Human Genetics and Genomics

Human Genetics and Genomics

Editor: Rosanna Mann

R CALLISTO
REFERENCE

www.callistoreference.com

Callisto Reference,
118-35 Queens Blvd., Suite 400,
Forest Hills, NY 11375, USA

Visit us on the World Wide Web at:
www.callistoreference.com

ISBN: 978-1-63239-808-6 (Hardback)

The publisher's policy is to use permanent paper from mills that operate a sustainable forestry policy. Furthermore, the publisher ensures that the text paper and cover boards used have met acceptable environmental accreditation standards.

Trademark Notice: Registered trademark of products or corporate names are used only for explanation and identification without intent to infringe.

Printed in the United States of America.

Cataloging-in-publication Data

Human genetics and genomics / edited by Rosanna Mann.
    p. cm.
Includes bibliographical references and index.
ISBN 978-1-63239-808-6
1. Human genetics. 2. Human genome. 3. Genomics. 4. Human biology. 5. Human chromosomes. I. Mann, Rosanna.
QH431 .H86 2017
599.935--dc23

Table of Contents

Permissions

List of Contributors

Index

Preface

Human genetics is study of heredity and inheritance as seen in human beings. It is interdependent on various other fields of science like genomics, molecular genetics, genetic counseling, population genetics, bio-chemical genetics, etc. This book unravels the recent studies in the field of human genetics and genomics. Also included in it is a detailed explanation of the various concepts and applications of this subject. The topics included in the text are of utmost significance and bound to provide incredible insights to readers. It is compiled in such a manner, that it will provide in-depth knowledge about the theory and practice of human genetics. This text will serve as a reference to a broad spectrum of readers, including biologists, genetic engineers, researchers, scientists and students involved with the field of human genetics across all levels.

After months of intensive research and writing, this book is the end result of all who devoted their time and efforts in the initiation and progress of this book. It will surely be a source of reference in enhancing the required knowledge of the new developments in the area. During the course of developing this book, certain measures such as accuracy, authenticity and research focused analytical studies were given preference in order to produce a comprehensive book in the area of study.

This book would not have been possible without the efforts of the authors and the publisher. I extend my sincere thanks to them. Secondly, I express my gratitude to my family and well-wishers. And most importantly, I thank my students for constantly expressing their willingness and curiosity in enhancing their knowledge in the field, which encourages me to take up further research projects for the advancement of the area.

Editor

Human Endogenous Retrovirus-K(II) Envelope Induction Protects Neurons during HIV/AIDS

Rakesh K. Bhat[1], Wallis Rudnick[1], Joseph M. Antony[1], Ferdinand Maingat[1], Kristofor K. Ellestad[1], Blaise M. Wheatley[2], Ralf R. Tönjes[3], Christopher Power[1]*

1 Department of Medicine (Neurology), University of Alberta, Edmonton, Alberta, Canada, 2 Department of Surgery (Neurosurgery), University of Alberta, Edmonton, Alberta, Canada, 3 Division of Medical Biotechnology, Paul-Ehrlich-Institut, Langen, Germany

Abstract

Human endogenous retroviruses (HERVs) are differentially expressed depending on the cell type and physiological circumstances. HERV-K has been implicated in the pathogenesis of several diseases although the functional consequences of its expression remain unknown. Human immunodeficiency virus (HIV) infection causes neuroinflammation with neuronal damage and death. Herein, we investigated HERV-K(II)/(HML-2) envelope (Env) expression and its actions in the brain during HIV/AIDS. HERV-K(II) Env expression was assessed in healthy brain tissues, autopsied HIV HIV− infected (HIV+) and uninfected (HIV−) brains and in neural cell cultures by real time RT-PCR, massively parallel (deep) sequencing, immunoblotting and immunohistochemistry. Neuronal and neural stem cells expressing HERV-K(II) Env were analyzed in assays of host responses including cellular viability, immune responses and neurobehavioral outcomes. Deep sequencing of human brain transcriptomes disclosed that RNA sequences encoded by HERV-K were among the most abundant HERV sequences detected in human brain. Comparison of different cell types revealed that HERV-K(II) *env* RNA abundance was highest in cultured human neurons but was suppressed by epidermal growth factor exposure. HERV-K(II) Env immunoreactivity was increased in the cerebral cortex from persons with HIV/AIDS, principally localized in neurons. Human neuronal cells transfected with HERV-K(II) Env exhibited increased *NGF* and *BDNF* expression. Expression of HERV-K(II) Env in neuronal cells increased cellular viability and prevented neurotoxicity mediated by HIV-1 Vpr. Intracerebral delivery of HERV-K(II) Env expressed by neural stem cells suppressed TNF-α expression and microglial activation while also improving neurobehavioral deficits in *vpr/RAG1*$^{-/-}$ mice. HERV-K(II) Env was highly expressed in human neurons, especially during HIV/AIDS, but in addition exerted neuroprotective effects. These findings imply that HERV gene products might exert adaptive effects in circumstances of pathophysiological stress, perhaps underlying the conservation of HERVs within the human genome.

Editor: Bruce W. Banfield, Queen's University, Canada

Funding: This study was supported by Canadian Institutes of Health Research to (CP) and none of the authors have financial interests related to the manuscript. The funders had no role in study design, data collection and analysis, decision to publish, or preparation of the manuscript.

Competing Interests: The authors have declared that no competing interests exist.

* Email: chris.power@ualberta.ca

Background

Human endogenous retroviruses (HERVs) represent approximately 8% of the human genome, which have been maintained through integration events over the past 50–100 million years [1,2,3]. In humans, endogenous retroviruses are not replication competent but can be engineered to replicate productively [4]. Endogenous retrovirus genes are inherited in a Mendelian manner in different species, usually remaining latent, but can become active depending on the individual cell type and host health status [5]. Although the human genome harbors a large number of endogenous retroviral sequences, their action(s) remain largely uncertain at present. We have shown previously that the human endogenous retrovirus (HERV)-W envelope protein, Syncytin-1, is highly expressed in glial cells within brain lesions of patients with multiple sclerosis and also contributes to endoplasmic reticulum stress [6,7].

HERV-K represents the most recent entry into the human genome and is also detected as multiple sub-types in humans [8]. There have been several disease associations with HERV-K [9,10,11,12]. The beta-retroviral HERV-K (HML-2), also referred to as the HERV-K(II) family, is considered to be among the youngest member of the HERVs and exhibits multiple polymorphic insertions, indicative of recent active replication in humans [8,13,14]. We previously showed that HERV-K(II) is one of the most transcriptionally active HERV families in brain and might be capable of generating virus-like particles [15]. Abnormal expression of HERV-K(II) proteins or transcripts has been associated with different pathological circumstances [16,17]. For example, induction of HERV-K *pol* transcript expression was reported in post-mortem brains from individuals with schizophrenia and other neuropsychiatric disorders [18,19,20]. HERV-K gene activation also occurs in different cancer cell lines and tumors [21]. Our group has previously shown an augmented expression of HERV-K *pol* transcripts in the brains of patients with neuroinflammatory disorders [22]. The high HERV-K Env amino terminal sequence conservation with Jaagsiekte sheep retrovirus (JSRV), which is contagious and causes lung cancer in sheep, suggests that the HERV-K Env might share similar properties in terms of receptor binding or modulating cellular entry [23,24]. However, it remains unclear if HERV-K genes exert pathogenic (or protective) effects.

During HIV/AIDS, HERV-K is highly expressed in blood although the determinants of its transcription and translation remain unclear [25,26]. Whether the increased expression of HERV-K in persons with HIV/AIDS requires specific pathophysiological triggers associated with HIV-1 infection is also uncertain. Given these circumstances we hypothesized that HERV-K envelope might exhibit increased expression in the brain during HIV infection. We observed differential expression of the HERV-K(II) envelope in the brain depending on the host neural cell type and disease state. Moreover, HERV-K(II) Env expression in neuronal cells was protective during *in vitro* and *in vivo* exposure to cytotoxic HIV-1 circumstances.

Results

HERV expression in healthy human brain

Although HERVs have been shown to be expressed in the human brain [20], their comparative expression levels have not been assessed to date using unbiased tools such as deep sequencing. The median number of HERV tags generated from human fetal (n = 3) and surgically resected (n = 2) brain RNA was 2738 tags per patient specimen by deep sequencing transcriptomic analysis while 31% belonged to the HERV-K family, which was only exceeded by the HERV-H family (57%) (**Figure 1A**). However, sequence tags were also assignable to HERV-W, -R, -E and –FRD. Fetal brains exhibited higher levels of HERV sequences for all HERV families (**Figure 1B**). Further analysis of the HERV-K tags revealed that LTR sequences were the most frequently detected tags among all patient brain specimens although *pol-gag* and *env* sequences were also detected (**Figure 1B**). LTR sequences dominated the total tag counts in all HERV families; for example, HERV-H, which contained the highest percentage tag frequency, displayed median tag frequencies of 0.5%, 17.6% and 81.8% for *env*, *gag-pol* and LTR, respectively. All host genes with transcript expression profiles were correlated with HERV-K(II) *env* tag abundance in the corresponding sample; based on sequence and bioinformatic analyses of differentially expressed host genes, there was substantial enrichment of host transcription-, translation- and cell cycle-related mRNAs associated with HERV-K(II) tag abundance (**Figure 1C**). Of interest, HERV-K *env* sequences are located throughout the human genome (**Figure S1A**) although the density of HERV-K LTR sequences was overall highest in specific chromosomes (**Figure S1B**). These findings highlighted the diverse expression of HERV genes in the human brain together with showing age-related expression and associations with fundamental host gene functions.

Ex vivo modulation of HERV-K(II) env expression

Stimulation of trophoblast cell fusion and differentiation by cyclic AMP (cAMP) has been associated with increased HERV-W *env* transcript and protein expression [27]. We investigated the effects of cAMP and epidermal growth factor (EGF) exposure on the expression profile of HERV-K *env* transcripts in different cell types. In this assay, HERV-K(II) *env* transcript levels were measured in U373 (human astrocytoma cell line), HFN (human fetal neurons), U937 (human leukemic monocyte lymphoma cell line) and HFA (human fetal astrocytes). HFNs showed the highest constitutive levels of HERV-K(II) *env* transcripts among all cell lines (**Figure 2A**). There was no effect of cAMP exposure on HERV-K(II) *env* transcript levels in HFNs, whereas there was decrease in HERV-K(II) *env* transcription in EGF-exposed HFNs

(**Figure 2B**). Both cAMP- and EGF-exposed U937 (**Figure 2C**) and HFA (**Figure 2D**) cells showed a reduction in HERV-K(II) *env* transcripts. These observations highlighted HERV-K(II) *env* expression was differentially regulated depending on the individual cell type and stimulus.

HERV-K(II) Env induces neurotrophin expression and neuronal process growth

Transfection of the human neuronal cell line, SK-N-SH, with an HERV-K(II) *env*-encoding vector (pHERV-Kenv) disclosed HERV-K(II) Env immunoreactivity at the predicted molecular weight (80 kDa) although HERV-K(II) immunoreactivity was not detected in cells transfected with a control vector (pGFP) (**Figure 3A**). In addition, transcripts encoded by *BDNF* and *NGF* were induced in HERV-K(II)-transfected SK-N-SH cells (**Figure 3B**) relative to cells transfected with the pGFP control vector. In addition, analyses of SK-N-SH cells transfected with the HERV-K(II) *env* containing vector showed increased levels of βIII-tubulin immunoreactivity compared to the control vector-transfected cells (**Figure 3C**). These findings implied that HERV-K(II) Env expression could be increased in human cells and might confer neurotrophic effects on neuronal cells.

HERV-K(II) Env expression in human brain during HIV/AIDS

Previous studies have suggested that HERV-K(II) *pol* transcripts are expressed in the brain in disease [11,20]. To investigate the *in vivo* abundance and specificity of HERV expression, we analyzed cerebral white matter from patients with HIV/AIDS (HIV+, n = 3) and uninfected persons with other diseases as controls (HIV−, n = 3) by transcriptomic deep sequencing. Massively parallel sequencing of samples produced a large number of short reads/tags, which were assembled into contigs (overlap length of 36–77 nucleotides). Overall 5,640,659–8,803,479 tags were obtained depending upon the individual sample, of these, 32.8% were mapped to human rRNA (one or more read per pair), 12.3% were aligned to human transcriptome, 34.0% to the human nuclear genome (but not to human transcriptome), 8.0% to human mitochondrion DNA (mtDNA), 0.1% to bacterial and viral sequences (but not to human genome or transcriptome) and 15% sequences were not found in sequence database queried. These studies revealed that the median HERV tag number specimen was 666 tags/specimen, of which 74% belonged to the HERV-K family; HERV-K tags were the most abundant tags detected in both clinical groups with the HIV+ specimens showing a higher HERV-K tag frequency than the HIV− group but HERV-W and HERV-H associated tags were more abundant in the HIV− group (**Figure 4A**). Sequence tags belonging to the HERV-W, -H and other HERVs were present in all specimens examined but the HIV HIV− group showed greater expression of these latter HERVs. Analysis of the relative abundance of all HERV-K *env* sequences tags showed no difference in tag numbers between the clinical groups. Comparison of the relative expression of different host genes in the HIV+ and HIV− groups' brains disclosed that tags of multiple groups of host genes implicated in a wide range of fundamental functions were enriched in the HIV+ group's brains based on gene ontology (GO) analyses (**Figure 4B**). These findings implied that there was differential expression of both HERV and host genes in the HIV+ and HIV− brain specimens.

To extend these findings, we focused on HERV-K(II) Env expression, which showed a significant increase in HERV-K(II) *env* transcript levels in cerebral cortical specimens from HIV+ patients

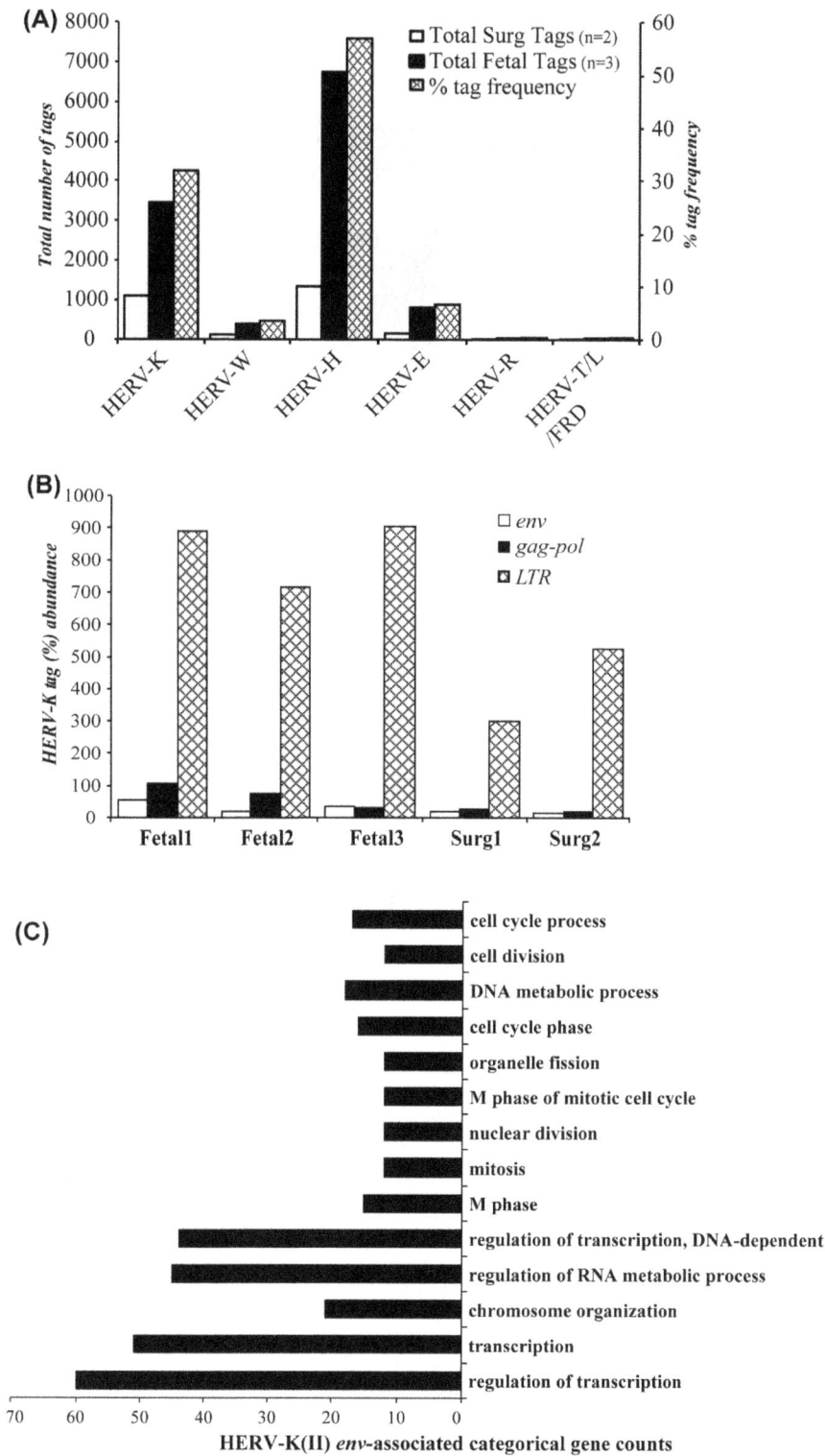

Figure 1. Deep sequencing analyses of HERVs in healthy brain. (**A**) Deep sequencing of the fetal and surgically resected (Surg) brain samples revealed that HERV-H exhibited the highest tag frequency and median number of tags followed by HERV-K. (**B**) When analyzing the HERV-K tags, LTR tags were most abundant, followed by *gag-pol* and then the *env* region tags (tags were normalized to respective gene lengths) (**C**) All host genes with transcript expression profiles correlated with HERV-K(II) *env* tag abundance (r$^2 \geq$ 0.5) were analyzed using the DAVID tools [58] for enriched gene ontology (GO) terms. Genes related to cell cycle functions and chromosomal organization were most strongly associated with HERV-K(II) *env* expression. With the use of DAVID bioinformatics resources [59], the predicted target genes were classified according to KEGG functional annotations to identify pathways that were actively regulated by HERV-K(II) *env* transcripts in brain tissue. The most over-represented GO term belonged to the transcriptional regulation and chromosome organization followed by different stages of cell cycle pathway. (Mann Whitney t test, *$p <$ 0.05, **$p <$ 0.01).

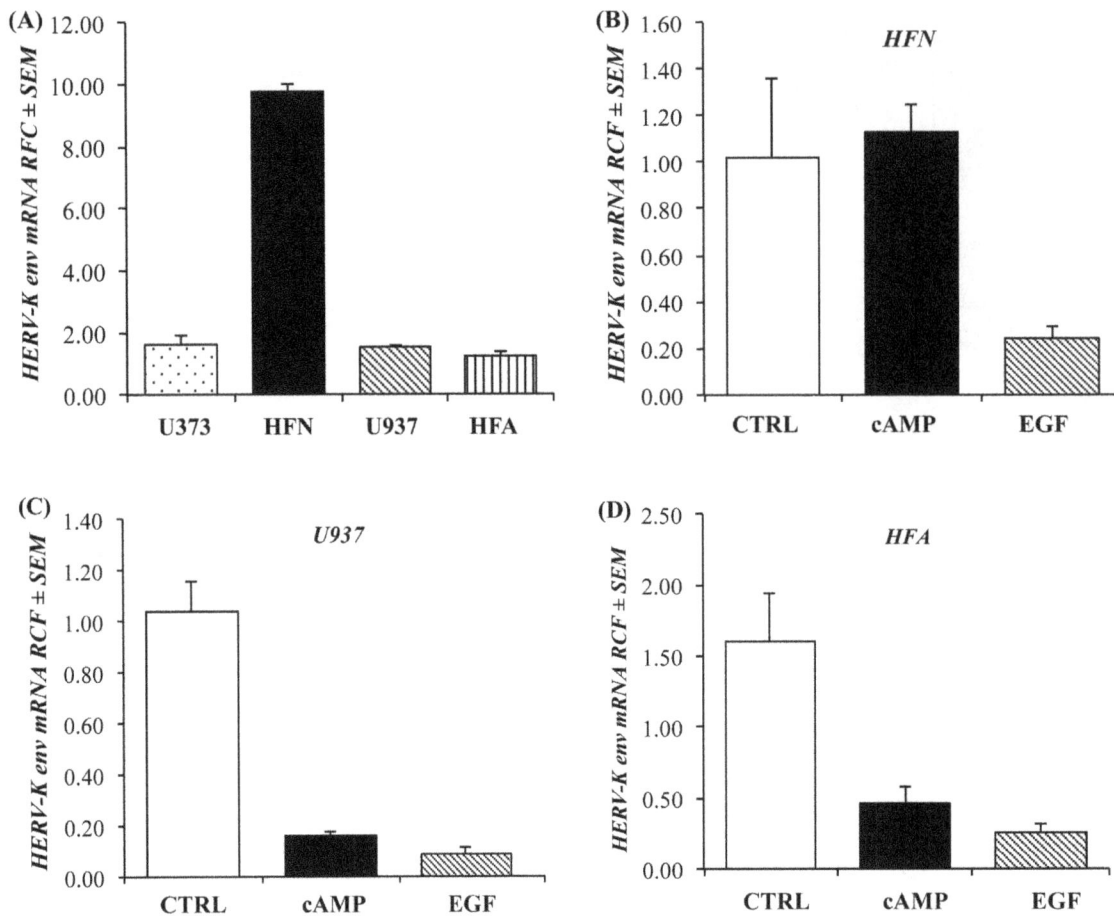

Figure 2. Activation of HERV-K(II) *env* **by cAMP and EGF in different human cell lines** (**A**) Individual cell lines displayed differential constitutive HERV-K(II) *env* expression profiles. (**B**) Upon treatment of human fetal neurons, db-cAMP did not have any effects on HERV-K(II) *env* expression but EGF down-regulated HERV-K(II) expression. (**C**) U937 and (**D**) HFA showed decreased in HERV-K(II) *env* expression upon both db-cAMP and EGF exposure. (n = 4 replicates per group across two independent experiments).

compared to matched white matter as well as cortex and white matter of HIV− patients using real time RT-PCR (**Figure 5A**). HERV-K(II) Env immunoreactivity in cerebral sections was minimal in HIV− patients (**Figure 5B**) but HIV+ brain sections displayed immunoreactivity in cells resembling cortical neurons, which was co-localized with MAP-2 immunoreactivity (**Figure 5B, insert**). Western blotting of cerebral cortex specimens from HIV− and HIV+ brains showed that HERV-K(II) Env expression was greater in the brains of HIV+ patients compared with HIV− patients (**Figure 5C**). Densitometry analyses of immunoblots showed that HERV-K Env expression was increased in the brains of HIV+ patients (**Figure 5D**). These findings suggested that HERV-K(II) Env was expressed in human cortical neurons, which was augmented during HIV infection.

HERV-K(II) Env prevents neuronal injury

As both BDNF and NGF are known to exert neurotrophic actions and were induced by HERV-K(II) Env over-expression in neuronal cells (**Figure 3B**), we investigated the contributions of HERV-K(II) Env to neuronal viability. Cell lines were transfected with pGFP or pHERV-Kenv and subsequently analysed for relative HERV-K(II) *env* transcript abundance, displaying variable expression depending on the individual cell line (**Figure S2A**). HERV-K(II) Env immunopositive cells were detected in ~5% of SK-N-SH transfected with pGFP, which rose to ~20% in

pHERV-Kenv-transfected cells (**Figure 6A**). Similarly, HERV-K(II) Env immunoreactivity in pGFP-transfected neuronal cells was minimally detected (**Figure 6B**) but exhibited robust immunoreactivity in HERV-K(II) *env*-transfected cells (**Figure 6C**) with cytosolic and plasma membrane immunoreactivity (**Figure 6D**). Transfection of the human neuronal cell line SK-N-SH with pHERV-Kenv resulted in increased transcript levels of *BDNF* and *NGF* transcripts compared to pGFP-transfected cells (**Figure 6E**). To evaluate cell viability with and without concurrent HERV-K(II) Env expression, the murine NG108 neuronal cell line was transfected with each vector and subsequently exposed to different neurotoxins (**Figure 6F**). Cell viability was found to be preserved differentially in the HERV-K(II) Env-transfected cells following exposure to staurosporine, the HIV-1 Vpr protein or NMDA relative to pGFP-transfected cells with ~100% and ~40% loss of pGFP-transfected cells following exposure to staurosporine and Vpr, respectively, relative to the pHERV-Kenv-transfected cells. These observations suggested that HERV-K(II) Env expression in neurons selectively prevented their injury upon exposure to different neurotoxic molecules.

HERV-K(II) Env is neuroprotective *In Vivo*

Because of the apparent neuroprotective effects identified above, the *in vivo* effects of HERV-K(II) Env in HIV-1 *vpr/RAG1*$^{-/-}$ mice were investigated by stereotaxically implanting

(A)

(B)

(C)

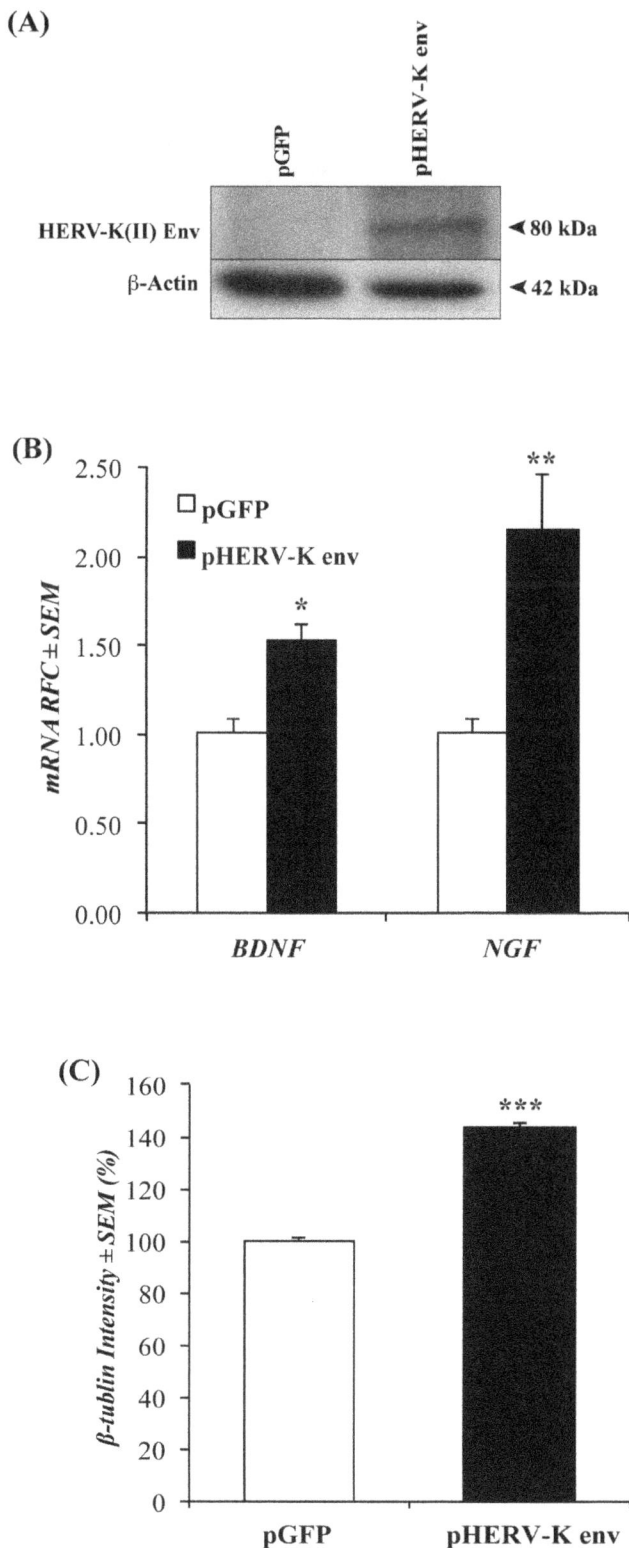

Figure 3. Over expression of HERV-K(II) Env exerts neurotrophic effects: (**A**) Transfection of the pHERV-Kenv plasmid into SK-N-SH cells showed HERV-K(II) Env immunoreactivity at the predicted molecular weight on western blot. (**B**) Upon treatment with supernatants from SK-N-SH cells transfected with pHERV-Kenv plasmid, HFN showed increases in *BDNF* and *NGF* transcript abundance compared to the control vector transfected cells. (n = 3, with technical quadruplicates) (**C**) βIII-tubulin expression in HFN following 24-hour exposure to supernatants from HFA-transfected with the pHERV-Kenv or the control

vector, showing an increase in βIII-tubulin immunoreactivity in cells exposed to HERV-K Env-transfected cells. (n = 2, with technical octuplicates) (Student t test, *p<0.05, **p<0.01).

neural stem cells (NSCs) that were transfected with both the pHERV-Kenv and pGFP vectors or only the pGFP vector into the striatum (**Figure 7A**). Molecular, neuropathological and neurobehavioral studies were subsequently performed. Immunoblots of NSCs transfected with each of the above vectors revealed detection of HERV-K Env immunoreactivity in cells transfected with pHERV-Kenv/pGFP vectors (**Figure 7B**). Analyses of host transcript levels in the brains of *vpr/RAG1*$^{-/-}$ animals demonstrated that animals implanted with cells expressing HERV-K(II) *env* exhibited showed diminished transcript levels of *TNF-α* (**Figure 7C**), together with increased levels of *IL-6* (**Figure 7D**) as well as a trend towards increased *BDNF* expression in transgenic animals (**Figure S3D**). The *GFP* transgene transcript levels expressed by transfected cells were similar in animals receiving cells with the pGFP or the pHERV-Kenv/GFP vectors (**Figure S3C**). Neuropathological analyses revealed that Nissl-positive neurons in the striatum displayed similar morphology and density in animals implanted with NSCs transfected with pGFP (**Figure 7E**) and pHERV-Kenv/pGFP (**Figure 7J**). Iba-1 immunoreactivity was more abundant on hypertrophied cells, resembling microglia in the striatum of animals receiving the pGFP-transfected NSCs (**Figure 7F**) compared to animals receiving cells expressing HERV-K(II) Env (**Figure 7H**). GFAP immunoreactivity in astrocytes was increased in the animals receiving the pHERV-Kenv/pGFP-transfected cells (**Figure 7L**). In keeping with previous studies from our group, cleaved caspase-6 immunoreactivity was increased on cells resembling astrocytes in the striatum of animals implanted with NSCs transfected with pGFP (**Figure 7H**) compared to those implanted with NSCs expressing HERV-K(II) *env* (**Figure 7M**). In contrast, BDNF immunostaining was abundant the striatum of animals implanted with NSCs expressing HERV-K(II) Env (**Figure 7M**) compared to those implanted with NSCs transfected with pGFP alone (**Figure 7I**).

In neurobehavioral studies, animals that received the pGFP vector-transfected cells exhibited greater neurobehavioral deficits in terms of rotary behavior at days 7 and 14 post-striatal implantation of transfected cells compared to animals receiving cells transfected with the pHERV-Kenv/pGFP vectors (**Figure 7K**), underscoring the potential neuroprotective properties exerted by HERV-K(II) Env.

Discussion

The present studies represent the first unbiased analysis of HERV transcript abundance in the human brain in both health and disease. HERV-K was among the most abundant HERVs identified; not surprisingly, the HERV-K LTR was the most frequently detected viral sequence. However, the HERV-K(II) Env was also observed in all brain specimens with the highest expression in neurons. Moreover, its expression was increased in the setting of HIV/AIDS in terms of both transcript and protein levels. Overexpression of HERV-K(II) Env in human neuronal cells induced neurotrophin expression and ensuing neuronal process extension while its *in vivo* expression in neural stem cells exerted beneficial effects in terms of reduced neuroinflammation (diminished microglial activation and *TNFα* expression) and improved neurobehavioral outcomes. Collectively, these observations point to a capacity for HERV-K(II) Env expression to be cell-

(A)

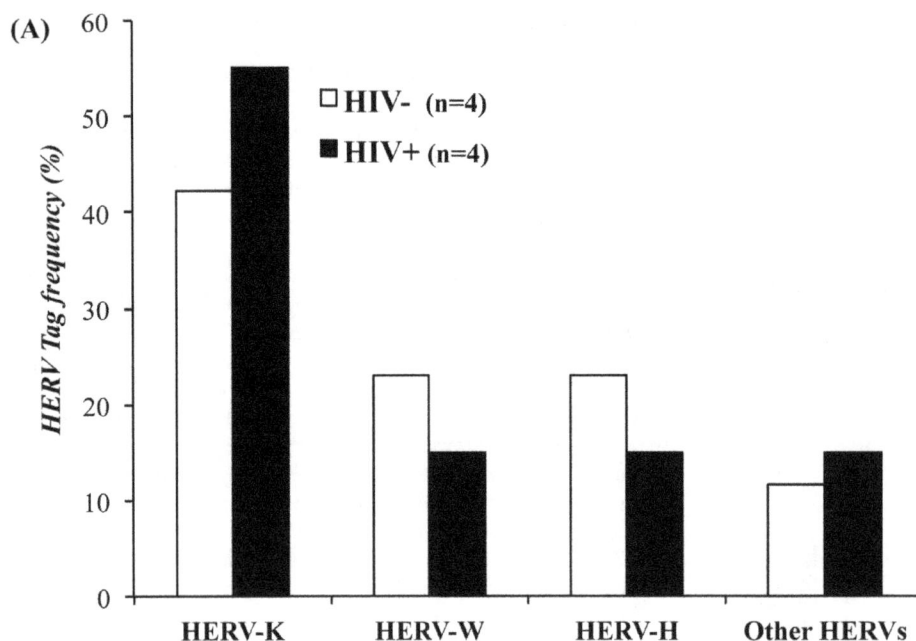

(B)

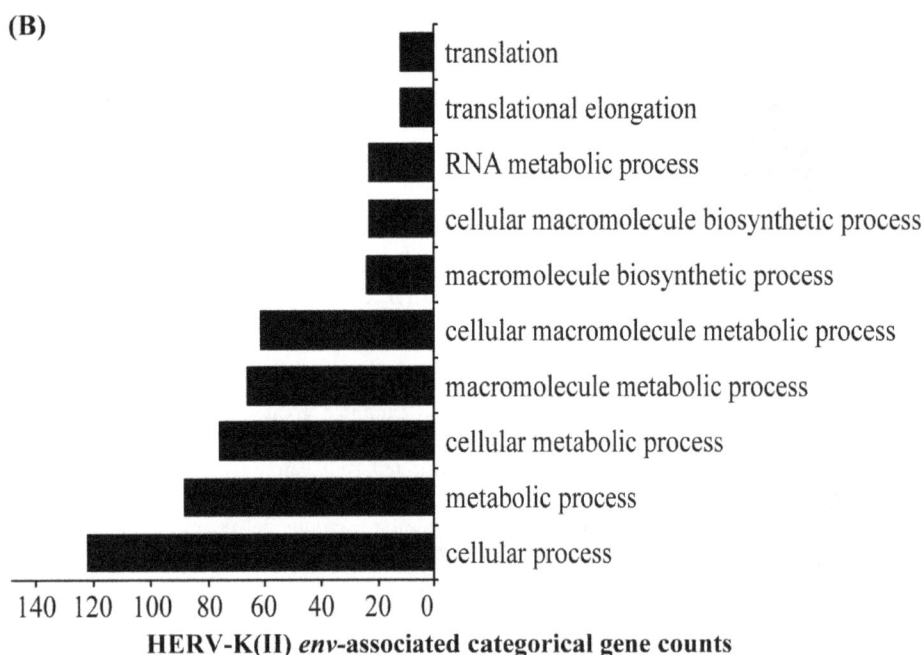

HERV-K(II) *env*-associated categorical gene counts

Figure 4. HERV transcripts in HIV− infected brain specimens. (**A**) Deep sequencing of the HIV− and HIV+ autopsied cerebral white matter revealed a higher tag frequency of HERV-K in both clinical groups compared to other HERVs. (**B**) With the use of the DAVID bioinformatics resources, the predicted target genes were classified according to KEGG functional annotations to identify pathways that were actively regulated by HERV-K(II) *env* transcripts in brain tissue.

type specific but also to respond to pathogenic stimuli in a manner that enhanced host fitness through preserved brain function.

Chronic neurodegeneration during HIV-1 infection of the nervous system remains a major clinical challenge, manifested as HIV− associated neurocognitive disorders, despite the burgeoning availability of highly active antiretroviral therapies [28]. The principal mechanisms by which HIV-1 injures the brain is through the release of virus-encoded proteins (e.g., Vpr, Tat, Env) or

induction and release of potential immunopathogenic host molecules (e.g., TNF-α) from infected or activated glial cells and leukocytes [29,30]. These secreted factors are toxic to neurons depending on the proximal concentrations, duration of exposure and host susceptibility factors (i.e. age), leading to apoptosis or necrosis but are also able to accentuate inflammation within the brain. Indeed, the present Vpr transgenic mouse, which selectively expresses Vpr protein in myeloid cells, exhibits a robust

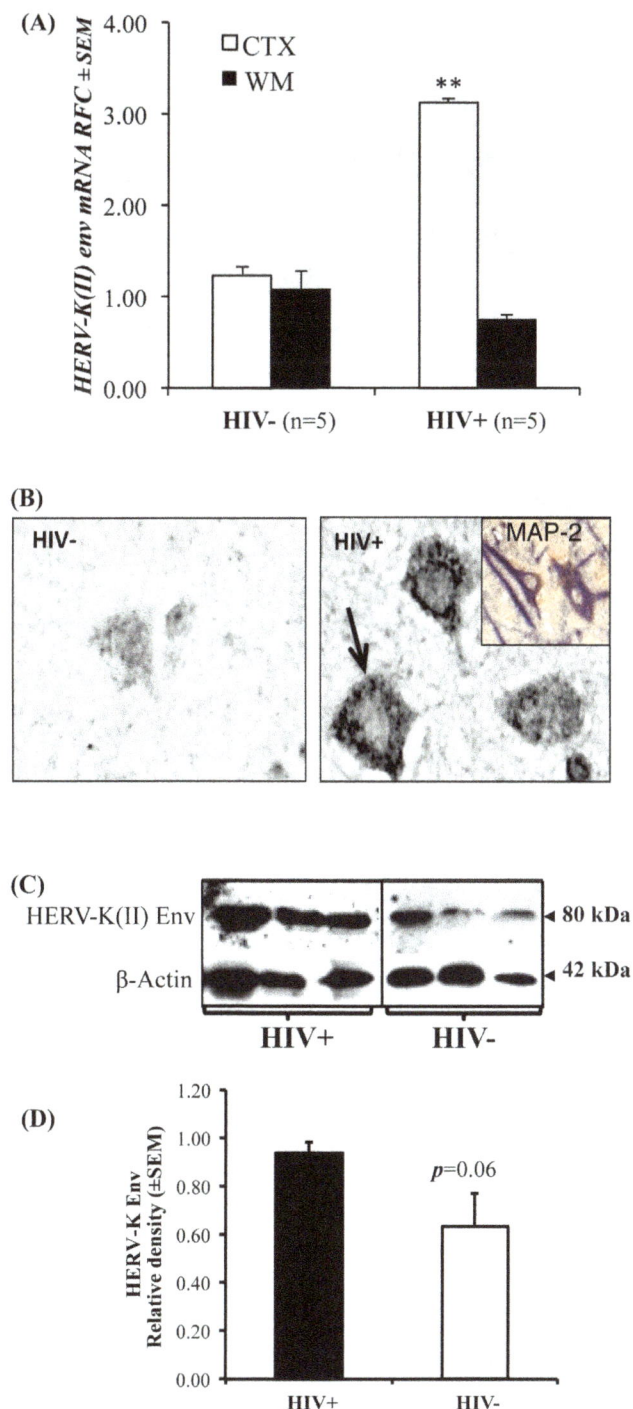

Figure 5. Brain expression of HERV-K(II) Env in HIV/AIDS: (**A**) HERV-K(II) *env* transcript analysis of HIV−) and HIV+ brains revealed high levels of HERV-K(II) *env* in cortex of HIV+ as compared to HIV− brains. (**B**) Immunohistochemical analyses of brain sections from HIV+ patients showed increased immunoreactivity of HERV-K(II) Env (arrow) protein in neurons as compared to the HIV− brain sections. HERV-K(II) Env protein expression co-localized in neurons expressing MAP-2 (insert: brown, MAP-2; blue, HERV-K Env). (**C**) In cerebral cortical specimens, HIV+ patients exhibited higher levels of HERV-K(II) Env detection than HIV− patients on immunoblotting of HERV-K(II) Env protein. (**D**) Quantitation of HERV-K(II) Env/β-actin band density on immunoblots (Original magnification: B-400X; insert, 200X). (Student t test, **$p < 0.01$).

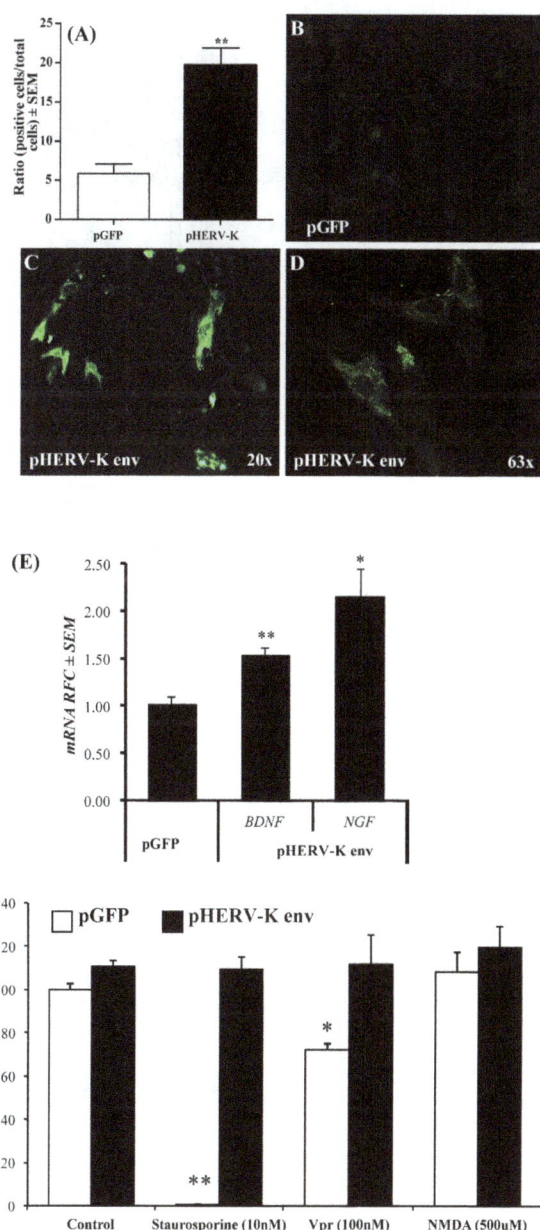

Figure 6. HERV-K(II) *env* transfection of neuronal cells was neuroprotective. (**A**) Analyses of SK-N-SH cells transfected with the pHERV-Kenv plasmid compared to the control (pGFP) showed that the efficiency of transfection was ∼20% (n = 3, with technical triplicates). (**B**) HERV-K(II) Env immunoreactivity was minimally detected in cells transfected with the control vector. pHERV-Kenv-transfected cells showed HERV-K(II) Env immunoreactivity at low (**C**) and high magnification (**D**). (**E**) Comparison of *BDNF* and *NGF* transcript levels in SKN-N-SH cells transfected with pGFP or pHERV-Kenv. (**F**) Exposure of pHERV-Kenv and control vector-transfected NG108 cells to staurosporine, HIV-1 Vpr or NMDA, showed that pHERV-Kenv-transfected cells were differentially protected depending on the neurotoxin. (Student t test, *$p < 0.05$, ***$p < 0.001$).

neurodegenerative phenotype defined by synapto-dendritic and neuronal loss coupled with worsened neurobehavioral performance on tasks of executive and motor functions [31]. These pathogenic circumstances represent a perturbed biological environment within the brain and thus induction of host molecules,

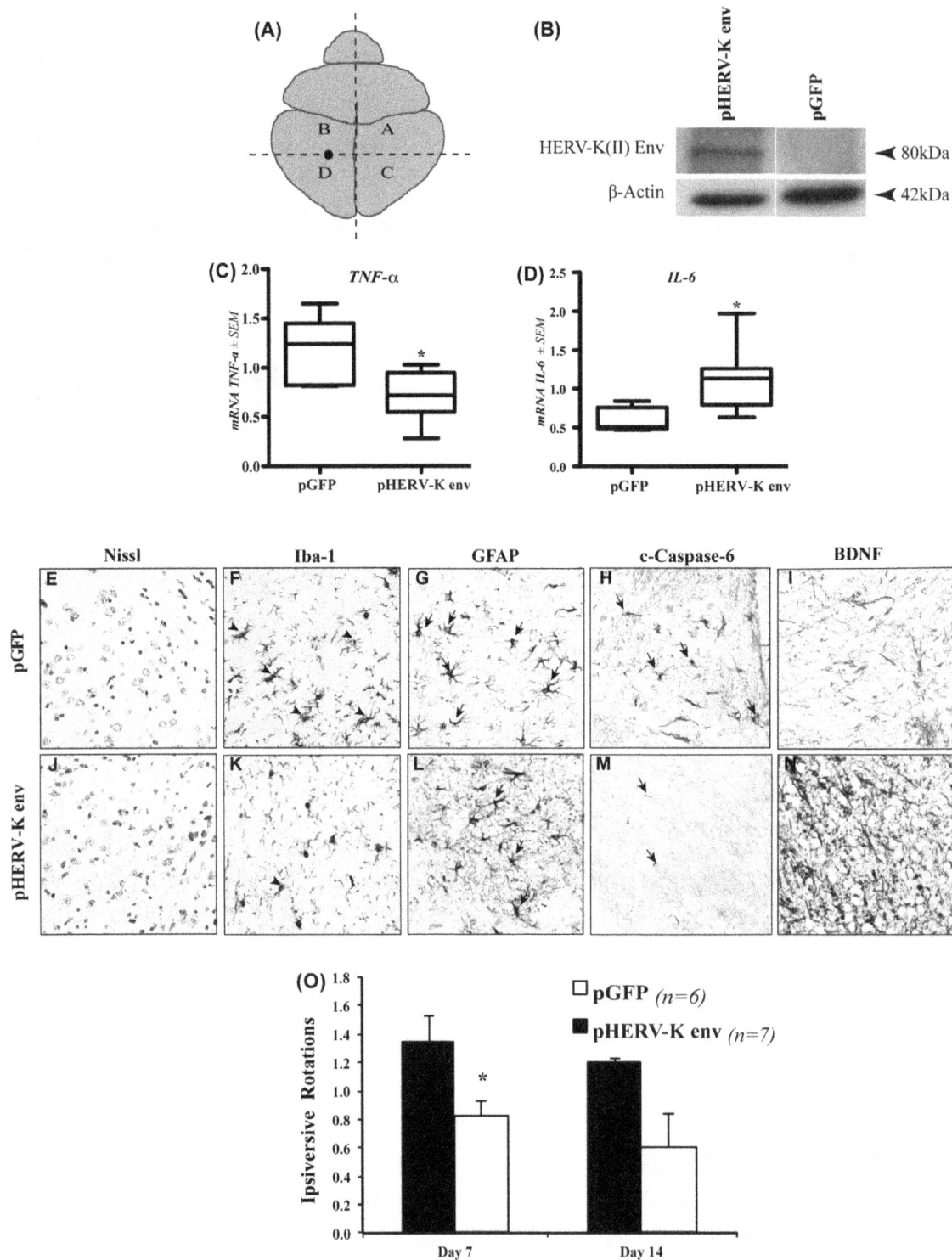

Figure 7. Neural stem cells expressing HERV-K(II) Env are protective in *vpr/RAG1*$^{-/-}$ animals. (A) Schematic of representation of C17.2 implantation site (marked by the ●) in Vpr/RAG1$^{-/-}$ mice. (B) Western blot showing HERV-K(II) Env immunoreactivity in transfected cells. (C) *TNF-α* expression was suppressed in the brains of animals implanted with cells expressing HERV-K(II) *env* while (C) *IL-6* was induced. Nissl staining showed similar striatal neuronal densities in animals implanted with cells transfected with either pGFP or pHERV-Kenv/pGFP (**E, I**). Immunohistochemistry revealed lower expression the microglia protein, Iba-1 (**K**) and higher expression levels astrocyte protein, GFAP (**L**) in HERV-K(II) *env* implanted brains compared to control vector (*pGFP*) implanted animals (**F, G**), respectively. Cleaved caspase-6 immunoreactvity was comparative reduced in striatum of animals receiving cells transfected with pHERV-Kenv/pGFP (**M**) but BDNF immunoreactivity was increased in the same animals (**N**) compared to controls (**H, I**). (**O**) At days 7 and 14, neurobehavioral deficits were greater in terms of ipsiversive rotations among the animals implanted with c17.2 cells transfected with the *pGFP* vector. (Original magnification: E–J, 400X) (Mann-Whitney test, *$p < 0.05$).

which could avert or restrict host injury, is a plausible (and desirable) response to HIV infection.

Induction of HERV-K expression during HIV/AIDS is a recognized phenomenon in blood and appears to be associated with disease progression during HIV/AIDS [32,33,34]. Conversely, HERV-K *pol* transcripts appear to be induced in central nervous system tissues from patients with amyotrophic lateral sclerosis with the reverse transcriptase protein principally localized in cortical neurons and associated with TDP-43 expression [11]. Similarly, we also observed HERV-K(II) Env expression chiefly in cortical neurons as well as in cultured human fetal neurons. Given that neurons are terminally differentiated cells, they require robust protective mechanisms to survive; the conserved ability to induce expression of an ancient virus-encoded protein could be a valuable evolutionary strategy. The ability of HERV-K(II) Env to mediate activation of the neurotrophins, BDNF and NGF, which are altered in HIV/AIDS implies HERV overexpression might have an intrinsic adaptive function by reducing the brain's susceptibility to neuronal injury. As mentioned above, other HERVs have been shown to be overexpressed in neurological diseases within different cell types including glia. For example, the HERV-W Env protein, Syncytin-1, is highly expressed in astrocytes in the brains of multiple sclerosis (MS) patients [7,35] [36,37,38] and mediates endoplasmic reticulum stress *in vitro* and *in vivo* in astrocytes [6]. In fact, Syncytin-1 overexpression in the central nervous system during MS is pathogenic, resulting in neuroinflammation with ensuing oligodendrocyte (but not neuronal) injury and death [6,38,39]. By contrast, the murine endogenous retroviral envelope proteins, Syncytin A and B, are not induced in animal models of MS (Power, *unpublished*) emphasizing the species-specificity and diversity of responses among different endogenous retroviruses.

While *in vivo* HERV-K induction in blood is a consistent feature of HIV/AIDS, its *in vitro* activation is more variable, perhaps reflecting the different cell types involved [40]. However, increased HERV-K expression in human cortical neurons was a constant feature in this study of HIV/AIDS as well as in a previous study of a neurodegenerative disease, amyotrophic lateral sclerosis [11]. Several mechanisms underlying the relative HERV-K(II) Env induction in neurons include local inflammation secondary to the primary disease process in which inflammatory molecules such as cytokines activate the HERV-K(II) LTR as suggested for other retroviruses. Alternatively, a loss of CpG methylation leading to increased provirus transcriptional activity might permit HERV-K(II) induction in the setting of neuronal de-differentiation or injury. An (secondary) opportunistic infection such as CMV might also activate retroviral gene expression, which frequently complicates HIV/AIDS, as suggested for other opportunistic infections [16,41]. In the case of HIV/AIDS, while neurons are not productively infected, the secretion of the HIV-1 viral proteins, Vpr or Tat, by infected myeloid/microglial cells could transactivate retroviral gene expression in nearby neurons. Nonetheless, whatever the process is by which HERV-K(II) Env expression is enhanced, the resulting effect was beneficial to host neurons in the present studies through the concurrent stimulation of neurotrophin expression and ensuing neuroprotective effects. Indeed, this neuroprotective phenomenon was particularly evident in murine neuronal cells following transfection of the pHERV-Kenv vector (**Figure 6F**), possibly through the improved efficiency of transfection of this cell type as well as the absence of any residual HERV-K(II) *env* expression, creating a dominant negative effect. While TNF-α is widely recognized as toxic factor acting on neurons through engagement of its p75 receptor, the actions of IL-6, which was induced by HERV-K(II) expression in the implanted brains of *vpr/RAG1*$^{-/-}$ mice, is less clear. IL-6 expression is

induced in many pathological circumstances but its downstream effects are divergent with both pathogenic and protective actions [42,43,44,45]. The current studies imply that conservation and expression of HERVs in specific organs or cells might contribute to host adaptation to pathogenic circumstances, which could be exploited as preventative or therapeutic strategies in the future.

Conclusions

The present studies demonstrate that HERV-K(II) Env was highly expressed in human neurons *in vitro* and *in vivo*, but was also induced in neurons during HIV/AIDS. Moreover, HERV-K(II) Env exerted protective effects on neuronal cells. These findings indicate that HERV gene-encoded proteins potentially mediate beneficial actions in circumstances of pathophysiological stress. Advantageous effects to host functions might underlie the conservation of HERVs within the human genome over time.

Materials and Methods

Human brain samples

Adult human brain (frontal lobe) specimens were collected at the time of autopsy or at the time of surgical resection for epilepsy with consent from HIV-1 sero-negative (uninfected) or -positive (HIV/AIDS) patients and stored at −80°C. All HIV/AIDS (HIV+) patients were AIDS-defined, as described previously [7,46,47,48]. Uninfected disease controls were comprised of different diseases (HIV−, sepsis, cancer, multiple sclerosis, stroke). Surgically-resected brain specimens were derived from patients undergoing surgery for removal of an epileptogenic focus; tissue specimens used herein were remote from the epileptogenic lesion. The use of brain tissues is part of an ongoing research study (Pro0002291) approved by the University of Alberta Human Research Ethics Board. Human brain fetal tissues were obtained from 15–19 week (elective) aborted fetuses with written consent approved by the University of Alberta Human Research Ethics Board (Biomedical-Pro00027660) from which neurons and astrocytes were prepared. The protocols for obtaining brain specimens comply with all federal and institutional guidelines with special respect for the confidentiality of the donor's identity and collected with consent.

Human fetal neural cell cultures

To establish human neuronal cultures, fetal brain tissues were collected and prepared on the same day; the meninges were removed, tissues mechanically minced and a single cell suspension was prepared by trituration through serological pipettes, followed by digestion for 30 min with 0.25% trypsin (Gibco, Burlington, ON) and 0.2 mg/mL DNase I (Roche Diagnostics, Mannheim, Germany) and passage through a 70 micron cell strainer (BD Biosciences, Mississauga, ON). Cells were washed 2 times with fresh medium, and plated in T-75 flasks coated with poly-L-ornithine (Sigma Aldrich, Oakville, ON) at $6-8\times10^7$ cells/75 mm2 flask in MEM supplemented with 10% FBS (Gibco), 2 mM L-glutamine (Gibco), 1 mM sodium pyruvate (Gibco), 1X MEM non-essential amino acids (Gibco), 0.1% dextrose (Sigma Aldrich), 100 U/mL Penicillin (Gibco), 100 µg/mL Streptomycin (Gibco), 0.5 µg/mL amphotericin B (Gibco) and 20 µg/mL gentamicin (Gibco). Cultures of neurons were additionally supplemented with 25 µM cytosine arabinoside (Sigma Aldrich, Oakville, ON) to prevent astrocyte growth.

Cell lines

Cell lines were obtained from the American Type Culture Collection (ATCC; www.atcc.org) and cultured according to standard mammalian tissue culture protocols and sterile techniques. Human neuroblastoma (SK-N-SH) and murine neuronal (NG108) cells were cultured as monolayer in Dulbecco's Modified Eagle Medium (DMEM). All media was supplemented with 10% FBS/100 units/ml penicillin/100 μg/ml streptomycin/2 mM L-glutamine. The RPMI medium was also supplemented with 1 mM sodium pyruvate/10 mM HEPES buffer. All tissue culture media and supplements were obtained from Invitrogen. Human fetal neurons (HFN), human fetal astrocytes (HFAs) and U937 cells [49] were cultured in 6 well plates and exposed to di-butyl cAMP (50 μg/ml) or epidermal growth factor (EGF (10 μg/ml) for 4 days. Following exposure, total RNA was extracted and relative mRNA levels of the different genes of interest were measured by a semi-quantitative reverse transcription PCR assay [7].

Neural cell transfection and implantation

Murine neural stem (C17.2) cells [50] and human or murine neuroblastoma (SK-N-SH or NG108, respectively) cells were grown in Dulbecco's modified Eagle's medium (DMEM) supplemented with 10% fetal calf serum, 5% horse serum, 2 mM glutamine, penicillin/streptomycin/fungizone (Invitrogen, 100x stock, 1/100 ml media) as previously described [51,52] in 25 cm^2 uncoated tissue culture flasks at 37°C. Half of the media was changed every 3–4 days and cells were split (1:20) weekly except when the cells were prepared for the implantation. Cells were transfected with a control plasmid (pBUD-GFP, a gift from Dr. David Vergote, University of Alberta) encoding enhanced green fluorescent protein (eGFP) or a HERV-K(II) Env encoding vector (pHERV-Kenv) [15], and henceforth these vectors were termed pGFP and pHERV-Kenv. For co-transfection, cells were grown overnight in medium in 6 well plates before transfecting with pGFP and pHERV-Kenv using lipofectamine reagents (Invitrogen) according to manufacturer's protocol. Selection of the positively transfected cells was performed over 3 months with puromycin (2.5 μg/ml) and Zeocin. At the time of implantation, near confluent undifferentiated cells were trypsinized with Trypsin-EDTA (0.05%), washed 2 times with phosphate-pHERV-Kenv buffered saline (PBS) and re-suspended in Hank's Balanced Salt Solution (HBSS) at a final concentration of 2×10^5 viable cells/μl [53].

Neurobehavioral studies

Vpr/RAG1$^{-/-}$ mice were generated by crossing Vpr transgenic mice which expressed HIV-1 Vpr under the control of the *c-fms* (M-CSF receptor) promoter, directing transgene expression chiefly in monocytoid cells [42], were crossed with RAG1$^{-/-}$ animals, as previously reported [31] and were used for the present *in vivo* studies (Research study AUP00000318 approved by the University of Alberta Animal Care and Use Committee for Health Sciences). Neurobehavioral deficits were assessed by the Ungerstadt assay [43,44]. Animals (4 weeks, n = 6–7) were anesthetized with Ketamine/Xylazine, ocular ointment was applied to their eyes to prevent drying and placed in a stereotaxic frame. The heads were cleaned with 70% ethanol, skin incised at the midline and a small cranial burr hole was made with a dental drill bit on a pre-marked skull area. The coordinates of implantation were 3.5 mm posterior, 2.5 mm lateral and 3 mm deep relative to the bregma resulting in an implantation site within the striatum (**Figure 7A**). 5 μl of transfected-cell suspension (HERV-K(II) *env*/*eGFP* or control, *eGFP*) containing ~1×10^6 viable cells were stereotactically implanted into the right striatum of each animal over 5 minutes.

The wound was closed with cyanoacrylate glue (Vectabond). Ipsiversive rotational behavior, which is indicative of neurological injury, was measured over 10 min after intraperitoneal injection of amphetamine (1 mg/kg) on days 4, 7, 14 and 21 following intrastriatal injection. Animals were sacrificed after 21 days followed by intracardial perfusion with saline, followed by PBS/4% paraformaldehyde. The brain was removed and the tissue anterior to the implantation site was frozen at−80°C while the posterior tissue was post-fixed in PBS/4% paraformaldehyde embedded in paraffin from which 10-μm sections were prepared for immunohistochemical analysis.

Neurotoxicity assays

HFN cells were cultured in 96-well flat bottom plates and exposed to either supernatants from SKN-N-SH cells transfected with plasmids encoding HERV-K(II) *env* (pHERV-Kenv) protein or control (pGFP). 48 h after treatment, cells were fixed, permeabilized and stained with mouse anti-β-tubulin antibody (1:800 dilution, Sigma-Aldrich) as previously described [54]. Neuronal injury was quantified by βIII-tubulin immunoreactivity using Odyssey Imager (LI-COR, Lincoln, NE). Diminished βIII-tubulin immunoreactivity was indicative of reduced cellular viability [55]. For assaying the cytotoxic effects of different neurotoxins, murine neuronal NG108 cells were stably co-transfected with pBUD-GFP or pHERV-Kenv plasmids. The cells were grown on 4 well chamber slides to ~60% confluency and then exposed to staurosporine (10 mg/ml), HIV-1 Vpr (100 nM) or NMDA (500 nM) for 24 hours [55,56]. After the incubation period, cells were fixed with 4% formalin, washed in PBS containing 0.1% Triton X-100 (Sigma-Aldrich), and blocked for 90 min at 4°C with LI-COR Odyssey Blocking Buffer (LI-COR, Lincoln, NE), following which antibodies to βIII-tubulin (1:1000) were applied to each well overnight and washed X3. A labeled secondary anti-mouse IgG antibody was applied for 1 hr, washed and then the relative immunoreactivity was assessed in each well [49,55].

Real-time RT-PCR

First-strand cDNA was synthesized by using aliquots of 1 μg of total RNA from cortex and basal ganglia, reverse transcriptase and random primers [7]. Specific genes were quantified by real-time PCR using i-Cycler MYIQ system (Bio-Rad, Mississauga, ON). cDNA prepared from total RNA of cultured cells was diluted 1:1 with sterile water and 5 μl were thereafter used per RT-PCR reaction. Semi-quantitative analysis was performed by monitoring in real time the increase of fluorescence of the SYBR Green dye on a Bio-Rad detection system, as previously reported [57] and expressed as relative fold change (RFC) compared to control. Oligonucleotide primers are provided in Table 1.

Immunohistochemistry

Formalin-fixed, paraffin-embedded sections of human brain tissue (frontal lobe) on glass slides were de-paraffinized and rehydrated using decreasing concentrations of ethanol. 4% PBS-buffered paraformaldehyde fixed mouse brains (left hemisphere) were embedded in paraffin and sections (5.0 μm thick) were cut with microtome onto glass slides (Reichert, Austria). The section slides were de-paraffinized in 2 changes of xylene for 5 minutes each followed by rehydration using a series of graded alcohols. Antigen retrieval was performed by boiling the slides in 0.01 M tri-sodium citrate buffer, pH 6, for 10 min followed by incubation with Levamisole to block endogenous alkaline phosphatase. Sections were then pre-incubated with 10% normal goat serum, 0.2% Triton X-100 overnight at 4°C to block nonspecific binding.

Table 1. Oligonucleotide primers used in Real-time RT PCR analyses.

Primer name	Sequence (5′ → 3′)	T$_a$ (°C)	Species
GAPDH forward	AGCCTTCTCCATGGTGGTGAAGAC	60	Human/Mouse
GAPDH reverse	CGGAGTCAACGGATTTGGTCG		
HERV-K (II) forward	CCTGCAGTCCAAAATTGGTT	55	Human
HERV-K (II) reverse	GGGGCAAGTTTTTCCCTTTAG		
hIL-6 forward	ACCCCTGACCCAACCACAAAT	58	Human
hIL-6 reverse	AGCTGCGCAGAATGAGATGAG		
hTNFα forward	CCCAGGGACCTCTCTCTAATCA	57	Human
hTNFα reverse	GCTACAGGCTTGTCACTCGG		
hIFN-β forward	CAGCAATTTTCAGTGTCAGAAGCT	57	Human
hIFN-β reverse	TCATCCTGTCCTTGAGGCAGTA		
hGFAP forward	GGACATCGAGATCGCCACCTACAG	60	Human
hGFAP reverse	CTCACCATCCCGCATCTCCACAGT		
hIL-1β forward	CCAAAGAAGAAGATGGAAAAGC	55	Human
hIL-1β reverse	GGTGCTGATGTACCAGTTGGG		
hBDNF forward	GAAAGTCCCGGTATCCAAAG	50	Human
hBDNF reverse	CCAGCCAATTCTCTTTTT		
hNGF forward	CCAAGGGAGCAGCTTTCTATCCTGG	60	Human
hNGF reverse	GGCAGTGTCAAGGGAATGCGAAGTT		
GFP forward	CCACAACATCGAGGACGGCA	55	pBud-GFP plasmid
GFP reverse	CGGGATCACTCTCGGCATGG		
mTNFα forward	ATGCTGGGACAGTGACCTGG	54	Mouse
mTNFα reverse	CCTTGATGGTGGTGCATGAG		
mIL-6 forward	ATGGATGCTACCAAACTGGAT	54	Mouse
mIL-6 reverse	TGAAGGACTCTGGCTTTGTCT		
mIFN-β forward	AAGAGTTACACTGCCTTTGCCATC	55	Mouse
mIFN-β reverse	CACTGTCTGCTGGTGGAGTTCATC		
mGFAP forward	GGACATCGAGATCGCCACCTACAG	55	Mouse
mGFAP reverse	CTCACCATCCCGCATCTCCACAGT		
mBDNF forward	AGTTCCACCAGGTGAGAAGA	55	Mouse
mBDNF reverse	GGTAATTTTTGTATTCCTCCAGCAGA		

To detect Iba-1, cleaved caspase-6, BDNF or GFAP immunoreactivity, slides were incubated overnight at 4°C with antibodies to MAP-2 (1:800; Sigma, USA), Iba-1 (1:1000, Waco), GFAP (1:5000, DAKO), BDNF (1:1000, eBioscience) and cleaved caspase-6 (1:500, gift from Dr. Andrea LeBlanc, McGill University), diluted in 5% normal goat serum, 0.2% Triton X-100. Mouse brain sections were also Nissl stained. A secondary alkaline phosphatase–conjugated goat anti–mouse or anti-mouse antibody (Jackson ImmunoResearch Laboratories) followed by NBT/BCIP substrate (Vector Laboratories) was used for single labeling. For double labeling with HERV-K(II) Env, human brain sections pretreated with 0.3% hydrogen peroxide to block endogenous peroxidases were incubated with rabbit polyclonal HERV-K(II) env antibody (1:200; Novus, USA), followed by biotinylated goat anti–rabbit antibody by avidin–biotin–peroxidase complexes (Vector Laboratories), then 3,3′-diaminobenzidine tetrachloride (Vector Laboratories) was applied.

Western blotting

Brain tissue or transfected cells were lysed with Laemmli buffer with 0.1% β-mercaptoethanol and boiled at 95°C for 10 minutes. Proteins from whole cell lysates were separated using polyacrylamide gel electrophoresis and protein fractions were transferred to a nitrocellulose membrane overnight (Bio-Rad, Mississauga, ON, CA). The membrane was blocked with 5% milk for 1 hour and labeled with monoclonal mouse anti-HERV-K(II) Env antibody (1:200; Novus, USA), overnight at 4°. The immunolabeled membrane was then probed with secondary peroxidase-conjugated goat anti-rabbit IgG (1:500: Jackson ImmunoResearch Laboratories, Inc., West Grove, PA, USA) for 2 hrs. Anti-βIII-actin antibodies were used to assess gel loading (1:1000) (Santa Cruz Biotechnology, Inc., Dallas, TX, USA). Membranes were developed with Pierce ECL Western blotting substrate (Fisher Scientific, Ottawa, ON, Canada) and exposed on film (Canon Canada, Inc., Mississauga, ON, Canada).

Deep sequencing and analyses of brain transcriptome

The high throughput brain transcriptome analysis was performed as described previously [7]. In brief ten micrograms of total RNA from fetal, surgical and clinical brain samples were used for the cDNA synthesis using ds-cDNA synthesis kit (Invitrogen) according to the manufacturer's instructions. The resulting ds-cDNA was cleaned and single end tag (SET) sequencing was performed using the Illumina Genome Analyzer per the manufacturer's instructions. Short read sequences (tags) obtained from the Illumina Genome Analyzer were mapped to the reference HERV mRNAs from the NCBI database (study accession number SRP032168).

Bioinformatic analyses

The sequence tags derived from deep sequencing of healthy surgical and clinical brain samples were unambiguously assigned to different HERV families and host genes and were analyzed for the abundance of different HERVs and host genes. Bioinformatic analysis was performed to gain insight into the functional aspects of host genes with expression levels highly correlated with those of HERV-K(II) *env*. To account for differences in sequence tags obtained from different sets of experiments and variation in starting materials (cDNA) was reconciled by normalizing different genes across all the samples (global normalization), which assumes the distribution of gene expression over different experiments was similar. All genes, which passed the filtering criteria (≥ 2 tags detected) and showed a high degree of correlation with respect to HERV-K(II) *env* transcript levels (Pearson $r^2 \geq 0.5$) in each sample, were analyzed in context of the BP_FAT gene ontology (GO) terms for overrepresented functional classes and tissue specificity examination using the DAVID tool (http://david.abcc.ncifcrf. gov). For clinical samples GO analysis was performed on the genes passing the above criteria as well as showing $> = 0.3$ fold change compared to controls.

Statistical analyses

Experimental variables were analyzed by Student t or Kruskal-Wallis tests for parametric or non-parametric continuous variables and the Chi-square test for categorical variables. *In vitro* data were tested by one-way ANOVA with Bonferroni *post hoc* tests. The level of significance was defined as $p < 0.05$.

Supporting Information

Figure S1   (A) Human-specific HERV-K insertion loci in the human genome. The red arrows indicated the chromosomal locations of truncated and full-length HERV-K elements i.e. Env, LTR and central ORFs. (B) A heatmap showing relative expression of different HERV-K elements on human genome.

Figure S2   Relative expression of HERV-K(II) *env* in *pHERV-Kenv* transfected HEK, SK-N-SH and C17.2 cell lines normalized to the control vector (*pGFP*) transfected-matched cell line.

Figure S3   Transcript levels $vpr/RAG1^{-/-}$ **mice implanted with HERV-K(II) env/eGFP expressing c17.2 cells.** Analyses of gene expression from implanted/non-implanted hemispheres of brain sections did not reveal significant differences in **(A)** *IFN-β*, **(B)** *GFAP*, **(C)** *GFP* and **(D)** *BDNF* transcript levels.

Acknowledgments

The authors thank Dr. Evan Snyder for kindly providing C17.2 cells.

Author Contributions

Conceived and designed the experiments: RB CP. Performed the experiments: RB WR JMA FM. Analyzed the data: RB FM KKE CP. Contributed reagents/materials/analysis tools: RRT BMW. Wrote the paper: RB CP.

References

1. Becker Y (1995) Endogenous retroviruses in the human genome–a point of view. Virus genes 9: 211–218.
2. Lander ES, Linton LM, Birren B, Nusbaum C, Zody MC, et al. (2001) Initial sequencing and analysis of the human genome. Nature 409: 860–921.
3. Bannert N, Kurth R (2004) Retroelements and the human genome: new perspectives on an old relation. Proc Natl Acad Sci U S A 101 Suppl 2: 14572–14579.
4. Paces J, Pavlicek A, Paces V (2002) HERVd: database of human endogenous retroviruses. Nucleic acids research 30: 205–206.
5. Weiss RA (2006) The discovery of endogenous retroviruses. Retrovirology 3: 67.
6. Deslauriers AM, Afkhami-Goli A, Paul AM, Bhat RK, Acharjee S, et al. (2011) Neuroinflammation and endoplasmic reticulum stress are coregulated by crocin to prevent demyelination and neurodegeneration. Journal of immunology 187: 4788–4799.
7. Bhat RK, Ellestad KK, Wheatley BM, Warren R, Holt RA, et al. (2011) Age-and disease-dependent HERV-W envelope allelic variation in brain: association with neuroimmune gene expression. PloS one 6: e19176.
8. Barbulescu M, Turner G, Seaman MI, Deinard AS, Kidd KK, et al. (1999) Many human endogenous retrovirus K (HERV-K) proviruses are unique to humans. Current biology: CB 9: 861–868.
9. Mallet F, Prudhomme S (2004) [Retroviral inheritance in man]. Journal de la Societe de biologie 198: 399–412.
10. Reynier F, Verjat T, Turrel F, Imbert PE, Marotte H, et al. (2009) Increase in human endogenous retrovirus HERV-K (HML-2) viral load in active rheumatoid arthritis. Scandinavian journal of immunology 70: 295–299.
11. Douville R, Liu J, Rothstein J, Nath A (2011) Identification of active loci of a human endogenous retrovirus in neurons of patients with amyotrophic lateral sclerosis. Annals of neurology 69: 141–151.
12. McCormick AL, Brown RH Jr, Cudkowicz ME, Al-Chalabi A, Garson JA (2008) Quantification of reverse transcriptase in ALS and elimination of a novel retroviral candidate. Neurology 70: 278–283.
13. Belshaw R, Dawson AL, Woolven-Allen J, Redding J, Burt A, et al. (2005) Genomewide screening reveals high levels of insertional polymorphism in the human endogenous retrovirus family HERV-K(HML2): implications for present-day activity. Journal of virology 79: 12507–12514.
14. Turner G, Barbulescu M, Su M, Jensen-Seaman MI, Kidd KK, et al. (2001) Insertional polymorphisms of full-length endogenous retroviruses in humans. Current biology: CB 11: 1531–1535.
15. Tonjes RR, Boller K, Limbach C, Lugert R, Kurth R (1997) Characterization of human endogenous retrovirus type K virus-like particles generated from recombinant baculoviruses. Virology 233: 280–291.
16. Contreras-Galindo R, Lopez P, Velez R, Yamamura Y (2007) HIV-1 infection increases the expression of human endogenous retroviruses type K (HERV-K) in vitro. AIDS Res Hum Retroviruses 23: 116–122.
17. Lower R, Lower J, Kurth R (1996) The viruses in all of us: characteristics and biological significance of human endogenous retrovirus sequences. Proceedings of the National Academy of Sciences of the United States of America 93: 5177–5184.
18. Yolken RH, Karlsson H, Yee F, Johnston-Wilson NL, Torrey EF (2000) Endogenous retroviruses and schizophrenia. Brain research Brain research reviews 31: 193–199.
19. Crow TJ (1987) Integrated Viral Genes as Potential Pathogens in the Functional Psychoses. Journal of Psychiatric Research 21: 479–485.
20. Christensen T (2010) HERVs in Neuropathogenesis. Journal of Neuroimmune Pharmacology 5: 326–335.
21. Rakoff-Nahoum S, Kuebler PJ, Heymann JJ, Sheehy ME, Ortiz GM, et al. (2006) Detection of T lymphocytes specific for human endogenous retrovirus K (HERV-K) in patients with seminoma. AIDS research and human retroviruses 22: 52–56.
22. Johnston JB, Silva C, Holden J, Warren KG, Clark AW, et al. (2001) Monocyte activation and differentiation augment human endogenous retrovirus expression: implications for inflammatory brain diseases. Annals of neurology 50: 434–442.
23. Maeda N, Palmarini M, Murgia C, Fan H (2001) Direct transformation of rodent fibroblasts by jaagsiekte sheep retrovirus DNA. Proceedings of the National Academy of Sciences of the United States of America 98: 4449–4454.

24. Palmarini M, Sharp JM, De las Heras M, Fan H (1999) Jaagsiekte sheep retrovirus is necessary and sufficient to induce a contagious lung cancer in sheep. Journal of virology 73: 6964–6972.

25. Contreras-Galindo R, Kaplan MH, Contreras-Galindo AC, Gonzalez-Hernandez MJ, Ferlenghi I, et al. (2012) Characterization of human endogenous retroviral elements in the blood of HIV-1-infected individuals. Journal of virology 86: 262–276.

26. Contreras-Galindo R, Kaplan MH, Markovitz DM, Lorenzo E, Yamamura Y (2006) Detection of HERV-K(HML-2) viral RNA in plasma of HIV type 1-infected individuals. AIDS research and human retroviruses 22: 979–984.

27. Frendo JL, Olivier D, Cheynet V, Blond JL, Bouton O, et al. (2003) Direct involvement of HERV-W env glycoprotein in human trophoblast cell fusion and differentiation. Molecular and Cellular Biology 23: 3566–3574.

28. McArthur JC, Steiner J, Sacktor N, Nath A (2010) Human immunodeficiency virus-associated neurocognitive disorders: Mind the gap. Annals of neurology 67: 699–714.

29. Herbein G, O'Brien WA (2000) Tumor necrosis factor (TNF)-alpha and TNF receptors in viral pathogenesis. Proceedings of the Society for Experimental Biology and Medicine Society for Experimental Biology and Medicine 223: 241–257.

30. Cheeran MC, Hu S, Sheng WS, Peterson PK, Lokensgard JR (2003) CXCL10 production from cytomegalovirus-stimulated microglia is regulated by both human and viral interleukin-10. Journal of virology 77: 4502–4515.

31. Acharjee S, Noorbakhsh F, Stemkowski PL, Olechowski C, Cohen EA, et al. (2010) HIV-1 viral protein R causes peripheral nervous system injury associated with in vivo neuropathic pain. FASEB journal: official publication of the Federation of American Societies for Experimental Biology 24: 4343–4353.

32. Garrison KE, Jones RB, Meiklejohn DA, Anwar N, Ndhlovu LC, et al. (2007) T cell responses to human endogenous retroviruses in HIV-1 infection. PLoS pathogens 3: e165.

33. Nixon DF, Townsend AR, Elvin JG, Rizza CR, Gallwey J, et al. (1988) HIV-1 gag-specific cytotoxic T lymphocytes defined with recombinant vaccinia virus and synthetic peptides. Nature 336: 484–487.

34. Sacha JB, Kim IJ, Chen L, Ullah JH, Goodwin DA, et al. (2012) Vaccination with cancer- and HIV infection-associated endogenous retrotransposable elements is safe and immunogenic. J Immunol 189: 1467–1479.

35. Antony JM, Izad M, Bar-Or A, Warren KG, Vodjgani M, et al. (2006) Quantitative analysis of human endogenous retrovirus-W env in neuroinflammatory diseases. AIDS research and human retroviruses 22: 1253–1259.

36. Antony JM, Zhu Y, Izad M, Warren KG, Vodjgani M, et al. (2007) Comparative expression of human endogenous retrovirus-W genes in multiple sclerosis. AIDS research and human retroviruses 23: 1251–1256.

37. Antony JM, Ellestad KK, Hammond R, Imaizumi K, Mallet F, et al. (2007) The human endogenous retrovirus envelope glycoprotein, syncytin-1, regulates neuroinflammation and its receptor expression in multiple sclerosis: a role for endoplasmic reticulum chaperones in astrocytes. Journal of immunology 179: 1210–1224.

38. Antony JM, van Marle G, Opii W, Butterfield DA, Mallet F, et al. (2004) Human endogenous retrovirus glycoprotein-mediated induction of redox reactants causes oligodendrocyte death and demyelination. Nature neuroscience 7: 1088–1095.

39. Antony JM, Deslauriers AM, Bhat RK, Ellestad KK, Power C (2011) Human endogenous retroviruses and multiple sclerosis: innocent bystanders or disease determinants? Biochimica et biophysica acta 1812: 162–176.

40. Fuchs NV, Kraft M, Tondera C, Hanschmann KM, Lower J, et al. (2011) Expression of the human endogenous retrovirus (HERV) group HML-2/

41. HERV-K does not depend on canonical promoter elements but is regulated by transcription factors Sp1 and Sp3. Journal of virology 85: 3436–3448.

41. van der Kuyl AC (2012) HIV infection and HERV expression: a review. Retrovirology 9: 6.

42. Carlson NG, Wieggel WA, Chen J, Bacchi A, Rogers SW, et al. (1999) Inflammatory cytokines IL-1 alpha, IL-1 beta, IL-6, and TNF-alpha impart neuroprotection to an excitotoxin through distinct pathways. Journal of immunology 163: 3963–3968.

43. Moonen G, Malgrange B, Rigo JM, Rogister B (1996) Neurotrophic factors: past and future. Acta neurologica Belgica 96: 203–218.

44. Yamada M, Hatanaka H (1994) Interleukin-6 protects cultured rat hippocampal neurons against glutamate-induced cell death. Brain research 643: 173–180.

45. Kristiansen OP, Mandrup-Poulsen T (2005) Interleukin-6 and diabetes: the good, the bad, or the indifferent? Diabetes 54 Suppl 2: S114–124.

46. St Hillaire C, Vargas D, Pardo CA, Gincel D, Mann J, et al. (2005) Aquaporin 4 is increased in association with human immunodeficiency virus dementia: implications for disease pathogenesis. J Neurovirol 11: 535–543.

47. Noorbakhsh F, Ramachandran R, Barsby N, Ellestad KK, LeBlanc A, et al. (2010) MicroRNA profiling reveals new aspects of HIV neurodegeneration: caspase-6 regulates astrocyte survival. FASEB journal: official publication of the Federation of American Societies for Experimental Biology 24: 1799–1812.

48. van Marle G, Henry S, Todoruk T, Sullivan A, Silva C, et al. (2004) Human immunodeficiency virus type 1 Nef protein mediates neural cell death: a neurotoxic role for IP-10. Virology 329: 302–318.

49. Na H, Acharjee S, Jones G, Vivithanaporn P, Noorbakhsh F, et al. (2011) Interactions between human immunodeficiency virus (HIV)-1 Vpr expression and innate immunity influence neurovirulence. Retrovirology 8: 44.

50. Ryder EF, Snyder EY, Cepko CL (1990) Establishment and characterization of multipotent neural cell lines using retrovirus vector-mediated oncogene transfer. Journal of neurobiology 21: 356–375.

51. Snyder EY, Deitcher DL, Walsh C, Arnold-Aldea S, Hartwieg EA, et al. (1992) Multipotent neural cell lines can engraft and participate in development of mouse cerebellum. Cell 68: 33–51.

52. Snyder EY (1995) Immortalized neural stem cells: insights into development; prospects for gene therapy and repair. Proceedings of the Association of American Physicians 107: 195–204.

53. Kim DE, Tsuji K, Kim YR, Mueller FJ, Eom HS, et al. (2006) Neural stem cell transplant survival in brains of mice: assessing the effect of immunity and ischemia by using real-time bioluminescent imaging. Radiology 241: 822–830.

54. Vivithanaporn P, Maingat F, Lin LT, Na H, Richardson CD, et al. (2010) Hepatitis C Virus Core Protein Induces Neuroimmune Activation and Potentiates Human Immunodeficiency Virus-1 Neurotoxicity. PLoS one 5.

55. Jones GJ, Barsby NL, Cohen EA, Holden J, Harris K, et al. (2007) HIV-1 Vpr causes neuronal apoptosis and in vivo neurodegeneration. Journal of Neuroscience 27: 3703–3711.

56. Silva C, Zhang K, Tsutsui S, Holden JK, Gill MJ, et al. (2003) Growth hormone prevents human immunodeficiency virus-induced neuronal p53 expression. Annals of neurology 54: 605–614.

57. Power C, Henry S, Del Bigio MR, Larsen PH, Corbett D, et al. (2003) Intracerebral hemorrhage induces macrophage activation and matrix metalloproteinases. Ann Neurol 53: 731–742.

58. DAVID Database for Annotation V, and Integrated Discovery. Available: http://david.abcc.ncifcrf.gov/.

59. Dennis G Jr, Sherman BT, Hosack DA, Yang J, Gao W, et al. (2003) DAVID: Database for Annotation, Visualization, and Integrated Discovery. Genome biology 4: P3.

2

An Examination of the Relationship between Hotspots and Recombination Associated with Chromosome 21 Nondisjunction

Tiffany Renee Oliver[1,2]*[9], **Candace D. Middlebrooks**[1][9], **Stuart W. Tinker**[1], **Emily Graves Allen**[1], **Lora J. H. Bean**[1], **Ferdouse Begum**[3], **Eleanor Feingold**[3,4], **Reshmi Chowdhury**[4], **Vivian Cheung**[5,6], **Stephanie L. Sherman**[1]

1 Department of Human Genetics, Emory University School of Medicine, Atlanta, Georgia, United States of America, 2 Department of Biology, Spelman College, Atlanta, Georgia, United States of America, 3 Department of Biostatistics, Graduate School of Public Health, University of Pittsburgh, Pittsburgh, Pennsylvania, United States of America, 4 Department of Human Genetics, Graduate School of Public Health University of Pittsburgh, Pittsburgh, Pennsylvania, United States of America, 5 Howard Hughes Medical Institute, University of Michigan, Ann Arbor, Michigan, United States of America, 6 Department of Human Genetics, University of Michigan, Ann Arbor, Michigan, United States of America

Abstract

Trisomy 21, resulting in Down Syndrome (DS), is the most common autosomal trisomy among live-born infants and is caused mainly by nondisjunction of chromosome 21 within oocytes. Risk factors for nondisjunction depend on the parental origin and type of meiotic error. For errors in the oocyte, increased maternal age and altered patterns of recombination are highly associated with nondisjunction. Studies of normal meiotic events in humans have shown that recombination clusters in regions referred to as hotspots. In addition, GC content, CpG fraction, Poly(A)/Poly(T) fraction and gene density have been found to be significant predictors of the placement of sex-averaged recombination in the human genome. These observations led us to ask whether the altered patterns of recombination associated with maternal nondisjunction of chromosome 21 could be explained by differences in the relationship between recombination placement and recombination-related genomic features (i.e., GC content, CpG fraction, Poly(A)/Poly(T) fraction or gene density) on 21q or differential hot-spot usage along the nondisjoined chromosome 21. We found several significant associations between our genomic features of interest and recombination, interestingly, these results were not consistent among recombination types (single and double proximal or distal events). We also found statistically significant relationships between the frequency of hotspots and the distribution of recombination along nondisjoined chromosomes. Collectively, these findings suggest that factors that affect the accessibility of a specific chromosome region to recombination may be altered in at least a proportion of oocytes with MI and MII errors.

Editor: Beth A. Sullivan, Duke University, United States of America

Funding: This work was supported by R01 HD38979, R01 HL083300 and UL1TR000454. The funders had no role in study design, data collection and analysis, decision to publish, or preparation of the manuscript.

Competing Interests: The authors have declared that no competing interests exist.

* E-mail: Toliver4@spelman.edu

[9] These authors contributed equally to this work.

Introduction

Trisomy 21, leading to Down Syndrome (DS), is the most common autosomal trisomy among live-born infants, occurring in approximately 1 in 700 live-births, and is caused mainly by the failure of chromosome 21 to properly segregate during oogenesis [1]. Increased maternal age and altered number and location of recombination events have been found to be associated with maternal meiotic errors involving chromosome 21 [2,3]. Specifically, the absence of recombination [4] or the presence of a single recombinant event near the telomere of 21q [2] are associated with maternal meiosis I (MI) errors and these associations appear to be independent of the age of the oocyte (i.e., maternal age at the time of birth of the infant with trisomy 21) [5]. Meiosis II (MII) errors appear to be driven by different age and recombination traits: MII errors are associated with the placement of a recombinant event near the centromere of 21q [2] and this association increases with increasing age of the oocyte [5].

Studies of normal meiotic events in humans show that the placement of recombination is not a random event. Rather, both cis and trans-acting factors have been found to be associated with the placement of recombination. Specifically, GC content, CpG fraction and Poly(A)/Poly(T) fraction have each been found to be significant predictors of placement of sex-averaged recombination events in the human genome [6]. In addition, sequence variation in the zinc-finger domain of the gene *Proline Rich Domain Containing 9* (PRDM9) has a major impact on the location of recombination in humans [7,8,9,10]. Specifically, allelic differences in the zinc finger binding domain of *PRDM9* explain approximately 80% of the heritable variation in "hotspot usage" " (i.e. the frequency in which recombination occurs within linkage disequilibrium (LD) or "historically"-defined hotspots) [8,11,12]. The observation that both cis and trans-acting factors are associated with the placement of recombination led us to question whether the altered patterns of recombination associated with nondisjunction of chromosome 21 could be explained by differences in the relationship between recombination and genomic features (i.e., GC content, CpG fraction, Poly(A)/Poly(T) fraction or gene density) on 21q or differential hot-spot usage. This paper presents the first analyses of the relationship between recombination rate and the quantity of genomic features or LD-defined hotspots specifically along chromosome 21 in oocytes with a normal meiotic outcome, a MI nondisjunction error or a MII nondisjunction error.

Materials and Methods

Ethics Statement

The work presented in this publication was approved by the Emory Univeristy Institutional Review Board. All participants in provided written consent which indicated that the individual (1) agreed for study personnel to proceed with the interview and (2) consented for biological specimens to be obtained from them and their child. All information obtained during participant interviews and related to sample collection were catalogued electronically and de-identified.

Trisomic Population

Families with an infant with full trisomy 21 were recruited through a multisite study of risk factors associated with chromosome mal-segregation [2,13,14]. Parents and the infant donated a biological sample (either blood or buccal) from which DNA was extracted. Only families in which DNA was available from both biological parents and the child with trisomy 21 were included, leading to a total of 297 maternal MI and 277 maternal MII cases of trisomy 21 (Table 1).

Trisomic Population Genotyping and Quality Control

Samples were genotyped at 1536 SNP loci on 21q by the Center for Inherited Disease Research using the Illumina Golden Gate Assay. The most centromeric SNP was rs2259403 (13,615,252 bp) and the most telomeric was rs7116 (46,909,248 bp). The average number of SNPs per 500 kb bin was 25.56 with a standard deviation of 25.91 with over 70% of cases exhibiting a recombinant having recombination breakpoints smaller than 1 Mb. Mendelian inconsistencies and sample mix-ups were identified using RelCheck among the trios. In addition, parental genotyping data were used to identify poorly performing SNPs.

Table 1. Population Sample Sizes.

Meiotic Outcome Group and Recombination Type	Number of samples
MI Single	222
MI Proximal	75
MI Distal	75
MII Single	202
MII Proximal	75
MII Distal	75
Normal Single	1272
Normal Proximal	342
Normal Distal	342

SNPs that met the following criteria were excluded from our analyses: minor allele frequency (MAF) <0.01, deviation from Hardy Weinberg Equilibrium (HWE) (p<0.01), heterozygosity > 0.60 or > 10% missingness. We also excluded SNPs on a family-by-family basis if >50% of the genotype data for a proband had low intensity levels. As it relates to our exclusion of SNPs with a heterozygosity rate of >0.60, while we understand that is a very conservative/stringent cutoff, we did indeed examine the distribution of cases by stage and origin upon changing the heterozygosity rate and we did not see any significant changes in stage (data not shown). In addition, for a significant majority of our cases, stage and origin had been previously determined using STR data and compared to what was identified with our SNP only data.

Determining Stage and Origin of Meiotic Chromosome Mal-Segregation

Individuals with trisomy 21 have three copies of chromosome 21 and thus display three alleles for each SNP genotyped on chromosome 21. In instances where trisomy 21 is caused by a maternal meiotic error, for each SNP examined, one of these alleles is inherited from dad, while the other two are inherited from mom. Maternal meiotic errors were confirmed upon determining that trisomic offspring inherited two alleles from mom and one from dad for SNPs genotyped on chromosome 21. Only cases of maternal origin were included in our analyses. Once the maternal origin of the meiotic error was established, markers located in the pericentromeric region (13,615,252 bp – 16,784,299 bp) of 21q were used to infer the stage of the meiotic error, MI or MII. If maternal heterozygosity was retained in the trisomic offspring, we concluded a MI error. If maternal heterozygosity was reduced to homozygosity, we concluded a MII error. In this assay, we cannot distinguish between the different types of underlying errors that might lead to an MII error. For example, sister chromatids that fail to separate during anaphase of MII or an error that is initiated in MI and not resolved properly in MII both lead to the contribution of sister chromatids to the gamete. Also, if sister chromatids prematurely separate in MI, some configurations will lead to both sister chromatids segregating to the same pole in MII. Lastly, when all informative markers in the parent of origin were reduced to homozygosity, the origin of nondisjunction was inferred to be a post-zygotic, mitotic error and excluded from the study.

Identifying the Location of Recombination – Trisomic Samples

After genotyping quality control measures were implemented and SNP data were combined with STR data from our previous studies [3], we defined the location of recombinant events. The breakpoints of a single recombinant event were defined by a minimum of either one STR or eight consecutive, informative SNPs flanking the recombination breakpoint. An exception to this rule occurred when the most proximal or most distal informative markers on 21q indicated the presence of recombinant event. In these instances, a minimum of either one STR or four consecutive, informative SNPs were required to define the breakpoints of recombination. The presence of a double recombinant event was defined by a minimum of either one STR or 8 consecutive, informative SNPs flanking the recombination breakpoint on each side for both events.

Euploid Population

SNP genotyping data for normally segregating chromosomes 21 were taken from families recruited for 1) the Autism Genetic Research Exchange (AGRE) (N = 743) [15], 2) the Framingham

Heart Study (FHS) (N = 764) [16] and 3) the GENEVA Dental Caries Study (N = 107) [17] (Table 1). All families were two-generation families with a minimum of three children. This was necessary to define specific recombination profiles for each parent child transmission.

Euploid Population Genotyping and Quality Control

The AGRE samples were genotyped for SNPs genome-wide using the Infinium(R) HumanHap550-Duo BeadChip. The AGRE data included genotypes at 520,017 markers genome-wide, however 11,473 markers were excluded from the analysis due to deviation HWE ($p < 10^{-7}$). After quality control measures were completed, there was genotype information for 7,810 SNPs on 21q for the AGRE dataset. The FHS samples were genotyped for SNPs genome-wide using the Genome-Wide Human SNP Array 5.0. The FHS data included genotypes at 500,568 markers. However, 22,000 markers were excluded from the analysis due to deviation from HWE ($p < 10^{-7}$). After quality control measures were completed, there was genotype information for 6,705 SNPs on 21q for the FHS dataset. The GENEVA samples were genotyped using the Illumina 610-Quad Array. The GENEVA dataset included genotypes at 620,901 SNPs. 58,610 markers were excluded from the analysis due to deviation from HWE ($p < 10^{-5}$), a MAF < 0.02. After quality control measures were completed, there was genotype information for 8,189 SNPs on 21q from the GENEVA population. All SNP locations were based on human NCBI Build 36 (hg18).

Identifying the Location of Recombination – Euploid Samples

For the AGRE, FHS and GENEVA datasets, genotype data from members of two-generation families with three or more children were used to infer the location of recombination along the maternal chromosome 21. Our approach and software are described in Chowdhury et al. [18]. Briefly, parental genotypes were used to identify informative markers. Then, using these markers, genotypes of the children were compared to identify alleles inherited identical-by-descent from the mothers and fathers. Between two sibs, a switch from sharing the same maternal allele to not sharing was scored as a maternal recombination event.

Examining the Relationship Between Genomic Features and Recombination

We used linear regression models to assess the relationship between the quantity of recombination and the quantity of each variable of interest found within regions across 21q. We divided 21q into 500 kb bins and calculated the quantity of each variable within a bin. We chose this bin size based on our level of refinement of recombination break-points. For the genomic features, we quantified the amount of each bin occupied by each genomic feature of interest, GC content, CpG content, Poly(A)/Poly(T) content were calculated as the proportion of each bin occupied by each feature however the number of genes per bin was calculated for gene density. Data on genomic features were based on the hg18 build of the human genome and retrieved from the following tables within the USCS Genome Browser: gc5Base, CpGIslandExt and rmsk (repeat master), UniGene_3 and RefGene. As for hotspots, we used the number of LD-defined hotspots, as defined by Myers et. al.[19] per bin as the predictor variable (Figure S1). The outcome variable was defined as the proportion of all chromosome 21 single or double recombinant events that occurred within the bin. As it is well known that single recombination events cluster in the telomeric and centromeric

regions of 21q for the MI and MII error groups (Fig. 1), respectively, we included bin location as a variable in our models as it may be a confounding variable. We stratified analyses by chromosomes with single and double recombinant events (Figs. 2 and 3, Table 1) as mechanisms of chromosome 21 nondisjunction may differ based on the number of recombinant events on 21q[2,3,5,14]. Univariate linear regression was then used to determine whether there was significant correlation between the quantity of each predictor variable and the proportion of recombination within a bin ($p \leq 0.05$).

General linear regression models were used to test for differences in the slopes of the regression models between comparison groups (MI or MII versus Controls) for each predictor. That is, to compare MI error to normal meiotic outcomes, we included the interaction term of comparison group by genomic feature within a bin. This type of model was also used to compare MII errors with controls. Once again, we included bin location as a covariate for the reason stated above.

Data Availability

Data on recombination along normally segregating chromosomes 21 came from three different studies, the AGRE, FHS and GENEVA. Access to data used in this analysis from the AGRE is publically available upon IRB approval or exemption. For more information please logon to https://research.agre.org. Data from the FHS and GENEVA Studies is now available via dbGaP, accession numbers phs000007.v23.p8 and phs000440.v1.p1 respectively. Genotypes used to determine the placement of recombination along nondisjoined chromosomes 21 will also be available via dbGAP.

Results

Association between genomic features along chromosome 21q and the proportion of recombination events

We first examined meiotic events with one detectable recombinant event on 21q (Table 2). In regression models that included both the specific genomic feature (i.e., GC content, CpG fraction, Poly(A)/Poly(T) fraction or gene density) and the location of the bin along 21q, only location, was found to be a significant predictor of the amount of recombination for the vast majority of features. This is consistent with previous work that has established altered placement of recombination as a significant risk factor for chromosome 21 nondisjunction [2,3]. There was one exception to this pattern: among the MII errors with a single recombinant, both location and GC content were significant predictors of the amount of recombination. This suggests that among MII single recombinant events, where the increased risk is associated with a pericentromeric recombinant, there may be a preference for recombination to occur in regions with elevated GC content and close to the centromere.

We then looked at meiotic events with two detectable recombinants and separated the analyses by the proximal and distal event. For proximal recombinant events (Table 3), GC and CpG content as well as bin location were found to be positively correlated with recombination among MI and MII errors; no association for these features was found among normal meiotic control recombinant events of this type (Table 3). Poly(A)/Poly(T) fraction was found to be inversely correlated with the amount of recombination among MI and MII errors and normal outcomes. Collectively these observations suggest that MI and MII proximal recombinant events occur in GC rich regions more often than statistically expected if there was no relationship between the

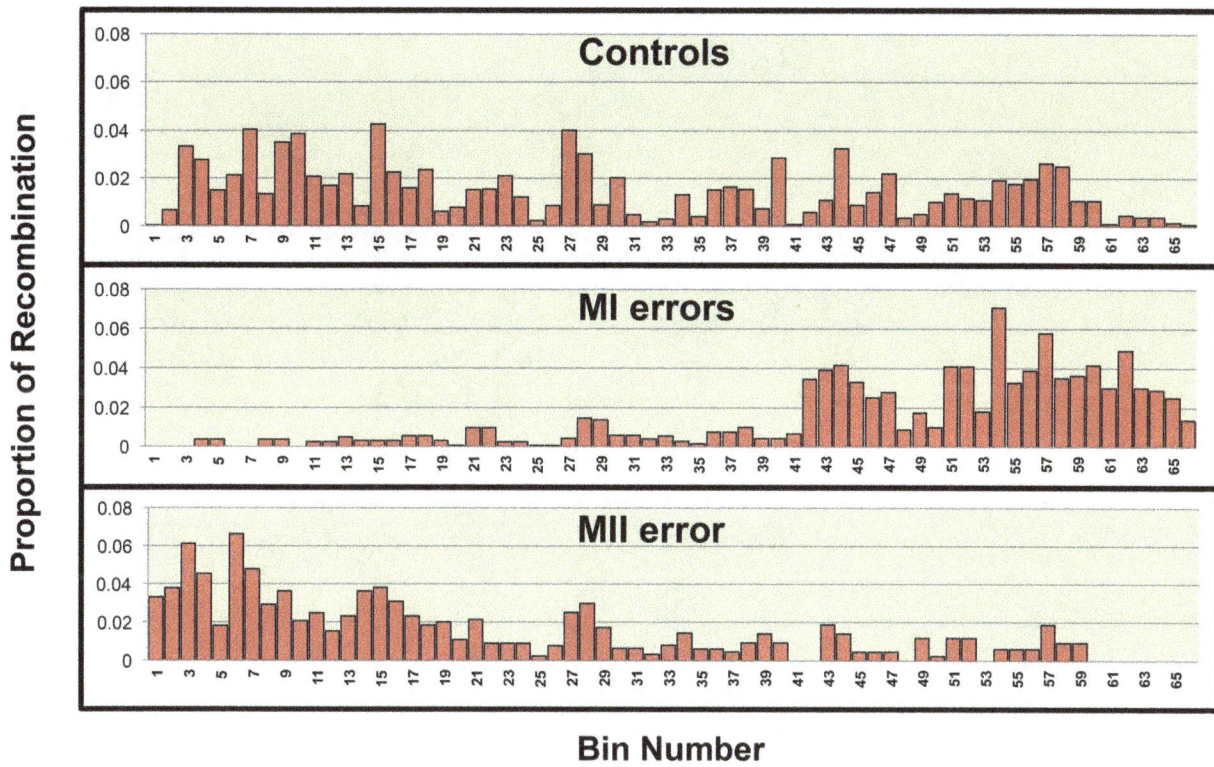

Figure 1. The distribution of single recombination events across the long arm of 21q by population. 21q was divided into 66 500 kb bins and the proportion of recombination in each bin from chromosomes with only one recombinant event is depicted above.

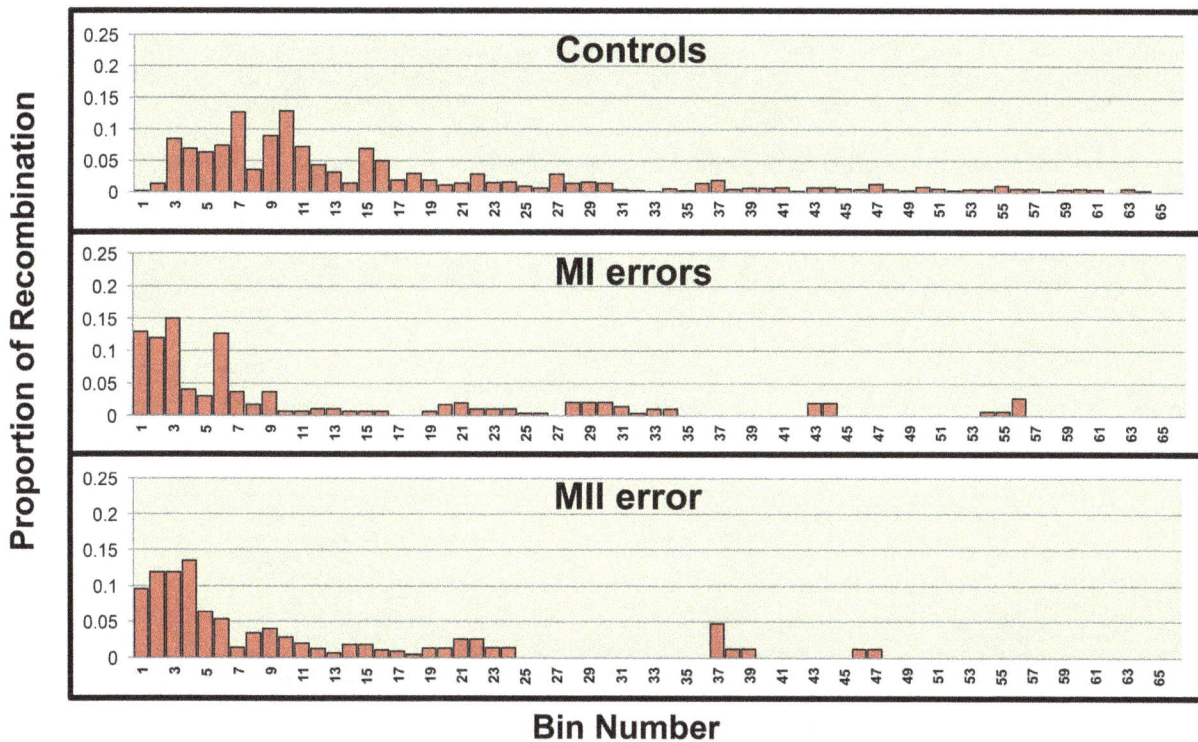

Figure 2. The distribution of the proximal recombinant of a double recombinant event across the long arm of 21q by population. 21q was divided into 66 500 kb bins and the distribution of Recombination in each bin from the proximal recombinant event of chromosomes displaying two recombinant events is depicted above.

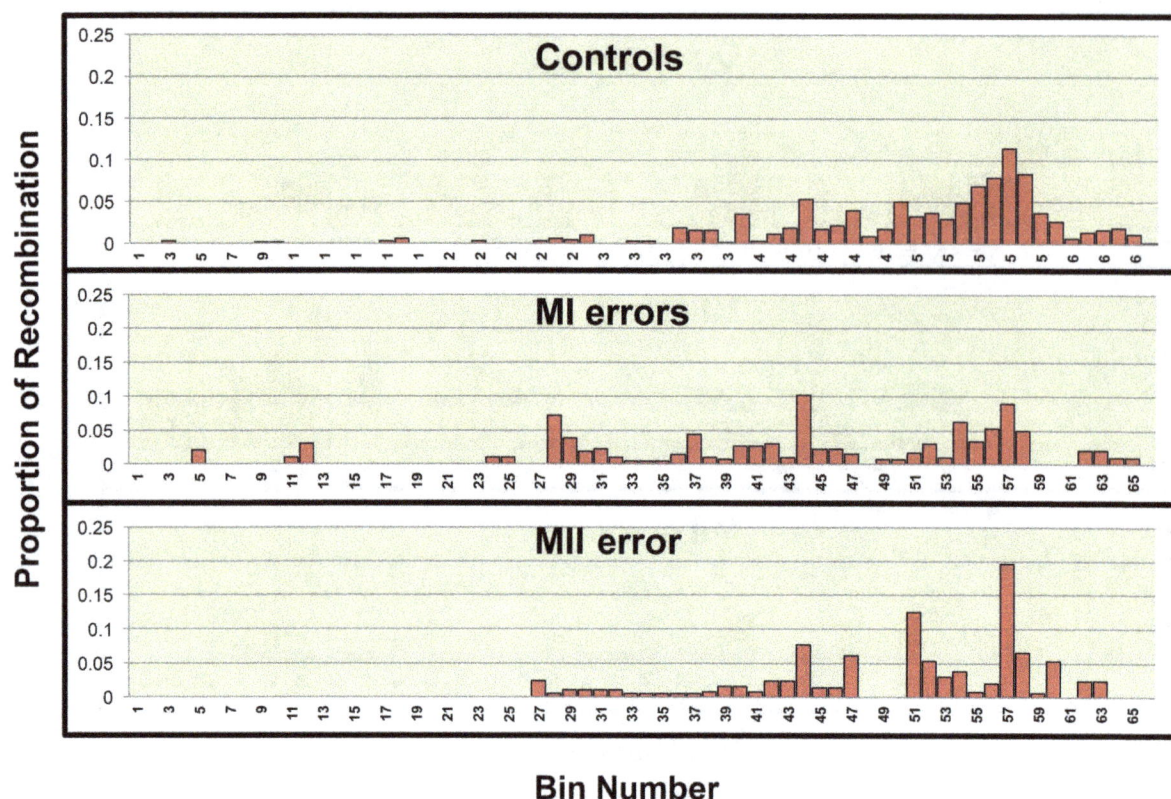

Figure 3. The distribution of the distal recombinant of a double recombinant event across the long arm of 21q by population. 21q was divided into 66 500 kb bins and the distribution of recombination in each bin from the distal recombinant event of chromosomes displaying two recombinant events is depicted above.

amount of recombination and GC (or CpG) content. We did not find any associations between genomic features and recombination among our MI and MII distal recombination events (Table 4).

Hotspot usage among normally disjoined chromosome 21 events

We examined LD-defined hotspots first among normally disjoining chromosomes (controls). We looked separately at those with one recombinant event and those with two recombinant events. Among those with one detectable event, we found a significant positive association between the number of hotspots per bin and the proportion of recombination per bin (p<.0001) (Table 5). Similarly, among those with two detectable events, we found that the proportion of proximal and distal recombinant events within a bin was significantly associated with LD-defined hotspots density (p = 0.001 and <.0001, respectively, Table 5). Thus, as expected, the amount of recombination per bin is positively correlated with historical hotspot density suggesting that historical hotspots are used for recombination along normally segregating chromosomes 21.

Table 2. Values of slopes/beta coefficients for GC, CpG, PolyAT and gene denisty for single recombinants stratified by meiotic outcome group.

Predictor Variable	Controls	MI	MII
GC	−0.0108	0.0377	**0.0856***
21q location	−0.0002	0.0006*	−0.0008*
CpG	−0.145	−0.3102	0.161
21q location	−0.0002	0.0008*	−0.0006*
Poly(A)/Poly(T)	−0.0354	−0.1311	−0.4959
21q location	−0.0002*	0.0006*	−0.0007*
Gene Density	−0.0005	−0.0041	0.0044
21q location	−0.0002*	0.0007*	−0.0006*

Beta values for each genomic feature adjusted for bin variable. Beta coefficients/slopes that are significantly different from zero are marked with an asterisk (p<0.05)*.

Table 3. Values of slopes/beta coefficients for GC, CpG, PolyAT and gene density for the proximal recombinant of a double recombinant event stratified by meiotic outcome group.

Predictor Variable	Controls	MI	MII
GC	0.167	0.366*	0.477*
21q location	−0.001*	−0.002*	−0.002*
CpG	0.36	0.891*	1.099*
21q location	−0.001	−0.001*	−0.001*
Poly(A)/Poly(T)	−1.739*	−3.237*	−3.612*
21q location	−0.001*	−0.001*	−0.002*
adjusted Gene Density	0.0137	0.019	0.005
21q location	−0.001*	−0.001*	−0.001*

Beta values for each genomic feature adjusted by bin variable. Beta coefficients/slopes that are significantly different from zero are marked with an asterisk (p<0.05)*.

Hotspot usage among nondisjoined chromosome 21 events due to MI errors

We first examined single recombinants along 21q. Similar to normally segregating chromosomes, we found hotspot density to be a positively correlated with the proportion of recombination within a bin (p = 0.0006, Table 5). In order to determine whether the strength of the relationship between recombination and hotspot density differed between control and MI single recombinant events, we next tested whether the strength of the association between the proportion of recombination and hotspot density among MI errors significantly differed from that of controls and found no evidence for different patterns (Fig. 4A, p = 0.43).

Among nondisjoining chromosomes with two detectable recombinants we separated analyses by the proximal and distal event. We did not detect a significant relationship between hotspot density and the proportion of recombination per bin for proximal recombinants (fig. 5A, table 5). For distal recombinant events, we found that recombination was significantly associated with LD-defined hotspot density (Table 5, p = 0.02), however for the patterns of association did not differ between MI and controls (Fig. 6A, p = 0.21).

Hotspot usage among nondisjoined chromosome 21 events due to MII errors

As for MII, we detected a significant positive correlation between hotspot density and the proportion of recombination

across 21q for single recombinants. The association patterns differed significantly from that of controls (Fig. 4B, p = 0.01), with MII single recombinant events being less correlated with hotspot density than controls. Among MII errors with two recombinant events, as with MI errors, we did not detect a significant correlation between the proportion of recombination per bin and the density of LD-defined hotspots in the proximal region (fig 5B, table 5). For MII distal events, there was a significant positive association between LD-defined hotspot density and the proportion of recombination per bin (Table 5, p = 0.02). The association patterns did not differ significantly between MII versus control events (Fig. 6B, p = 0.69).

Discussion

Association between genomic features along chromosome 21 and the proportion of recombination events along 21q

In our analysis of the relationships between our genomic features of interest and the proportion of recombination per bin, we found several genomic features to be associated with recombination, although these results were not consistent among recombination types (single, double proximal or distal event). Based on the lack of patterns, we were unable to draw any significant conclusions. We do note that our large sample of normal maternal meiotic events (n = 1,272) for 21q did not show many of the relationships found in the study of Kong et al.[20].

Table 4. Values of slopes/beta coefficients for GC, CpG, Poly(A)/Poly(T) and gene density for the distal recombinant of a double recombinant event stratified by meiotic outcome group.

Predictor Variable	Controls	MI	MII
GC	−0.039	−0.098	0.037
21q location	0.001*	0.001*	0.001*
CpG	−0.705*	−0.573	−0.558
21q location	0.001*	0.001*	0.001*
Poly(A)/Poly(T)	−0.215	0.195	−0.794
21q location	0.001*	0.0004*	0.001*
Gene Density	0.002	0.001	−0.008
21q location	0.001*	0.0004*	0.001*

Beta values for each genomic feature adjusted by bin variable. Beta coefficients/slopes that are significantly different from zero are marked with an asterisk*.

Table 5. Beta coefficient/slope values for hotspots variable adjusted by bin variable and stratified by meiotic outcome group and number of recombinants on chr21.

Recombination Type	Predictor Variable	Controls	MI	MII
Single Recombination	Hotspot count	0.002*	0.002*	0.001*
	Bin location	−0.0002*	0.0007*	−0.0006*
Double Recombination -Proximal	Hotspot count	0.003*	−0.0005	0.0002
	Bin location	−0.001*	−0.0009	−0.001
Double Recombination - Distal	Hotspot count	0.0034*	0.002*	0.0028*
	Bin location	0.0007*	0.0004*	0.0007*

Beta coefficients/slopes that are significantly different from zero are marked with an asterisk*.

We attribute this to a difference in the study design, not to the sample size, as both studies had comparable numbers of meiotic events. First, we restricted our analysis to 21q, whereas the original associations were found through the analysis of the entire genome. Second, the Kong et al. study the sex-averaged associations based on 628 paternal and 629 maternal meiotic outcomes; we only examined maternal recombination events. Taken together, a study of sex-specific, chromosome-specific associations of genomic features and recombination may provide further insights into the control of recombination.

Hotspot usage among nondisjoined chromosome 21 events

Our findings with regard to LD-defined historical hotspots differ between our meiotic outcomes groups and provide some insight into recombination-associated nondisjunction. First, we gain confidence that our analyses are able to identify associations with hotspot usage, as our findings from normally disjoining chromosomes 21 are consistent with expectation. That is, using our

sample of normal meiotic events, our statistical analysis showed the expected pattern of increased recombination in the LD-defined hotspots for single events and double recombinant events on 21q. As it relates to MI errors, our analysis of single recombinants indicated an association of recombination with the distribution of LD-defined hotspots along 21q, similar to controls, suggesting that these events occur preferentially near or within LD-defined hotspots. This is interesting as our previous studies have shown that the average location of MI single recombinant events is approximately 10 Mb closer to the telomere of 21q than normal single recombinant events [21]. As a result, it does not appear that the altered patterns of recombination associated with MI errors can be explained by differential hotspot usage.

We found different patterns of association for MII single recombinant events compared with those for MI-single recombinants events and controls. Specifically, we found that the proportion of single recombinants across 21q per bin is significantly correlated with LD-defined hotspots; however, this association is not as strong as it is in controls. From our most

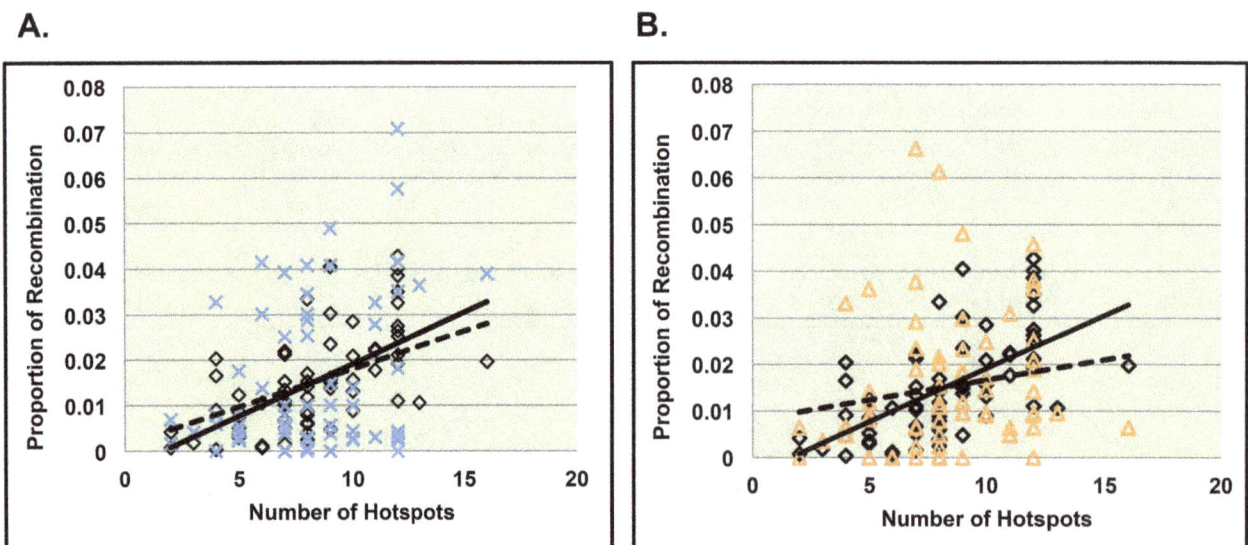

Figure 4. Comparison of the relationship between hotspot usage between MI and MII cases and Controls. Figure 4A and 4B represent MI and MII cases respectively with only one recombinant event on 21q. The solid line represents the relationship between the number of hotspots per bin and the proportion of recombination per bin along normally segregating chromosomes 21. The dotted line represents the relationship between the number of hotspots per bin and the proportion of recombination per bin along chromosomes 21 from MI errors (figure 4A) and MII errors (figure 4B).

A.

B.

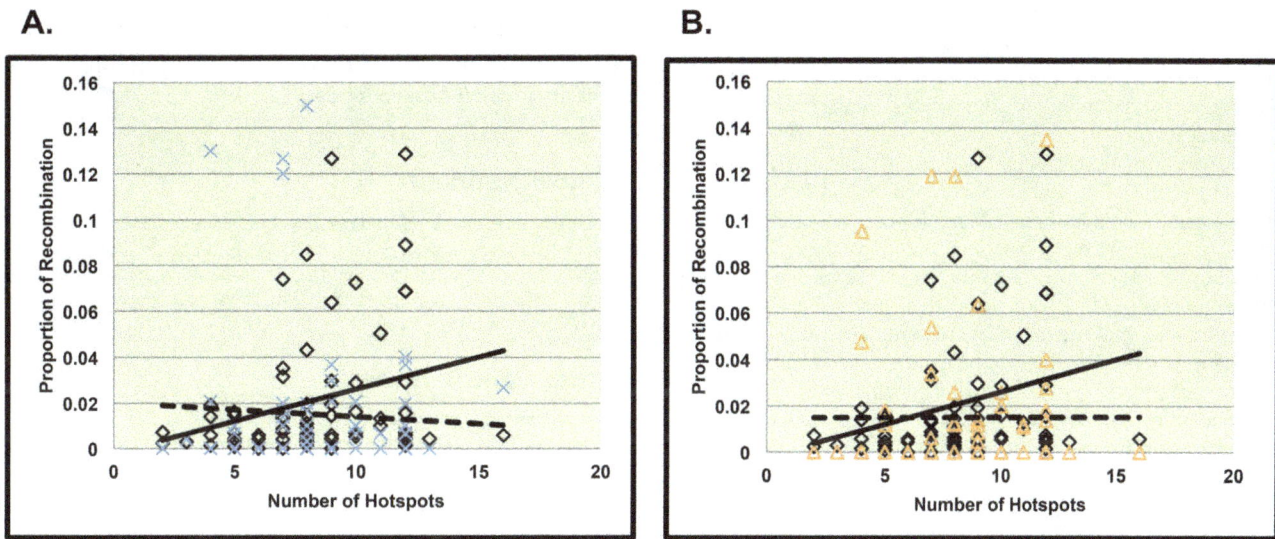

Figure 5. Comparison of slopes between MI or MII errors and controls for the proximal recombinant of double recombinant events. Figures 5A and 5B represent data from the proximal recombinant event of chromosomes displaying two recombinant events on 21q. The solid line represents the relationship between the number of hotspots per bin and the proportion of recombination per bin along normally segregating chromosomes 21. The dotted line represents the relationship between the number of hotspots per bin and the proportion of recombination per bin along chromosomes 21 from MI errors (figure 5A) and MII errors (figure 5B).

recent work, recombination along 21q among MII errors is more proximally located: average location 22.60 Mb on 21q compared with 27.53 Mb on 21q among normal events [21]. Potentially factors characteristic of pericentromeric DNA such as chromatin structure or epigenetic modifications may affect the accessibility of a specific chromosome region to recombination in at least a proportion of oocytes with meiotic errors.

In our analysis of double recombinants events, we found similar results with respect LD-defined hotspots among MI and MII errors. We detected a significant relationship between LD-defined hotspots for the distal recombinant events among doubles, but not the proximal events. Furthermore, the lack of evidence for an association in the proximal region differed from that in controls where an association was detected (i.e., significant interaction). Oliver et al. [21]. found that the unusual pericentromeric proximal

A.

B.

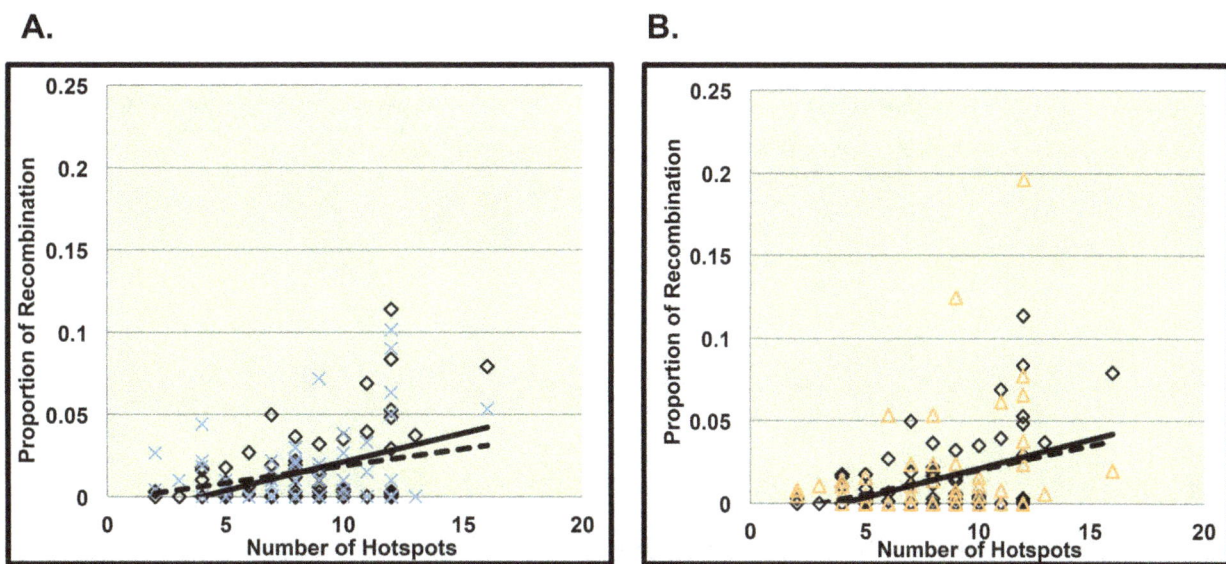

Figure 6. Comparison of slopes between MI or MII errors and controls for the distal recombinant of double recombinant events. Figures 6A and 6B represent data from the distal recombinant event of chromosomes displaying two recombinant events on 21q. The solid line represents the relationship between the number of hotspots per bin and the proportion of recombination per bin along normally segregating chromosomes 21. The dotted line represents the relationship between the number of hotspots per bin and the proportion of recombination per bin along chromosomes 21 from MI errors (figure 6A) and MII errors (figure 6B).

events imposed a risk for MII nondisjunction and were associated with increased maternal age, but this pattern was not found among MI errors. Further work is needed to synthesize these results with those based on location of the events along nondisjoined chromosomes 21.

We do not provide consistent evidence that genomic features present at the site of recombination or differential hotspot usage are implicated in the nondisjunction of chromosome 21. However, altered patterns of recombination on 21q have long-been identified to be associated with an increased risk for chromosome 21 nondisjunction. Thus we believe that either the absence or altered physical placement of recombination may be more important as it relates to the risk for chromosome 21 nondisjunction. Moving forward we plan to take a genome-wide approach in efforts to identify genetic factors implicated in the altered patterns of recombination associated with chromosome 21 nondisjunction.

Supporting Information

Figure S1 Distribution of Hotspots along 21q. Hotspot counts for each of the 66 bins across 21q.

Acknowledgments

We would like to thank our lab personnel, recruiters and the families that participated in this study. We would also like to thank Dr. Mary Marazita [M.M] for contributing data on the location of recombination along properly segregating chromosomes. This data was derived from the GENEVA dental caries study.

Author Contributions

Conceived and designed the experiments: TO SS CM EF LB. Performed the experiments: TO CM. Analyzed the data: TO CM SS EF LB EGA ST. Contributed reagents/materials/analysis tools: VC FB RC SS. Wrote the paper: TO CM EF SS VC.

References

1. Sherman SL, Allen EG, Bean LH, Freeman SB (2007) Epidemiology of Down syndrome. Ment Retard Dev Disabil Res Rev 13: 221–227.
2. Lamb NE, Feingold E, Savage A, Avramopoulos D, Freeman S, et al. (1997) Characterization of susceptible chiasma configurations that increase the risk for maternal nondisjunction of chromosome 21. Hum Mol Genet 6: 1391–1399.
3. Oliver TR, Feingold E, Yu K, Cheung V, Tinker S, et al. (2008) New insights into human nondisjunction of chromosome 21 in oocytes. PLoS Genet 4: e1000033.
4. Warren AC, Chakravarti A, Wong C, Slaugenhaupt SA, Halloran SL, et al. (1987) Evidence for reduced recombination on the nondisjoined chromosomes 21 in Down syndrome. Science 237: 652–654.
5. Oliver TR, Tinker SW, Allen EG, Hollis N, Locke AE, et al. (2011) Altered patterns of multiple recombinant events are associated with nondisjunction of chromosome 21. Hum Genet.
6. Holmquist GP, Ashley T (2006) Chromosome organization and chromatin modification: influence on genome function and evolution. Cytogenet Genome Res 114: 96–125.
7. Berg IL, Neumann R, Lam KW, Sarbajna S, Odenthal-Hesse L, et al. (2010) PRDM9 variation strongly influences recombination hot-spot activity and meiotic instability in humans. Nat Genet.
8. Hinch AG, Tandon A, Patterson N, Song Y, Rohland N, et al. (2011) The landscape of recombination in African Americans. Nature 476: 170–175.
9. Berg IL, Neumann R, Sarbajna S, Odenthal-Hesse L, Butler NJ, et al. (2011) Variants of the protein PRDM9 differentially regulate a set of human meiotic recombination hotspots highly active in African populations. Proc Natl Acad Sci U S A 108: 12378–12383.
10. Kong A, Thorleifsson G, Gudbjartsson DF, Masson G, Sigurdsson A, et al. (2010) Fine-scale recombination rate differences between sexes, populations and individuals. Nature 467: 1099–1103.
11. Baudat F, Buard J, Grey C, Fledel-Alon A, Ober C, et al. (2010) PRDM9 is a major determinant of meiotic recombination hotspots in humans and mice. Science 327: 836–840.
12. Fledel-Alon A, Leffler EM, Guan Y, Stephens M, Coop G, et al. (2011) Variation in human recombination rates and its genetic determinants. PLoS One 6: e20321.
13. Freeman SB, Allen EG, Oxford-Wright CL, Tinker SW, Druschel C, et al. (2007) The National Down Syndrome Project: design and implementation. Public Health Rep 122: 62–72.
14. Lamb NE, Freeman SB, Savage-Austin A, Pettay D, Taft L, et al. (1996) Susceptible chiasmate configurations of chromosome 21 predispose to nondisjunction in both maternal meiosis I and meiosis II. Nat Genet 14: 400–405.
15. Weiss LA, Shen Y, Korn JM, Arking DE, Miller DT, et al. (2008) Association between microdeletion and microduplication at 16p11.2 and autism. N Engl J Med 358: 667–675.
16. Dawber TR, Meadors GF, Moore FE Jr (1951) Epidemiological approaches to heart disease: the Framingham Study. Am J Public Health Nations Health 41: 279–281.
17. Polk DE, Weyant RJ, Crout RJ, McNeil DW, Tarter RE, et al. (2008) Study protocol of the Center for Oral Health Research in Appalachia (COHRA) etiology study. BMC Oral Health 8: 18.
18. Chowdhury R, Bois PR, Feingold E, Sherman SL, Cheung VG (2009) Genetic analysis of variation in human meiotic recombination. PLoS Genet 5: e1000648.
19. Myers S, Bottolo L, Freeman C, McVean G, Donnelly P (2005) A fine-scale map of recombination rates and hotspots across the human genome. Science 310: 321–324.
20. Kong A, Gudbjartsson DF, Sainz J, Jonsdottir GM, Gudjonsson SA, et al. (2002) A high-resolution recombination map of the human genome. Nat Genet 31: 241–247.
21. Oliver TR, Tinker SW, Allen EG, Hollis N, Locke AE, et al. Altered patterns of multiple recombinant events are associated with nondisjunction of chromosome 21. Hum Genet.

3

In*Ce*DB: Database of Human Long Noncoding RNA Acting as Competing Endogenous RNA

Shaoli Das[1][◐], Suman Ghosal[1][◐], Rituparno Sen[2], Jayprokas Chakrabarti[1,2]*

1 Computational Biology Group, Indian Association for the Cultivation of Science, Kolkata, West Bengal, India, **2** Gyanxet, Kolkata, West Bengal, India

Abstract

Long noncoding RNA (lncRNA) influences post-transcriptional regulation by interfering with the microRNA (miRNA) pathways, acting as competing endogenous RNA (ceRNA). These lncRNAs have miRNA responsive elements (MRE) in them, and control endogenous miRNAs available for binding with their target mRNAs, thus reducing the repression of these mRNAs. In*Ce*DB provides a database of human lncRNAs (from GENCODE 19 version) that can potentially act as ceRNAs. The putative mRNA targets of human miRNAs and the targets mapped to AGO clipped regions are collected from TargetScan and StarBase respectively. The lncRNA targets of human miRNAs (up to GENCODE 11) are downloaded from miRCode database. miRNA targets on the rest of the GENCODE 19 lncRNAs are predicted by our algorithm for finding seed-matched target sites. These putative miRNA-lncRNA interactions are mapped to the Ago interacting regions within lncRNAs. To find out the likelihood of an lncRNA-mRNA pair for actually being ceRNA we take recourse to two methods. First, a ceRNA score is calculated from the ratio of the number of shared MREs between the pair with the total number of MREs of the individual candidate gene. Second, the P-value for each ceRNA pair is determined by hypergeometric test using the number of shared miRNAs between the ceRNA pair against the number of miRNAs interacting with the individual RNAs. Typically, in a pair of RNAs being targeted by common miRNA(s), there should be a correlation of expression so that the increase in level of one ceRNA results in the increased level of the other ceRNA. Near-equimolar concentration of the competing RNAs is associated with more profound ceRNA effect. In In*Ce*DB one can not only browse for lncRNA-mRNA pairs having common targeting miRNAs, but also compare the expression of the pair in 22 human tissues to estimate the chances of the pair for actually being ceRNAs. **Availability:** Downloadable freely from http://gyanxet-beta.com/lncedb/.

Editor: Sandro Banfi, Telethon Institute of Genetics and Medicine, Italy

Funding: The authors received financial support from the intramural funds of their host institutes. SD, SG and JC received financial support from the Indian Association for the Cultivation of Science. RS was hosted by Gyanxet, an independent non-profit body. Funds for the website were contributed by Gyanxet. The fund giving organizations, namely, the Indian Association for the Cultivation of Science and Gyanxet, had no role in the design, data collection and analysis, the decision to publish, or in the preparation of the manuscript.

Competing Interests: Rituparno Sen is an employee of Gyanxet, which hosts the database InCeDB, together with several freely downloadable software. There are no further patents, products in development or marketed products to declare. The database is freely accessible from http://gyanxet-beta.com/lncedb [8].

* E-mail: j.chakrabarti@gyanxet.com

◐ These authors contributed equally to this work.

Introduction

A major part of the transcriptome of higher eukaryote does not code for any protein, but functions as regulatory RNAs. Long noncoding RNAs (lncRNAs) are long transcripts (generally more than 200 bases) that in many ways are similar to protein coding transcripts, except for the lack of a meaningful coding sequence (CDS) or open reading frame (ORF) [1–2]. In the last decade it has become possible to annotate a greater number of lncRNA transcripts. Consequently the study of them has gained much significance as many of them have been linked with epigenetic, transcriptional and post-transcriptional regulation of gene expression [3]. Despite having generally a lower level of concentration than protein coding transcripts, lncRNAs exhibit more tissue specific expressions [4–5]. A vast set of lncRNA transcripts are differentially expressed during development where many of them play critical roles [6–9]. LncRNAs are now known to have a major involvement in cancer [10–11]. Though, till now, a majority of the lncRNAs have been linked with epigenetic modulation of gene expressions [12–13], they can also regulate gene expression by

transcriptional or post transcriptional modes [14–15]. LncRNAs can influence post-transcriptional regulation by interfering with the miRNA pathways, by acting as competing endogenous RNAs (ceRNAs) [16]. In recent years it has been discovered that endogenous RNAs (mRNAs, pseudogenes, long noncoding RNAs or circular RNAs) compete with each other for the limited pool of cellular microRNAs (miRNAs) and thus affect the competing RNA's level. These RNAs have miRNA responsive elements (MRE), i.e., the miRNA binding sites in them, and act as miRNA sponges to control endogenous miRNAs available for binding with their target mRNAs, thus reducing the repression of these mRNAs [17–24]. This phenomenon adds a significant new dimension to the miRNA mediated regulation of gene expression in cells. ceRNAs are important regulators in cell cycle control and tumor suppression (e.g. PTEN-P1 blocking miR-19b and miR-20a from binding to PTEN tumor suppressor [17–19]), modulating self-regulation in hepatocellular carcinoma (HCC) (HULC lncRNA acts as ceRNA of the protein coding gene PRKACB that induces activation of CREB which in turn is involved in upregulation of

HULC [20]) as well as in developmental stages (e.g. linc-MD1 blocking miR-133 from binding to transcription factors involved in myogenic differentiation [21] and H19 blocking the miRNA let-7 to affect muscle differentiation in vitro [22]). Circular RNAs have recently been shown to be involved in pathways of cancer and many other diseases [23,24].

Due to the availability of huge lncRNA datasets from recent GENCODE versions [25], 13870 lncRNA genes in GENCODE 19, it has become imperative to uncover the potential functions of these transcripts. In the light of new findings on ceRNAs and lncRNA-miRNA interactions, we developed a database, lnCeDB, of human lncRNAs that can potentially act as ceRNAs (based on the newly available GENCODE 19 annotated lncRNAs). Recently, databases describing lncRNA-miRNA interactions, like miRCode [26], Diana-lncBase [27], lncRNome [28] and StarBase v2.0 [29], have become available. But none of them documents miRNA interactions with lncRNAs annotated past GENCODE 17. We used lncRNA-mRNA interaction pairs from miRCode database of miRNA targets for lncRNAs in GENCODE 11 [26], and for the newly enlisted lncRNAs in GENCODE 19, we predicted seed-matched miRNA targets using our algorithm. We mapped these putative miRNA-lncRNA interactions into the Ago-interacting regions within lncRNAs, collected from a recent study [30]. In lnCeDB, the users can also browse for miRNA targets on recently available GENCODE 19 lncRNAs not available from other databases. Moreover, the objective of lnCeDB is not just describing lncRNA-miRNA interactions, but providing researchers with a database of human lncRNAs that can potentially act as ceRNAs to protein coding genes. The chances of an lncRNA-mRNA pair for actually being ceRNA depend not only on the fact that they are targeted by common miRNA(s), but also other factors like relative concentrations of individual component ceRNAs and the number of shared MREs. lnCeDB is built by taking into consideration these varied and complex features.

A previously published database of ceRNAs, ceRDB, provides data of mRNAs that can putatively act as ceRNAs, but it does not have information about lncRNAs [31]. It should also be noted that unlike the ceRDB database, that used putative miRNA-mRNA interactions predicted by TargetScan [32], we include mRNA-miRNA and miRNA-lncRNA interactions predicted from AGO CLIP-Seq data [30,33]. The user can limit the target search within regions of AGO interaction, significantly reducing false-positive target. Another recently published database, StarBase v2.0 include ceRNA pairs predicted from available AGO PAR-CLIP datasets [29]. However, the use of only the PAR-CLIP data for prediction of ceRNA pairs limits the result set to only a few cell lines where the AGO PAR-CLIP was performed. As mentioned earlier, our dataset includes, but is not limited to predictions from just the AGO CLIP-SEQ data. This gives the user a broader set of probable ceRNA pairs in many human tissues, and also the option to narrow down the search to only the AGO interacting regions as available from AGO CLIP-SEQ data.

As mentioned earlier, the chances of an lncRNA-mRNA pair actually being ceRNA depend not just on the fact that they are targeted by common miRNA(s), but also on the number of miRNA responsive elements (MRE), number of distinct miRNAs targeting both transcripts and the concentration and cellular levels of the competing RNAs. A study by Ala et al provides a comprehensive view on the ceRNA network and the possible outcome of perturbation in the components (miRNA and competing RNA levels) of the network [34]. The authors showed that the relative expressions of competing RNAs play a vital part in determining the ceRNA effect. While, for a pair of ceRNAs, it is seen that the competing RNA with higher expression has greater

ceRNA effect on the other competing RNA, it has also been observed that competing RNAs with near-equal expression exhibit more robust ceRNA effect than other ceRNA pairs having largely different expressions [34]. Thus, the information on the concentration levels of the two RNAs making the ceRNA pair is very crucial. Also, to determine the potential cross-regulation of a ceRNA pair, it is very important to check the co-expression of shared miRNAs along with the ceRNA pair. One major drawback of the existing ceRNA databases, other than lnCeDB, is that they do not offer the option to the users to check the co-expression of the ceRNA pair and the shared miRNAs. Following from the observation by Ala et al, to estimate the chances of an lncRNA-mRNA pair for actually being ceRNAs in particular tissues, lnCeDB offers users the possibility to browse for lncRNA-mRNA pairs targeted by common miRNAs (sorted by the number of shared miRNAs of the pair) and compare the expression of the pair in 22 human tissues (from RNA-Seq expression data from Cabili et al [35]). Moreover, lnCeDB also provides users with the information on the shared miRNAs co-expressed in each of the 22 different tissues. This feature is not offered by any other ceRNA database.

To assess the likelihood of an lncRNA-mRNA pair to act as ceRNAs, we provide two different methods. In the first approach, we calculate the P-value for each ceRNA pair by hypergeometric test, similar to the study by Sumazin et al [19] and StarBase v2.0 [29], considering the number of shared miRNAs between the pair against the total number of miRNAs targeting each component in the pair, i.e. the lncRNA and the mRNA. But in the second approach, unlike other ceRNA databases that predict the likelihood for being ceRNAs by the number of shared miRNAs [19,29], we calculate a ceRNA score for each probable ceRNA pair by taking into consideration the number of shared MREs against the total number of MREs for the candidate lncRNA. A major drawback of the other ceRNA databases [29] is that they calculate the likelihood of a pair of genes to act as ceRNA by considering only the number of shared miRNAs between the pair. But there is a certain importance to the number of shared MREs compared to the number of shared miRNAs between the ceRNA pair. Thus, a candidate lncRNA having 100 MREs for 1 shared miRNA would be considered lesser than a candidate lncRNA with 2 MREs for 2 shared miRNAs in some of the existing databases. We believe that the number of shared MREs would be more appropriate instead of the number of shared miRNAs between the ceRNA pair. And that makes lnCeDB different from the existing databases on ceRNA. In lnCeDB, the ceRNA score, along with the provision for checking relative expressions of the ceRNA pair over different tissues, offer users a better assessment of the potential ceRNAs. We believe, therefore, that lnCeDB addresses the finer aspects of post transcriptional gene expression regulation in human.

Results

lnCeDB contains dataset for human genome wide lncRNAs that potentially can act as ceRNAs. Presently we have miRNA targets on 22286 lncRNAs out of 25978 lncRNA transcripts in GENCODE 19. miRNA targets for lncRNAs in GENCODE 11 were collected from miRCode database and targets for newly enlisted lncRNAs in GENCODE 19 were predicted by our algorithm (see section 2). Among the 22286 lncRNAs with putative miRNA target sites, 15842 lncRNAs were predicted to be working as potential ceRNAs. Figure 1 shows a statistics of the database contents in lnCeDB.

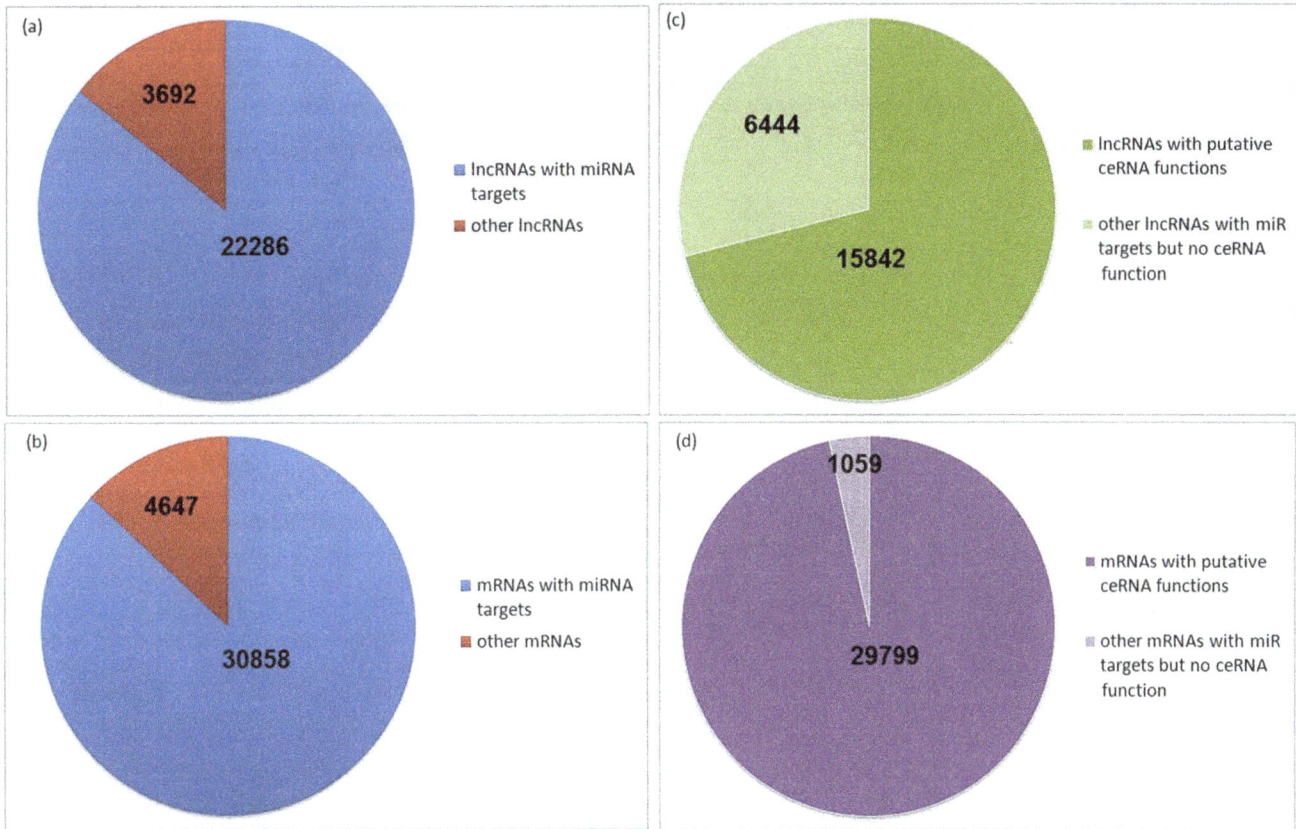

Figure 1. A statistics of the database contents in ln_Ce_DB. (a) Fraction of GENCODE 19 lncRNA transcripts with putative miRNA targets. (b) Fraction of mRNAs with predicted miRNA targets. (c) Fraction of lncRNAs with predicted ceRNA function compared to all lncRNAs with putative miRNA targets. (d) Fraction of mRNAs with predicted ceRNA function compared to all mRNAs with putative miRNA targets.

lncRNA-miRNA interactions mapping into Ago interaction sites within lncRNA loci

Among the putative lncRNA-miRNA interactions, 72 interaction pairs were mapped to the Ago-interaction sites within lncRNA transcript loci which contained 41 miRNAs targeting 27 lncRNA transcripts. ceRNAs specifically corresponding to these PAR-CLIP supported miRNA target interactions are browsable in ln_Ce_DB.

Putative mRNA-miRNA interactions from TargetScan

Conserved human miRNA targets on mRNAs were collected from TargetScan [32]. The current version stores miRNA targets on 30858 mRNA transcripts for human.

mRNA-miRNA interactions derived from Ago interacting regions within mRNA

miRNA targets on 6090 mRNAs were collected from StarBase; these targets were predicted from CLIP-Seq data by five most widely used miRNA target prediction tools (see section 2).

Expression data for a ceRNA pair over 22 tissues

Users can compare the cellular levels of the competing RNAs to estimate the chances of the pair for actually being ceRNAs by viewing the tissue specific expression of a ceRNA pair over 22 human tissues. Presently we have tissue specific expressions for 7017 lncRNA genes and 17701 protein coding genes.

Experimentally verified lncRNAs acting as ceRNAs

A small number of lncRNAs are reported and experimentally verified as ceRNAs. Experimentally validated lncRNA ceRNAs are identified in ln_Ce_DB. The reported lncRNA-ceRNAs include lncRNA HULC acting as ceRNA of the gene PRKACB [20], lincRNA MD1 working as ceRNA of the gene MAML1 [21] and linc-RoR acting as ceRNA of the pluripotency associated transcription factors SOX2 and NANOG [36]. Table 1 lists the experimentally verified lncRNA ceRNAs in ln_Ce_DB.

Utility

Users can browse for ceRNA candidates for a protein coding gene (by gene symbol, gene id or refseq accession) and/or lncRNA gene (by gene name, ensemble gene id or ensemble transcript id) and/or miRNA name. lncRNA-mRNA pairs targeted by common miRNAs are displayed as potential ceRNAs. The list of ceRNAs is sorted by the number of shared miRNAs. The miRNA targets for a particular lncRNA or mRNA can also be browsed in our database by choosing a particular lncRNA/mRNA from the ceRNA list. By choosing a particular record for a ceRNA pair, users can view the RNA-Seq expressions of the lncRNA-mRNA pair over 22 human tissues in form of a heat map (if the expression data for the corresponding lncRNA and mRNA is available) and a bar chart with actual FPKM values appearing in tool tip text. The users can also view the co-expressed miRNAs shared between the ceRNA pair in each of the 22 tissues. Figure 2 shows a flow diagram for the usability of ln_Ce_DB.

Table 1. Experimentally verified lncRNA ceRNAs in ln*Ce*DB.

lncRNA acting as ceRNA	Competing protein coding gene	Shared miRNA	ceRNA score	Reference
HULC (Highly Upregulated in Liver Cancer)	PRKACB	miR-372	0.026 (p-value = 0.001)	[20]
lincRNA MD1	MAML1	miR-133	0.022 (p-value = 0.02)	[21]
H19	Targets of hsa-let-7	Let-7	-	[22]
Linc-RoR (Regulator of Reprogramming)	SOX2 and NANOG	miR-145	0.038 (p-value = 0.008)	[36]
PTCSC3	Targets of miR-574-5p in thyroid cancer cell line	miR-574-5p	-	[37]

Discussion

As competing endogenous RNAs are crucial new determinants of gene expression regulation [38], new data sources are needed. Following the availability of huge dataset of annotated lncRNA transcripts from GENCODE project, the possible functions of the transcripts have to be addressed. In ln*Ce*DB, we have tried to explore the potential sponge activity of recently annotated lncRNAs in the miRNA mediated gene regulation networks at human genome wide scale. Unlike other ceRNA databases, ln*Ce*DB includes but is not limited to miRNA targets on protein coding and lncRNA transcripts predicted from Ago interaction sites within them. The advantage is that it reduces the false positive target detection in our miRNA-target interaction dataset and enhances the reliability of the prediction. At the same time, the dataset is not limited to predictions from only a few cell lines where AGO PAR-CLIP were performed. Also, in ln*Ce*DB we considered, for the first time, that ceRNA activity largely depends on the relative concentration of the components of a ceRNA network, i.e. the pair of competing RNAs and also the miRNAs they compete for. The provision for checking the tissue specific expression for a potential ceRNA pair (whenever available) along with the co-expression of shared miRNAs, gives the user a higher chance of identifying the most likely ceRNA candidates in a tissue of interest.

One interesting example of a putative ceRNA pair identified by ln*Ce*DB is the lncRNA maternally expressed 3 (MEG3) and a transcript (the longer isoform) of the protein coding gene Myeloid Cell Leukemia Sequence 1 (MCL1). MEG3 is a maternally expressed imprinted gene encoding a number of alternatively spliced lncRNA transcripts. It interacts with the tumour suppressor P53, and is supposedly a tumour suppressor itself. MEG3 is expressed in many normal tissues including breast, colon, liver, ovary but its expression is lost in many tumour cells. Interestingly, it has been reported that MEG3 is targeted by miRNAs [39]. MCL1 is a member of the BCL2 family and it has three isoforms. The longer isoform is anti-apoptotic whereas the shorter isoforms are pro-apoptotic. The ceRNA pair MEG3-MCL1 putatively shares 16 common miRNAs including miR-28, miR-181d, miR-520a, miR-520b and miR-876-3p and show comparable high co-expressions in breast and colon, especially in colon (MEG3 (66.2): MCL1 (37.9)). The co-expression pattern of MEG3 and MCL1, along with the co-expressed shared miRNAs indicates that MEG3 may act as a ceRNA to MCL1 in colon. Interestingly, the MEG3 gene locus has been reported to be hypermethylated in colorectal cancer cells [40] indicating the possible perturbation of MEG3 lncRNA expression in colorectal carcinoma. Furthermore, in a colorectal cancer cell, the anti-apoptotic MCL1 has been reported to be regulated (in vitro) by a number of miRNAs, including miR-876-3p [41] which we predicted to be shared by MEG3 and MCL1. Together, these observations suggest that there may be a disruption of the potential MEG3-MCL1 ceRNA network in colorectal cancer cells as opposed to the normal colon cells. This observation, however, needs to be validated by further investigations. This example shows the importance of ln*Ce*DB over other ceRNA databases as no other ceRNA database allows the users to check the co-expression patterns of the competing RNAs and the shared miRNAs in different tissue types. Some other interesting observations from ln*Ce*DB are MEG3 and CMPK1 as potential ceRNA pair with near-equal expression signature in colon and ovary and MALAT1-PRKACB potential ceRNA pair in liver.

We believe this database will help researchers in deciphering the larger and more complex scenario of miRNA mediated gene regulatory networks in human in the real world of ceRNAs.

Materials and Methods

Data Source

We collected predicted miRNA target sites on human mRNA transcripts from TargetScan. The miRNA-mRNA interactions mapped to the AGO interacting regions were collected from StarBase database [33]. StarBase houses datasets of mRNA-miRNA interactions predicted from AGO CLIP-Seq data (AGO interacting regions) by widely used miRNA target prediction tools like TargetScan [32], PITA, PicTar, RNA22 and miRanda. The lncRNA-miRNA interaction dataset (for lncRNAs in GENCODE 11 version) was downloaded from miRCode database [26]. For newly enlisted lncRNAs in GENCODE 19 version, we used our miRNA-target prediction algorithm for finding seed matched miRNA target sites on lncRNAs. The dataset for Ago interacting regions within lncRNA transcript loci was collected from Jalali et al [30]. Tissue specific expressions (RNA-Seq) for lncRNAs and mRNAs in 22 human tissues were used from the dataset of Cabili et al [35]. Tissue specific miRNA expressions (QRT-PCR) were collected from miRNA body-map [42].

Prediction of miRNA target sites on lncRNAs

We used a 25 base window on the target to run a modified version of the Smith-Waterman alignment (as used in the miRanda algorithm [43]) to find complementary alignment of the 25 base target region with the miRNA. For reducing runtime, at the first step we searched for 6-mer seed region (position 2-7 on miRNA) complementarities on the target. If a match was found, then a 25 base window around the seed-matched site was used for further alignment. For prediction of different types of miRNA target sites (6-mer, 7-mer, 7-merA1 and 8-mer), we changed the seed region definition in each case [43]. We considered transcripts with perfect seed complementarity as well as one base mismatch tolerance (in position 2- 8 or 2-7) in the seed region with 3′ compensatory complementarity (at position 13-18).

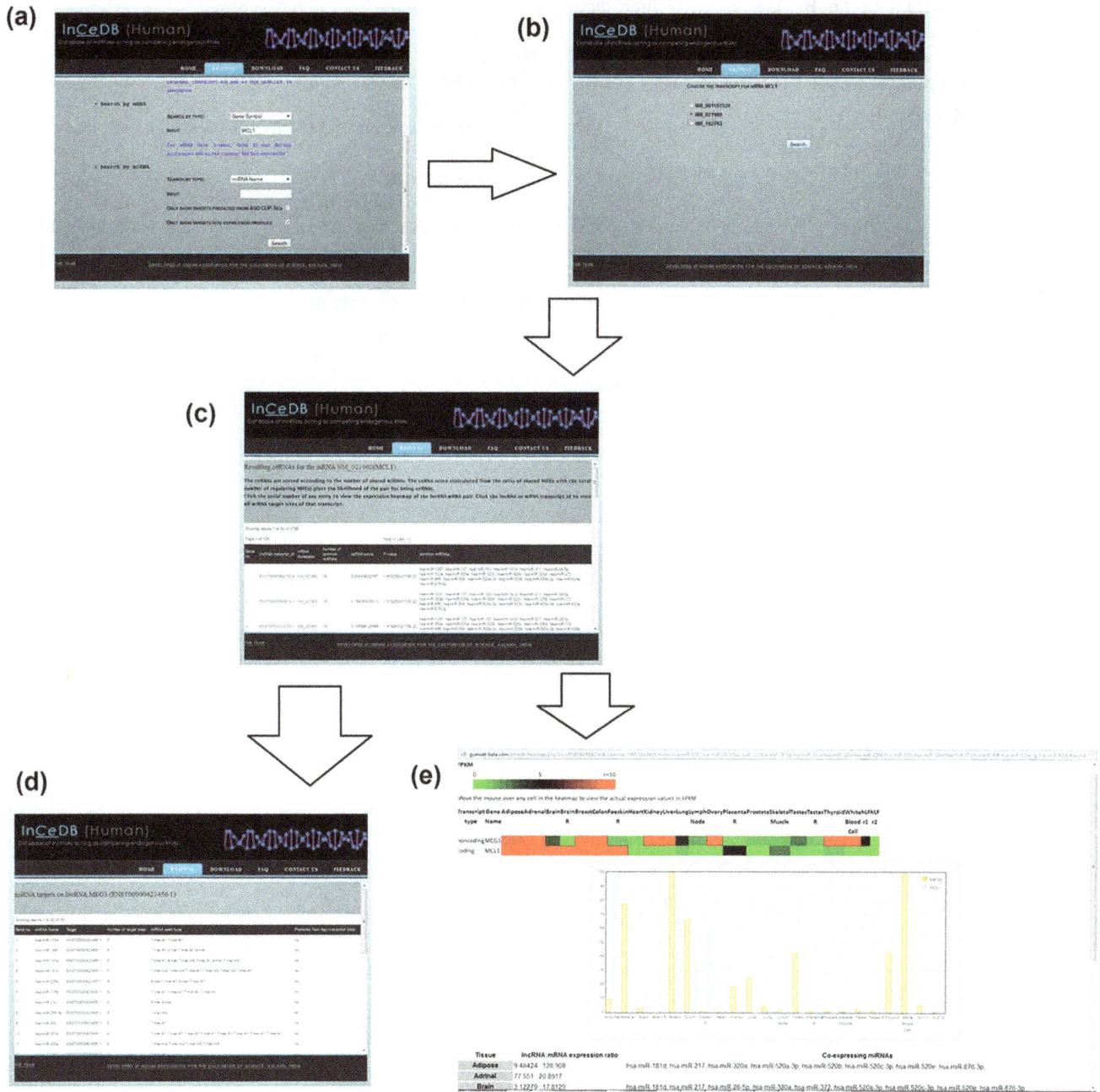

Figure 2. The flow diagram for browsing ceRNAs in In*Ce*DB. (a) InCeDB is searched by the gene symbol MCL1. (b) An intermediate page shows the different transcripts of the gene MCL1. (c) Upon choosing a transcript, the results page shows the potential lncRNAs working as ceRNA for the chosen transcript of MCL1. (d) By clicking on a lncRNA or mRNA id in the ceRNA table, the user views all miRNA targets on the chosen transcript. (e) By clicking on a serial number in the ceRNA table, the user views the expression heatmap and bar chart for the ceRNA pair along with co-expressed shared miRNAs in 22 human tissues.

Consideration for target site conservation

We treated conserved and non-conserved target sites separately. We searched for target region conserved among human, chimp, mouse, rat and dog. Genome wide conservation data generated using multiz 46-way alignment (for 46 vertebrate species) was downloaded from UCSC genome browser [44]. Genomic regions (within human genome) conserved within human, chimp, mouse, rat and dog were then extracted and mapped within the coordinates of human mRNAs (downloaded from UCSC genome browser) to get the location of the conserved regions within human

mRNAs. Conserved regions of length 8 bases or more were only considered.

The ceRNA score

We generated the ceRNA score of an lncRNA-mRNA pair targeted by common miRNAs to measure the likelihood of a lncRNA to act as ceRNA to a protein coding gene. The ceRNA score was calculated keeping in mind the observation by Ala et al [34] that genes that shared a large number of distinct miRNAs, as opposed to genes targeted by a small number of shared miRNAs, compared to the total number of miRNAs targeting the individual genes, exhibit more profound ceRNA

$$ceRNAscore = \frac{\text{the number of MREs for the distinct shared miRNAs between the pair}}{\text{the total number MREs for all distinct miRNAs targeting the lncRNA}} \quad (1)$$

Calculating the probability of ceRNA pair to cross-regulate each other

We implemented another measure to assess the likelihood of a ceRNA pair to regulate each other via shared miRNAs. This approach was similar to what has been used in the study of Sumazin et al [19] and in StarBase v2.0 [29]. We calculated the p-value for each potential ceRNA pair by hypergeometric test considering the number of shared miRNAs between a ceRNA pair against the number of miRNAs targeting individual components of the ceRNA pair. The p-value was measured as:

$$p = \sum_{i=m_c}^{\min(m_p, m_n)} \frac{\binom{m_n}{i}\binom{M_T - m_n}{m_p - i}}{\binom{M_T}{m_p}} \quad (2)$$

Where,

M_T = Total number of miRNAs in the human genome

m_p = Number of miRNAs interacting with the mRNA (protein-coding)

m_n = Total number of miRNAs interacting with the lncRNA (non protein-coding)

m_c = Number of miRNAs shared between the ceRNA pair

Implementation

The miRNA target finding algorithm was implemented in JAVA and the web interface of the database ln*Ce*DB was developed using PHP-mySql.

Author Contributions

Conceived and designed the experiments: SD SG JC. Analyzed the data: SD SG. Wrote the paper: SD SG JC. Implemented the programs necessary for performing the study: SD RS. Supervised the whole work: JC.

References

1. Ponting CP, Oliver PL, Reik W (2009) Evolution and functions of long noncoding RNAs, Cell 136: 629–641.
2. Ponjavic J, Ponting CP, Lunter G (2007) Functionality or transcriptional noise? Evidence for selection within long noncoding RNAs. Genome Res. 17: 556–565.
3. Mercer TR, Dinger ME, Mattick JS (2009) Long non-coding RNAs: insights into functions. Nat Rev Genet. 10: 155–159.
4. Pang KC, Frith MC, Mattick JS (2006) Rapid evolution of noncoding RNAs: lack of conservation does not mean lack of function. Trends Genet. 22: 1–5.
5. Marques AC, Ponting CP (2009) Catalogues of mammalian long noncoding RNAs: modest conservation and incompleteness. Genome Biol. 10: R124.
6. Amaral PP, Mattick JS (2008) Noncoding RNA in development. Mamm Genome. 19: 454–492.
7. Meola N, Pizzo M, Alfano G, Surace EM, Banfi S (2012) The long noncoding RNA Vax2os1 controls the cell cycle progression of photoreceptor progenitors in the mouse retina. RNA 18: 111–123.
8. Paralkar VR, Weiss MJ (2011) A new 'Linc' between noncoding RNAs and blood development. Genes Dev. 25: 2555–2558.
9. Ghosal S, Das S, Chakrabarti J (2013) Long noncoding RNAs: new players in the molecular mechanism for maintenance and differentiation of pluripotent stem cells. Stem Cells Dev. 22:2240–2253.
10. Aguilo F, Zhou MM, Walsh MJ (2011) Long noncoding RNA, polycomb, and the ghosts haunting INK4b-ARF-INK4a expression. Cancer Res. 71: 5365–5369.
11. Lv XB, Lian GY, Wang HR, Song E, Yao H, et al. (2013) Long Noncoding RNA HOTAIR Is a Prognostic Marker for Esophageal Squamous Cell Carcinoma Progression and Survival. PLoS One. 8: e63516.
12. Zhao J, Sun BK, Erwin JA, Song JJ, Lee JT (2008) Polycomb proteins targeted by a short repeat RNA to the mouse X chromosome. Science 322: 750–756.
13. Ng SY, Johnson R, Stanton LW (2011) Human long noncoding RNAs promote pluripotency and neuronal differentiation by association with chromatin modifiers and transcription factors. EMBO J. 31: 522–533.
14. Hawkins PG, Morris KV (2010) Transcriptional regulation of Oct4 by a long noncoding RNA antisense to Oct4-pseudogene 5. Transcription 1: 165–175.
15. Tripathi V, Ellis JD, Shen Z, Song DY, Pan Q, et al. (2010) The nuclear-retained noncoding RNA MALAT1 regulates alternative splicing by modulating SR splicing factor phosphorylation. Mol Cell 39: 925–938.
16. Salmena L, Poliseno L, Tay Y, Kats L, Pandolfi PP (2011) A ceRNA hypothesis: the Rosetta Stone of a hidden RNA language? Cell 146: 353–358.
17. Tay Y, Kats L, Salmena L, Weiss D, Tan SM, et al. (2011) Coding-independent regulation of the tumor suppressor PTEN by competing endogenous mRNAs. Cell 147: 344–357.
18. Karreth FA, Tay Y, Perna D, Ala U, Tan SM, et al. (2011) In vivo identification of tumor- suppressive PTEN ceRNAs in an oncogenic BRAF-induced mouse model of melanoma. Cell 147: 382–395.
19. Sumazin P, Yang X, Chiu HS, Chung WJ, Iyer A, et al. (2011) An extensive microRNA-mediated network of RNA-RNA interactions regulates established oncogenic pathways in glioblastoma. Cell 147: 370–381.
20. Wang J, Liu X, Wu H, Ni P, Gu Z, et al. (2010) CREB up-regulates long non-coding RNA, HULC expression through interaction with microRNA-372 in liver cancer. Nucleic Acids Res. 38: 5366–5383.
21. Cesana M, Cacchiarelli D, Legnini I, Santini T, Sthandier O, et al. (2011) A long noncoding RNA controls muscle differentiation by functioning as a competing endogenous RNA. Cell 147: 358–369.
22. Kallen AN, Zhou XB, Xu J, Qiao C, Ma J, et al. (2013) The imprinted H19 lncRNA antagonizes let-7 microRNAs. Mol Cell. 52:101–112.
23. Hansen TB, Kjems J, Damgaard CK (2013) Circular RNA and miR-7 in Cancer. Cancer Res. 73, 5609–5612.
24. Ghosal S, Das S, Sen R, Basak P, Chakrabarti J (2013) Circ2Traits: A comprehensive database for circular RNA potentially associated with disease and traits. Front Genet. Doi: 10.3389/fgene.2013.00283.
25. Derrien T, Johnson R, Bussotti G, Tanzer A, Djebali S, et al. (2012) The GENCODE v7 catalog of human long noncoding RNAs: Analysis of their gene structure, evolution, and expression. Genome Res 22: 1775–1789.
26. Jeggari A, Marks DS, Larsson E (2012) miRcode: a map of putative microRNA target sites in the long non-coding transcriptome. Bioinformatics. 28: 2062–2063.
27. Paraskevopoulou MD, Georgakilas G, Kostoulas N, Reczko M, Maragkakis M, et al. (2013) DIANA-LncBase: experimentally verified and computationally predicted microRNA targets on long non-coding RNAs. Nucleic Acids Res. 41(Database issue): D239–45.
28. Bhartiya D, Pal K, Ghosh S, Kapoor S, Jalali S, et al. (2013) lncRNome: a comprehensive knowledgebase of human long noncoding RNAs. Database (Oxford). 2013:bat034.
29. Li JH, Liu S, Zhou H, Qu LH, Yang JH (2014) starBase v2.0: decoding miRNA-ceRNA, miRNA-ncRNA and protein-RNA interaction networks from large-scale CLIP-Seq data., Nucleic Acids Res. 42:D92–D97.
30. Jalali S, Bhartiya D, Lalwani MK, Sivasubbu S, Scaria V (2013) Systematic transcriptome wide analysis of lncRNA-miRNA interactions. PLoS One. 8: e53823.
31. Sarver AL, Subramanian S. (2012) Competing endogenous RNA database. Bioinformation. 8: 731–733.
32. Grimson A, Farh KK, Johnston WK, Garrett-Engele P, Lim LP, et al. (2007) MicroRNA Targeting Specificity in Mammals: Determinants beyond Seed Pairing. Molecular Cell. 27: 91–105.

33. Yang JH, Li JH, Shao P, Zhou H, Chen YQ, et al. (2011) starBase: a database for exploring microRNA-mRNA interaction maps from Argonaute CLIP-Seq and Degradome-Seq data. Nucleic Acids Res. 39(Database issue):D202–D209.

34. Ala U, Karreth FA, Bosia C, Pagnani A, Taulli R, et al. (2013) Integrated transcriptional and competitive endogenous RNA networks are cross-regulated in permissive molecular environments. Proc Natl Acad Sci U S A. 110: 7154–7159.

35. Cabili MN, Trapnell C, Goff L, Koziol M, Tazon-Vega B, et al. (2011) Integrative annotation of human large intergenic noncoding RNAs reveals global properties and specific subclasses. Genes Dev. 25: 1915–1527.

36. Wang Y, Xu Z, Jiang J, Xu C, Kang J, et al. (2013) Endogenous miRNA sponge lincRNA-RoR regulates Oct4, Nanog, and Sox2 in human embryonic stem cell self-renewal. Dev Cell. 25: 69–80.

37. Fan M, Li X, Jiang W, Huang Y, Li J, et al. (2013) A long non-coding RNA, PTCSC3, as a tumor suppressor and a target of miRNAs in thyroid cancer cells. Exp Ther Med. 5: 1143–1146.

38. Tay Y, Rinn J, Pandolfi PP (2014) The multilayered complexity of ceRNA crosstalk and competition. Nature. 505:344–352.

39. Braconi C, Kogure T, Valeri N, Huang N, Nuovo G, et al. (2011) microRNA-29 can regulate expression of the long non-coding RNA gene MEG3 in hepatocellular cancer., Oncogene. 30:4750–4756.

40. Menigatti M, Staiano T, Manser CN, Bauerfeind P, Komljenovic A, et al. (2013) Epigenetic silencing of monoallelically methylated miRNA loci in precancerous colorectal lesions, Oncogenesis. 2:e56.

41. Lam LT, Lu X, Zhang H, Lesniewski R, Rosenberg S, et al. (2010). A microRNA screen to identify modulators of sensitivity to BCL2 inhibitor ABT-263 (navitoclax). Mol Cancer Ther. 9:2943–2950.

42. Mestdagh P, Lefever S, Pattyn F, Ridzon D, Fredlund E, et al. (2011) The microRNA body map: dissecting microRNA function through integrative genomics., Nucleic Acids Res. 39:e136.

43. Enright AJ, John B, Gaul U, Tuschl T, Sander C, et al. (2003) MicroRNA targets in Drosophila. Genome Biol. 5: R1

44. Karolchik D, Hinrichs AS, Kent WJ (2012) The UCSC Genome Browser. Curr Protoc Bioinformatics. 40: 1.4.1–1.4.33.

4

HIVE-Hexagon: High-Performance, Parallelized Sequence Alignment for Next-Generation Sequencing Data Analysis

Luis Santana-Quintero[1], Hayley Dingerdissen[1,2], Jean Thierry-Mieg[3], Raja Mazumder[2]*, Vahan Simonyan[1]*

1 Center for Biologics Evaluation and Research, US Food and Drug Administration, Rockville, Maryland, United States of America, 2 Department of Biochemistry and Molecular Biology, George Washington University Medical Center, Washington, DC, United States of America, 3 National Center for Biotechnology Information, U.S. National Library of Medicine, National Institutes of Health, Bethesda, Maryland, United States of America

Abstract

Due to the size of Next-Generation Sequencing data, the computational challenge of sequence alignment has been vast. Inexact alignments can take up to 90% of total CPU time in bioinformatics pipelines. High-performance Integrated Virtual Environment (HIVE), a cloud-based environment optimized for storage and analysis of extra-large data, presents an algorithmic solution: the HIVE-hexagon DNA sequence aligner. HIVE-hexagon implements novel approaches to exploit both characteristics of sequence space and CPU, RAM and Input/Output (I/O) architecture to quickly compute accurate alignments. Key components of HIVE-hexagon include non-redundification and sorting of sequences; floating diagonals of linearized dynamic programming matrices; and consideration of cross-similarity to minimize computations.

Availability: https://hive.biochemistry.gwu.edu/hive/

Editor: Tom Gilbert, Natural History Museum of Denmark, Denmark

Funding: This research was supported in part by the Food and Drug Administration Medical Countermeasures Initiative and in part by Intramural Research Program of the NIH, National Library of Medicine. The funders had no role in study design, data collection and analysis, decision to publish, or preparation of the manuscript.

Competing Interests: The authors have declared that no competing interests exist.

* E-mail: mazumder@gwu.edu (RM); Vahan.Simonyan@fda.hhs.gov (VS)

Introduction

Sequence alignment is the critical first step of sequence analysis [1,2], after which the alignment results are used as a source of data for numerous downstream analyses (e.g., the genetic content of short reads, pathway analysis, and etc). Before proceeding to the description of the optimized, ultra-fast alignment algorithm implemented in the High-performance Integrated Virtual Environment (HIVE), the following section describes the task of alignment and conventional methods currently used to solve it.

Given

- There exists a set of "Reference Genomes" numbered *1...r...N* with sizes of G_r and cumulative size of $R = \Sigma G_r$ bases.
- There exists a set of "Short Reads" from *1...s...S,* each one having a length of *L*.

Task

- Define an alignment as $A(s,r) = (s_1 r_1),...(s_i r_j),... (s_{As} r_{Ar})$ where $(s_i r_j)$ signifies the correspondence between *i*-th letter of the short sequence and *j*-th letter of the reference sequence. *As* and *Ar* correspond to the length of the alignment with respect to the corresponding sequence or reference.

- Define a set of "Scoring Parameters" *P* defining the benefit and cost factors for matches, mismatches, insertions and deletions between bases $(s_i r_j)$ of the short read and reference genomes
- Define an additive "Score of Alignment" as the sum of scoring factors $SA(s,r) = \Sigma(P_l)$ where **1** is chosen based on the match, mismatch, insertion or deletion of the sequence positions of *s* and *r*.

Solve

- Find an optimal alignment *A(s,r)* between short sequence *s* and reference genome *r* such that *SA(s,r)* is no smaller ($\geq$) than any other *SA(s',r')* where *s'* is not equal ($\neq s$) to *s* and/ or *r' ≠ r*.

(Notice that dynamic programming alignments are only optimal relative to additive scoring schemes.

If, for example, we considered a triple deletion less costly than three separate single letter deletions, the Smith Waterman algorithm, which assumes additive costs, may fail to find the best solution.).

The simplistic approach of comparing every short read position to every other genomic position, even without mismatches allowed, has a complexity of *O(S×L×R)* in big O notation. Such an approach has no technical value for realistic sizes of genomes (in Giga-bases $R\sim 10^9$) and high throughput sequence (HTS)-scale sequence read files (600 Giga-bases $S\times L>10^{12}$) for a single run. A more realistic approach is to detect highly scoring regions of candidate positions by finding exact matches of sequence seeds of length K (K-mers), up to K = 14, either by hashing techniques[3–5] or by other indexed methods like full-text minute-space (FM) indexes [6,7] as described below:

- **K-mer seed based hash indexes.** The reference genome is precompiled into a hash table where every K-mer's occurrence is maintained in the hash container [8]. Candidate alignment positions are then detected by looping through all hash elements corresponding to each seed of length K occurring in the short read.

- **FM-index based substring search methods.** The reference genome is compiled into a compressed suffix array container using the Burrows-Wheeler transform [9]. Lookup operations are then implemented through backward iterators searching for the sequence patterns within suffix array [10].

The speed and computational complexity of detecting candidate positions are comparable for both approaches and either can be suitable depending on the exact situation. The K-mer hash compilation stage is usually much faster than FM index building, but the hash table is also significantly larger in memory than an FM-index array [11,12]. The result of the first stage of lookups is a list of exact matches of certain lengths. The next step generally involves extension of the preliminary matches using a heuristic extension algorithm (BLAST-like) [13] or a dynamic programming method (Needleman-Wunsch [14] or Smith-Waterman [15,16]). The key considerations of extending seed alignment to obtain the optimal alignment include:

- **Extension of the seed candidate alignment.** The detected seeds' exact matches are extended in both directions with or without mismatches, insertions and deletions. This step is typically very fast and is of $\sim O(L)$ average performance.

- **Optimal alignment.** Dynamic programming techniques are usually performed in a rectangular matrix where alternative trajectories of alignments are considered concurrently. Each cell represents an alignment between two sequences at a given position. The best possible trajectory across cells is determined by cumulative alignment scores from left to right [17]. Such techniques are generally of $\sim O(L^2)$ [18,19] performance and, having square dependent memory footprint, are not cheap with respect to memory and CPU-clock.

Depending on the approach used, one may impose certain requirements on alignment algorithms to ensure the reliability of the computational results: **optimality** –demands that no better alignment is possible for the specified region of the supplied reference genome and short read; **fuzziness** – a small number of errors are acceptable within an allowed error density; **quantifiable** – each alignment can be assigned a number and score for the purpose of comparing two alternative alignments; **customizable** – behavior of alignment can be finely controlled by a set of parameters; **robustness** – small changes in parameters should lead to small changes in alignment results; **reproducible** – should arrive to the same alignment despite the stochastic nature of the algorithm's initial detection of seeds' exact matches.

The HIVE-hexagon aligner applies modified versions of the aforementioned approaches in conjunction with a new suffix-based approach to the removal of duplicate data and strategic sorting to optimize the alignment process: only the best candidate alignments undergo the computationally expensive stage of optimal alignment.

Results

The implementation of novel and traditional approaches to the alignment task in HIVE-hexagon promotes competitive performance when compared to current industry standards. The overall alignment pipeline employed by HIVE-hexagon can be seen in Figure 1. Short read data sets are non-redundified such that only a unique copy of any given read is subjected to the alignment while appropriate counts and indexes are maintained for all such reads. Remaining reads are then sorted by sequence similarity for efficient lookup in later stages, and both reads and reference sets are distributed among the computational cloud. Parallelized reference sets are compiled into a bloom/hash table such that each read thread undergoes K-mer query against each reference, followed by extended inexact alignments for all identified matches. If an inexact alignment does not meet score requirements, self-similarity can be used to filter neighboring reads based on the implied similarity of proximal data as ensured by earlier sorting. Thus, the number of actual alignments calculated is drastically reduced. Finally, the optimal, floating diagonal adaptation of the Smith-Waterman algorithm is computed for all candidate reads that score above the specified threshold in the inexact alignment stage.

This paradigm greatly benefits HIVE-hexagon with respect to computational speed and sensitivity: the time saved allows HIVE-hexagon to take on more costly computations to achieve greater sensitivity without sacrificing overall alignment speed. To support this claim, we have performed validation and benchmarking procedures to compare the HIVE-hexagon alignment algorithm to similar software packages used in the industry. For this paper, we have chosen Bowtie and BWA for comparison since these two tools have been readily adopted for high throughput sequence alignments.

In the first category of tests we only compare algorithms without considering parallelization, computer stress factors, and performance or I/O characteristics of computations. In the second category we compare the alignment platform as a whole with competitive applications running in singular and parallel modes (where possible).

Sample Notation

We have chosen.

a) **Influenza (IZ) sample.** Multi-segmented RNA virus which is accurately represented by its H5N1 genome and the mutations and divergence of the sample were well categorized from previous Sanger sequencing data;

b) *Mycoplasma hyorhinis* **(MH) sample.** Bacterial sample of known origin with a known set of multiple repeats determined by Sanger sequencing;

c) *Homo sapiens* **(HS) whole genome sequencing sample.** Eukaryotic DNA-seq sample from Sequence Read Archive out of which artificial reads were generated from 2 reference segments (chromosomes) X and Y Human Genome v19 Build 37.3;

d) **mixture of 15 viruses (VM)** out of which artificial reads are generated from 2 of the genomes Human adenovirus C (gi|9626158|ref|NC_001405.1) and Dengue virus 1 (gi|9626685|ref|NC_001477.1);

e) and, similarly, a **mixture of 10 bacterial genomes (BM)** from which artificial reads are generated for 2 of the genomes

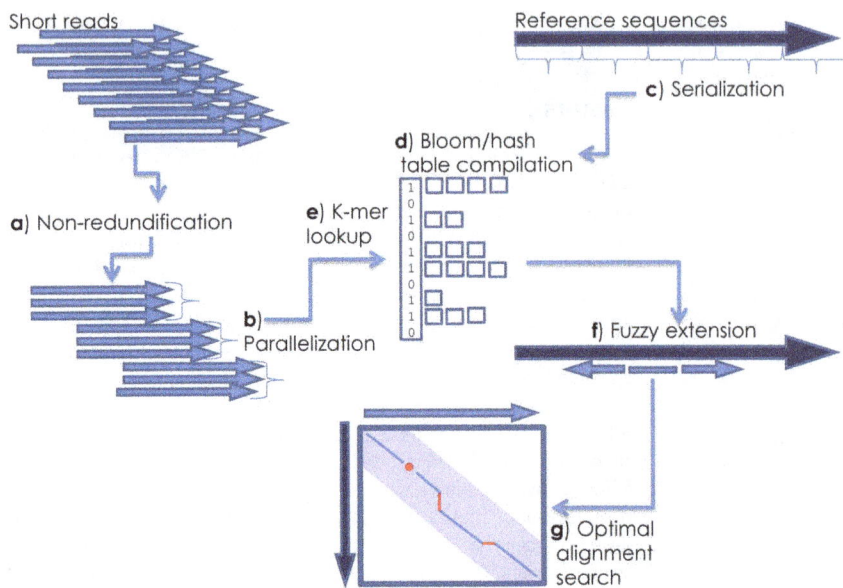

Figure 1. Workflow for HIVE-hexagon alignment utility. Overall alignment schema for HIVE-hexagon: short reads are non-redundified (**a**) and parallel portions (**b**) are sent to distributed cloud for computation. Reference genomes are then split into smaller pieces (**c**) and compiled into bloom/hash table (**d**). Every parallel execution thread performs a K-mer lookup against every reference sequence (**e**) then extends matches via inexact alignments (**f**) and performs a subsequent optimal alignment search on remaining candidates (**g**).

Shigella dysenteriae Sd197 (gi|82775382|ref|NC_007606.1) and Streptococcus pneumonia ST556 chromosome (gi|387787130|ref|NC_017769.1).

Full information regarding mixed samples and concentrations can be found in Table S1.

Inputs and Arguments

We have generated artificial reads from original reference sequences with no error (prefix **AR0-**), 1% error (**AR1**) and 5% error (**AR5**) or taken the original submissions produced by a sequencing hardware (**OR**). To define a consistent notation for our samples within this publication we also signify the number of reads as a suffix for the sample name. Thus, in this notation the sample AR5-IZ-1M would mean 1 million reads artificially generated from the Influenza reference with 5% random errors; similarly OR-HS-100M indicates a large, original sequence file of human origin with 100 million reads.

All HIVE-hexagon runs were performed with optimized advanced parameters including K-mer extension minimal length percent of 75, K-mer extension mismatch allowance % of 15, optimal alignment search of only identities and seed K-mer length of 14. Basic parameters were set to a minimum match length of 36 and 15 percent mismatches allowed, reporting the first match found to have the highest score of alignment. All Bowtie runs were performed with additional argument −e 600 both inside and outside of HIVE. All BWA runs were performed with additional argument −n 15 both inside and out of HIVE. These parameters were chosen to mimic the sensitivity of those tools with HIVE-hexagon as much as possible. Alignment results are reported as the percentage of unaligned reads. Detailed count information for all runs can be viewed in Table S2.

Proof of concept. Error free sequence files were artificially generated using the generateSeqs script (Table S3) with 1 million reads each generated from sequence data originating from influenza, mycoplasma and human samples. This test acts as a

proof of concept since we expect to detect 100% of artificial error free reads when aligned to the appropriate reference. As seen in Table 1, we fully aligned all error-free reads for influenza and mycoplasma runs. A very small number of error-free human reads were left unaligned by all tools: 16 by HIVE-hexagon, 150 by Bowtie and 147 by BWA. The exclusion of some artificial reads may be due to the over-optimization of heuristic algorithms with regard to seed over-representation which can have a degrading effect on alignments. This can have a drastic impact on low-complexity read alignments as evidenced by the provided example alignments for human samples with no noise. Low-complexity sequences generate thousands of hit candidate positions, making comprehensive alignment costly and disadvantageous due to the required increase in computational time without much added benefit. Additionally, specifying a smaller seed length for the determination of candidate regions results in a larger number of positions to be considered for extension. Each one base shortening of the seed results in four times as many computations in the candidate discovery stage. Thus, the huge decrease in the amount of unmapped reads observed for more divergent samples can be explained by the gain in extra sensitivity provided by shorter seed specification.

Sensitivity check for single species. Three sets of artificial reads were generated from the same influenza sample with varying degrees of error. A higher induced error rate simulates the real-life scenario of divergent sequences. Thus, this test shows if and how sensitivity varies in alignments between increasingly divergent sequences. As expected, an increasing number of alignments are missed across all tools as error or noise (indicating sequence divergence) increases. The higher sensitivity of a method for divergent sequences is critical for detection pipelines where adventitious agents present in small quantities can have adverse effects on biological products safety. An inability to detect a significant amount of sequence hits when the reference sequence is not well known can render Next-Gen based techniques useless. HIVE-hexagon has been specifically optimized to improve

Table 1. Validity and Sensitivity Comparison of Alignment Tools in Native Environments.

TEST	PURPOSE	SAMPLE	HIVE-HEXAGON	BOWTIE	BWA
Proof of concept for single species	Influenza mapping	AR0-IZ-1M	0.0000	0.0000	0.0000
	Mycoplasma mapping	AR0-MH-1M	0.0000	0.0000	0.0000
	Human mapping	AR0-HS-1M	0.0016	0.0150	0.0147
Sensitivity check for single species	Proof of concept	AR0-IZ-1M	0.0000	0.0000	0.0000
	Sensitivity level 1 check	AR1-IZ-1M	0.0013	0.5171	0.4930
	Sensitivity on more divergent sample	AR5-IZ-1M	0.4645	16.7000	21.3228
Sensitivity check for samples with	Mixture proof of concept for viruses: Capability to separate different references	AR0-VM-1M	0.0000	0.0000	0.0000
many species: Viral genomes	Capability to separate different references with greater sensitivity	AR1-VM-1M	0.0310	0.4376	0.6207
	Capability to separate different/divergent references with great sensitivity	AR5-VM-1M	0.6079	16.7687	21.2656
Sensitivity check for samples with	Mixture proof of concept for bacteria: Capability to separate different references	AR0-BM-1M	0.0000	0.0000	0.0000
many species: Bacterial genomes	Capability to separate different references with greater sensitivity	AR1-BM-1M	1.1817 0.0002*	0.2963	0.4078
	Capability to separate different/divergent references with great sensitivity	AR5-BM-1M	7.3305 0.3539*	16.2000	20.3869
Sensitivity check for large genomes:	Proof of concept for large, human genome	AR0-HS-1M	0.0016	0.0150	0.0147
Human data	Capability to separate different references with greater sensitivity	AR1-HS-1M	5.8383	18.8320	18.4592
	Sensitivity on more diverse sample from a large genome	AR5-HS-1M	15.3918	62.8203	62.4555

*with repeat and transposition search sub-algorithm on.

sensitivity and clearly outperforms both Bowtie and BWA in this respect.

Sensitivity check for samples with many species. As mentioned above, read sets with variable error rates were generated from select genomes within viral mixture and bacterial mixture samples. Alignment of the read files created with zero error to the entire mixture sample allows us to demonstrate HIVE-hexagon's ability to separate multiple references within one sample. Subsequent comparison to files with error rates tests the sensitivity with which HIVE-hexagon can separate and map a query to the correct reference in a mixed file when the query has an increasing degree of divergence from its reference. Separation and mapping follows the same principle here as in the single species check such that it is increasingly difficult (and therefore more alignments are missed) as divergence increases.

The viral mixture shows HIVE-hexagon once again to surpass performance of other tools. The bacterial mixture results demonstrate a more complex mode of alignment, ultimately showing HIVE-hexagon to be more sensitive for files with or without divergence. Because the chosen bacterial mixture contains species having significant numbers of repeats, HIVE-hexagon was run both with and without a specific argument forcing careful detection of such repeats. In the repeat and transposition detection mode, HIVE-hexagon misses significantly fewer alignments while being only 10–15% slower compared to the non-repeat mode. Once again, HIVE demonstrated much higher sensitivity than Bowtie and BWA while being more time-optimal.

Sensitivity check for large genomes. The large genome sensitivity concept is similar to the mixture approach because of the nature of eukaryotic (human) genome references as compilations of separate reference files of various segments (chromosomes, genes, etc.). Thus, HIVE-hexagon's ability to separate references is essential to its utility in human mappings. Furthermore, human genomes and sequence data are much larger, on average, than bacterial and viral counterparts and the stress to memory and IO in algorithms is significantly greater. Regardless, results follow the established trends with HIVE-hexagon missing fewer alignments than both Bowtie and BWA. This test showcases the viability of HIVE-hexagon as a faster and more sensitive tool for eukaryotic contiguous alignments.

Performance/scalability Testing

Performance dependency on the size of the genome. The hashing of extra-large eukaryotic genomes can be a time consuming step. An FM index may take 2–3 hours to compile for BWA and Bowtie on modern x86 CPUs, but once finished, results can be maintained for application to future computations. As a native component of the HIVE system, HIVE-hexagon is required to allow users to subset and superset genomic reference sequences in an arbitrary manner during alignments and therefore cannot maintain permanent indexes of precompiled references: HIVE-hexagon recompiles the reference sequence K-mer hash tables every time a computation is initiated. Although it takes up to 8 minutes to recompile a eukaryotic size genome, this is considered an affordable tradeoff between convenience and functionality given the approximately 1–1.5 hours it takes to perform an alignment of 100 million sequences on such genomes. For smaller genomes this step takes only a few seconds and does not play a significant role in our performance considerations.

As mentioned before, low complexity regions and over-represented repeated subsequences can strongly diminish alignment algorithm performance. Large reference genomes tend to be more prone to such regions; it takes super-linearly disproportional time to map reads to such genomes compared to compact and dense genomes. With simple low complexity read filters, low complexity reference region masking and over-represented seed masking, the performance of HIVE-hexagon alignment of 100 million reads to human genome can boost from ~1–1.5 hours to 35–40 minutes. The sensitivity of biologically dense sequence alignments deteriorates only slightly (usually less than 0.1% in real samples) and only in long, low complexity stretches of the reference genome.

Thus, the total time consumed by HIVE-hexagon may be from 15 minutes to 1.5 hours for 100 million reads against a 3GB human genome depending on the set of parameters. Comparable and still less sensitive runs using BWA and Bowtie take about 1–1.5 hours assuming the references have been precompiled (2–3 hours).

Performance dependency on the coverage and the number of reads. Due to its non-redundification feature, HIVE-hexagon consistently outperforms BWA and Bowtie linearly proportional to the coverage. For example: alignment of a real dataset containing 100 million reads (out of which only 16 million were unique) to a small influenza genome (12KB) resulted in 6x time-saving (2–3 min) for HIVE-hexagon when compared to itself without utilizing the non-redundification algorithm (~15 min). Similar computation against 25 pico- and entero-viruses using HIVE-hexagon took around 15–17 minutes (larger cumulative genome) for a metagenomic poli-virus environmental sample dataset of the same size where the unique count was roughly similar to the total sequence count. BWA and Bowtie both take about 20–25 minutes for the same datasets, although generally slightly less sensitive.

Performance dependency on the execution environment. All sequence aligners (HIVE-hexagon, BWA, Bowtie or others) implemented within HIVE infrastructure run in a parallel execution environment. The time and benchmarks compared in the two previous sections were those of the applications run from within the system. The actual running times for standalone applications can be significantly longer, and are roughly linearly proportional to the level of parallelism used to run. For example, computational alignment of a human genome of 100 million sequences which takes ~1–1.5 hours from within HIVE with a level of parallelism of 20 takes roughly 22–30 hours when run in a standalone, single thread mode on a comparable computer. HIVE-hexagon, specifically designed for a parallel execution environment like that provided by HIVE, additionally benefits from parallel data storage and decreased data mobility, achieving returns beyond what the algorithm itself provides.

Table S4 lists differential hit counts obtained for both Bowtie and BWA when run in their native, external environments compared to results when run inside HIVE's parallelized environment.

Follow-up Testing and Benchmarking

It is clear from this preliminary validation and performance testing that HIVE-hexagon has all the characteristics necessary to become an industry staple for alignments on any species. HIVE has already embarked on a number of studies and collaborations using the HIVE-hexagon aligner and other HIVE-developed tools and workflows [20–22], so the quality and integrity of the system and each of its tools are of the highest priority.

A number of the previously mentioned performance-enhancing ideas have been borrowed from the AceView Magic [23] alignment algorithm, and further integration of these tools is intended to continue; we therefore avoided inclusion of AceView Magic in this direct comparison but will follow-up with full

benchmarking in the next publication. We plan to conduct more extensive comparisons between HIVE-hexagon and all currently available, comparable alignment algorithms within the next few months to better determine HIVE-hexagon's respective performance and to identify any weaknesses or needs for improvement. Additionally, we intend to continue integrating other external tools directly into the native HIVE execution environment. Furthermore, we aim to use the underlying HIVE-hexagon algorithm to implement robust multiple sequence alignment, recombinant and clone discovery utilities.

Methods

Aligning short reads to reference genomes results in a high degree of coverage, such that every genetic position is mapped by a large number of base-calls. It is not unusual to see 10,000–1,000,000× uniform coverage on shorter genomes like viruses and bacteria and 5–100× coverage on eukaryotic genomes with a lesser degree of uniformity [24]. The ratio of short read count to the number of reference positions is a heuristic measure of unique short read redundancy. Errors and noise introduced into short read sequences by sequencing chemistry or processing pipelines can reduce the actual redundancy rate by randomly introducing differences between similar DNA/RNA molecules. However, given the small systemic error rates produced by present day technologies [25–27], actual redundancy rates can be tens to sometimes hundreds or thousands for highly expressed regions of reference genomes. Innovative identification and usage of redundancy and cross-similarity between reads can be beneficial for bioinformatics pipelines, minimizing the storage memory footprint and bio-analytics complexity by removing repeats, and therefore allowing a higher rate of vertical compression and avoiding unnecessary repeated computations of identical/redundant reads.

Bioinformatics pipelines performing alignment and mapping computations are frequently heuristic in nature and, with memory access patterns driven by the data itself, often need to access large memory chunks randomly. For such large datasets, the "memory hungry" processes with unorganized "page hits" can create a bottleneck in computations where the vast majority of time is lost on memory pages caching into the CPU. The HIVE-hexagon algorithm proposes a novel method of linearized data organization to minimize memory usage of underlying matrixes and structures used for programming alignment algorithms, optimize memory page hits and boost algorithmic performance by taking advantage of modern CPU architecture.

The overall pipeline for the HIVE-hexagon algorithm proposed within this publication consists of a few major steps (see Figure 1):

Non-redundification

The non-redundification of the sequence space is accomplished by building a sequence suffix tree [28] for text with the four letter alphabet A = a, c, g, t (see Figure 2). The use of a tree-like structure to solve sorting problems is not new in the field of bioinformatics[29–31]; however, previous efforts have reported problems related to the prohibitive size of suffix trees when applied to longer strings in larger quantities, increasing the string universe from hundreds to thousands or millions.

Our simple tree algorithm minimizes the number of comparisons against new sequences base-four logarithmically as they are added into the tree. Computationally, the average scenario behaves as a task of ~$O(S*log_4L)$ complexity; the worst case scenario, when all sequences are identical, will require an all-to-all comparison and therefore will increase processing time to

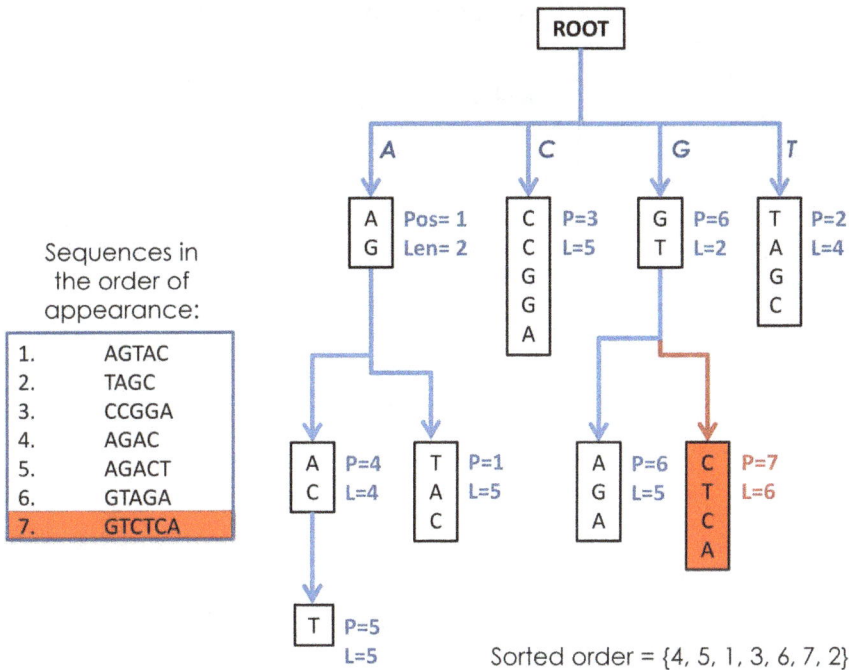

Figure 2. Non-redundification. This tree representation of 6 sequences is composed by linking all sequences' tails (suffixes) to the parent nodes which represent the longest prefixes common to each branch. The tree contains all information about the 6 sequences, including position and length, at each node. Once the tree is complete, we can traverse the tree in a left maze order (always taking a left path) to obtain the sorted list of elements as: S = 4,5,1,3,6,7,2, which refers to the sorted list AGAC, AGACT, AGTAC, CCGGA, GTAGA, GTCTCA, TAGC.

~$O(S*L)$, but the corresponding storage needs are minimal. Conversely, a very diverse set of reads is computationally cheaper but has higher storage demands to accommodate the increased quantity of unique records.

Redundant sequences are discarded, but counters are associated with the unique sequences to signify that a similar sequence (in this case, an exact copy) has already been stored in the tree. Thus, each sequence in the tree at any given time is unique and, upon addition of a new sequence to the tree, it is a simple process to identify whether the sequence is a repeat or whether a new branch should be created.

The actual implementation of the algorithm has a built-in parallelization schema such that, depending on the memory limitations imposed on the algorithm from the execution environment, a decision is made to split data into 4, 16, or 4^K portions. Each thread will pick up sequences starting with a prefix which is the 2-na representation of its thread index in K-mer space. For example, in 16 part execution mode the portions will include the prefixes: AA..., AC..., AG..., AT..., CA..., CC..., CG..., CT..., TA..., TC..., TG..., TT... Splitting the tree parsers in this manner removes the need for a joining step where trees must be collapsed. Since no overlaps are possible in each portion and within portions, there is no need to repeat the non-redundification step and re-sort the sequences.

Another piece of valuable information stored at each sequence node is the cross-similarity coefficient which shows the degree of similarity between prefixes of consecutive sequences in the tree. This is used for explicit cross-similarity optimization of the alignment algorithm described later in this paper.

Sorted Paths

Once all sequences are digested into the tree, it is easy to traverse the tree by visiting every node exactly once in a pre-

determined order using a single traversal iteration function which operates on any given node. The function returns the node itself (which represents a sequence) or it continues the traversal through its children nodes from left to right if such nodes exist. This simple recursive procedure will, in turn, generate a final version of the file with unique and sorted sequences.

By maintaining an iterator structure on a bound, self-referenced tree this operation is of constant time access and the whole tree can be generated within $O(S)$ operations where S is the count of non-redundant sequences.

K-mer Hashing

HIVE-hexagon compiles a dictionary of K-mer occurrences (seeds) in the reference set of genomes, where the K-mer is defined as a shorter subset of $r_1,...r_i... r_K$. The compiled result is a hashed bucket list (Figure 3d) where each bucket represents the positions of its seed's occurrences in a reference genome. This step has a complexity of $O(R)$. In a genomic alphabet of four letters (A, T, G, C), the K-mer table consists of 4^K elements, each occupying so many integers in memory to hold the list of occurrence positions and the hash back reference. In a conventional hash-table implementation, one would need to store backward references to hash indexes. However, in the HIVE-hexagon implementation, the K-mers themselves are considered indexes in 2-na representation of sequence space where each nucleotide is represented by a 2-bit value (A = 00 = 0, C = 01 = 1, G = 10 = 2, T = 11 = 3). By considering sequences as indexes we remove the need to maintain the sparse hash table back-references and avoid hash collisions using an over-exaggerated hash table. There is some penalty for having to maintain significantly sparse arrays for small genomes, but the benefit outweighs the cost, especially for larger genomes where the hash table is almost fully occupied.

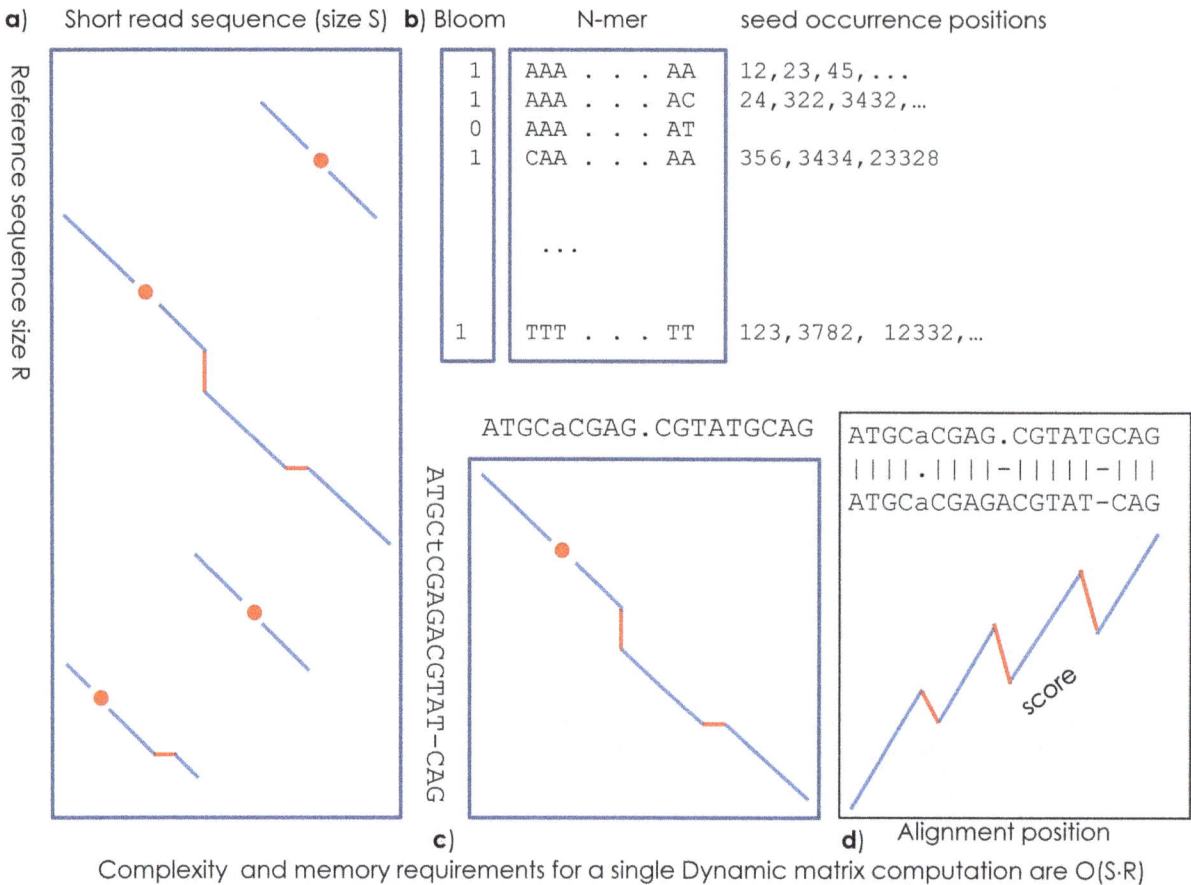

Figure 3. Optimal alignment search optimization schema. (**a**) Dynamic programming matrix Needleman-Wunsch or Smith-Waterman algorithms use a two dimensional rectangular matrix where cells represent the cumulative score of the alignment between short read (horizontal) and reference genome(vertical). (**b**) HIVE-hexagon maintains a bloom lookup table where each K-mer is represented only by a single bit signifying the presence or absence of that K-mer and 2-na hash table where a sequence's binary numeric representation is used as an index. (**c, d**) The fuzzy extension algorithm allows accurate definition of the alignment frame and squeezes the window where the high scoring alignments might be discovered.

The list of occurrence positions in a bucket list requires at least $2 \times R$ cells of integers to refer to the index of the reference genome and to a position on the genome where the particular seed has occurred. Thus, the memory footprint for a seed-hash table is roughly in the order of $\sim 4^K + \sim R$ integers. Contemporary (2013) computers can realistically hold a dictionary of up to 14-mers without sacrifice to the execution environment. K-mers larger than this are typically problematic, causing too great a stress to memory and, in parallel execution environments, diminishing performance benefits of hashing by memory swapping. Additionally, long K-mers require a sacrifice in sensitivity with over $1/K \sim 7\%$ error. HIVE-hexagon implements a double-hashing schema where lookup for K-mers larger than 14 is done by double-lookups of K-mers with $K<14$ in consecutive continuous positions.

Lookup Step

For every short read, HIVE-hexagon retrieves the K-mers sequentially and matches them to a seed-dictionary to obtain the list of occurrences of each particular K-mer on a reference sequence as potential candidates of alignment position. A genome of size R has in average $R/4^K$ occurrences of candidates for every K-mer. Increasing K results in fewer candidate positions where each has a higher chance of being a true alignment, thus increasing the speed of computations. However, an increase in K

also has the potential of increasing the footprint of the memory as $\sim 4^K + \sim R$. For L positions of the sequence there are $L-K$ (usually $K<<L$) positions to be looked up using the dictionary. Thus, the complexity of lookup step is $O(S \times L)$ and the memory footprint measures as $\sim R/4^K \times S \times L$ with proportionality coefficient dependent on the relatedness of the reference sequence and the number of successful alignments.

A significant number of hash lookups are misses, and from the perspective of the CPU the lookup results in a memory page load on a random access basis and is therefore costly and to be avoided. HIVE-hexagon maintains a bloom lookup table (Figure 3d) [32] where each K-mer is represented by only a single bit, signifying the presence or absence of that K-mer in the table. Due to its more compact size, the lookups from the bloom table are much less costly from a paging viewpoint and result in significant time-saving. The bloom table itself occupies a single bit for each hash element instead of full size 64 bit integer; therefore, the additional cost of the bloom implementation is only $\sim 1-2\%$ more memory compared to the original bloom-less variant of the algorithm.

The HIVE-hexagon implementation of the lookup table ignores seeds which are overexpressed as defined by a count greater than a given threshold: overexpressed seeds are usually present in low complexity regions in eukaryotic genomes and, although not always, frequently do not represent biological relevance. However, if the actual biological genomic region does have such overexpressed seeds adjacent to those with normal level of expression, the alignment algorithm will be able to capture the exact seed match in subsequence k-mers and then extend as much as needed regardless of the seed's expression level. If the entire region is made of overexpressed seeds, HIVE-hexagon will exclude alignments above a certain number of findings.

HIVE-hexagon can maintain seeds of K-mers on every genomic position, but it can also skip positions based on expression levels and preset parameters. This decreases the memory required for bucket list storage and decreases the pool of candidate positions without a great sacrifice to sensitivity for cases where the reference is well known. In viruses and bacteria, divergence can achieve large values and so this technique may decrease the sensitivity due to the fact that an exactly-matched seed has been skipped. For eukaryotic genomes, however, the reference tends to be more stable, so we frequently observe continuous exact matches longer than that of the K-mer size chosen for HIVE-hexagon.

Bracketing and Fuzzy Extension

Once K-mer seeding has detected a candidate region, HIVE-hexagon determines the frame of possible alignment windows around the exact seed match positions. The presence of a short exact match only hints at the potential alignment: the actual alignment must be computed by other means. Extension methodologies (like that in BLAST) do not need an exact reference frame of alignment, but dynamic programming methods require a strictly defined matrix where the optimal alignment will be computed. Underestimation of the frame may result in loss of sensitivity whereas overestimation may result in vertical extension (see Figure 3a) of the matrix, slowing down computations by increasing the absolute size of the matrix, and therefore the memory and the number of computations to be performed.

HIVE-hexagon implements both double-sided extension of the seed and a dynamic programming matrix. The fuzzy extension algorithm is similar to cost-based dynamic programming methods except that it runs in a very small floating window along the bidirectional extension front using integer arithmetic. This approach not only allows more accurate definition of the alignment frame (Figure 3c), but also filters a significant number of accidental K-mer hits. The number and the density of mismatches, insertions and deletions allowed during extension are customizable and, by default, correlated with relevant parameters for the optimal alignment algorithm used downstream.

Optimal Alignment

Dynamic programming methods of alignment [33] matrix evaluation usually include the computation of all alternative matches, mismatches, insertions and deletions (Figures 3 and 4). Each cell's value is computed as the best score from all alternate trajectories leading to that cell either as a match/mismatch (diagonal), insertion (vertical) or deletion (horizontal). All possible cumulative scores are computed across those trajectories and the highest value $SA(s,r) = \Sigma(P_k)$ is reported as a potential alignment score along with the trajectory $A(s,r)$ leading to it. The usual strategy involves computing the dynamic matrix values and backward pointers from the top left corner down to the bottom right corner. Backward pointers are then propagated in the opposite direction starting from the maximal score position to re-identify the trajectory which generated best local or global alignment.

The first, most obvious level of NW/SW optimization implemented in HIVE-hexagon is to avoid computation of the whole matrix and concentrate only on the diagonal region (Figure 4a) where the expected alignment usually lies because the extension algorithm applied in HIVE-hexagon ensures the accuracy of the frame positioning. Using a diagonal of constant width allows translation of computational complexity of $O(L \times G_r)$ into $O(L \times w)$ where w is the constant width of the diagonal and does not scale with the size of the selected reference segment.

Alignments with multiple insertions or deletions can be problematic for such optimizations. Although not usually an issue for short read alignments, a significant number of multiple in-dels is a critical problem (Figure 4b) for longer contig alignments or mutual alignments of reference genes. Unlike existing analogues [34,35], HIVE-hexagon allows the diagonal to float along the two sides of the highest scoring path (Figure 4c), thus allowing the generation of longer, multiple in-del-containing alignments to take advantage of the dynamic matrices diagonalization method.

To even further reduce the amount of computations in memory, HIVE-hexagon maintains a variable width diagonal (Figure 4d). Cells located at a greater distance from the optimal diagonal have an associated cost of insertions and deletions. Thus, alignments involving these cells will have a limit to the maximum possible score and minimal number of insertions, deletions and mismatches. Using this information, HIVE-hexagon estimates each particular cell's potential to generate a successful alignment within required thresholds. If the maximum possible score implies no potential, that position will be ignored and no memory will be allocated for it nor will computations be performed for trajectories through that cell. For this reason, the diagonal itself has "holes": black, ignored spots, hence the association with a particular kind of Armenian string cheese which looks like a string with variable thickness along its profile with possible holes on the sides.

Additional significant optimization implemented in HIVE-hexagon is the linearization of the final dynamic matrix. The benefit of storing a linear matrix $\sim w \times L$ instead of the rectangular matrix $L \times G_r$ where memory requirements are tens or hundreds of times larger is significant. In a real parallel execution environment where hundreds of processes compete for memory pages, such optimization has a huge potential of improving the actual execution speed.

Additional Features Contributing to Competitive HIVE-hexagon Performance

The alignment further benefits from the consideration and implementation of the following:

- **Alignment of unique sequences only.** By aligning every redundant sequence only once and maintaining the count of redundancies, HIVE-hexagon already achieves significant improvements in computational speed without additional algorithmic modifications. The speed benefit resultant from non-redundification is above linear due to the decrease in stress to memory pages, caching and swapping.

- **Sorting and implicit cross-similarity.** Further lazy optimization is achieved implicitly due to cross-similarity and sorting of the short reads computed during non-redundification of short read sequences.

Seed lookup involves reading particular locations from the bloom table, from the hash table, and therefore from the computer's memory. Modern CPUs optimize memory-reading by double-layer caching of memory pages. Random access to memory is usually slower than sequential access with additional hits to pages already in the cache.

In an attempt to align the first sequence, the CPU is forced to load a certain number of pages (containing seeds, reference

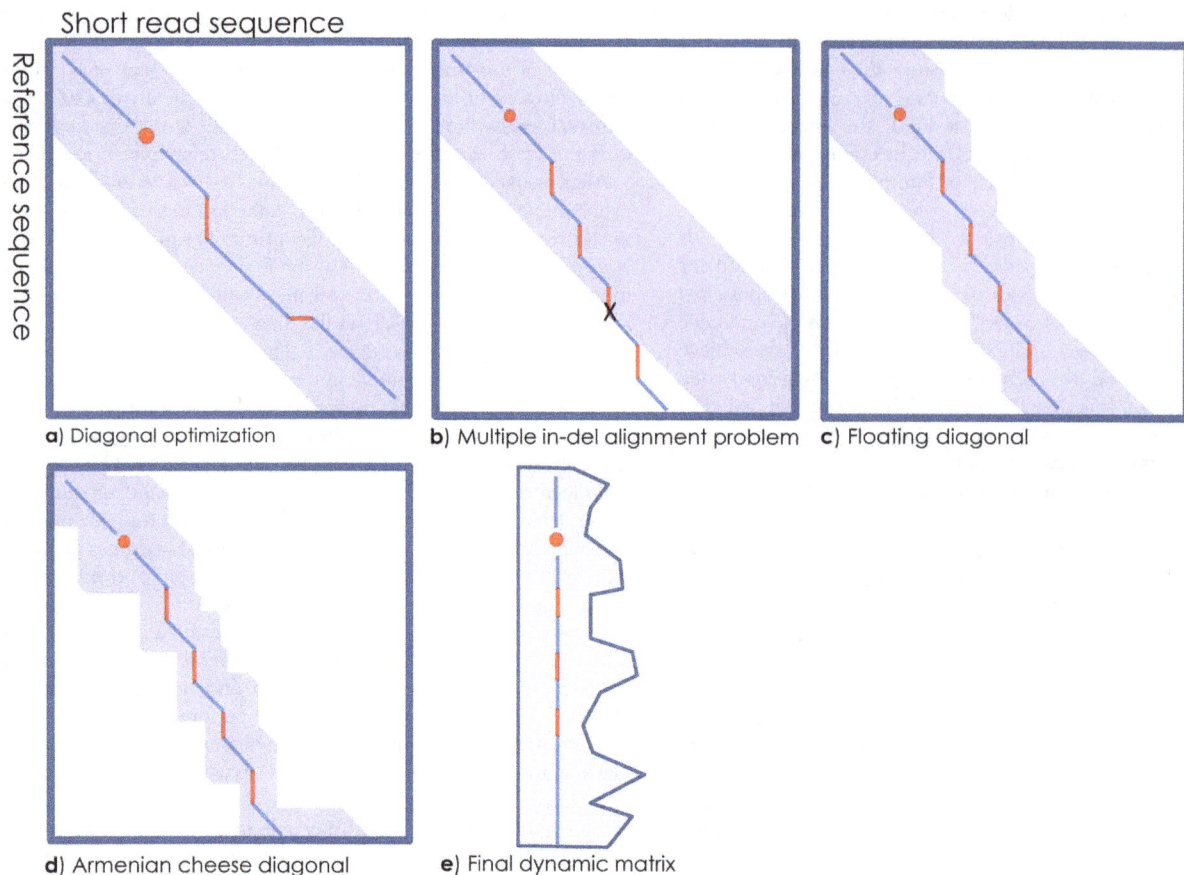

Figure 4. Dynamic programing matrix linearization schema. HIVE-hexagon implements a floating diagonal approach where the diagonal of the computation is maintained along the two sides of the current highest scoring path of the matrix. (**a**) We assume the optimal alignment will be along the diagonal. (**b**) Multiple insertions or deletions can result in the optimal path traversing outside the defined diagonal belt. (**c**) By defining a constant width to the diagonal and pre-computing cell scores line by line, the limits of the remaining diagonal matrix can float along with the likely optimal scoring path. (**d**) Furthermore, pre-computation of cells in the diagonal line by line allows the exclusion of regions which cannot possibly contain the highest scoring path. (**e**) The resulting minimized, dynamic diagonal matrix is linearized to further simplify the process and minimize the memory footprint required.

genome, etc.) into the cache. If the next sequence is very dissimilar to the first, the pages loaded are different for each, thus resulting in cache saturation with potential of dumping memory pages. However, if the sequences are very similar, the seeds hit the same set of positions in memory and the CPU does not need to reload new pages. HIVE-hexagon greatly benefits from the implicit optimization of short read sorting prior to alignment. We see up to 2-fold improvement of the speed solely due to sorting.

- **Explicit cross-similarity.** Even deeper optimization can be achieved explicitly by considering cross-similarity among the short reads. If two sequences are similar up to a certain number of characters, the alignments of those to the reference genome are expected to be similar.

Let us assume the first sequence has hit the K-mer table in a large number of candidate positions, out of which only a few have passed the extension step filter (described above) and even fewer have resulted in a real alignment for a given score threshold. The following sequence is nearly identical to the first and can therefore skip all candidate positions where the extension attempt for the first failed with a wide margin of error. Since HIVE maintains all self-similarities between consecutive sequences, it is possible to quantitatively predict if the best potential alignment score between

a sequence and the candidate position is within the required error threshold. Thus, only a few successful hits or near-successful mis-hit positions are typically considered for real optimal alignment search (Figure 5).

The pattern of cross-similarity in non-redundant sequence files is oscillatory, frequently in a zig-zag pattern where the "zigs" (horizontals) are changes in a base early in the sequence and "zags" (diagonals) are changes at the end of the sequences. Given the inexact nature of the alignments, each consecutive sequence has to be considered in more places than assumed by the cross-similarity, so the efficiency of zig-zag is not a full 100%.

We observe up to 2–4x fold performance boost depending on the parameters defined in the alignment algorithm, assuming the choice of the reference is accurate. For these cases we notice almost no sacrifice to sensitivity (<0.01%), but for cases with reference sequence further from the subject we saw some degradation (1–2%) of sensitivity with strong usage of cross-similarity.

This parameter is optional in HIVE-hexagon and is not recommended to be used in its current stage for references distant from the subject, or for references containing multiple repeats and transpositions. We do recommend using this option to benefit from

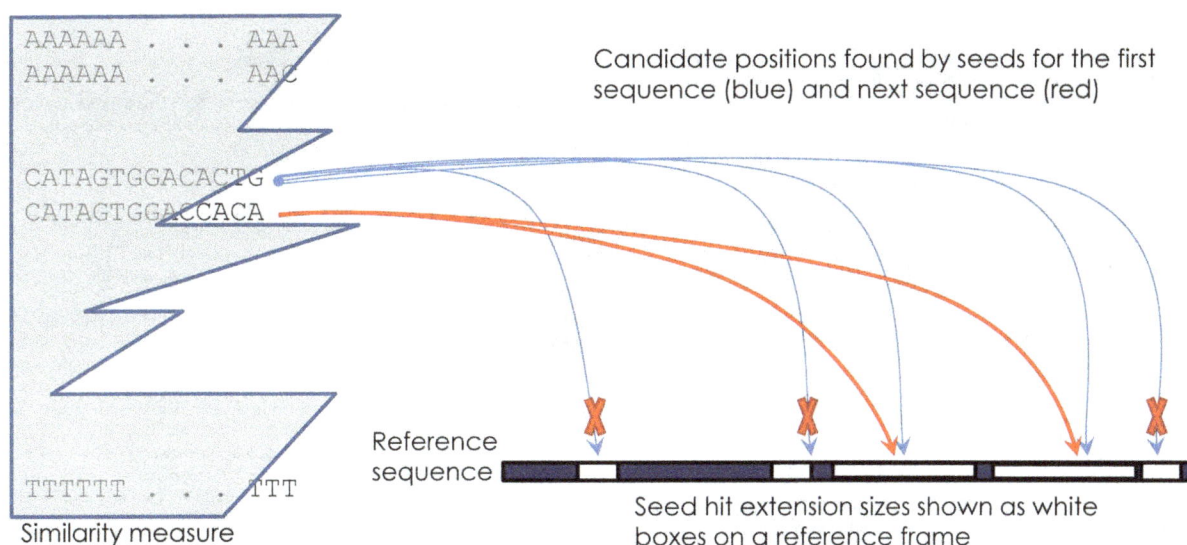

Figure 5. Cross-similarity decreases the pool of optimal alignment candidates. In this example, the first sequence CATAGTGGACACTG has generated 5 possible hit candidate positions (blue arrows), but only three of those have failed to extend or align long enough (marked by **X**). The next sequence, CATAGTGGACCACA, having a significant length of prefix similar to the prior one, does not need to consider all candidate positions, and can specifically exclude from consideration those for which the extension attempt failed with a wide margin of error for the prior, similar sequence.

significant improvements of the speed for more accurate references. We expect further development to take place on this subject.

Discussion

In this article, we have discussed the challenges associated with NGS data analysis, with special emphasis on alignment, and provided improvised and novel approaches to overcoming these challenges. Specifically, we have introduced new and significantly improved algorithms developed by HIVE which greatly reduce the memory footprint required by alignment of NGS data and, therefore, decrease the overall time needed for the alignment process.

A great quality of these HIVE algorithms is that the decrease in computational cost, memory requirement and time for processing is not accompanied by a sacrifice in the quality of the approach or the results. In fact, being native to the massively parallelized HIVE environment, the overall speed increase afforded by the infrastructure alone allows these algorithms to perform at a higher sensitivity than other industry algorithms of similar functionality while still outperforming in terms of time required for the computation.

Experiments have shown HIVE-hexagon is more sensitive and faster than current industry standard alignment algorithms due to scalability, high parallelizability, non-redundification, dynamic matrix linearization and implicit and explicit cross-similarity usage. There is already a great deal of interest surrounding HIVE-hexagon, and HIVE plans to continue developing this and other new tools to further promote the advancement of NGS technologies and the larger field of genomics.

Supporting Information

Table S1 Genomes, Mixes & Conc. Lists the components and accession IDs of all sequences used in the validation and testing section of the paper.

Table S2 All Run Counts. A detailed spreadsheet containing all counts of alignments (and unaligned sequences) used to summarize the validation and testing section.

Table S3 generateSeqs. Provides the text of a short script written to generate the random subset of reads used in the validation and testing.

Table S4 Externalvs.HIVE Bowtie&BWA. Shows the compared alignment counts of external tools when run both inside the HIVE environment and in their native external environments.

Table S5 Access Data in HIVE. Provides access information and instruction for reviewers and other interested individuals to see and replicate the results presented herein.

Acknowledgments

The authors wish to thank Konstantin Chumakov, PhD, Associate Director for Research, Office of Vaccines Research and Review, FDA CBER, for providing a deep insight and great advice regarding the nature and biology of the next-generation sequencing, in addition to provision of datasets; Carolyn A. Wilson, PhD, Associate Director for Research, Office of the Center Director, FDA CBER, for continuous support to the HIVE project and our research goals; Thomas Maudru FDA CBER, and Cristopher Kiem FDA CBER for their support for hardware/software and administrative infrastructure of HIVE project.

Implementation
blueHIVE technology group bluehivescience@gmail.com
Contact us
https://hive.biochemistry.gwu.edu/contact.php

Author Contributions

Conceived and designed the experiments: VS RM JT. Performed the experiments: VS JT LS. Analyzed the data: VS HD. Contributed reagents/materials/analysis tools: LS VS JT. Wrote the paper: HD VS.

References

1. Torri F, Dinov ID, Zamanyan A, Hobel S, Genco A, et al. (2012) Next Generation Sequence Analysis and Computational Genomics Using Graphical Pipeline Workflows. Genes (Basel) 3: 545–575.

2. Pabinger S, Dander A, Fischer M, Snajder R, Sperk M, et al. (2013) A survey of tools for variant analysis of next-generation genome sequencing data. Brief Bioinform.

3. Altschul SF, Gish W, Miller W, Myers EW, Lipman DJ (1990) Basic local alignment search tool. J Mol Biol 215: 403–410.

4. Delcher AL, Kasif S, Fleischmann RD, Peterson J, White O, et al. (1999) Alignment of whole genomes. Nucleic Acids Res 27: 2369–2376.

5. Li H, Homer N (2010) A survey of sequence alignment algorithms for next-generation sequencing. Brief Bioinform 11: 473–483.

6. Langmead B, Trapnell C, Pop M, Salzberg SL (2009) Ultrafast and memory-efficient alignment of short DNA sequences to the human genome. Genome Biol 10: R25.

7. Ferragina PM G (2000) Opportunistic data structures with applications; Redondo Beach, California. 390–398.

8. Dua S, Chowriappa P (2013) Data Mining for Bioinformatics. Boca Raton, FL CRC Press. Taylor & Francis Group, LLC.

9. Burrows M, Wheeler DJ (1994) A block-sorting lossless data compression algorithm.

10. Sun WK (2010) Algorithms in Bioinformatics: A Practical Introduction. Boca Raton, FL Chapman & Hall/CRC Press. Taylor & Francis Group, LLC.

11. Wang W, Wei Z, Lam TW, Wang J (2011) Next generation sequencing has lower sequence coverage and poorer SNP-detection capability in the regulatory regions. Sci Rep 1: 55.

12. Homer N, Merriman B, Nelson SF (2009) BFAST: an alignment tool for large scale genome resequencing. PLoS One 4: e7767.

13. Schuler GD, Altschul SF, Lipman DJ (1991) A workbench for multiple alignment construction and analysis. Proteins 9: 180–190.

14. Needleman SB, Wunsch CD (1970) A general method applicable to the search for similarities in the amino acid sequence of two proteins. J Mol Biol 48: 443–453.

15. Smith TF, Waterman MS (1981) Identification of common molecular subsequences. J Mol Biol 147: 195–197.

16. Smith TF, Waterman MS, Fitch WM (1981) Comparative biosequence metrics. J Mol Evol 18: 38–46.

17. Shen S, Tuszynski JA (2008) Theory and Mathematical Methods for Bioinformatics. Berlin, Heidelberg: Springer-Verlag.

18. Gotoh O (1982) An improved algorithm for matching biological sequences. J Mol Biol 162: 705–708.

19. Waterman MS (1984) Efficient sequence alignment algorithms. J Theor Biol 108: 333–337.

20. Karagiannis K, Simonyan V, Mazumder R (2013) SNVDis: a proteome-wide analysis service for evaluating nsSNVs in protein functional sites and pathways. Genomics Proteomics Bioinformatics 11: 122–126.

21. Dingerdissen H, Motwani M, Karagiannis K, Simonyan V, Mazumder R (2013) Proteome-wide analysis of nonsynonymous single-nucleotide variations in active sites of human proteins. FEBS J 280: 1542–1562.

22. Lam PV, Goldman R, Karagiannis K, Narsule T, Simonyan V, et al. (2013) Structure-based comparative analysis and prediction of N-linked glycosylation sites in evolutionarily distant eukaryotes. Genomics Proteomics Bioinformatics 11: 96–104.

23. Thierry-Mieg D, Thierry-Mieg J (2006) AceView: a comprehensive cDNA-supported gene and transcripts annotation. Genome Biol 7 Suppl 1: S12 11–14.

24. Zagordi O, Daumer M, Beisel C, Beerenwinkel N (2012) Read length versus depth of coverage for viral quasispecies reconstruction. PLoS One 7: e47046.

25. Kircher M, Heyn P, Kelso J (2011) Addressing challenges in the production and analysis of illumina sequencing data. BMC Genomics 12: 382.

26. Voelkerding KV, Dames SA, Durtschi JD (2009) Next-generation sequencing: from basic research to diagnostics. Clin Chem 55: 641–658.

27. McElroy KE, Luciani F, Thomas T (2012) GemSIM: general, error-model based simulator of next-generation sequencing data. BMC Genomics 13: 74.

28. Bieganski P, Riedl J, Cartis JV, Retzel EF (1994) Generalized suffix trees for biological sequence data: applications and implementation; Hawaii.

29. Valimaki N, Gerlach W, Dixit K, Makinen V (2007) Compressed suffix tree–a basis for genome-scale sequence analysis. Bioinformatics 23: 629–630.

30. Soares I, Goios A, Amorim A (2012) Sequence comparison alignment-free approach based on suffix tree and L-words frequency. ScientificWorldJournal 2012: 450124.

31. Makinen V, Navarro G, Siren J, Valimaki N (2010) Storage and retrieval of highly repetitive sequence collections. J Comput Biol 17: 281–308.

32. Bloom BH (1970) Space/time trade-offs in hash coding with allowable errors. Commun ACM 13: 422–426.

33. Holmes I, Durbin R (1998) Dynamic programming alignment accuracy. J Comput Biol 5: 493–504.

34. Chao KM, Pearson WR, Miller W (1992) Aligning two sequences within a specified diagonal band. Comput Appl Biosci 8: 481–487.

35. Lopez R, Silventoinen V, Robinson S, Kibria A, Gish W (2003) WU-Blast2 server at the European Bioinformatics Institute. Nucleic Acids Res 31: 3795–3798.

5

SeqEntropy: Genome-Wide Assessment of Repeats for Short Read Sequencing

Hsueh-Ting Chu[1,2], William W.L. Hsiao[3,4], Theresa T.H. Tsao[5], D. Frank Hsu[6], Chaur-Chin Chen[7], Sheng-An Lee[8]*, Cheng-Yan Kao[5]*

1 Department of Biomedical informatics, Asia University, Taichung, Taiwan, 2 Department of Computer Science and Information Engineering, Asia University, Taichung, Taiwan, 3 British Columbia Public Health Microbiology and Reference Laboratory, Vancouver, British Columbia, Canada, 4 Department of Pathology and Laboratory Medicine, University of British Columbia, Vancouver, British Columbia, Canada, 5 Department of Computer Science and Information Engineering, National Taiwan University, Taipei, Taiwan, 6 Department of Computer and Information Science, Fordham University, New York, New York, United States of America, 7 Department of Computer Science, National Tsing Hua University, Hsinchu, Taiwan, 8 Department of Information Management, Kainan University, Taoyuan, Taiwan

Abstract

Background: Recent studies on genome assembly from short-read sequencing data reported the limitation of this technology to reconstruct the entire genome even at very high depth coverage. We investigated the limitation from the perspective of information theory to evaluate the effect of repeats on short-read genome assembly using idealized (error-free) reads at different lengths.

Methodology/Principal Findings: We define a metric $H^{(k)}$ to be the entropy of sequencing reads at a read length k and use the relative loss of entropy $\Delta H^{(k)}$ to measure the impact of repeats for the reconstruction of whole-genome from sequences of length k. In our experiments, we found that entropy loss correlates well with de-novo assembly coverage of a genome, and a score of $\Delta H^{(k)} > 1\%$ indicates a severe loss in genome reconstruction fidelity. The minimal read lengths to achieve $\Delta H^{(k)} < 1\%$ are different for various organisms and are independent of the genome size. For example, in order to meet the threshold of $\Delta H^{(k)} < 1\%$, a read length of 60 bp is needed for the sequencing of human genome (3.2×10^9 bp) and 320 bp for the sequencing of fruit fly (1.8×10^8 bp). We also calculated the $\Delta H^{(k)}$ scores for 2725 prokaryotic chromosomes and plasmids at several read lengths. Our results indicate that the levels of repeats in different genomes are diverse and the entropy of sequencing reads provides a measurement for the repeat structures.

Conclusions/Significance: The proposed entropy-based measurement, which can be calculated in seconds to minutes in most cases, provides a rapid quantitative evaluation on the limitation of idealized short-read genome sequencing. Moreover, the calculation can be parallelized to scale up to large euakryotic genomes. This approach may be useful to tune the sequencing parameters to achieve better genome assemblies when a closely related genome is already available.

Editor: Aaron Alain-Jon Golden, Albert Einstein College of Medicine, United States of America

Funding: No current external funding sources for this study.

Competing Interests: The authors have declared that no competing interests exist.

* E-mail: cykao@csie.ntu.edu.tw (CYK); shengan@gmail.com (SAL)

Introduction

The development of the next generation sequencing technologies (NGS) raised the hope to conduct true haplotype analysis of human genome [1] and for rapid full genome sequencing and typing of various organisms. The 1000 Genomes Project, launched in 2008, began to sequence one thousand human genomes with SGS platforms [2]. In the first phase of the project, the goal was to generate low coverage whole genome shotgun sequencing of 185 individuals. These data were produced in order to validate millions of published genetic variations including single nucleotide polymorphisms (SNPs), insertions and deletions (indels), and other structural variants. Soon after the announcement of the project, another group of scientists started the Genome 10 K project in 2009 which aims to "assemble a genomic zoo" by sequencing the genomes of vertebrate animals [3]. These studies help us understand the correlation between genotypes and phenotypes if large-scale genome shotgun sequencing could be unambiguously and accurately assembled.

Recently, Alkan *et al.* published their analysis of the short-read sequencing data generated from the whole genomes of a Han Chinese individual and a Yoruban individual [4]. In contrary to the initial optimistic view of using NGS technologies to reconstitute the whole genome, it showed a severe deficiency in disambiguating certain genomic regions with short reads. Compared to reference human genome, more than 400 mega-base-pairs (Mbps) of common repeats are missing. As a consequence, it is still a challenge to perform accurate haplotype analysis even though a massive amount of genome sequencing data from multiple individuals is currently available. In another study by Kingsford *et al.*, they explored the effect of repeats in prokaryotic genome assembly using de Bruijn graphs and derived an upper bound of contig sizes for a large number of prokaryotic genomes based on simulated short-reads of different lengths [5]. They concluded that while most genes (>98%) can be recovered in contigs derived from reads as short as 100 bps, even reads as long as 1000 bps are not sufficient to produce a complete prokaryotic genome in most cases.

In this paper, we investigated the impact of read length with a different quantitative analysis. We define the entropy of nucleotide fragments (H) and use the loss in entropy to measure the influence of repeats on genome assembly. The repeat problem has plagued the assembly process since the first generation of sequencers [6,7]. Regardless of the sequence assembler used, the de novo assembly of sequencing data will collapse identical repeats if the length of repeated segments is greater than the read length, resulting in incomplete genome reconstruction. As a result, with the read length limitations imposed by sequencing platforms, repetitive regions will not be reconstructed. In information theory, the entropy score is an index to measure the disorder in a system. Thus we use the definition of k-substring entropy to represent the expected value of the quantified information contained in the reads of length k produced by the sequencing procedure. We apply this measurement to both prokaryotic and eukaryotic genomes, including the human genome. We demonstrated the usefulness of the score as a measurement of the repeat structures of the genomes and proposed how it can be used to aid genome sequencing efforts from the perspective of read lengths and repetitiveness of target genomes.

Materials and Methods

An Idealized Model of Short Read Sequencing

In Figure 1, we illustrated a simplified model of the process to generate high coverage sequence data using a modern sequencing platform such as the Illumina/Solexa system. We considered fixed read length systems for this study since such systems currently provide the cost-effectiveness required for large-scale sequencing and have wider market adoptions. For our current purpose, we simplified the sequencing into two major steps. In the first step, a target DNA sequence is broken into smaller fragments. The fragments are filtered by size and then form a sequencing library. The second step is the parallel sequencing of the ends of these fragments. Current parallelized sequencing technologies are based on various sequencing-by-synthesis methods which can produce a massive number of reads with high redundancy. In this model, we assume that both of the steps are random. That is, the produced sequencing reads can be from any position in the DNA sequence with equal probability although various factors can contribute to sequence sampling bias in real life - resulting in uneven coverage and gaps in sequence assembly [8]. In addition, the reads can come from both the forward strand T_F and the reverse strand T_R of the target DNA sequence T. For convenience, we just use T to present both strands T_F and T_R in the following sections.

The Computation of k-substring Entropy Loss as a Quantitative Measurement of Repeats

If we can filter out all sequencing errors, we can see in Figure 1 that the reads generated are substrings of the target DNA sequence. Let x be a substring of length k from DNA sequence $T = t_1t_2t_3...t_m$ where each t_j, $1 \leq j \leq m$, is one of the nucleotides {A, C, G, T}.

In other words, for short read sequencing with reads of length k, the reads could be denoted as

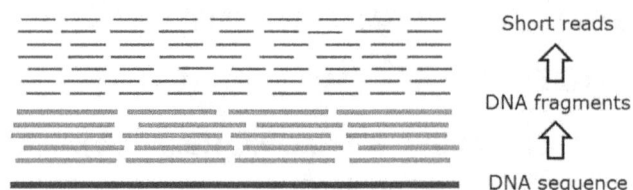

Figure 1. Model of typical short read sequencing. (a) The target sequence is randomly broken into fragments and filtered by their lengths to form a sequencing library. (b) The end or ends of the DNA fragments are sequenced in parallel to generate a massive set of short reads. We assumed the sequencing is random so that each position is more or less covered by equal numbers of fixed-length reads.

$$x_i = t_{i+}t_{i+1}...t_{i+k-1} \text{ where } 1 \leq i \leq m-k+1. \quad (1)$$

Let S_k be the collection of all possible substrings $x_i = t_{i+1} t_{i+2} ...t_{i+k}$ with length k where each $t_j \in \{A, C, G, T\}$.

Define

$$p(x_i) = \frac{c(x_i)}{m-k+1} \quad (2)$$

where $c(x_i)$ is the number of occurrences of x_i in the sequence T.

The Shannon entropy H of S is defined as

$$H^{(k)} = -\sum_{x_i} p(x_i)\log_{10}p(x_i) \quad (3)$$

where k is the length of substrings [9].

In particular, if any substring $x_i = t_{i+1} t_{i+2} ...t_{i+k}$ of length k is unique. Then,

$$p(x_i) = \frac{1}{m-k+1} \quad (4)$$

If any substring of length k is unique, there is no repeat whose length is greater than or equal to k. In this case, it achieves a maximum of entropy

$$I^{(k)} = -\sum_{i=1}^{m-k+1} \frac{1}{m-k+1}\log_{10}\frac{1}{m-k+1} \quad (5)$$
$$= \log_{10}(m-k+1).$$

In almost all of the applications, we have

$$H^{(k)} < I^{(k)} \text{ for any } k > 1. \quad (6)$$

We define the relative entropy loss of length k as

$$\Delta H^{(k)} = \frac{I^{(k)} - H^{(k)}}{I^{(k)}}. \quad (7)$$

If there are a large number of substrings $\{x_i\}$, we can divide the substrings into independent sets $T_1, T_2, ..., T_m$ according their prefixes. Then we can rewrite (3) as

$$H^{(k)} = - \sum_{T_j} \sum_{x_{ij}} p(x_{ij}) \log_{10} p(x_{ij}) \qquad (8)$$

Equation 8 represents an approach to divide up the input into smaller sets and to process them in parallel. This tactic allows one to process a large genome in a timely manner using modest computing resources.

Results

Deficiency in Sequence Coverage Caused by Sequence Repeats is Strongly Correlated to Loss in k-substring Entropy

We investigated the impact of repeat structures on genome assembly and the correlation of sequence coverage and k-substring entropy using the SeqEntropy program that we developed by comparing the sequence coverage results obtained from the SHARCGS de-novo assembler paper [10] and our entropy measurements. Table 1 lists the BAC insert sequences derived from *Arabidopsis thaliana* and *Drosophila melanogaster* (fruit fly). The reconstructed sequence coverage of assembly was quoted from the original SHARCGS paper. The result showed a strong correlation between the ratio of reconstructed coverage and entropy loss. Most BAC sequences have small deficiency from the reconstructed sequence coverage except the following three sequences: AC009243, AC092242 and AC007329 (the bolded rows in Table 1). These three sequences have significantly incomplete coverage such that the sequence coverage by SHARCGS assembled contigs is less than 90%. All of these three sequences lose more than 1% entropy at read length of 30 bp. Consequently, we propose that >1% entropy loss results in poor assembly results using de novo assemblers. For comparison, we also calculated the percentage of simple repeats in these BACs and showed the results in the last column of Table 1. The proposed relative entropy loss scores correlate well with the percentages of repeats. When both are high, we observed poor coverage of assembly, as shown in the bolded rows.

Evaluation of the Limitation of Short-read Sequencing for Animal Genomes

We applied the entropy measurement to analyze the limitation of short-read sequencing for different organisms. We selected five model animals (Table 2) with genome sizes ranging from $10^7 \sim 10^9$ bp. In order to calculate the entropy scores for large eukaryotic genomes on a desktop computer (Intel i7-3820 CPU and 8G RAM), we applied the principle behind Equation 8 above and divided up the input sequences into 256 (4^4) or 1024 (4^5) subsets based on the prefix (4mer or 5mer, respectively) of the sequences. The subsets from each genome are than run sequentially on the desktop computer. We had to run the processes sequentially due to memory constraint (only 8 Gb of RAM was available on the desktop computer) The total run-time for each organism is reported in Table 2. While large genomes such as that of human took a long time (295 hours) to complete, with some modifications to the program, we can run each subset in parallel on a computing cluster to reduce the run time significantly.

Figure 2 depicts the relative entropy losses at different read lengths if idealized sequences are used for the organisms. Human requires read length of 60 bp and zebra fish requires read length of

100 bp to overcome the 1% entropy loss threshold. The genome size of zebra fish is less than half of human genome. It indicates that zebra fish genome is more repetitive than human genome. Moreover, the nematode, *C. elegans*, requires very short read (less than 30 bp) to avoid 1% entropy loss whereas fruit fly (*D. melanogaster*) requires more than 320 bp. Our analysis shows that genome assembly of many other organisms using short reads may be more challenging than human genome assembly.

The relationship between entropy loss and read length explains the limitation of short-read sequencing technology illustrated by Alkan *et al.* [4]. In the early experiments of the 1000 Genomes Project (like SRA ID: ERX000020), the read length of sequencing data is 36 bp and the curve of relative entropy loss for human genome in Figure 2 indicates more than 2% entropy loss at the read length of 36 bp. As a result, it is almost impossible to retrieve a perfect whole genome assembly from those WGS experiments. On the other hand, Sundquist *et al.* showed the sequencing of *D. melanogaster* still achieved worse genome sequence coverage than that of human chromosomes at read lengths of 200 bp [11]. Our proposed quantitative model illustrates the deficiency of sequence coverage for *D. melanogaster* comes from the richness of repeats in its genome.

Evaluation of Bacterial Whole Genome Sequencing at Different Read Lengths

The *Escherichia coli* strain MG1655 whole genome shotgun sequencing datasets SRX000429 and SRX000430 generated using Illumina Genome Analyzer are commonly used as performance benchmark of short read sequencing [12]. The complete genome sequence of the same strain (NCBI REFSEQ ID: NC_000913) had been well characterized since 1997 [13]. Therefore, we can compare the result of the de novo assembly using the Illumina reads to the completed reference genome by mapping the contigs to the reference genome sequence.

To explore the reliability of the de novo assembly result, we computed the entropy at read length of 36 bp for the *E. coli* genome sequence NC_000913. We listed the entropy loss of the *E. coli* genome sequence along with some other prokaryotic whole genome sequences in Table 3. It shows that the entropy loss for the *E. coli* genome sequenced at read length of 36 bp is 0.22%. Compared this with the results obtained in Table 1, a relative entropy loss of 0.22% corresponds to about 2% genome coverage loss and suggests the difficulty in achieving a perfect genome coverage. Most of the assembly results without pair information by publicly available de novo assemblers can only achieve a genome-wide coverage of around 98% for the *E. coli* short reads dataset SRX000429 (https://wiki.nbic.nl/index.php/ Raw_results_of_NGS_de_novo_assembly). With the help of longer reads or paired-end reads information available since the two 36 bp *E. coli* datasets were generated, the de novo assembly can achieve a better genome coverage than 98%. To approximate the effects of longer reads and paired-end reads, we calculate the relative entropy losses at k-substring length of 500 bp and 1000 bp (Table 3). However, based on the entropy losses, we think it is still very difficult to develop an automated de novo assembler to reach lossless assembly using data generated from current sequencing platforms. Manual inspections and the use of long range mapping information are necessary in most of the genome assembly. In general, our analysis indicates that smaller prokaryotic genomes with fewer repeats have less entropy loss compared to larger genomes. However, it's the repeat structure and not the genome size that plays a determining role in entropy loss. *Bacteroides thetaiotaomicron* VPI-5482 has a larger genome but much fewer long repeats than *E. coli*. [14]. As a consequence, *B. thetaiotaomicron*'s

Table 1. Comparison of coverage and entropy loss of BAC sequences[a].

Clone (BAC no.)	Sequence Length(bp)	Sequence Coverage (%)	Coverage Deficiency (%)	H³⁰	I³⁰	ΔH³⁰	% of repeat
AC011809	108,767	99.98	0.02	5.335862	5.337411	0.03%	0.46%
AC002328	109,171	99.61	0.39	5.335947	5.339022	0.06%	0.99%
AC064879	109,180	99.68	0.32	5.335971	5.339058	0.06%	0.97%
AC023673	109,367	99.91	0.09	5.338616	5.339801	0.02%	0.38%
AC011713	109,694	98.85	1.15	5.332009	5.341098	0.17%	2.91%
AC009243	110,565	100.00	0.00	5.344373	5.344534	0.00%	0.02%
AC022520	110,611	99.63	0.37	5.341812	5.344714	0.05%	0.92%
AC018460	**110,619**	**66.48**	**33.52**	**4.99782**	**5.344746**	**6.49%**	**41.15%**
AC007764	111,222	99.84	0.16	5.346664	5.347107	0.01%	0.13%
AC000348	111,566	98.78	1.22	5.339937	5.348449	0.16%	2.81%
AC092191	80919	99.52	0.48	5.205896	5.208925	0.06%	1.01%
AC185533	95808	98.46	1.54	5.271617	5.2823	0.20%	3.41%
AC018485	**99441**	**82.86**	**17.14**	**5.094117**	**5.298469**	**3.86%**	**19.86%**
AC018478	103809	99.90	0.10	5.315052	5.317144	0.04%	0.56%
AC092242	111023	100.00	0.00	5.346276	5.346329	0.00%	0.01%
AC018482	**113821**	**87.42**	**12.48**	**5.261927**	**5.357142**	**1.78%**	**19.27%**
AC185534	119461	99.42	0.58	5.37266	5.378151	0.10%	1.75%
AC092399	122013	99.92	0.08	5.386815	5.387333	0.01%	0.17%
AC007837	123647	99.90	0.10	5.391958	5.393112	0.02%	0.28%
AC007329	126140	99.99	0.01	5.401638	5.401783	0.00%	0.05%

Sequence coverage percentages as listed in Dohm *et al* [10].
[a]The programs for the computation are available at: http://sourceforge.net/projects/seqentropy/files/SeqEntropy-demo-20130203.zip.
[b]The columns H30, I30,ΔH30 are computed by our program "SeqReadEntropy" using read length of 30 bp.
[c]The column "% of repeat" is computed by our program "SeqReadRepeat" using read length of 30 bp.

entropy loss at read length 36 bps is more similar to that of a much smaller prokaryotic genome. Using entropy losses calculated at various read lengths, we can capture the repeat structure of a genome.

The third generation sequencing platforms use single-molecule technologies and other nanotechnologies [15]. The new methods claim to produce longer reads (>3000 bp). For instance, the reads for sequencing the *E. coli* genome could reach an average length of more than 3000 bps [16]. We showed the entropy loss at read length of 500 and 1000 bps for a few prokaryotic whole genome sequences. At the read length of 500 bps, *Mycoplasma genitalium*, an obligate parasitic bacterium with a highly reduced genome, has an insignificant entropy loss (Table 3).

In light of the new and emerging sequencing technologies, we applied the entropy calculation to 2725 prokaryotic chromosomes and plasmids which were downloaded from the NCBI FTP site (ftp://ftp.ncbi.nih.gov/genomes/Bacteria/- downloaded June 2012). We computed the relative entropy losses at read lengths of 125 bp, 500 bp and 3000 bp – roughly correspond to the read lengths of Illumina, Roche 454 pyrosequencing and PacBio SMRT platforms, respectively. We showed the distribution of the relative entropy loss scores in Figure 3. While there are some replicons, especially certain highly repetitive plasmids, with entropy loss scores >1.0% (see Table S1), the vast majority (>99%) of the prokaryotic replicons would lose less than 1.0% entropy with current sequencing technologies. These results are

Table 2. Five animal genomes for entropy measurement.

Organism	Genome size	Version[a]	Computation time[b]
Yeast (*S. cerevisiae*)	1.2×10^7	sacCer3	1.3 minutes
Nematode (*C. elegans*)	1.0×10^8	ce10	33 minutes
Fruit fly (*D. melanogaster*)	1.3×10^8	dmel_r5.42	42 minutes
Zebrafish (*D. rerio*)	1.4×10^9	danRer7	66 hours
Human (*H. sapiens*)	3.2×10^9	hg19, GRCh37.p5	295 hours

[a]The whole genome sequences were downloaded from http://hgdownload.cse.ucsc.edu/for the organisms: *S. cerevisiae*, *C. elegans*, *D. rerio*, and *H. sapiens* and ftp://ftp.flybase.net/for *D. melanogaster*.
[b]The computation time of entropy measurement was recorded for read length 100 bp on a PC with Intel i7-3820 CPU and 8G RAM.

Relative entropy loss at different read lengths

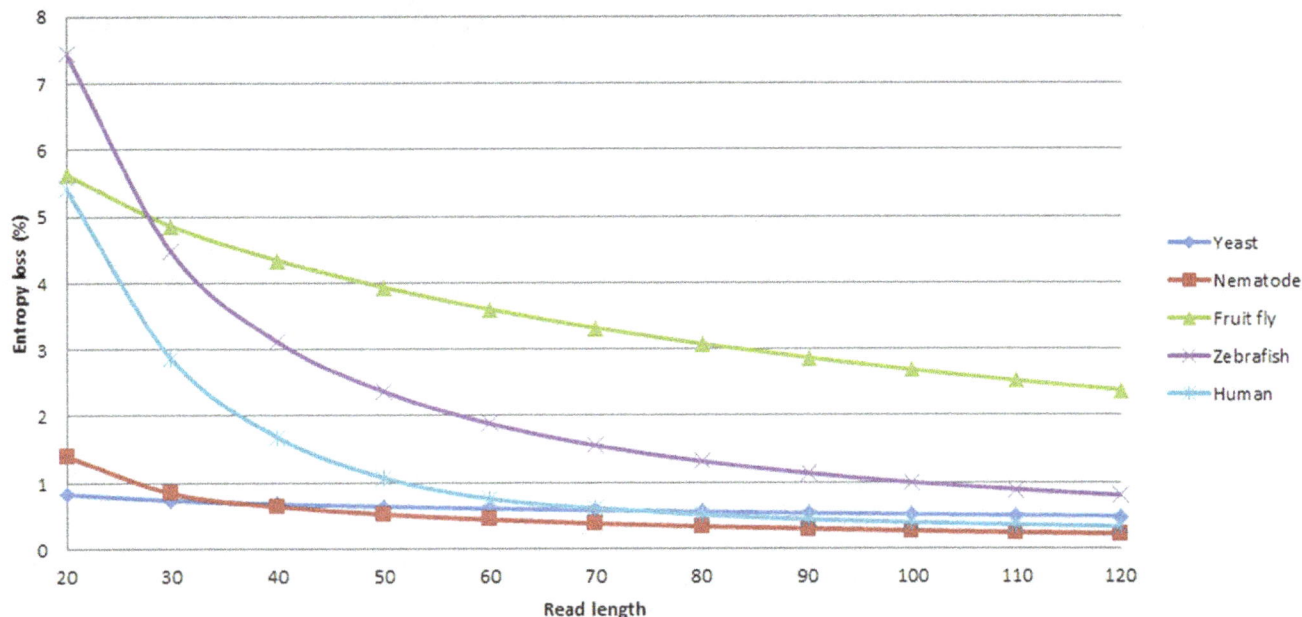

Figure 2. Entropy losses at different read lengths for different organisms. In the five organisms, the genomes of zebra fish (*D. rerio*) and fruit fly (*D. melanogaster*) will lose more entropy regardless of any read length used for sequencing. In particular, the fruit fly loses >2% of entropy loss even with read length of 120 bp. It will be <1% of entropy loss at read length of 230 bp. On the other hand, the genomes of Yeast (*S. cerevisiae*) and Nematode (*C. elegans*) have minor entropy loss even with very short reads. The detail results of entropy measurements are listed in **Table 5**.

confirmed by the fact that the vast majority of the prokaryotic genomes can be reasonably reconstructed into long contigs with current shotgun sequencing technologies. In figure 4, we zoomed in to show only the results from replicons with entropy loss <1% in order to further differentiate the effect of read length. The entropy loss scores decrease as the read length increases. At read length of 125 bp, even if the sequencing reads cover the genomes evenly and completely, de novo assemblers without utilizing pair-end information would still not be able to fully reconstruct the genome due to the presence of repeats in ~80% of the cases (2140/2725 with $\Delta H^{125} > 0$). On the other hand, with read length of 3000 bp, 58% of the replicons surveyed could be reconstructed completely (i.e. no loss of entropy due to repeats). We list the top 10 genomes (excluding plasmids) with the largest entropy losses at read length of 125 bp, 500 bp and 3000 bp in **Table 4** and the complete results of the entropy loss scores are listed in **Table S1**. There are 41 chromosomes with $\Delta H^{125} > 1\%$ and 6 chromosomes with $\Delta H^{500} > 1\%$. Many of these outliers have been noted in the

literature to have highly repetitive genomes due to various evolutionary forces [17,18,19,20].

Taken together, these observations mean the third generation sequencing technologies can theoretically provide complete genome sequences for most prokaryotic organisms if the error rate is controlled or corrected by other short read sequencing technologies. On the other hand, bigger eukaryotic genomes, which are likely to have more complex repeat structure, would benefit less from the longer reads produced by third generation sequencing technologies.

Discussion

The evaluation of k-substring entropy shows that the genomes of different organisms may have distinct repeat structures that impose limitation on sequencing at a certain read length regardless of their genome sizes. Using the entropy measurement, we can estimate an ideal read length for a given genome sequencing project by trying to minimize the entropy loss. Kingsford *et al.* previously introduced a de Bruijn graph method to evaluate the influence of repeat structure on sequence assembly [21]. However, their analysis was only applied to prokaryotic genomes which are relatively small (most <10 megabases). For the analysis of large genomes such as the mammalian or plant genomes, which can be a gigabases long, a program requires huge amount of memory to record all distinct substrings in the genome. Our algorithm is designed to handle genomes at any scale. With eq. (8), we can separate the substrings into different subsets by their prefix of length m ($m = 5$ for human genome and $m = 4$ for zebra fish genome) and calculate the entropy measurement of each set individually. As a result, the proposed program can be run on a desktop computer (8 Gb of RAM) for genomes of an arbitrary size at the cost of time. If completely parallelized, the calculation

Table 3. Entropy loss of selected prokaryotic whole genomes with reads of lengths 36, 500 and 1000 bps.

Seq. no	Organism	Sequence Length(bp)	ΔH^{36}	ΔH^{500}	ΔH^{1000}
NC_000913	E. *coli* K-12	4,639,675	0.22%	0.09%	0.04%
NC_004663	B. *thet* VPI-5482	6,260,361	0.15%	0.09%	0.05%
NC_008525	P.*pent* ATCC 25745	1,832,387	0.16%	0.11%	0.08%
NC_000908	M.*geni* G37	580,076	0.11%	0.00%	0.00%

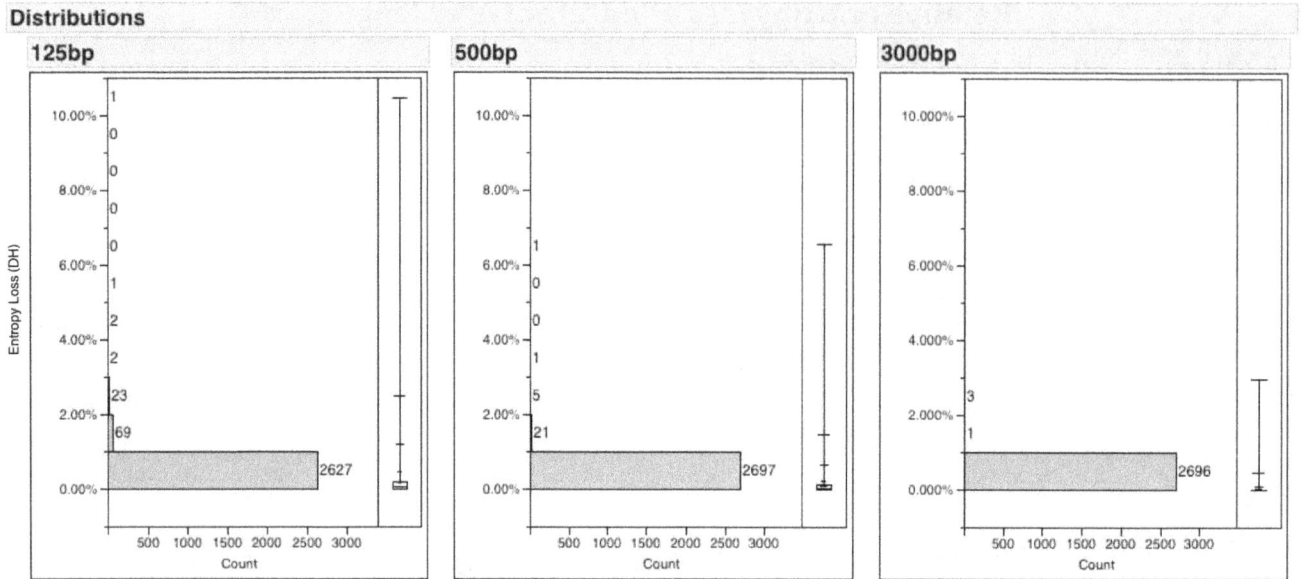

Figure 3. Histograms and quartile box plot of relative entropy losses in 2725 prokaryotic replicons. The x-axis shows the number of replicons in each bin while the y-axis shows the % entropy loss (ΔH). The quartile box plot displays the mean (diamond shape), the medium (50%) the first (25%) and the third (75%) quartiles (the boxes), and the entire range (the whiskers). The vast majority of the replicons lost <1% entropy regardless of the read length.

can be done on a large cluster to reduce the run time significantly. For a typical prokaryotic genome, it takes a few seconds to a few minutes to calculate the entropy loss at a given read length. The processing time is read length dependent and for read length <1000 bp, the calculation takes <1 min using a single CPU core. This length should be sufficient for most of the sequencing technologies in the near future with a few exceptions (e.g. Pacific

Biosciences SMRT platform). For read length of 3000 bp, it takes about 20 min to calculate the entropy for one bacterial genome.

This task of entropy measurement can be performed on a preliminary assembly or on a reference genome from a closely related organism. Since read length depends on the sequencing platforms and the protocols used, it is often not possible to alter. Experimentally, it is easier to construct paired-end libraries with different insert sizes. We can use pair-end libraries of different

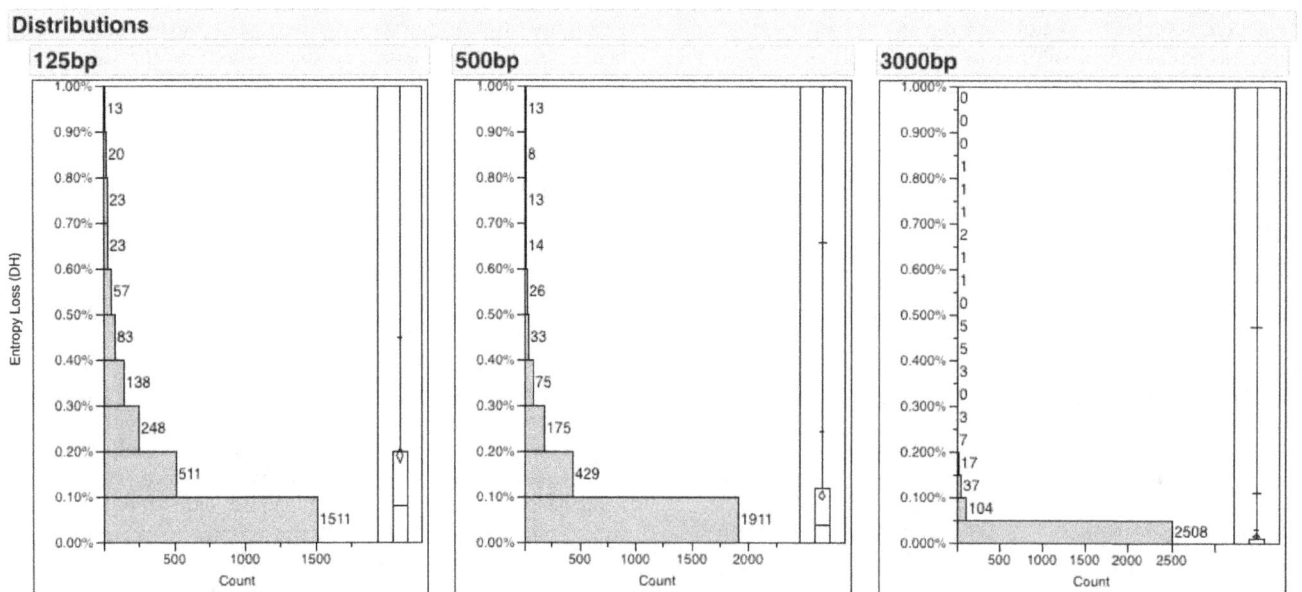

Figure 4. Histograms and quartile box plot of entropy losses in 2725 prokaryotic replicons truncated at 1% entropy loss in order to see the finer breakdown. The x-axis shows the number of replicons in each bin while the y-axis shows the % entropy loss (ΔH). The quartile box plot displays the mean (diamond shape), the medium (50%) the first (25%) and the third (75%) quartiles (the boxes), and the entire range (the whiskers). It is clear that as read length increases, the entropy loss decreases. As a result, a higher number of replicons have ΔH <1.0%.

Table 4. Prokaryotic chromosomes with largest entropy losses at read lengths of 125, 500 and 3000 bp.

Genome ID	ΔH^{125}	Genome ID	ΔH^{500}	Genome ID	ΔH^{3000}
Bordetella pertussis Tohama I	1.79136%	Bordetella pertussis Tohama I	0.898868%	Mycoplasma agalactiae	0.367427%
Xanthomonas oryzae pv. oryzae PXO99A	1.82280%	Mycoplasma fermentans M64 chromosome	0.934389%	Dehalococcoides ethenogenes 195	0.394745%
Wolbachia sp. wRi	1.98404%	Acinetobacter baumannii SDF	0.934389%	Mycoplasma fermentans M64	0.415779%
Aliivibrio salmonicida LFI1238 chrom 1	2.04042%	Mycoplasma mycoides subsp. mycoides SC str. PG1	0.940714%	Orientia tsutsugamushi Boryong	0.416469%
Shigella boydii CDC 3083-94	2.09366%	Wolbachia endosymbiont of Culex quinquefasciatus Pel	1.028366%	Methylobacillus flagellatus KT	0.42687%
Shigella dysenteriae Sd197	2.40508%	Aliivibrio salmonicida LFI1238 chrom 1	1.153796%	Wolbachia sp. wRi	0.43966%
Acinetobacter baumannii SDF	2.64465%	Wolbachia sp. wRi	1.285340%	Alteromonas macleodi	0.465815%
Orientia tsutsugamushi str. Iked	2.67584%	Aliivibrio salmonicida LFI1238 chrom 2	1.334486%	Bartonella tribocorum CIP 105476	0.484289%
Mycoplasma mycoides subsp. mycoides SC str. PG1	2.75548%	Orientia tsutsugamushi str. Ikeda	1.570541%	Streptococcus agalactiae NEM316	0.49117%
Orientia tsutsugamushi Boryong	4.62902%	Orientia tsutsugamushi Boryong	2.753110%	Candidatus Phytoplasma mali	0.693043%

[a]The complete entropy computations of 2725 prokaryotic replicons are listed in **Table S1**.

insert sizes to mimic the effect of longer reads. Wetzel *et al.* recently demonstrated that the assembly outcome can be improved drastically by tuning mate-pair sizes (i.e. adjusting the average insert size of the pair-end library) to match the dominant repeat types [22]. Furthermore, they showed that "short" inserts that are between 4 to 6 times the actual read lengths perform better than long inserts that are a few kilobases long. This is because short inserts that barely span the repeats are more effective at resolving local ambiguities than long inserts. Their work is based on idealized de Bruijn graph reconstruction of a genome (see Kingsford *et al* 2010 for the method). This process is computationally and memory intensive. As a result, it is not easily scaled up to handle large eukaryotic genomes. On the contrary, our method can estimate the ideal read length and is fast and highly scalable as we have demonstrated in the previous sections.

We propose that SeqEntropy can be run with different read length parameters to detect the minimum entropy loss. Based on the work of Wetzel *et al*, we propose that the mate-pair sizes of the sequence library be slightly longer than the theoretical ideal read length detected by SeqEntropy. While the insert size is tuned, the actual read length is still based on available funding and sequencing platforms, allowing minimal interruption to existing and on-going sequencing projects. As more genomes are being sequenced in an automated fashion, the ability to tune the sequencing parameters to achieve better assemblies is highly desirable. We propose that entropy loss can be used to provide an accurate and objective estimate for the optimal sequence length.

Table 5. The relative entropy losses of five animal genomes at different read lengths.

ReadLen k	ΔH^k of Yeast	ΔH^k of Nematode	ΔH^k of Fruit fly	ΔH^k of Zebrafish	ΔH^k of Human
20	**0.831217%**	1.397796%	5.627856%	7.44469%	5.406936%
30	0.736196%	**0.844718%**	4.852753%	4.461654%	2.858026%
40	0.683882%	0.642178%	4.337883%	3.11601%	1.684438%
50	0.644228%	0.522634%	3.932514%	2.363446%	1.072922%
60	0.611417%	0.440682%	3.599801%	1.886473%	**0.655126%**
70	0.582789%	0.380003%	3.319438%	1.556083%	0.482622%
80	0.557695%	0.332944%	3.078671%	1.315029%	0.379953%
90	0.536264%	0.295774%	2.868949%	1.134327%	0.313174%
100	0.517069%	0.265546%	2.684404%	**0.996147%**	0.2661318%
110	0.499602%	0.240705%	2.520669%	0.887384%	0.2306789%
120	0.483514%	0.219947%	2.374214%	0.799316%	0.2028979%
320	–	–	1.007846%	–	–
330	–	–	**0.973252%**	–	–

[a]The bold numbers show the relative entropy loss values and the corresponding minimal read lengths at which the relative entropy losses are below 1% for different animal genomes.

Acknowledgments

We would like to thank Drs Chun-Fan Chang (Chinese Culture University) and Chen-Hsiung Chan (Tzu Chi University) for useful comments regarding new sequencing technologies.

References

1. Consortium TIHGS (2005) A haplotype map of the human genome. Nature 437: 1299–1320.
2. Durbin RM, Abecasis GR, Altshuler DL, Auton A, Brooks LD, et al. (2010) A map of human genome variation from population-scale sequencing. Nature 467: 1061–1073.
3. Scientists GKCo (2009) Genome 10K: a proposal to obtain whole-genome sequence for 10,000 vertebrate species. J Hered 100: 659–674.
4. Alkan C, Sajjadian S, Eichler EE (2011) Limitations of next-generation genome sequence assembly. Nat Methods 8: 61–65.
5. Kingsford C, Schatz M, Pop M (2010) Assembly complexity of prokaryotic genomes using short reads. BMC Bioinformatics 11: 21.
6. Tammi MT, Arner E, Britton T, Andersson B (2002) Separation of nearly identical repeats in shotgun assemblies using defined nucleotide positions, DNPs. Bioinformatics 18: 379–388.
7. Shapiro JA, von Sternberg R (2005) Why repetitive DNA is essential to genome function. Biol Rev Camb Philos Soc 80: 227–250.
8. Brown J Jr, Brown TA (2006) Genomes 3: Garland Publishing, Incorporated.
9. Schneider T (1995) Information Theory Primer.
10. Dohm JC, Lottaz C, Borodina T, Himmelbauer H (2007) SHARCGS, a fast and highly accurate short-read assembly algorithm for de novo genomic sequencing. Genome Res 17: 1697–1706.
11. Sundquist A, Ronaghi M, Tang H, Pevzner P, Batzoglou S (2007) Whole-genome sequencing and assembly with high-throughput, short-read technologies. PLoS One 2: e484.
12. Simpson JT, Wong K, Jackman SD, Schein JE, Jones SJ, et al. (2009) ABySS: a parallel assembler for short read sequence data. Genome Res 19: 1117–1123.
13. Blattner FR, Plunkett G 3rd, Bloch CA, Perna NT, Burland V, et al. (1997) The complete genome sequence of Escherichia coli K-12. Science 277: 1453–1462.
14. Xu J, Bjursell MK, Himrod J, Deng S, Carmichael LK, et al. (2003) A genomic view of the human-Bacteroides thetaiotaomicron symbiosis. Science 299: 2074–2076.
15. Flusberg BA, Webster DR, Lee JH, Travers KJ, Olivares EC, et al. (2010) Direct detection of DNA methylation during single-molecule, real-time sequencing. Nat Methods 7: 461–465.
16. Rasko DA, Webster DR, Sahl JW, Bashir A, Boisen N, et al. (2011) Origins of the E. coli strain causing an outbreak of hemolytic-uremic syndrome in Germany. N Engl J Med 365: 709–717.
17. Bischof DF, Vilei EM, Frey J (2006) Genomic differences between type strain PG1 and field strains of Mycoplasma mycoides subsp. mycoides small-colony type. Genomics 88: 633–641.
18. Cho NH, Kim HR, Lee JH, Kim SY, Kim J, et al. (2007) The Orientia tsutsugamushi genome reveals massive proliferation of conjugative type IV secretion system and host-cell interaction genes. Proc Natl Acad Sci U S A 104: 7981–7986.
19. Nakayama K, Yamashita A, Kurokawa K, Morimoto T, Ogawa M, et al. (2008) The Whole-genome sequencing of the obligate intracellular bacterium Orientia tsutsugamushi revealed massive gene amplification during reductive genome evolution. DNA Res 15: 185–199.
20. Cerveau N, Leclercq S, Leroy E, Bouchon D, Cordaux R (2011) Short- and long-term evolutionary dynamics of bacterial insertion sequences: insights from Wolbachia endosymbionts. Genome Biol Evol 3: 1175–1186.
21. Kingsford C, Schatz MC, Pop M (2010) Assembly complexity of prokaryotic genomes using short reads. BMC Bioinformatics 11: 21.
22. Wetzel J, Kingsford C, Pop M (2011) Assessing the benefits of using mate-pairs to resolve repeats in de novo short-read prokaryotic assemblies. BMC Bioinformatics 12: 95.

Author Contributions

Problem definition: CYK CCC. Conceived and designed the experiments: HTC. Performed the experiments: HTC WLH. Analyzed the data: CCC DFH. Contributed reagents/materials/analysis tools: CYK SAL. Wrote the paper: HTC WLH TTT DFH SAL CCC.

6

Tandem Repeats and G-Rich Sequences Are Enriched at Human CNV Breakpoints

Promita Bose[1], Karen E. Hermetz[1], Karen N. Conneely[1,2], M. Katharine Rudd[1]*

1 Department of Human Genetics, Emory University School of Medicine, Atlanta, Georgia, United States of America, **2** Department of Biostatistics and Bioinformatics, Emory University School of Public Health, Atlanta, Georgia, United States of America

Abstract

Chromosome breakage in germline and somatic genomes gives rise to copy number variation (CNV) responsible for genomic disorders and tumorigenesis. DNA sequence is known to play an important role in breakage at chromosome fragile sites; however, the sequences susceptible to double-strand breaks (DSBs) underlying CNV formation are largely unknown. Here we analyze 140 germline CNV breakpoints from 116 individuals to identify DNA sequences enriched at breakpoint loci compared to 2800 simulated control regions. We find that, overall, CNV breakpoints are enriched in tandem repeats and sequences predicted to form G-quadruplexes. G-rich repeats are overrepresented at terminal deletion breakpoints, which may be important for the addition of a new telomere. Interstitial deletions and duplication breakpoints are enriched in *Alu* repeats that in some cases mediate non-allelic homologous recombination (NAHR) between the two sides of the rearrangement. CNV breakpoints are enriched in certain classes of repeats that may play a role in DNA secondary structure, DSB susceptibility and/or DNA replication errors.

Editor: Brian P. Chadwick, Florida State University, United States of America

Funding: This work was supported by a grant from the National Institutes of Health (grant number MH092902 to MKR) and a grant from the March of Dimes (grant number #12-FY11-203 to MKR). The content is solely the responsibility of the authors and does not necessarily represent the official views of the National Institutes of Health or the March of Dimes. The funders had no role in study design, data collection and analysis, decision to publish, or preparation of the manuscript.

Competing Interests: The authors have declared that no competing interests exist.

* Email: katie.rudd@emory.edu

Introduction

Genomic CNV is a major cause intellectual disability, autism spectrum disorders, epilepsy, and psychiatric disorders. Large, pathogenic CNVs are located throughout the human genome and include tens to hundreds of genes [1,2]. Though most germline CNVs have different chromosome breakpoints [3–6], it is possible that breakpoint regions share common DNA features that make them susceptible to double-strand breaks (DSBs). This is true of chromosome rearrangements in leukemia that vary in location, but share repetitive DNA and/or DNAse hypersensitive sites at breakpoint cluster regions (BCRs) [7–9].

Classic studies of chromosomal fragile sites have revealed that DNA sequence and structure can influence chromosome breakage. Fragile sites were originally identified as breaks and gaps in metaphase chromosomes, induced under conditions of DNA replication stress [10,11]. Mapping and sequence analyses of fragile sites have uncovered repetitive classes of DNA at many loci. Trinucleotide repeats and tandemly repeated minisatellites underlie many rare fragile sites [12,13]. The FRA7E and FRA16D common fragile sites are made up of AT-rich repeats [14], and yeast studies have shown that the FRA16D AT-rich dinucleotide repeat intrastrand pairs to form a secondary structure that stalls DNA replication [15]. Though there is no single DNA sequence responsible for fragile sites in the human genome, in general fragile sites are made up of repetitive DNA that may form secondary structures.

Some studies of germline CNV breakpoints have attempted to identify common DNA sequences that, like fragile sites, contribute to chromosome breakage [16–21], but the search for breakage-prone DNA at CNV boundaries is challenging for a number of reasons. First, more than half of the human genome is made up of repetitive DNA, so finding a repeat at or near a CNV breakpoint may be a circumstantial finding unrelated to breakage. Second, DNA resection after chromosome breakage can lead to a CNV breakpoint that is kilobases (kb) away from the initial DSB [22,23]. Thus, studies that only focus on the sequence directly adjacent to the post-repair junction will miss some DSB sites. Finally, since chromosome breaks are caused by heterogeneous factors, it is necessary to analyze breakpoints from a large cohort of annotated CNVs to find DNA motifs that are significantly enriched at breakpoints.

Using our large dataset of fine-mapped and sequenced CNV breakpoints from patients with neurodevelopmental disorders, we applied several motif and repeat discovery tools to search for DNA sequences enriched at CNV breakpoints compared to control regions of the genome. We broadened the breakpoint regions to include flanking sequence and account for DNA resection. To search for a common breakage-associated motif, we analyzed patient breakpoint regions and control sequences using Multiple EM for Motif Elicitation (MEME) and nested motif independent component analysis (NestedMICA). We also searched for repetitive DNA with Tandem Repeats Finder, QuadParser, and RepeatMasker. This large-scale analysis revealed an enrichment of tandem repeats and potential G-quadruplex sequences at human CNV breakpoints, providing insight into DNA sequences susceptible to DSBs.

Materials and Methods

CNV breakpoint and control sequences

We previously fine-mapped and/or sequenced CNV breakpoints from 116 individuals with abnormal clinical cytogenetic testing results [5,6]. Patients have unique deletions and duplications that alter the copy number of different genes, so they do not share a common phenotype. In general, individuals with large pathogenic CNVs exhibit developmental delay, intellectual disability, autism spectrum disorders, and/or congenital anomalies. We analyzed 48 terminal deletions, 41 inverted duplications adjacent to terminal deletions, 11 translocations, 10 interstitial deletions, four interstitial duplications and two terminal duplications. Terminal deletions and terminal duplications have one breakpoint per rearrangement. Translocations, interstitial deletions and interstitial duplications have two breakpoints per rearrangement. For 18q-71c's translocation, we only identified the chromosome 18 breakpoint; the other breakpoint on chromosome 4 is cryptic [5]. Thus, there are a total of 140 breakpoints in 116 individuals.

We calculated 4-kb windows surrounding 140 CNV breakpoints and downloaded the corresponding DNA sequence from the NCBI 36.1/hg18 build of the human genome assembly using the Table Browser from the UCSC Genome Browser (http://genome. ucsc.edu/). Four-kb CNV breakpoint regions were unique and did not overlap with one another. We also randomly selected 4-kb control sequences from the same genome build. We concatenated the coordinates of chromosomes 1–22, X and Y that make up the 3,080,419,480 bp in the haploid human genome. Next we used a random number generator to select 10,000 numbers between one and 3,080,419,480. We added four kb to each number to produce start and stop coordinates for 10,000 regions and downloaded the associated DNA sequences from the UCSC Genome Browser. We excluded control regions with "N" bases that correspond to sequencing gaps, resulting in 9243 ungapped 4-kb sequences. From these, we randomly selected 140 sequences 20 times to make up 20 datasets of 140 control sequences. We saved 140 CNV breakpoint sequences and 140 control sequences per dataset in FASTA format for analysis.

MEME and NestedMICA

We searched for common motifs in the CNV breakpoint and 20 control datasets using MEME [24] and NestedMICA [25] with default parameters to find a single ungapped 50-basepair (bp) motif. We ran NestedMICA motif inference tool (NMinfer) and NestedMICA motif scanner module (NMscan) with a cutoff of -15 to determine the number of motifs per sequence. MEME and NestedMICA programs were executed on the Emory Human Genetics Computing Cluster (HGCC). We used NMinfer to identify the 50-bp NestedMICA motif in all 21 datasets and MochiView [26] to align the motifs to the sequences in the CNV breakpoint and control datasets.

Repeat searches

To identify *Alu* repeats in the CNV breakpoint and control datasets, we ran RepeatMasker using default settings [27]. We identified tandem repeats and G-quadruplex sequences using

Tandem Repeats Finder (TRF) [28] and QuadParser [29], respectively. All three programs were executed with default parameters. We used custom scripts to calculate the lengths of non-overlapping tandem repeats. We calculated the GC content of sequences using the geecee program (http://mobyle.pasteur.fr/ cgi-bin/portal.py#forms::geecee) within the European Molecular Biology Open Software Suite (EMBOSS) [30].

We used chi-squared goodness-of-fit tests to test whether each type of repeat was proportionally distributed across CNV types (i.e., independent of CNV type). We used two-sided binomial tests to test each type of CNV breakpoint for enrichment with each type of repeat. Since we performed 18 of these enrichment tests (testing six CNV types for enrichment for three repeat types), we performed Bonferroni adjustment for the 18 tests and used a p-value cutoff of $.00278 = .05/18$ to assess significance.

Results

Human CNV breakpoints

Our goal was to identify DNA sequence motifs that are overrepresented in human CNV breakpoint regions compared to control regions of the genome. We analyzed the breakpoint sequences of pathogenic CNVs ascertained from 116 children with phenotypes including intellectual disability, developmental delay, congenital abnormalities, and autism spectrum disorders. We excluded recurrent CNVs mediated by NAHR between segmental duplications. The 140 breakpoints from 116 CNVs have been fine-mapped by high-resolution array comparative genome hybridization (CGH) [5,6]. Thirty-two out of 116 CNV junctions have been sequenced, resolving the breakpoints to the bp. Most of the sequenced junctions were simple with little or no microhomology at the breakpoint junctions; three had more complex junctions with short insertions 10–16 bp long [5,6] (Table S1). The other 84 CNV junctions were fine-mapped with custom microarrays that had, on average, one oligonucleotide probe per 200 bp. Oligonucleotide spacing is not uniform throughout the genome due to repetitive sequences that confound unique probe design. Thus, the mean and median resolutions of breakpoints are 468 bp and 101 bp, respectively (Table S1).

To characterize a diverse collection of CNV breakpoints, we included breakpoints from 48 terminal deletions, 41 inverted duplications adjacent to terminal deletions, 11 translocations, 10 interstitial deletions, four interstitial duplications and two terminal duplications. Interstitial deletions and duplications have two breakpoints in the same chromosome arm, and translocations have two breakpoints in different chromosomes. Terminal deletions and duplications have a single breakpoint. Inverted duplications adjacent to terminal deletions are a specific type of CNV where the deletion and duplication form as part of one chromosome rearrangement. In this case, the terminal deletion is the site of the initial DSB [5,6], so we only included that breakpoint in the analysis. Since our CNV dataset is enriched in terminal deletions and duplications, breakpoints are overrepresented towards chromosome ends. The mean and median distances from the CNV breakpoint to the end of the chromosome are 4.8 Mb and 2.3 Mb, respectively (Table S2). It is important to recognize that all CNVs in this study extend beyond the terminal segmental duplications that make up human subtelomeres. Thus, none of the CNV breakpoints lie in subtelomeric segmental duplications.

Array CGH and junction sequencing can resolve chromosome breakpoints to a relatively small region; however, that may not correspond to the exact DSB site. After the initial DSB, 5' to 3' DNA resection can lead to a CNV breakpoint that is up to 1.3 kb away [22,23]. We included additional sequence around each of the 140 breakpoints to account for DNA resection and array resolution. Breakpoint regions are based on the normal locus in the reference genome (before breakage), not the patient's CNV (after breakage). In the case of sequenced breakpoint junctions, we added sequence two kb proximal and two kb distal of the breakpoint. For breakpoints that were resolved by array but not sequencing, we added two kb proximal and distal to the midpoint between the abnormal and normal probes that defined the CNV breakpoint. Thus, each breakpoint region is four kb and centered around the post-repair chromosome breakpoint. We downloaded the 140 4-kb breakpoint regions from the reference genome assembly (NCBI Build 36.1/hg18).

As a comparison, we analyzed 4-kb control regions from the human genome. We compiled 140 control sequences 20 times to make up 20 control datasets (see Methods). In the following experiments, we compared the motifs in the CNV breakpoint dataset to those in the 20 control datasets. We applied motif-finding tools to search for DNA motifs with the potential to form secondary structures susceptible to DSBs. Since most of our breakpoints are fine-mapped, but not sequenced, we focused on long repeats that span much of the breakpoint region. We did not analyze short motifs reported at other chromosome breakpoints (e.g., 6-8-bp translin target sites) due to the imprecision of most CNV junctions in our study.

Common motif search

It is possible that CNVs are caused by breakage in a common DNA sequence motif present in many or all CNV breakpoints. To look for common motifs among the 140 CNV breakpoints, we performed MEME [24] and NestedMICA [25] searches. We queried the top 50-bp motif in the CNV breakpoint dataset using default parameters for both programs. MEME and NestedMICA output almost identical motifs; only one bp is different between the two consensus sequences, at position 25 (Figure 1). We performed NestedMICA searches with the same parameters in the 20 control datasets and found motifs that were very similar to each other, and very similar to the motif present in the CNV breakpoint dataset (Figure S1). Further review of this common motif revealed that it is part of the *Alu* repeat sequence. This 50-bp motif is present twice in the ~300-bp *Alu* consensus (Figure 2) and includes a 26-bp core sequence that is a hotspot within *Alu*s involved in gene rearrangements [31].

It is possible that although the 50-bp motif is present in both breakpoint and control datasets, it is enriched in CNV break-

points. To explore this, we examined the number of motifs in CNV breakpoints and control regions. Using the 50-bp Nested-MICA motif detected in the CNV breakpoint dataset, we visualized the motif alignments with MochiView [26] and counted the number of motifs in all 21 datasets. The number of 4-kb sequences with at least one motif is not enriched in the CNV breakpoint dataset (68) versus controls (56–85), and the total number of motifs in the CNV breakpoint dataset (161) is not greater than the number of motifs in the control datasets (137–209) (Figures S2 and S3, Table S3). Since *Alu*s are the most abundant mobile element in the human genome [32], it is not surprising that we find a substring of the *Alu* sequence as the most common 50-bp sequence in the CNV breakpoint dataset and the 20 control datasets; however, this motif is not overrepresented in breakpoint regions compared to control regions of the genome.

Repeats enriched at CNV breakpoints

Diverse types of repetitive sequences may be involved in DSBs that give rise to CNVs. This is the case for fragile sites, which are made up of various classes of satellite DNA, dinucleotide and trinucleotide repeats. In addition, different types of repetitive DNA predicted to form secondary structures underlie many BCRs in tumor genomes [7–9,16,33,34]. To investigate repetitive DNA sequences involved in germline CNVs, we searched for tandem repeats and predicted G-quadruplex DNA in the CNV breakpoint and control datasets.

We used Tandem Repeats Finder with default settings to identify tandem repeats in the 21 datasets. Tandem repeats are not based on a consensus sequence, rather they are defined by two or more duplicated sequences arrayed head-to-tail. Since tandem repeats may overlap one another (Figure 2), we counted the total number of tandem repeats as well as the total non-overlapping bp occupied by at least one tandem repeat in each 4-kb sequence. In the CNV breakpoint dataset, 104 out of 140 breakpoint sequences had at least one tandem repeat (Table S4). For the 104 sequences with tandem repeats, 25–4000 bp were occupied by tandem repeats, and the mean and median amounts of sequence including at least one tandem repeat were 330 bp and 133 bp, respectively. The 20 control datasets had 71–95 out of 140 sequences with at least one tandem repeat (mean = 82; median = 83). For those control sequences with at least one tandem repeat, the mean and median numbers of bp occupied by a tandem repeat per 4-kb sequence were 156 bp and 63 bp, respectively. Since all 21 datasets had the same number of bp analyzed (4 kb * 140 sequences = 560 kb), we can compare the number of tandem

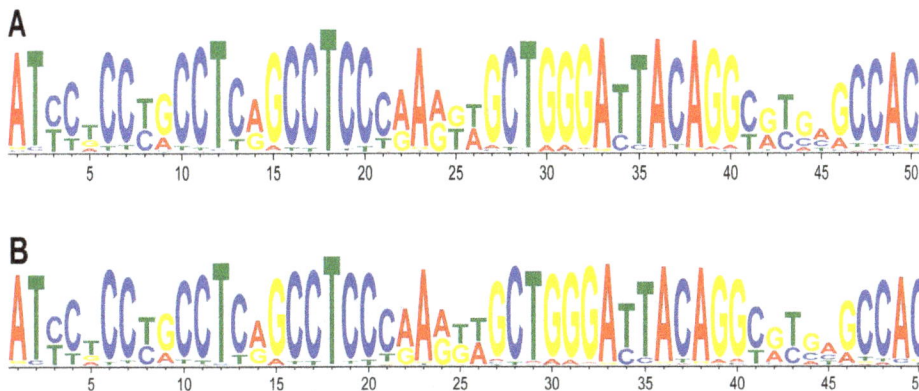

Figure 1. Common 50-bp motif in CNV breakpoint dataset. Logo plots for (**A**) MEME and (**B**) NestedMICA motifs show nearly identical consensus sequences.

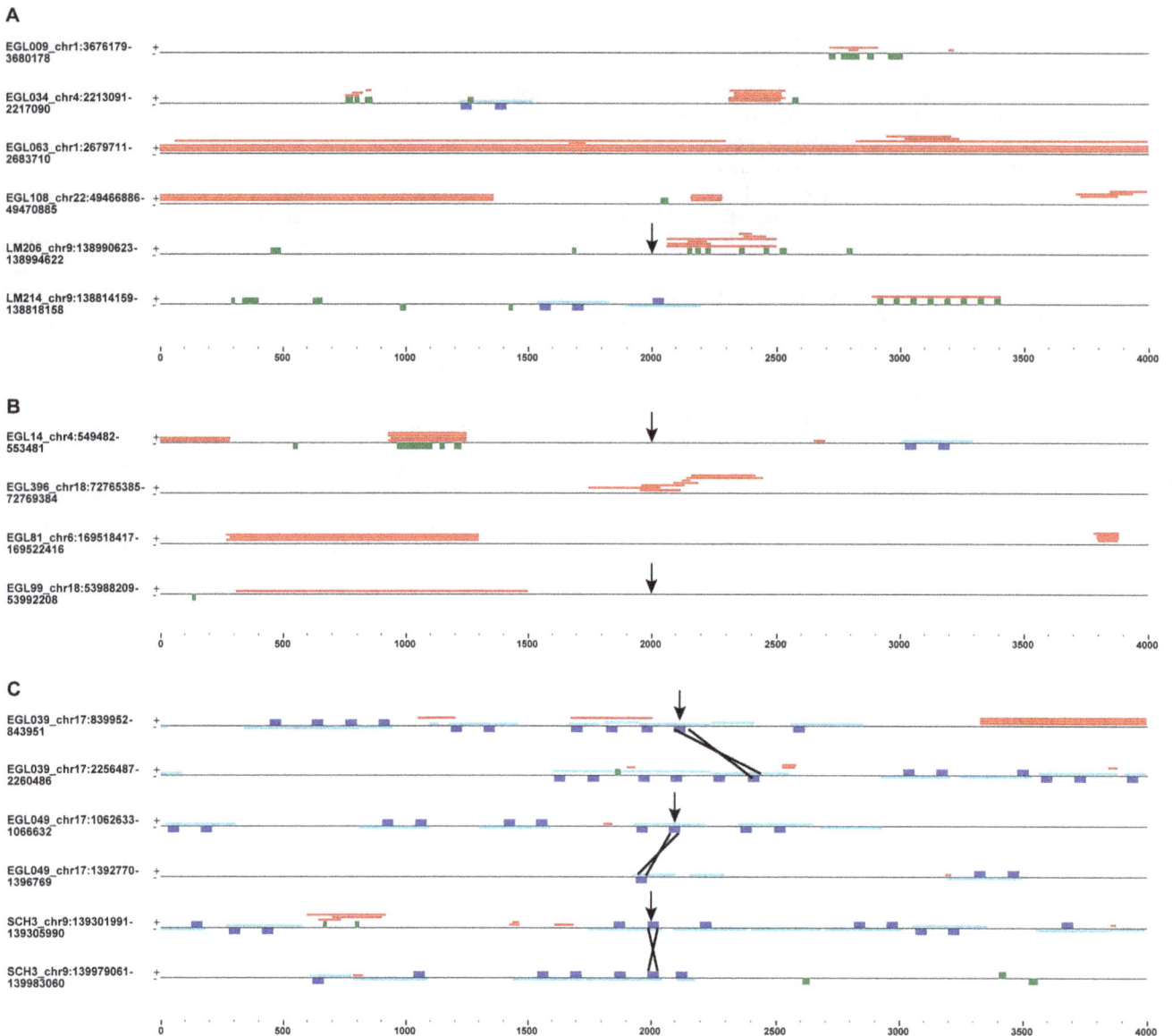

Figure 2. Repeats in 4-kb CNV breakpoint regions. The location of tandem repeats (red), G-quadruplexes (green), *Alus* (light blue) and 50-bp motif (dark blue) sequences are shown for a subset of terminal deletion (**A**), inverted duplication adjacent to terminal deletion (**B**) and interstitial deletion (**C**) breakpoint regions. Repeats on the positive (+) and negative (−) strands are shown on the top and bottom of the black line, respectively. Arrows point to sequenced breakpoint junctions. The breakpoint region from EGL108 underlies terminal deletions and one interstitial duplication. Black Xs show *Alu-Alu* recombination sites for sequenced junctions (**C**).

repeats without adjusting for the size of the dataset. The CNV breakpoint dataset had 318 tandem repeats, whereas the control datasets had 133–254 tandem repeats (Figure 3). Thus, the CNV breakpoint regions are enriched in the number and density of tandem repeats compared to the control regions.

Tandem repeats include other classes of repetitive sequences, including triplet repeats and satellite DNA. It is possible that the enrichment in tandem repeats in the CNV breakpoint dataset is due largely to one particular type of duplicated sequence. Instead, we found that the tandem repeats in the breakpoint regions vary in repeat unit size, repeat array size, AT- and GC-content. We concatenated the tandem repeats in 104 CNV breakpoints to assemble the non-overlapping tandem repeat sequences and avoid counting segments of breakpoint regions more than once. Non-overlapping tandem repeat regions have mean and median GC

percentages of 44% and 49%, respectively. Thus, both AT-rich and GC-rich tandem repeats are present at CNV breakpoints.

We also investigated sequences predicted to form G-quadruplexes in CNV breakpoint and control datasets. Sequences that contain four tracts of at least three guanines separated by other bases can form G-quadruplexes by intrastrand pairing between the four G-rich tracts. Such G-rich sequences can assemble G-quadruplex structures *in vitro* [35] and cause chromosome breakage and genomic instability *in vivo* [36,37]. We searched for the G-quadruplex consensus sequence, $G_{3+}N_{1-7}G_{3+}N_{1-7}G_{3+}N_{1-7}G_{3+}$, using the QuadParser program [29]. Sixty-eight out of 140 CNV breakpoint regions have at least one G-quadruplex, whereas 38 to 52 control regions in the 20 control datasets have at least one G-quadruplex (mean and median $= 42$). There are 201 G-quadruplexes in the CNV breakpoint dataset and 47–78 G-

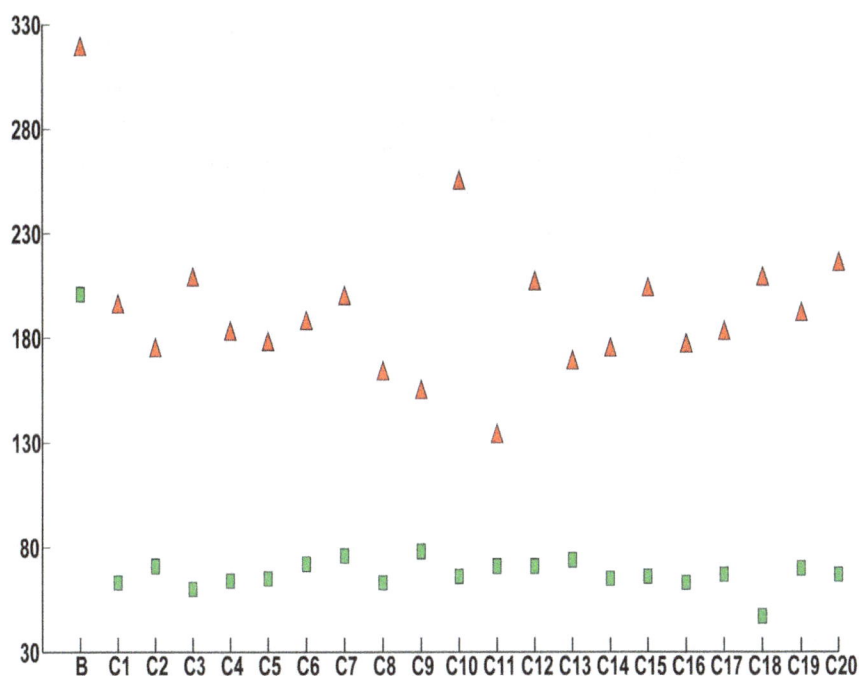

Figure 3. Number of repeats per dataset. The total number of tandem repeats (red triangles) and G-quadruplex (green rectangles), sequences per CNV breakpoint (B) and control (C1-C20) datasets are plotted.

quadruplexes in the control datasets (Figure 3; Table S5). Thus, CNV breakpoints are enriched in the G-quadruplex consensus sequence.

It is possible that the enrichment in G-quadruplex sequences in the CNV breakpoint datasets stems from an increase in overall GC-richness in CNV breakpoints. The human genome is organized into GC-rich and AT-rich isochores [38], and the genome-wide average of GC content is 41% [39]. The mean GC content of the 20 control datasets ranged from 39.8–41.6%, whereas the mean GC content of the CNV breakpoint regions was 47.2% (Table S6). CNV breakpoints are 133 kb to 75 Mb from the nearest telomere; the mean and median distances are 4.8 Mb and 2.3 Mb, respectively (Table S2). Thus, many breakpoints lie within the terminal chromosome band that is known to be elevated in GC content [40,41]. Since terminal deletion and duplication CNV breakpoints lie closer to chromosome ends than other CNV breakpoints, we would expect them to be GC-rich. However, breakpoint regions from all six types of CNV had a higher GC percentage than the genome average (Table 1). Therefore, the GC enrichment in CNV breakpoints is not due to only a subset of terminal chromosome rearrangements.

Repeats enriched in different types of CNV

We analyzed the enrichment of tandem repeats, G-quadruplexes and *Alu*s in the breakpoints from the six types of CNVs (Table 1). Interstitial and terminal duplications are underrepresented in the 140 breakpoints (n = 10), whereas terminal deletions and inverted duplications adjacent to terminal deletions make up more than half of the 140 breakpoints. Tandem repeats, G-quadruplexes and *Alu*s were not distributed proportionally across CNV types according to chi-square goodness-of-fit tests (1e-15< p<.0024). Terminal deletion breakpoint regions have an average GC content of 50.1% and are enriched in G-quadruplexes (p = 9.2e-6). Inverted duplication terminal deletion breakpoints were slightly depleted for G-quadruplexes (p = .0019) and *Alu*s

(p = 2.4e-7) (Table 1). Breakpoints from interstitial duplications and deletions were enriched in *Alu*s (p = 3.9e-10 and 3.0e-4, respectively). The enrichment of motifs at certain types of CNV breakpoints was striking and points to specific repetitive DNA being involved in various types of chromosome rearrangements.

Discussion

Our analysis of 140 CNV breakpoints revealed an enrichment in tandem repeats and G-quadruplexes. It is possible that some of these sequences assemble secondary structures that are susceptible to DNA replication errors or DSBs. Tandem repeats have been described at other CNV breakpoints and are predicted to form a range of secondary structures [17–19]. In our CNV breakpoints, we find both AT-rich and GC-rich tandem repeats. Additional studies of DNA secondary structure and chromosome fragility are necessary to pinpoint the factors required for DSBs in particular classes of tandem repeats.

Breaks in repetitive sequences may facilitate particular types of chromosome rearrangements. DSBs that give rise to terminal deletions may be repaired by synthesis of a new telomere at the deletion breakpoint [5,42–44]. Breaks that occur in or resect to G-rich DNA are ideal substrates for telomerase to prime a new telomere sequence, (5′-TTAGGG-3′)n. In addition, G-rich sequences with the ability to form secondary structures are susceptible to DSBs. Thus, sequences that underlie terminal deletion breakpoints may be G-rich due to the propensity for DSBs plus the likelihood of recovering a chromosome break repaired by a new telomere. We sequenced 13 of 48 terminal deletion junctions to pinpoint the post-repair CNV junction. None of the 13 terminal deletion junctions lies in a G-quadruplex or tandem repeat; however, LM206's chromosome 9q terminal deletion junction is 65 bp proximal of a cluster of G-rich tandem repeats and G-quadruplexes (Figure 2). It is tempting to speculate that a DSB in the G-rich repeat region resected 65 bp and was the

Table 1. Repeats enriched and depleted in CNV breakpoints. The GC content and number of breakpoints are listed for the six CNV types.

CNV type	%GC	Breakpoints	Tandem repeats	G-quads	*Alu*s
Terminal deletion	50.1	48	122	99**	62
Inverted duplication with terminal deletion	43.6	41	70	39*	36*
Translocation	46.1	21	40	24	42
Interstitial deletion	50.0	20	60	30	72**
Interstitial duplication	45.4	8	25	9	28**
Terminal duplication	41.0	2	1	0	2
Total			140	318	201	242

The number of tandem repeats, G-quadruplexes, and *Alu*s per type of CNV are shown. Significant depletion (*) and enrichment (**) for repeats were determined by binomial test p-values <.00278 (.05/18, based on Bonferroni adjustment for 18 tests).

site of telomere addition. In this case, the new telomere lies directly adjacent to a G-rich sequence at the breakpoint, 5′-GGGGCGGAGGGGCCGAAGCTGGCTGGTGG-3′ [5].

Though *Alu* repeats were not enriched in the entire CNV breakpoint dataset compared to control datasets, they were enriched in interstitial deletion and duplication breakpoint regions (Table 1). *Alu*s that recombine to form interstitial deletions and duplications are oriented in the same direction, share high sequence homology (typically >85% identical), and crossover at a homologous site within the *Alu*s [5,45–48]. NAHR generates a hybrid *Alu* at the breakpoint that merges the two sides of the CNV, which is detectable by breakpoint sequencing. We sequenced three of the ten interstitial deletion junctions and none of the four interstitial duplication junctions. Sequence analysis revealed that EGL039, EGL049 and SCH3 breakpoints are the product of recombination between two highly identical *Alu* repeats [5] (Figure 2). In all three cases, the sequenced *Alu-Alu* breakpoints lie within a 50-bp motif. Other interstitial deletion breakpoints may have *Alu*s nearby, but are not the product of *Alu-Alu* NAHR. EGL094's sequenced interstitial deletion junction is the product of non-homologous end-joining (NHEJ) between two breakpoints that are not in *Alu*s [5]. However, EGL094's proximal breakpoint region has a cluster of five *Alu*s in four kb (Figure S2). Other studies of chromosome breakpoints have also found an enrichment of *Alu*s at interstitial deletion junctions [16,20].

CNV breakpoint regions in our study were significantly more GC-rich than the genome average. A previous study of germline and somatic breakpoints suggested that deletion breakpoints were AT-rich, whereas translocation breakpoints were GC-rich [16]. The deletion and translocation breakpoints in our study are both GC-rich, with GC contents of 50% and 46%, respectively (Table 1). This difference is likely due to the chromosome rearrangements selected for the two studies: Abeysinghe *et al.* examined a large cohort of mostly somatic chromosome rearrangements, whereas our study included only germline chromosome rearrangements. In addition, 87/140 (87%) of the CNV breakpoints in our study occur in the last 10 Mb of chromosomes, which are more GC-rich than the genome average. There are likely different biases in GC content for deletions, depending on the origin of the deletion (germline vs. somatic) and the location of the deletion breakpoints. In their large-scale analysis of 663,446 breakpoints from diverse cancer genomes, De and Michor found an enrichment of tandem repeats and *Alu*s [34]. Thus, some classes of DNA repeats are shared between germline and somatic breakpoints.

Repeat density may also play a role in chromosome breakage. For example, both of EGL039's breakpoints are made up of several *Alu*s and tandem repeats. EGL063's terminal deletion breakpoint is entirely covered by tandem repeats across the 4-kb region, and LM206's terminal deletion breakpoint has overlapping tandem repeats and G-quadruplexes (Figure 2). In some cases, G-quadruplexes are part of the tandem repeat structure, rather than separate sequences (see EGL99 and LM214 breakpoints). In other breakpoints, different types of repeats are dispersed across the 4-kb region (see EGL034, EGL108 and EGL81 breakpoints). Repeats at CNV breakpoints could have an additive effect, whereby more repeats lead to a greater propensity for chromosome breakage and/or recombination. On the other hand, there may be only one repeat per locus that is responsible for chromosome rearrangement.

Our analysis of CNV breakpoint regions revealed an enrichment in tandem repeats and sequences predicted to form G-quadruplexes. Furthermore, particular classes of repeats are overrepresented at breakpoints of different types of CNV. Thus, when interpreting mechanisms of CNV formation, it is important to consider the DNA at breakpoints as well as the resulting chromosome rearrangement. Functional analysis of individual DNA motifs will delineate the sequences responsible for gross chromosomal rearrangement [49–51]. In addition, motif mining of even larger CNV breakpoint datasets from diverse CNV classes will tell us more about the factors required for CNV formation.

Supporting Information

Figure S1 Logo plots of top 50-bp motif detected in each control dataset (C1-C20) by NestedMica.

Figure S2 Repeats in 4-kb CNV breakpoint regions (html). The location of tandem repeats (red), G-quadruplexes (green), *Alu*s (light blue) and 50-bp motif (dark blue) sequences are shown for each of the 140 CNV breakpoint regions.

Figure S3 Repeats in 4-kb control regions (html). 140 control regions from control dataset C3 are shown as an example of repeat content in control sequences.

Table S1 CNV breakpoint resolution. For 140 breakpoint regions, minimum, maximum, mean, and median breakpoint resolution is 0 bp, 7,542 bp, 468 bp, and 101 bp, respectively.

The method of breakpoint mapping and the type of rearrangement is described. Insertion length is listed for three sequenced junctions with insertions.

Table S2 Distance from breakpoint region to the nearest chromosome end.

Table S3 50-bp NestedMica (NM) motifs in the CNV breakpoint region dataset (B) and control datasets (C1-C20). The number of sequences with at least one NM motif and the number of NM motifs per dataset are listed.

Table S4 Tandem repeats (TR) in the CNV breakpoint region dataset (B) and control datasets (C1-C20). The number of sequences with at least one TR and the number of TRs per dataset are listed.

Table S5 G-quadruplex consensus sequences ($G_{3+}N_{1-7}$ $G_{3+}N_{1-7}G_{3+}N_{1-7}G_{3+}$) in the CNV breakpoint region dataset

(B) and control datasets (C1-C20). The number of sequences with at least one G-quadruplex and the number of G-quadruplexes per dataset are listed.

Table S6 GC content of CNV breakpoint region dataset (B) and control datasets (C1-C20). For each dataset, minimum, maximum, mean, and median percent GC were calculated.

Acknowledgments

We thank Cheryl Strauss for editorial assistance and members of the Rudd laboratory for helpful discussions.

Author Contributions

Conceived and designed the experiments: PB MKR. Performed the experiments: PB KEH. Analyzed the data: PB KEH KNC MKR. Contributed to the writing of the manuscript: MKR.

References

1. Cooper GM, Coe BP, Girirajan S, Rosenfeld JA, Vu TH, et al. (2011) A copy number variation morbidity map of developmental delay. Nature genetics 43: 838–846.
2. Kaminsky EB, Kaul V, Paschall J, Church DM, Bunke B, et al. (2011) An evidence-based approach to establish the functional and clinical significance of copy number variants in intellectual and developmental disabilities. Genetics in medicine 13: 777–784.
3. Itsara A, Cooper GM, Baker C, Girirajan S, Li J, et al. (2009) Population analysis of large copy number variants and hotspots of human genetic disease. Am J Hum Genet 84: 148–161.
4. Rudd MK, Keene J, Bunke B, Kaminsky EB, Adam MP, et al. (2009) Segmental duplications mediate novel, clinically relevant chromosome rearrangements. Hum Mol Genet 18: 2957–2962.
5. Luo Y, Hermetz KE, Jackson JM, Mulle JG, Dodd A, et al. (2011) Diverse mutational mechanisms cause pathogenic subtelomeric rearrangements. Human molecular genetics 20: 3769–3778.
6. Hermetz KE, Newman S, Conneely KN, Martin CL, Ballif BC, et al. (2014) Large inverted duplications in the human genome form via a fold-back mechanism. PLoS genetics 10: e1004139.
7. Zhang Y, Rowley JD (2006) Chromatin structural elements and chromosomal translocations in leukemia. DNA repair 5: 1282–1297.
8. Wang G, Vasquez KM (2006) Non-B DNA structure-induced genetic instability. Mutation research 598: 103–119.
9. Sinclair PB, Parker H, An Q, Rand V, Ensor H, et al. (2011) Analysis of a breakpoint cluster reveals insight into the mechanism of intrachromosomal amplification in a lymphoid malignancy. Human molecular genetics 20: 2591–2602.
10. Sutherland GR (2003) Rare fragile sites. Cytogenet Genome Res 100: 77–84.
11. Glover TW, Arlt MF, Casper AM, Durkin SG (2005) Mechanisms of common fragile site instability. Hum Mol Genet 14 Spec No. 2: R197–205.
12. Yu S, Mangelsdorf M, Hewett D, Hobson L, Baker E, et al. (1997) Human chromosomal fragile site FRA16B is an amplified AT-rich minisatellite repeat. Cell 88: 367–374.
13. Hewett DR, Handt O, Hobson L, Mangelsdorf M, Eyre HJ, et al. (1998) FRA10B structure reveals common elements in repeat expansion and chromosomal fragile site genesis. Mol Cell 1: 773–781.
14. Zlotorynski E, Rahat A, Skaug J, Ben-Porat N, Ozeri E, et al. (2003) Molecular basis for expression of common and rare fragile sites. Mol Cell Biol 23: 7143–7151.
15. Zhang H, Freudenreich CH (2007) An AT-rich sequence in human common fragile site FRA16D causes fork stalling and chromosome breakage in S. cerevisiae. Mol Cell 27: 367–379.
16. Abeysinghe SS, Chuzhanova N, Krawczak M, Ball EV, Cooper DN (2003) Translocation and gross deletion breakpoints in human inherited disease and cancer I: Nucleotide composition and recombination-associated motifs. Human mutation 22: 229–244.
17. Gajecka M, Pavlicek A, Glotzbach CD, Ballif BC, Jarmuz M, et al. (2006) Identification of sequence motifs at the breakpoint junctions in three t(1;9)(p36.3;q34) and delineation of mechanisms involved in generating balanced translocations. Hum Genet 120: 519–526.
18. Gajecka M, Gentles AJ, Tsai A, Chitayat D, Mackay KL, et al. (2008) Unexpected complexity at breakpoint junctions in phenotypically normal individuals and mechanisms involved in generating balanced translocations t(1;22)(p36;q13). Genome Res 18: 1733–1742.
19. Yatsenko SA, Brundage EK, Roney EK, Cheung SW, Chinault AC, et al. (2009) Molecular mechanisms for subtelomeric rearrangements associated with the 9q34.3 microdeletion syndrome. Hum Mol Genet 18: 1924–1936.
20. Vissers LE, Bhatt SS, Janssen IM, Xia Z, Lalani SR, et al. (2009) Rare pathogenic microdeletions and tandem duplications are microhomology-mediated and stimulated by local genomic architecture. Hum Mol Genet 18: 3579–3593.
21. Conrad DF, Bird C, Blackburne B, Lindsay S, Mamanova L, et al. (2010) Mutation spectrum revealed by breakpoint sequencing of human germline CNVs. Nat Genet 42: 385–391.
22. Richardson C, Jasin M (2000) Frequent chromosomal translocations induced by DNA double-strand breaks. Nature 405: 697–700.
23. Simsek D, Jasin M (2010) Alternative end-joining is suppressed by the canonical NHEJ component Xrcc4-ligase IV during chromosomal translocation formation. Nature structural & molecular biology 17: 410–416.
24. Bailey TL, Williams N, Misleh C, Li WW (2006) MEME: discovering and analyzing DNA and protein sequence motifs. Nucleic acids research 34: W369–373.
25. Down TA, Hubbard TJ (2005) NestedMICA: sensitive inference of over-represented motifs in nucleic acid sequence. Nucleic Acids Res 33: 1445–1453.
26. Homann OR, Johnson AD (2010) MochiView: versatile software for genome browsing and DNA motif analysis. BMC biology 8: 49.
27. Smit AF (1996) The origin of interspersed repeats in the human genome. Curr Opin Genet Dev 6: 743–748.
28. Benson G (1999) Tandem repeats finder: a program to analyze DNA sequences. Nucleic Acids Res 27: 573–580.
29. Huppert JL, Balasubramanian S (2005) Prevalence of quadruplexes in the human genome. Nucleic Acids Res 33: 2908–2916.
30. Rice P, Longden I, Bleasby A (2000) EMBOSS: the European Molecular Biology Open Software Suite. Trends Genet 16: 276–277.
31. Rudiger NS, Gregersen N, Kielland-Brandt MC (1995) One short well conserved region of Alu-sequences is involved in human gene rearrangements and has homology with prokaryotic chi. Nucleic acids research 23: 256–260.
32. Batzer MA, Deininger PL (2002) Alu repeats and human genomic diversity. Nat Rev Genet 3: 370–379.
33. Popescu NC (2003) Genetic alterations in cancer as a result of breakage at fragile sites. Cancer letters 192: 1–17.
34. De S, Michor F (2011) DNA secondary structures and epigenetic determinants of cancer genome evolution. Nature structural & molecular biology 18: 950–955.
35. Burge S, Parkinson GN, Hazel P, Todd AK, Neidle S (2006) Quadruplex DNA: sequence, topology and structure. Nucleic Acids Res 34: 5402–5415.
36. Kruisselbrink E, Guryev V, Brouwer K, Pontier DB, Cuppen E, et al. (2008) Mutagenic capacity of endogenous G4 DNA underlies genome instability in FANCJ-defective C. elegans. Curr Biol 18: 900–905.
37. Ribeyre C, Lopes J, Boule JB, Piazza A, Guedin A, et al. (2009) The yeast Pif1 helicase prevents genomic instability caused by G-quadruplex-forming CEB1 sequences in vivo. PLoS Genet 5: e1000475.
38. Costantini M, Clay O, Auletta F, Bernardi G (2006) An isochore map of human chromosomes. Genome research 16: 536–541.
39. Consortium IHGS (2001) Initial sequencing and analysis of the human genome. Nature 409: 860–921.

40. Rudd MK (2007) Subtelomeres: Evolution in the Human Genome. Encyclo-pedia of Life Sciences: John Wiley & Sons, Ltd.

41. Rudd MK (2014) Human and Primate Subtelomeres. In: Louis EJ, Becker MM, editors. Subtelomeres: Springer. pp. 153–164.

42. Lamb J, Harris PC, Wilkie AO, Wood WG, Dauwerse JG, et al. (1993) De novo truncation of chromosome 16p and healing with (TTAGGG)n in the alpha-thalassemia/mental retardation syndrome (ATR-16). Am J Hum Genet 52: 668–676.

43. Flint J, Craddock CF, Villegas A, Bentley DP, Williams HJ, et al. (1994) Healing of broken human chromosomes by the addition of telomeric repeats. Am J Hum Genet 55: 505–512.

44. Ballif BC, Yu W, Shaw CA, Kashork CD, Shaffer LG (2003) Monosomy 1p36 breakpoint junctions suggest pre-meiotic breakage-fusion-bridge cycles are involved in generating terminal deletions. Hum Mol Genet 12: 2153–2165.

45. Lehrman MA, Schneider WJ, Sudhof TC, Brown MS, Goldstein JL, et al. (1985) Mutation in LDL receptor: Alu-Alu recombination deletes exons encoding transmembrane and cytoplasmic domains. Science 227: 140–146.

46. Pousi B, Hautala T, Heikkinen J, Pajunen L, Kivirikko KI, et al. (1994) Alu-Alu recombination results in a duplication of seven exons in the lysyl hydroxylase gene in a patient with the type VI variant of Ehlers-Danlos syndrome. Am J Hum Genet 55: 899–906.

47. Sen SK, Han K, Wang J, Lee J, Wang H, et al. (2006) Human genomic deletions mediated by recombination between Alu elements. American journal of human genetics 79: 41–53.

48. Beck CR, Garcia-Perez JL, Badge RM, Moran JV (2011) LINE-1 elements in structural variation and disease. Annual review of genomics and human genetics 12: 187–215.

49. Narayanan V, Mieczkowski PA, Kim HM, Petes TD, Lobachev KS (2006) The pattern of gene amplification is determined by the chromosomal location of hairpin-capped breaks. Cell 125: 1283–1296.

50. Kim HM, Narayanan V, Mieczkowski PA, Petes TD, Krasilnikova MM, et al. (2008) Chromosome fragility at GAA tracts in yeast depends on repeat orientation and requires mismatch repair. Embo J.

51. Shishkin AA, Voineagu I, Matera R, Cherng N, Chernet BT, et al. (2009) Large-scale expansions of Friedreich's ataxia GAA repeats in yeast. Molecular cell 35: 82–92.

7

Exome Sequencing from Nanogram Amounts of Starting DNA: Comparing Three Approaches

Vera N. Rykalina[1,2,3⦾], Alexey A. Shadrin[1,3⦾], Vyacheslav S. Amstislavskiy[1], Evgeny I. Rogaev[4], Hans Lehrach[1], Tatiana A. Borodina[1,2]*

1 Max-Planck Institute for Molecular Genetics, Berlin, Germany, 2 AlacrisTheranostics GmbH, Berlin, Germany, 3 Freie Universität Berlin, Berlin, Germany, 4 Vavilov Institute of General Genetics, Moscow, Russia

Abstract

Hybridization-based target enrichment protocols require relatively large starting amounts of genomic DNA, which is not always available. Here, we tested three approaches to pre-capture library preparation starting from 10 ng of genomic DNA: (i and ii) whole-genome amplification of DNA samples with REPLI-g (Qiagen) and GenomePlex (Sigma) kits followed by standard library preparation, and (iii) library construction with a low input oriented ThruPLEX kit (Rubicon Genomics). Exome capture with Agilent SureSelect^{XT2} Human AllExon v4+UTRs capture probes, and HiSeq2000 sequencing were performed for test libraries along with the control library prepared from 1 μg of starting DNA. Tested protocols were characterized in terms of mapping efficiency, enrichment ratio, coverage of the target region, and reliability of SNP genotyping. REPLI-g- and ThruPLEX-FD-based protocols seem to be adequate solutions for exome sequencing of low input samples.

Editor: Tom Gilbert, Natural History Museum of Denmark, Denmark

Funding: This work was supported by the 7th framework programme of the European Union (ADAMS) [242257, FP7-HEALTH-2009]. The funders had no role in study design, data collection and analysis, decision to publish, or preparation of the manuscript.

Competing Interests: The authors declare that no competing interests exist. Though two authors (VR and TB) are currently employed by a commercial company Alacris Theranostics GmbH. Results presented in the current manuscript are not IP of the company and have no relationship to the research performed at Alacris Theranostics GmbH or to the services provided by this company. The presence of the company's name in the list of affiliations merely reflects the fact that during the preparation of the manuscript affiliations of two of the authors have changed.

* Email: t.borodina@alacris.de

⦾ These authors contributed equally to this work.

Introduction

Whole exome sequencing (WES) is currently one of the main applications in next generation sequencing and this trend will likely continue in the near future. WES requires much less sequencing volume, allows higher throughput, and requires less computational resources than whole genome sequencing (WGS). WES is about ten times cheaper than WGS (calculated for the following WES settings: library preparation with Agilent SureSelect Human All Exon v4 kit, sequencing on Illumina HiSeq2000 platform in paired end 2×100 bp sequencing mode, with 12 GB output of filtered data; and the following WGS settings: library preparation with Illumina TruSeq library preparation kit, sequencing on Illumina HiSeq2000 platform in paired end 2×100 bp sequencing mode, with 150 GB output of filtered data). In many applications these advantages make WES a feasible alternative to WGS in terms of the price/results ratio. Multiple publications have demonstrated the impact of WES in identifying causative variants of Mendelian diseases [1–7]. WES is also performed to analyze complex traits, to both reveal trait-associated regions and screen for individual variations contributing to the trait manifestation [8–10].

Currently the preferred method for WES library preparation is hybridization-based enrichment of whole genome sequencing libraries [11]. Corresponding commercial products are available for example from Agilent, NimbleGen, and Illumina. Existing exome enrichment kits differ in total size of target region, and the number, length and nature (DNA or RNA) of the capture probes, as well as minor issues in laboratory procedures; however the principle of the protocol is the same. The procedure begins with preparing a whole genome library – random genomic DNA fragments flanked with common adapters. The library is amplified with several PCR cycles and mixed with a set of artificial biotinilated probes corresponding to the target region. Library molecules with inserts at least partly containing fragments of the target region hybridize to the capture probes. Capture probes, both free and hybridized to library molecules, are collected by their biotin groups using streptavidin-coated magnetic beads. Library molecules are then washed off the beads and amplified to a concentration appropriate for sequencing.

Hybridization-based exome enrichment kits require comparatively large amount of starting genomic DNA, i.e. 1–3 μg. For comparison, a WGS library of good complexity may be prepared from just 10 ng of genomic DNA. However, in many cases even such amounts are not available, for example in small size samples such as clinical biopsies. Another example is DNA collections in population analysis and genome-wide associated studies (GWAS) laboratories. Researchers try to use the material carefully, since it is hard to assemble such collections and samples may be required for several projects.

Cost-efficiency of exome sequencing makes it very attractive to be applied for low quantity samples. Several approaches have been

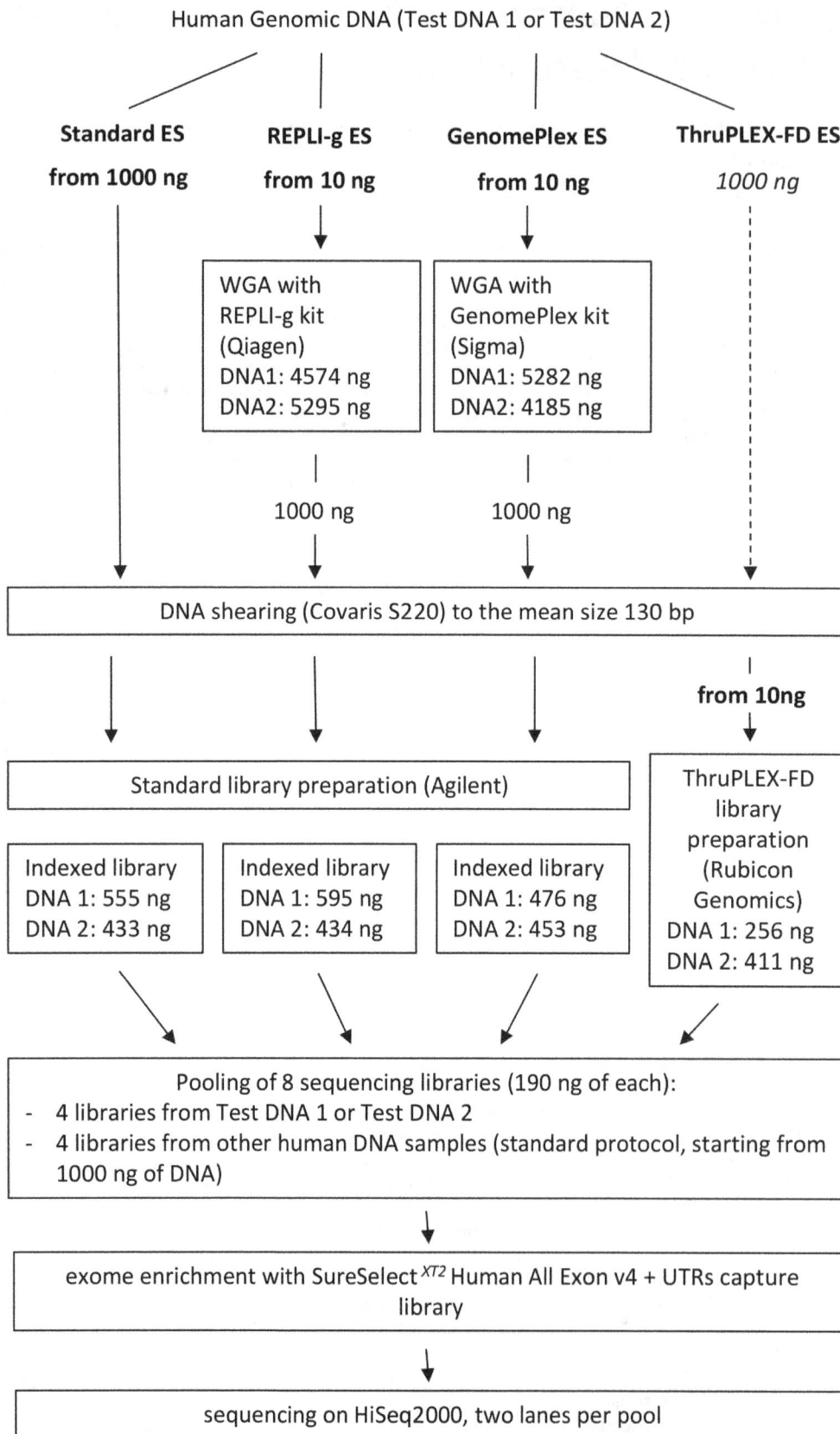

Figure 1. The experimental scheme. Two DNA samples (Test DNA 1 and Test DNA 2) were subjected to four exome sequencing (ES) protocols performed in parallel: control (Standard ES) and three modified (REPLI-g ES, GenomePlex ES and ThruPLEX-FD ES). Common steps performed in parallel for several protocols are shown by text boxes spanning the corresponding number of protocol columns.

Table 1. Features of the target region.

Number of segments	Total length of target region (Mb)	Mean length of the segment (bp)	Median length of the segment (bp)	Maximum length of the segment (bp)	Minimum length of the segment (bp)
199268	70.37	353.1	203	21747	114

suggested to overcome the sample amount requirement and a number of publications describe exome sequencing performed with sub-microgram amounts of starting DNA.

One approach is to increase the amount of starting material by whole genome amplification. In this case the WES library is prepared from the recommended amount of material and no changes to the protocol itself are necessary [12,13].

Another approach is to optimise the whole genome library preparation that precedes the hybridization-based enrichment (pre-capture library). The standard procedure involves DNA shearing followed by three enzymatic steps, with two purifications and amplification in between. During library preparation loss of material occurs due to the wide distribution of fragment sizes after DNA shearing, taking aliquots for quality and quantity evaluations, and purification steps. It is possible to reduce these losses and adjust the standard protocol for smaller starting amounts [14].

An alternative transposon-based library preparation method (Nextera technology, Illumina) gives enough material for hybridization-based enrichment staring with just 50 ng of genomic DNA. The efficiency of this method is explained by obtaining fragmented adapter-flanked genomic DNA in just one tagmentation reaction (material losses are minimized) and by a more narrow size distribution of fragments than obtained by e.g. ultrasonic shearing. However this method is sensitive to the fragmentation of starting DNA and produces more biased coverage of genomic regions [15].

The amount of input material for hybridization-based enrichment steps can also be reduced, as described in the MSA-Cap method suggested by Kosarewa et al. [14]. Except for optimising the steps prior to hybridization, the authors used the fact that commercial protocols are not working at the border of sensitivity; they used half the recommended amount of library for hybridization. In addition, they optimised the post hybridization procedure, decreasing the concentration of the library solution required for Illumina sequencing. All together, they managed to decrease the starting amount for exome sequencing using the Agilent SureSelect capture method from 3 µg to 50 ng.

During the ADAMS FP7 EU project we faced the necessity of performing exome sequencing and target sequencing of GWAS selected regions with low amounts of samples. Our partners in the ADAMS consortium have large DNA collections, often stored for a long time, of different quality. We had to look for a strategy to prepare libraries for exome sequencing starting from about 10 ng of genomic DNA.

To prepare the WES library we tested three commercially available systems for pre-capture library preparation: REPLI-g (Qiagen), GenomePlex (Sigma) and ThruPLEX-FD (Rubicon Genomics). Exome enrichment was performed with Agilent SureSelect^{XT2} Human AllExon v4+UTRs capture probes sets. Sequencing was performed on the Illumina HiSeq2000 platform. We compared three test sequencing data sets with data obtained from the recommended starting amount of material, using evaluation parameters commonly used to characterize exome sequencing: mapping efficiency, enrichment efficiency, coverage uniformity, and single nucleotide variants (SNV) detection efficiency.

Materials and Methods

DNA samples

Protocols compared in this study were independently tested on two human genomic DNA samples. Test DNA 1 was purchased from Bioline (Human Genomic DNA, BIO-35025). Test DNA 2 was isolated from peripheral blood of an anonymous blood donor using phenol-chloroform method. Original DNA purity and integrity was confirmed by gel electrophoresis. Concentrations were determined on a Qubit fluorometer using a dsDNA BR kit (Invitrogen, Q32853).

The blood sample for Test DNA 2 used for this work is one of the samples collected specifically for the ADAMS FP7 project, mentioned in the Funding section. All samples for this project were taken with written informed consent and all the data was anonymized. This particular sample comes from the group of one of the co-authors, Prof. Evgeny Rogaev. Prof. Rogaev got the approval of the Local Ethical Committee of Vavilov Institute of General Genetics of Russian Academy of Sciences for the ADAMS FP7 project. Prof. Rogaev did not collect blood himself and did not contact the donor, but he has access to the donor-identifying information.

Whole genome amplification

Whole genome amplification (WGA) was performed using the commercially available GenomePlex Complete Whole Genome Amplification Kit (Sigma, Cat. No. WGA2) and REPLI-g Mini Kit (Qiagen, Cat. No. 150023) following the manufacturers' protocols. Starting amounts of genomic DNA for WGA reactions was 10 ng in all cases.

The **REPLI-g** protocol includes denaturation of DNA and isothermal amplification. About 15 minutes hands-on time are required at the beginning. Total procedure time is determined by the recommended amplification duration: 10–16 hours.

DNA was first denatured: 5 µl of Buffer D1 were added to 5 µl of 2 ng/µl DNA solution and the sample was incubated at room temperature for 3 minutes. Denaturation was stopped by adding 10 µl of neutralization buffer N1. Then 30 µl of REPLI-g Master Mix were added to the sample, and amplification was carried out at 30°C for 16.5 hours. REPLI-g Mini DNA Polymerase was inactivated by heating at 65°C for 3 minutes.

The **GenomePlex** WGA protocol includes (i) fragmentation, (ii) preparation of OmniPlex library and (iii) amplification steps. No intermediate purifications are required. The procedure takes about 5 hours and requires about 30 minutes hands-on time. DNA was first fragmented: 1 µl of 10x Fragmentation Buffer was added to 10 µl of 1 ng/µl DNA solution and the sample was incubated at 95°C for 4 minutes and immediately cooled on ice.

Fragmented DNA was converted into OmniPlex library, fragments flanked with common sequences. First, 2 µl of 1x Library Preparation Buffer and 1 µl of Library Stabilization Buffer

Table 2. Alignment statistics.

| Library Preparation method | Number of raw reads (Mb of seq) | Percentage of duplicates (% of raw reads) | Percentage of high-confident* reads mapped to hg19 (% of raw reads) | Percentage of high-confident reads mapped uniquely to hg19 (% of raw reads reads) | | | Mate is on the same chromosome and has proper orientation (% of total) | Percentage of high-confident reads mapped uniquely to FR*** (% of raw reads) | Percentage of high-confident reads mapped uniquely to TR (% of raw reads) |
				Total	Mate is mapped (% of total)	Mate is on the same chromosome (% of total)			
Standard ES									
Test DNA 1	95033226 (9598)	20.21	75.62	71.82	99.51	99.26	99.25	50.20	47.97
Test DNA 2	173021034 (17475)	25.07	71.31	68.02	99.79	99.73	99.73	49.06	47.34
GenomePlex ES									
Test DNA 1	65957628 (5740)**	18.21	69.54	63.09	96.58	95.94	95.21	31.87	30.40
Test DNA 2	70253046 (6252)**	20.58	63.54	57.49	95.39	94.09	93.58	31.95	31.00
ThruPLEX-FD ES									
Test DNA 1	60302154 (6091)	34.43	61.02	57.83	99.04	97.45	97.41	40.26	38.25
Test DNA 2	81220550 (8203)	45.38	50.10	47.54	98.93	97.08	97.03	31.73	30.28
REPLI-g ES									
Test DNA 1	89106596 (8999)	20.70	75.59	71.90	99.66	99.53	99.53	50.99	49.03
Test DNA 2	146075078 (14754)	25.30	71.64	68.38	99.61	99.54	99.34	49.72	48.04

*high confident reads-reads with probability of wrong mapping lower than 0.05 according to their MAPQ score (MAPQ>13).

**some of GenomePlex ES library reads contained sequences of the primer used for whole genome amplification. These common segments were cut out before the alignment. As a result, 13.8% of and 11.9% of nucleotides were removed from the reads of the Test DNA 1 and Test DNA 2 libraries, respectively.

***FR-flanking regions (FR), which include 100 bp from both ends of the targeted sequences.

Table 3. Coverage statistics for the target region.

Amplification method	Mean coverage	Coverage depth (% of bases in TR)							
		0	1–10	11–20	21–30	31–40	41–50	51–60	61+
Standard ES									
Test DNA 1	20.80	1.56	30.75	29.95	17.49	9.12	4.70	2.48	3.83
Test DNA 2	20.07	2.46	34.32	26.60	15.65	8.82	4.95	2.81	4.05
GenomePlex ES									
Test DNA 1	19.59	8.85	39.18	21.20	11.53	6.55	3.89	2.44	5.92
Test DNA 2	17.79	11.14	39.75	19.97	10.95	6.25	3.71	2.29	5.06
ThruPLEX-FD ES									
Test DNA 1	19.90	1.27	26.61	32.28	20.85	10.51	4.71	2.06	1.62
Test DNA 2	19.07	1.56	26.95	31.34	21.94	11.57	4.60	1.40	0.41
REPLI-g ES									
Test DNA 1	20.92	1.99	30.96	28.39	17.17	9.48	5.11	2.79	3.98
Test DNA 2	20.01	3.41	36.06	25.67	14.58	8.08	4.56	2.61	4.65

Analysis was performed on subsets of reads uniquely mapped to the target region and having approximately equal total amounts of bases (~17×10^8 bases).

were added to the sample after fragmentation. The tube was incubated at 95°C for 2 minutes and cooled on ice. Then 1 μl of Library Preparation Enzyme was added and the sample was placed in a thermal cycler and incubated as follows: 16°C for 20 minutes; 24°C for 20 minutes; 37°C for 20 minutes; 75°C for 5 minutes; 4°C hold.

The OmniPlex library was diluted with 47.5 μl of water, 7.5 μl of 10x Amplification Master Mix and 5 μl of WGA DNA Polymerase and placed in thermal cycler. Following initial denaturation at 95°C for 3 minutes, 14 amplification cycles were performed: 94°C for 15 seconds; 65°C for 5 minutes.

WGA reactions were purified with Agencourt AMPure XP beads (Beckman Coulter, Cat. No. A63881). Concentrations of amplified DNA were measured with Qubit dsDNA BR kit. Both GenomePlex and REPLI-g amplification procedures showed ~230-fold amplification.

Pre-capture sample processing

For each DNA sample, four barcoded libraries were prepared for further exome capture (Figure 1). Three libraries were prepared starting from 1000 ng of original or WGA-material according to the sample preparation guidelines of the Agilent SureSelect^{XT2} Target Enrichment System for Illumina Multiplexed Sequencing Protocol (version A, January 2012). One library was prepared starting from 10 ng of sheared original DNA using ThruPLEX Fragmented DNA Prep kit (ThruPLEX-FD) from Rubicon Genomics (Cat. No. R40012).

DNA shearing. The first step in both Agilent SureSelect^{XT2} and ThruPLEX-FD library preparation is DNA fragmentation, which was performed on a Covaris S220 system. Shearing of 1000 ng was carried out in 50 μl of 1x TE buffer using the following instrument parameters: duty cycle 10%, intensity 5, cycles per burst 200, time 6 cycles per 60 seconds. The resulting size distribution was checked using the Agilent 2100 BioAnalyserDNA1000 assay. For all samples, shearing worked very consistently and the size distribution peak was around 130 bp.

In our experience shearing of small amounts of genomic DNA on Covaris S220 is associated with noticeable loss of material. For example, after fragmentation of 20 ng of genomic DNA, only 10 ng are detected. Most probably this is related to the adhesion of DNA molecules to the glass walls of the tubes used for shearing (microTUBE AFA Fiber tubes, Covaris, Cat. No. 520045). To avoid incorrect evaluation of the ThruPLEX-FD kit, we took as starting material 10 ng of already sheared original DNA.

Agilent SureSelect^{XT2} library preparation. Libraries from original, GenomePlex- and REPLI-g- amplified DNA were prepared with SureSelect^{XT2} Reagent kit (Agilent, Cat. No. G9621A) following the manufacturer's instructions. The procedure comprises standard steps for Illumina DNA library preparation: end-repair of fragmented DNA, A-tailing, adapter ligation and amplification. Purifications between steps were carried out with Agencourt AMPure XP beads. Library preparation takes about 6 hours, and about 2 hours hands-on time. The libraries' yield, estimated with Qubit dsDNA HS kit (Invitrogen, Q32853), varied between 476 and 595 ng for Test DNA 1 and between 433 and 453 ng for Test DNA2.

ThruPLEX-FD library preparation. The ThruPLEX–FD procedure comprises three steps: (i) template preparation, (ii) library synthesis and (iii) amplification and indexing. No intermediate purifications are required; reagents are subsequently added to the same tube. The procedure takes in total 3.5 hours with 45 minutes hands-on time. From 10 ng of starting material we obtained 256 ng for the Test DNA1 library and 411 ng for the Test DNA 2 library (measured with Qubit dsDNA HS kit).

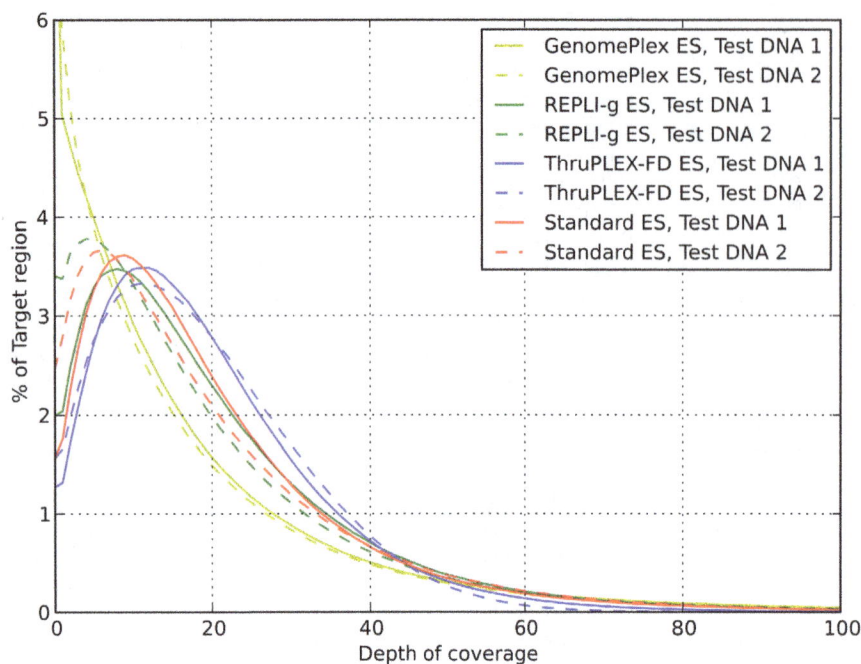

Figure 2. Per-base sequencing depth distribution on the target region.

ThruPLEX-FD libraries were prepared following the manufacturer's protocol. First, 3 µl of Template Preparation pre-mix were added to 10 µl of 1 ng/µl DNA solution and the sample was incubated in a thermal cycler under following conditions: 22°C for 25 minutes; 55°C for 20 minutes, 22°C for 5 minutes. Immediately 2 µl of Library Synthesis pre-mix were added to the tube and the sample was incubated at 22°C for 40 minutes and cooled at 4°C for 5 minutes. Again immediately, 58 µl of Library Ampli-fication pre-mix and 2 µl of one Indexing Reagent (1–12) were added to the sample and the tube was incubated in a thermal cycler using the following amplification program: 72°C for 3 minutes, 85°C for 2 minutes, 98°C for 2 minutes, 4 cycles of (98°C for 20 seconds, 67°C for 20 seconds, 72°C for 40 seconds), 7 cycles of (98°C for 20 seconds, 72°C for 50 seconds). Ready barcoded libraries were purified with Agencourt AMPure XP beads.

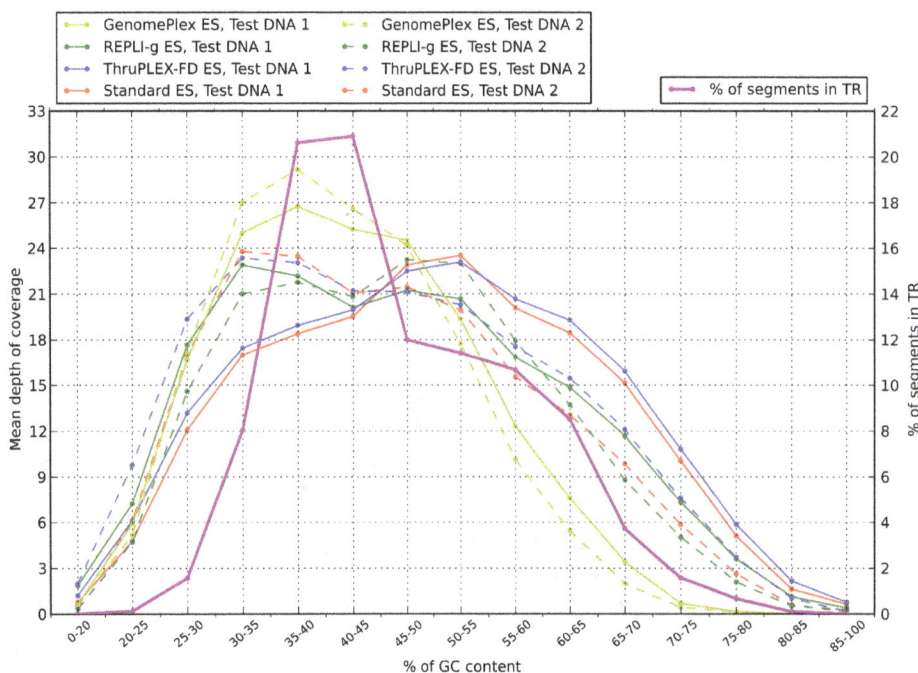

Figure 3. Coverage distribution along the target regions with different percentages of GC bases.

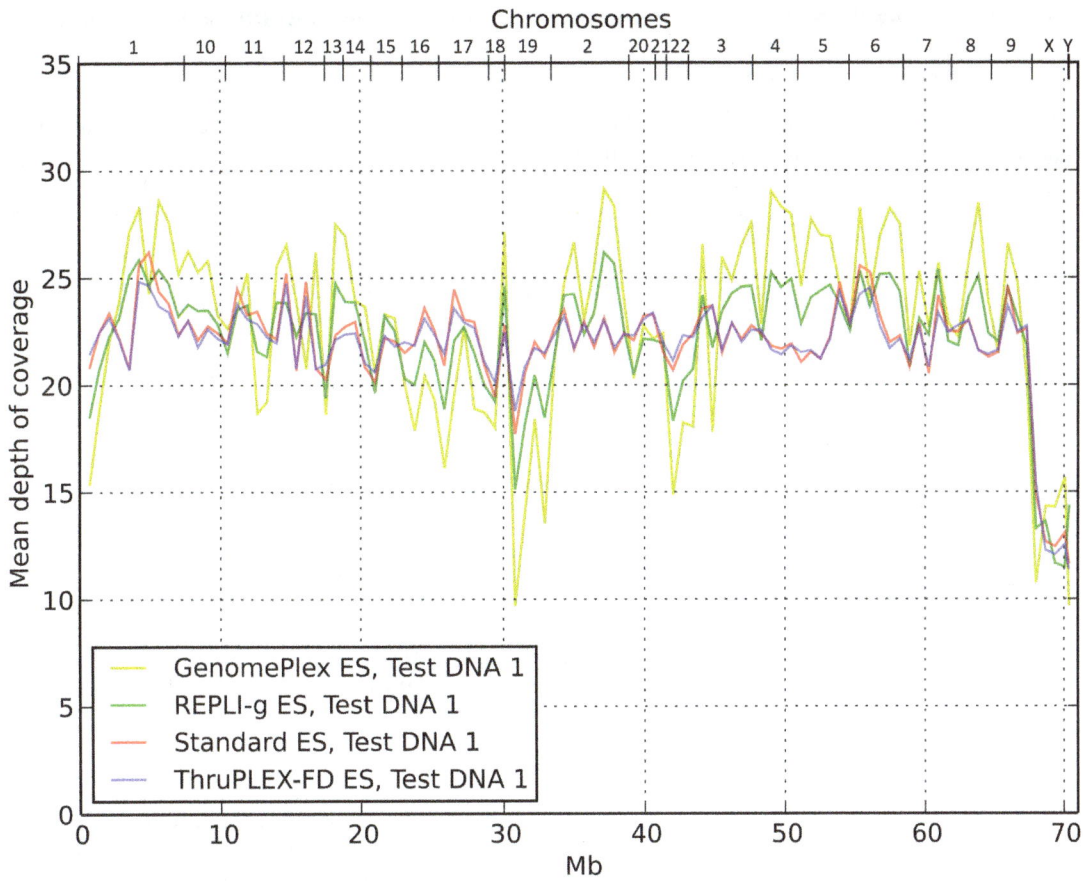

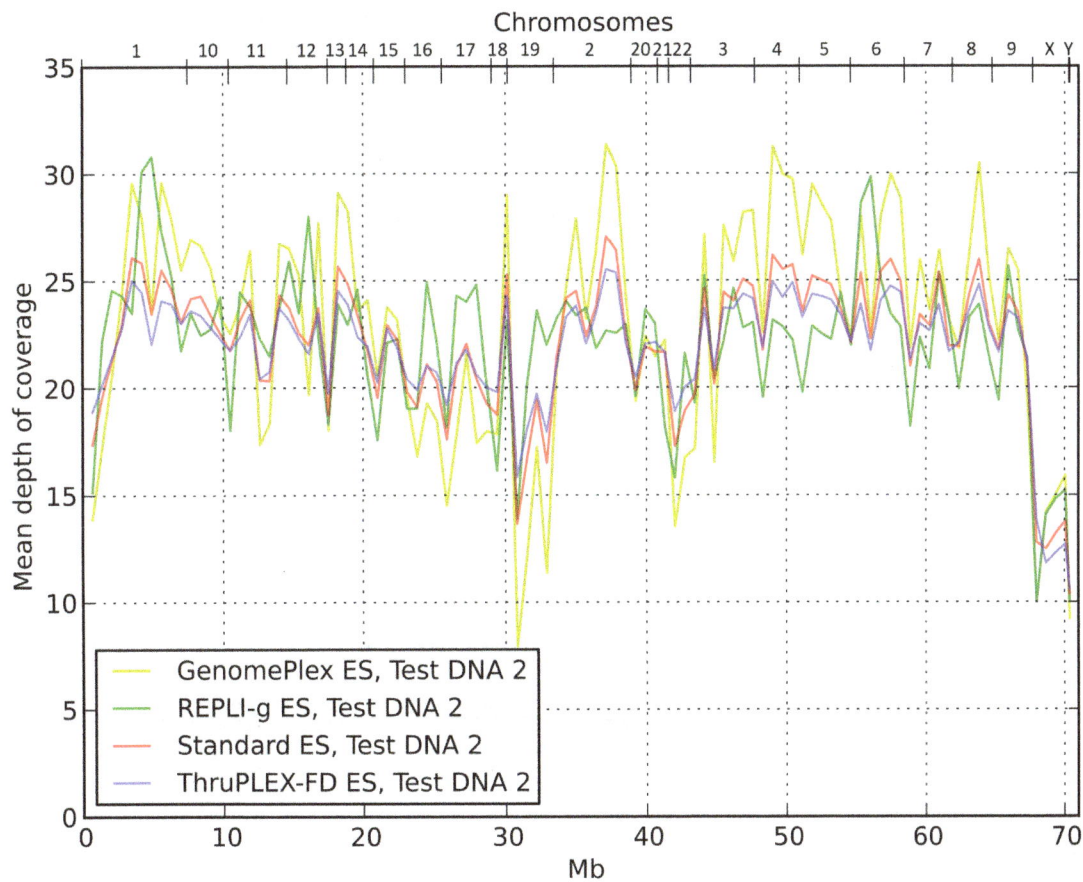

Figure 4. Profiles of coverage depth along the target region for Test DNA 1 (upper panel) and Test DNA 2 (lower panel) WES libraries.

Exome enrichment

Exome capture was performed with Agilent SureSelect^{XT2} Human All Exon v4 + UTRs capture probes set (Agilent, Cat. No. 5190-4671). The SureSelect^{XT2} Target Enrichment System is designed for exome capture of eight pooled libraries.

Libraries from Test DNA 1 and Test DNA 2 were prepared and processed within a two-month interval. To prepare library pools for exome enrichment, four test DNA libraries were pooled with four libraries from other human genomic DNA samples. In both cases those were genomic DNA samples isolated from peripheral blood of anonymous donors and converted into libraries with the standard Agilent SureSelect^{XT2} protocol starting from 1000 ng of DNA. 190 ng of each library were taken for pooling. In both cases the pool volume was larger than the 7 μl required for the following enrichment step, and the pools were concentrated using a SpeedVac concentrator.

Hybridization of pooled libraries to the capture probes and removal of non-hybridized library molecules were carried out according to the Agilent SureSelect^{XT2} Target Enrichment System for Illumina Multiplexed Sequencing Protocol (version A, January 2012).

Library molecules fished out by hybridization were amplified. Excessive amplification can increase sequence bias and cause PCR artefacts, so prior to amplification of the whole sample, a small aliquot (1 μl) of captured library was first tested in qPCR. The number of cycles corresponding to the mid-exponential phase of the amplification curve and adjusted to the volume of the captured library was taken for amplification. For both pools with Test DNA 1 and Test DNA 2 libraries post-hybridization PCR was performed with 11 cycles.

Sequencing

Sample dilution, flowcell loading and sequencing were performed according to the Illumina specifications. Pools with Test DNA 1 and Test DNA 2 libraries were each sequenced on two lanes of the HiSeq2000 platform as paired-end 101-bp reads.

Data processing and statistics

All sequencing data are submitted to the European Nucleotide Archive (ENA study accession number PRJEB6077).

Fastq files were generated with Illumina BCL2FASTQ Conversion Software (version 1.8.2).

Some GenomePlex ES library reads contain sequences of the primer used for whole genome amplification. These technical segments were mostly removed before alignment. However since

this primer has partly degenerate sequence, it is difficult to completely remove it from the reads.

Bowtie 2 (version 2.1.0) [16] was used to align the reads on human genome reference assembly (build hg19 GRCh37). The genome was downloaded from the UCSC Genome Browser website (http://genome.ucsc.edu/) [17].

The BED file with description of the target region of the SureSelectV4+UTRs kit was provided by the manufacturer. Brief specification of it is shown in Table 1.

After alignment, potential PCR duplicates were removed with Picard MarkDuplicates (version 1.91) [18]. Subsequently, using a home-made script, all reads with a probability of wrong mapping higher than 0.05 according to their mapping quality score (MAPQ<14) were filtered out from each library, the remaining reads were further filtered to remove the ambiguous reads (not uniquely aligned). Then from high-confident uniquely mapped reads, only reads overlapping the target region (TR) were selected with BEDTools (version 2.17.0) [19]. We also estimated the overlapping of high-confident reads with the flanking regions (FR), which include 100 bp from both ends of the targeted sequences.

SNVs and small INDELs were called using SAMtools' mpileup (version 0.1.19) [20].

Results

Strategies for pre-capture library preparation

Our aim was to select a strategy to perform exome sequencing starting from 10 ng of human genomic DNA. For exome capture we chose the Agilent SureSelect^{XT2} method. The XT2 modification of the SureSelect strategy is designed for hybridization-based enrichment of indexed libraries and seemed attractive in two aspects. First, it allows one to pool up to 8 libraries prior to capture, and is thus cheaper per sample while providing higher throughput. Second, the recommended starting DNA amount is 1 μg and the amount of whole genome library required for hybridization is 190 ng, instead of 3 μg and 500 ng, respectively, for single library processing.

Since hybridization efficiency depends on the concentration of the participating molecules, we wanted to modify the WES library preparation procedure before the hybridization step to obtain the amount of input material for hybridization as recommended by the manufacturer. We selected two approaches: (i) whole genome amplification of initial DNA, to obtain enough input material for a standard Agilent SureSelect protocol, and (ii) preparation of the pre-capture library starting with 10 ng of input material using an optimized protocol, and adopting the Agilent SureSelect protocol at the hybridization step.

Table 4. Pearson correlation coefficient with coverage profile of Standard ES strategy and average deviation from Standard ES coverage profile.

Strategy	Pearson correlation with Standard ES coverage profile	Average deviation from Standard ES coverage profile
ThruPLEX-FD ES, Test DNA1 (DNA2)	0.986 (0.980)	0.318 (0.681)
REPLI-g ES, Test DNA1 (DNA2)	0.809 (0.755)	1.478 (1.950)
GenomePlex ES, Test DNA1 (DNA2)	0.589 (0.947)	3.149 (2.233)

Test DNA 1

Rg 124	Rg-Tp 88	Rg-Tp-Gp 196	Rg-Gp 57
St-Rg 119	St-Rg-Tp 1177	St-Rg-Tp-Gp 10265	St-Rg-Gp 429
St 134	St-Tp 59	St-Tp-Gp 298	St-Gp 42

Total: 13249
St: 12523
Tp: 12191
Rg: 12455
Gp: 11471

Tp 77	Tp-Gp 31	Gp 153

Test DNA 2

Rg 159	Rg-Tp 97	Rg-Tp-Gp 219	Rg-Gp 68
St-Rg 128	St-Rg-Tp 1143	St-Rg-Tp-Gp 10883	St-Rg-Gp 466
St 147	St-Tp 71	St-Tp-Gp 321	St-Gp 54

Total: 14104
St: 13213
Tp: 12911
Rg: 13163
Gp: 12221

Tp 138	Tp-Gp 39	Gp 171

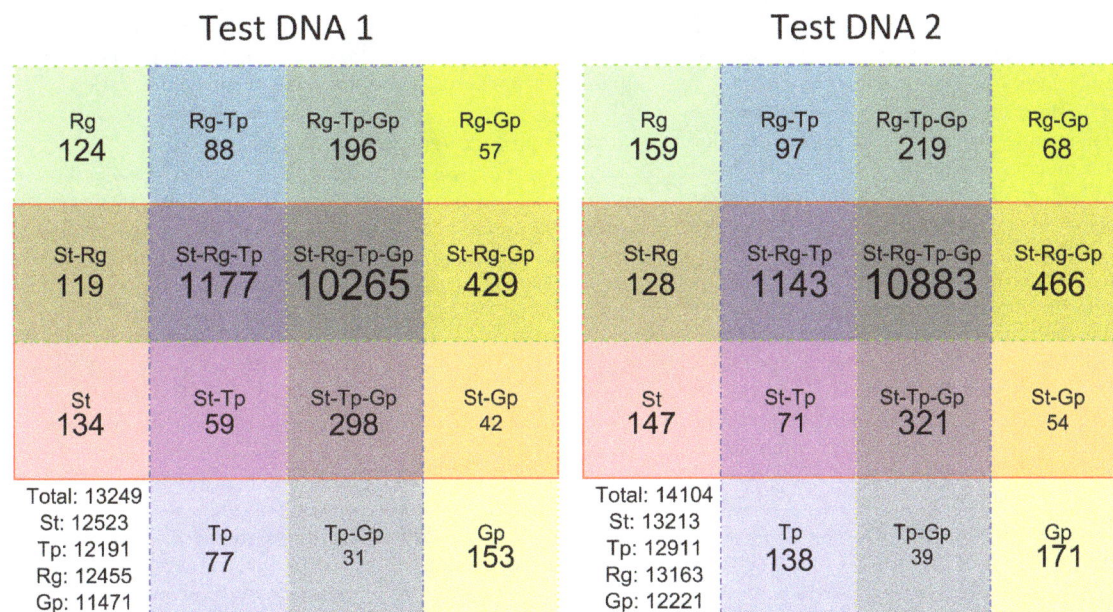

Figure 5. Sharing of genetic variations between strategies depicted in a Venn diagram. Only variation with minimum depth of coverage of 20x and minimum quality of 13 were taken into account in all four strategies. The names of the samples are abbreviated: Standard ES = St; ThruPLEX-FD ES = Tp; REPLI-g ES = Rg; GenomePlex ES = Gp. The lower left tile presents the overall statistics, where "Total" indicates the number of all unique SNVs found in the region of interest, i.e. the union of SNV sets found by each strategy.

For the first approach we chose two commercially available WGA kits: REPLI-g from Qiagen and GenomePlex from Sigma.

The REPLI-g kit uses isothermal genome amplification, called Multiple Displacement Amplification (MDA), which involves random hexamers binding to denatured DNA followed by strand displacement synthesis at a constant temperature using the Phi 29 polymerase [15]. Additional priming events can occur on each displaced strand, leading to a network of branched DNA structures. Phi 29 polymerase does not dissociate from the genomic DNA template, allowing the generation of DNA fragments up to 100 kb without sequence bias.

The GenomePlex Complete Whole Genome Amplification Kit uses the proprietary amplification OmniPlex technology [21]. The protocol involves isothermal primer extension on randomly fragmented genomic DNA. Oligonucleotides for primer extension have self-inert degenerate 3′ ends and universal 5′ tails. Extension is performed with polymerase with strand displacement activity, so, as in MDA amplification, brunched DNA structures are formed. This whole genome amplification step produces multiple, comparatively short fragments with common ends. These fragments are further amplified by PCR amplification using common tails, resulting in DNA fragments with an average size of 400 bp.

Both REPLI-g and GenomePlex technologies were reported to enable accurate genotyping [22,23], which is also important for targeted sequencing. REPLI-g amplified DNA was also previously used for exome sequencing [12,13]. GenomePlex has not yet been tried for exome sequencing, and we expected it to perform worse than REPLI-g for target enrichment since it is known that PCR-based WGA is prone to biased representation of different genomic regions. However we decided to compare and characterize this kit, since it is capable of amplifying denatured and degraded DNA. Moreover, if the performance of GenomePlex was acceptable, we knew it would be possible to incorporate NGS adapter sequences during the PCR step of the WGA, and thus simplify library preparation.

For the second approach, we decided to check an alternative library preparation method that had recently appeared on the market: the ThruPLEX-FD Kit from Rubicon Genomics. This kit focuses on NGS library preparation starting from small amounts; enzymatic reactions are optimized and performed in a single tube, without intermediate purifications. Background is minimized, e.g.

Table 5. Comparison of SNVs found in regions with coverage > = 20 in both Standard ES and one of three tested strategies.

Strategy	% of TR covered with depth≥20 for both tested and Standard ES libraries	Shared SNVs	Exclusive SNVs present only in Standard ES	Exclusive SNVs Present only in tested ES	Discovery rate (found by tested ES)/(Total found)
REPLI-g ES, Test DNA1 (DNA2)	34.92 (32.91)	18890 (17561)	1408 (1091)	1281 (1093)	0.9348 (0.9447)
ThruPLEX-FD ES, Test DNA1 (DNA2)	36.95 (37.80)	19836 (19722)	1592 (1803)	1152 (1273)	0.9295 (0.9209)
GenomePlex ES, Test DNA1 (DNA2)	26.27 (29.32)	13010 (14180)	1661 (2226)	803 (789)	0.8927 (0.8705)

Only high-confidence (probability of false positive detection <0.05) SNVs were taken into account.

ligation adapters are destroyed after ligation. The ThruPLEX-FD Kit was also reported to be applicable for degraded DNA samples.

For each of two human genomic DNA samples, Test DNA 1 and Test DNA 2, we performed four protocols of pre-capture library preparation for exome sequencing (Fig. 1): one for recommended starting amounts, and three for 10 ng of initial DNA. Test DNA 1 was high quality placenta DNA (Bioline). Test DNA 2 was DNA isolated from peripheral blood. Both samples were of good quality: RNA-free, and not fragmented.

The standard exome sequencing (ES) protocol was performed completely according to the Agilent recommendations, starting with 1000 ng of initial material and serving as a reference for the modified protocols. REPLI-g ES and GenomePlex ES protocols started with 10 ng of genomic DNA. DNA was first subjected to WGA. 1000 ng of amplified DNA were processed in parallel to the Standard ES sample according to the Agilent library preparation protocol. The ThruPLEX-FD ES protocol started with 10 ng of genomic DNA preliminary sheared to ~130 bp. 190 ng of amplified library were incorporated into the Agilent protocol starting from the hybridization-based enrichment step. Protocols details, as well as comments on intermediate results, particular protocol characteristics and performance are presented in the Methods section.

Sequencing data from four Test DNA 1 and four Test DNA 2 libraries allowed us to compare low-amount libraries to the recommended input library in the most important quantitative and qualitative parameters: enrichment efficiency, uniformity of coverage of the target region, and accuracy of SNV detection.

Comparison of sequencing statistics

The alignment statistics for the two sets of libraries grouped by the DNA of origin include basic library properties: mapping efficiency, duplication levels, and enrichment efficiency (Table 2). Generally, there are inter-sample differences, but differences between the library preparation methods are reproduced in both cases. REPLI-g ES library characteristics are very similar to those of the control Standard ES library. GenomePlex ES and ThruPLEX-FD ES libraries demonstrate poorer results.

The number of raw reads varies between the libraries. We cannot exclude that this is not related to some preference during hybridization. In our experience when exome enrichment is performed on pooled libraries, low-complexity samples are underrepresented in the final library. Moreover ThruPLEX-FD ES libraries, which have smaller numbers of reads, also show lower complexity. However, libraries prepared by the standard protocol (starting from 1000 ng of genomic DNA) also vary considerably in the number of reads (see Figure S1).

ThruPLEX-FD ES libraries show a considerably higher percentage of duplexes: 34% (45%) for Test DNA 1 (Test DNA 2). In comparison, Standard ES and REPLI-g ES libraries have ~20.5% (25%) duplexes for test DNAs. The GenomePlex ES library looks slightly better with 18% (21%) duplexes.

GenomePlex ES and especially ThruPLEX-FD ES libraries have noticeably less reads mapped to the genome: 70% (64%) and 61% (50%) versus ~76% (71.5%) for the other two protocols. The GenomePlex ES library shows about two-fold more non-uniquely mapped reads −6.5% (6%), than the other three protocols: 3.2–3.8% (2.6–3.3%). Also GenomePlex ES libraries have slightly more read pairs with improper position of the mate read.

With ThruPLEX-FD ES and GenomePlex ES libraries less reads mapped uniquely to the target region (TR). However for ThruPLEX-FD ES libraries, uniquely mapped TR reads constitute 66% (64%) of the total uniquely mapped reads, a proportion close to Standard ES (67% (70%)) and REPLI-g ES (68% (70%))

libraries. In contrast, for GenomePlex ES only 48% (54%) of uniquely mapped reads fall into the TR. Thus the enrichment efficiency is noticeably worse for GenomePlex ES libraries.

Target region coverage

To compare coverage features of the target region, such as evenness and depth of coverage, it is crucial to have the comparable amount of sequencing data (Mb) for each library. The data should also be independent from the total number of reads and the enrichment efficiency. Therefore, we made a sub-selection of high-confident reads uniquely mapped to TR, down-sampled with Picard DownsampleSam [18]. Except for Genome-Plex ES libraries, approximately 17×10^6 reads for each library were extracted. In the GenomePlex ES libraries reads were on average shorter due to the removal of the common primer sequence. Hence we took more reads (approximately 19×10^6 reads for each GenomePlex ES library), in order to approximately equalize the total number of bases of the extracted sequence among all samples ($\sim 17 \times 10^8$ bases). All the results presented here were obtained for these sets of comparable amounts of sequencing data uniquely mapped to the target region (Table 3).

Since we selected approximately equal amounts of sequencing data from each sample, the expected values of mean coverage should be very similar, which what we see in Table 3.

Evenness of coverage is an important characteristic of sequencing data for exome analysis. Finding genetic variations is a common task in exome analysis and the more uniform the coverage, the less amount of sequencing data is required to reliably detect variations and be able to use more straightforward and reliable statistical techniques.

Ideally the target region should be uniformly represented in the sequencing data. However sequence-dependent performance at many stages of the sequencing protocol, related to the composition and structure of the DNA, distort the sequences' representation in the original DNA [24]. For example, it is known that PCR amplification introduces a bias in standard Illumina library preparation depending on the GC composition of the sequence [25]. Both extremely GC-poor and extremely GC-rich loci are often underrepresented or even absent [26]. In the case of target enrichment, capture preferences add to the sequencing bias [27]. As a result, the region being sequenced will have segments that are either over- or under-represented, or completely absent in the sequencing reads.

The uniformity of coverage in the test libraries compared to the standard protocol was assessed and compared (Table 3, Fig. 2). GenomePlex ES libraries differ from the others: 9% (11%) of the TR are not covered at all for Test DNA 1 (Test DNA 2) and only 52% (48%) of the TR have coverage >10. ThruPLEX-FD ES libraries look the best: they demonstrate coverage >10 for 72% (71%) of the TR, while Standard ES libraries have coverage >10 for 68% (63%), and REPLI-g ES libraries for 67% (60%) of the TR.

For the mean coverage according to the GC content, ThruPLEX-FD ES libraries look most similar to controls for both DNA samples (Fig. 3). GenomePlex ES libraries show considerable under-representation of sequences with GC content ≥55%. Interestingly, except for GenomePlex ES libraries, all libraries show inter-sample differences in GC content profile.

We next asked whether the profiles of coverage depth along the target region differ between the libraries (Fig. 4). To obtain the profiles of coverage depth, segments of the target region were assembled together. Segments within single chromosomes were concatenated according to their coordinates on the chromosome. Chromosomes were concatenated in numerical order, as indicated in the upper horizontal axes. Each point of coverage depth profile represents coverage averaged over 700000 adjacent bases in the concatenated TR (i.e. a total of 101 points in each profile).

The profiles resemble each other closely, particularly the profiles of Standard ES and ThruPLEX-FD ES libraries. Profiles of the GenomePlex ES libraries have greater amplitudes. Visual impression is confirmed by the values of Pearson correlation and average deviation of the profiles relative to the Standard ES profile (Table 4). ThruPLEX-FD ES libraries coverage profile and depth are most close to those of the Standard ES libraries. ThruPLEX-FD ES approach also demonstrates smaller difference between the two test samples than the other protocols.

The inter-sample differences observed for the tested protocols might have different reasons. They may be partly caused by the natural variability of a laboratory procedure due to e.g. pipetting and measurement errors. A protocol may be not sufficiently optimized. Also, even a well set up protocol may be sensitive to certain properties of the samples, e.g. purification method used, fragmentation, etc. Small number of samples tested and absence of technical replicas do not allow us to compare the consistency of the approaches.

Genotype calling

Identification of genetic variations is usually the main aim of exome sequencing. We called variations from sequencing data obtained with all tested protocols and compared the results. For the comparison we selected only variations with minimum depth of coverage of 20x and minimum quality of 13 (i.e. the probability of false positive detection is less than 0.05) in all performed strategies. Positions covered with depth ≥ 20 in all methods constitute 22.1% (23.5%) of the TR for Test DNA 1 (Test DNA 2). Results of this comparison for both Test DNA 1 and Test DNA 2 are presented in Figure 5 in the form of Venn diagram. Each tile represents unique SNVs for a strategy or combination of strategies. For both Test DNA1 and Test DNA 2 all strategies revealed a similar number of SNVs and more than 77% (10265 of 13249; 10883 of 14104) of all detected SNVs are shared by all four methods.

GenomePlex ES libraries reveal more unique SNVs than the other three strategies. The lower right tile in each square diagram (Fig. 5; Gp-tile) represents the number of SNVs that are unique to GenomePlex ES, while Rg- St- and Tp-tiles represent the number of unique SNVs for the other ES strategies. Intersections of GenomePlex ES with other methods contain less SNVs than intersections that do not include GenomePlex ES. However, it is clear that these differences are not significant compared to the number of SNVs shared by all strategies.

Test libraries were also pair-wise compared to the control Standard ES library (Table 5). In all cases there are SNVs detected in just one library in the pair (exclusive SNVs). As a consequence of the higher similarity of coverage profiles of Standard ES and ThruPLEX-FD ES, the part of the target region with ≥ 20 coverage for both libraries is larger than for other methods. SNV statistics are very similar between ThruPLEX-FD ES and REPLI-g libraries.

Discussion

Hybridization-based enrichment requires a certain concentration of target DNA to provide good hybridization efficiency in a reasonable time. It is also necessary that the library undergoing hybridization is complex enough, and the target region is more or less uniformly represented, otherwise it will not be uniformly represented in the enriched library. Simply increasing the number of PCR cycles for amplifying the pre-capture library does not work, since over-amplified material has a distorted proportion of amplicons relative to initial PCR templates. In addition, highly duplicated library molecules compete with capture probes during hybridization, which leads to a lower output of enriched library, and when capture is performed on a pool of libraries, to under-representation of sequences of low complexity within the pool.

The task of exome sequencing starting from small amounts of DNA can be reformulated as having large enough quantities of sufficiently complex library before hybridization, to obtain enough complex library afterwards. We set out to characterize different protocols designed to address this problem both in terms of exome sequencing and handling of low quantity samples. We assessed the quality of test WES libraries against those using a standard protocol by comparing mapping characteristics, coverage of the target region, and efficiency of SNV detection.

The REPLI-g based protocol turned out to be the best in terms of resulting library complexity, with duplicate levels and mapping parameters being the same as for the standard protocol. The representation of certain genomic regions was distorted differently in the REPLI-g ES libraries compared to the Standard ES libraries (Fig. 4). However from the amount of sequencing data corresponding to ~20x coverage of the target region, 98% (96.6%) of the target region was represented in REPLI-g ES library samples for Test DNA 1 (Test DNA 2), which is very close to 98.4% (97.5%) in the Standard ES library.

The ThruPLEX-FD library preparation kit looks very interesting, since it closely resembles the procedure for standard library preparation; thus we expected GC amplification biases and sequencing data distribution along the target region to be similar to the standard procedure, as is seen in Figures 3 and 4. The weak point of the ThruPLEX-FD ES library is the lower library complexity, and as a result fewer unique reads map to the target region. However, for the two example samples target region representation was still the same as for the standard ES library, which means library complexity is still within the acceptable range. The less uniform coverage of the target region by the REPLI-g ES library is compensated by the higher percent of reads mapped to the target region than with the ThruPLEX-FD ES library. We ultimately think that both protocols are comparable in the amount of sequencing data required for certain coverage.

The GenomePlex ES libraries showed good complexity. The lower mapping efficiency for the GenomePlex ES strategy is probably related to the presence of the residual WGA-primer sequence in a considerable number of sequencing reads. The most problematic feature of the GenomePlex ES strategy is strong under-representation of GC-rich sequences and uneven coverage of the target region.

In terms of SNV detection, ThruPLEX-FD ES and REPLI-g ES were similar to each other and the standard protocol. GenomePlex ES library sequencing data revealed fewer SNVs and had more exclusive, protocol-specific SNVs. Sequencing data handling was also most problematic for GenomePlex, requiring specific approaches to remove the WGA-primer sequences.

With ThruPLEX-FD, care should be taken when amplifying the pre-capture libraries. It is important to keep the number of cycles as low as possible, and find a balance between obtaining sufficient amounts of material for hybridization, and not over-amplifying to create problems with complexity. In our experience this optimal number of PCR cycles varies between 9 and 11, despite the same amount (10 ng) of starting material. So we recommend following the Rubicon Genomics protocol and performing real-time PCR amplification, in order to follow the amplification curve and remain in the exponential phase.

WGA protocols are safer in terms of material loss, which might be of importance for precious samples. Both REPLI-g and GenomePlex WGA kits reproducibly generated around 5 µg of DNA from 10 ng, providing the possibility to repeat exome library preparation if necessary. In the ThruPLEX-FD protocol, all the material went into the library. Moreover, ThruPLEX-FD requires DNA shearing, which is associated with considerable losses for small amounts of starting material. For example, after fragmentation of 20 ng of genomic DNA using Covaris, only 10 ng are detected afterwards. This is probably related to DNA molecules adhering to the glass walls of the tubes used for shearing. We recommend incubating 0.5x BSA solution in the tubes for several minutes before shearing to reduce DNA binding during shearing.

ThruPLEX-FD ES and GenomePlex ES protocols may be advantageous for partly degraded DNA, since it is recommended that REPLI-g WGA be performed on DNA of good quality.

Apart from the described test experiments we performed REPLI-g ES protocol for three more samples. Sequencing, enrichment and coverage characteristics of the obtained datasets are presented in the Material S1: alignment statistics (Table S1 in Material S1), coverage statistics (Table S2 in Material S1), per-base sequencing depth distribution on the target region (Figure S2 in Material S1), dependence of the coverage on the GC content of the target region (Figure S3 in Material S1) and profiles of coverage depth along the target region (Figure S4 in Material S1). For the altogether five samples the REPLI-g ES protocol demonstrated good consistency.

All the three tested protocols for exome library preparation from 10 ng of starting DNA were successful. Definitely more samples need to be analyzed to make reliable conclusions about the reproducibility of the approaches as well as about superiority of one of them. This study demonstrated that the tested protocols are in general suitable for WES and revealed the parameters which have the tendency to differ between the protocols. GenomePlex ES showed more differences to the standard protocol than REPLI-g ES and ThruPLEX-FD ES protocols. When applying GenomePlex for amplification of original DNA, one should be aware of underrepresentation of GC rich regions in the final library. REPLI-g ES and ThruPLEX-FD ES protocols look like more or less equivalent alternatives, producing sequencing data comparable to the data obtained from 1000 ng of genomic DNA using the standard method. So far REPLI-g ES and ThruPLEX-FD ES protocols seem well-suited for target sequencing library preparation from nanogram amounts of starting material.

Supporting Information

Figure S1 Number of sequencing reads obtained per sample in the Agilent SureSelect^{XT2} All Exon assay. Sample pooling was performed before exome enrichment. Colors mark the protocol used to prepare the pre-capture library. Blue: standard Agilent protocol (data available for the two samples analyzed in this report, Standard ES (1) and Standard ES (2), as well as for 8 other samples), red: REPLI-g ES, green: GenomePlexES, violet: ThruPLEX-FD ES. The original DNA (Test DNA 1 or Test DNA 2) is indicated as a number in brackets.

Material S1 Three additional human DNA samples were processed according to the REPLI-g ES protocol starting from 10 ng. This supplement presents the data on sequencing statistics, enrichment efficiency and coverage uniformity for those three samples.

Acknowledgments

We would like to thank Rubicon Genomics for offering us a ThruPLEX-FD kit at a reduced price.

Author Contributions

Conceived and designed the experiments: TB VR HL. Performed the experiments: VR ER. Analyzed the data: AS VA. Contributed reagents/materials/analysis tools: ER. Wrote the paper: TB AS VR.

References

1. Biesecker LG, Shianna KV, Mullikin JC (2011) Exome sequencing: the expert view. Genome Biol 12: 128.
2. Ng SB, Buckingham KJ, Lee C, Bigham AW, Tabor HK, et al. (2010) Exome sequencing identifies the cause of a mendelian disorder. Nat Genet 42: 30–35.
3. Ng SB, Bigham AW, Buckingham KJ, Hannibal MC, McMillin MJ, et al. (2010) Exome sequencing identifies MLL2 mutations as a cause of Kabuki syndrome. Nat Genet 42: 790–793.
4. Li MX, Kwan JS, Bao SY, Yang W, Ho SL, et al. (2013) Predicting mendelian disease-causing single nucleotide variants in exome sequencing studies. PLoS Genet 9: e1003143.
5. Zhang X, Guo BR, Cai LQ, Jiang T, Sun LD, et al. (2012) Exome sequencing identified a missense mutation of EPS8L3 in Marie Unna hereditary hypotrichosis. J Med Genet 49: 727–730.
6. Wang K, Kim C, Bradfield J, Guo Y, Toskala E, et al. (2013) Whole-genome DNA/RNA sequencing identifies truncating mutations in RBCK1 in a novel Mendelian disease with neuromuscular and cardiac involvement. Genome Med 5: 67.
7. Woo HM, Park HJ, Baek JI, Park MH, Kim UK, et al. (2013) Whole-exome sequencing identifies MYO15A mutations as a cause of autosomal recessive nonsyndromic hearing loss in Korean families. BMC Med Genet 14: 72.
8. McClellan J, King MC (2010) Genetic heterogeneity in human disease. Cell 141: 210–217.
9. Wagner MJ (2013) Rare-variant genome-wide association studies: a new frontier in genetic analysis of complex traits. Pharmacogenomics 14: 413–424.
10. Panoutsopoulou K, Tachmazidou I, Zeggini E (2013) In search of low frequency and rare variants affecting complex traits. Hum Mol Genet 22: R16–21.
11. Mamanova L, Coffey AJ, Scott CE, Kozarewa I, Turner EH, et al. (2010) Target-enrichment strategies for next-generation sequencing. Nat Methods 7: 111–118.
12. Hou Y, Song L, Zhu P, Zhang B, Tao Y, et al. (2012) Single-cell exome sequencing and monoclonal evolution of a JAK2-negative myeloproliferative neoplasm. Cell 148: 873–885.

13. Lepere C, Demura M, Kawachi M, Romac S, Probert I, et al. (2011) Whole-genome amplification (WGA) of marine photosynthetic eukaryote populations. FEMS Microbiol Ecol 76: 513–523.
14. Kozarewa I, Rosa-Rosa JM, Wardell CP, Walker BA, Fenwick K, et al. (2012) A modified method for whole exome resequencing from minimal amounts of starting DNA. PLoS ONE 7: e32617.
15. Dean FB, Hosono S, Fang L, Wu X, Faruqi AF, et al. (2002) Comprehensive human genome amplification using multiple displacement amplification. Proc Natl Acad Sci U S A 99: 5261–5266.
16. Langmead B, Salzberg S (2012) Fast gapped-read alignment with Bowtie 2. Nature Methods 9: 357–359.
17. International Human Genome Sequencing Consortium (2001) Initial sequencing and analysis of the human genome. Nature 409: 860–921.
18. Picard website. Available: http://picard.sourceforge.net. Accessed 2014 June 10.
19. Quinlan AR, Hall IM (2010) BEDTools: a flexible suite of utilities for comparing genomic features. Bioinformatics 26: 841–842.
20. Li H, Handsaker B, Wysoker A, Fennell T, Ruan J, et al. (2009) The Sequence Alignment/Map format and SAMtools. Bioinformatics 25: 2078–2079.
21. Langmore JP (2002) Rubicon Genomics, Inc. Pharmacogenomics 3: 557–560.
22. Barker DL, Hansen MS, Faruqi AF, Giannola D, Irsula OR, et al. (2004) Two methods of whole-genome amplification enable accurate genotyping across a 2320-SNP linkage panel. Genome Res 14: 901–907.
23. Giardina E, Pietrangeli I, Martone C, Zampatti S, Marsala P, et al. (2009) Whole genome amplification and real-time PCR in forensic casework. BMC Genomics 10: 159.
24. Benjamini Y, Speed TP (2012) Summarizing and correcting the GC content bias in high-throughput sequencing. Nucleic Acids Res 40: e72.
25. Kozarewa I, Ning Z, Quail MA, Sanders MJ, Berriman M, et al. (2009) Amplification-free Illumina sequencing-library preparation facilitates improved mapping and assembly of (G+C)-biased genomes. Nat Methods 6: 291–295.

26. Aird D, Ross MG, Chen WS, Danielsson M, Fennell T, et al. (2011) Analyzing and minimizing PCR amplification bias in Illumina sequencing libraries. Genome Biol 12: R18.

27. Oyola SO, Otto TD, Gu Y, Maslen G, Manske M, et al. (2012) Optimizing Illumina next-generation sequencing library preparation for extremely AT-biased genomes. BMC Genomics 13: 1.

8

Hybrid Lentivirus-phiC31-int-NLS Vector Allows Site-Specific Recombination in Murine and Human Cells but Induces DNA Damage

Nicolas Grandchamp[1,2,3], Dorothée Altémir[1,2], Stéphanie Philippe[1,2,3], Suzanna Ursulet[1,2,3], Héloïse Pilet[1,2,3], Marie-Claude Serre[4], Aude Lenain[5], Che Serguera[6], Jacques Mallet[1], Chamsy Sarkis[1,2]*

1 Unit of Biotechnology and Biotherapy, Centre de recherche de l'Institut du Cerveau et de la Moelle Epinière, Pierre-and-Marie-Curie University/Institut National de la Santé et de la Recherche Médicale, Paris, France, 2 NewVectys, Villebon-sur-Yvette, France, 3 Biosource, Paris, France, 4 Laboratoire de Virologie Moléculaire et Structurale, Gif-sur-Yvette, France, 5 Commissariat à l'Energie Atomique, Laboratoire de Radiobiologie et Oncologie, Fontenay-aux-Roses, France, 6 Molecular Imaging Research Center - Modélisation des biothérapies, Fontenay-aux-Roses, France

Abstract

Gene transfer allows transient or permanent genetic modifications of cells for experimental or therapeutic purposes. Gene delivery by HIV-derived lentiviral vector (LV) is highly effective but the risk of insertional mutagenesis is important and the random/uncontrollable integration of the DNA vector can deregulate the cell transcriptional activity. Non Integrative Lentiviral Vectors (NILVs) solve this issue in non-dividing cells, but they do not allow long term expression in dividing cells. In this context, obtaining stable expression while avoiding the problems inherent to unpredictable DNA vector integration requires the ability to control the integration site. One possibility is to use the integrase of phage phiC31 (phiC31-int) which catalyzes efficient site-specific recombination between the *attP* site in the phage genome and the chromosomal *attB* site of its *Streptomyces* host. Previous studies showed that phiC31-int is active in many eukaryotic cells, such as murine or human cells, and directs the integration of a DNA substrate into pseudo *attP* sites (*pattP*) which are homologous to the native *attP* site. In this study, we combined the efficiency of NILV for gene delivery and the specificity of phiC31-int for DNA substrate integration to engineer a hybrid tool for gene transfer with the aim of allowing long term expression in dividing and non-dividing cells preventing genotoxicity. We demonstrated the feasibility to target NILV integration in human and murine *pattP* sites with a dual NILV vectors system: one which delivers phiC31-int, the other which constitute the substrate containing an *attB* site in its DNA sequence. These promising results are however alleviated by the occurrence of significant DNA damages. Further improvements are thus required to prevent chromosomal rearrangements for a therapeutic use of the system. However, its use as a tool for experimental applications such as transgenesis is already applicable.

Editor: Yuntao Wu, George Mason University, United States of America

Funding: This work was supported by grants from European FP6 (INTEGRA NEST-Adventure contract #29025), AFM (Association Française contre les Myopathies), and Rétina France. The funders had no role in study design, data collection and analysis, decision to publish, or preparation of the manuscript. Co-authors NG, DA, SP, SU and HP are employed by NewVectys SAS. NewVectys SAS provided support in the form of salaries for authors NG, DA, SP, SU and HP, but did not have any additional role in the study design, data collection and analysis, decision to publish, or preparation of the manuscript. The specific roles of these authors are articulated in the 'author contributions' section.

Competing Interests: NG, DA, SP, SU and HP are employed by NewVectys SAS. CS and JM own shares of NewVectys SAS. There are no patents, products in development or marketed products to declare.

* Email: chamsy.sarkis@newvectys.com

Background

Gene transfer technologies are essential for genetics studies and gene therapies. However, major challenges remain to be addressed. A major issue is the lack of control over the site of DNA integration in the host genome which leads to unpredictable gene expression level and potentially undesirable mutagenesis of important cellular genes [1]. Recent strategies to tackle this challenge are relying on the use of genome editing tools such as ZFNs [2–7], TALENs [8–13] or more recently CRISPR-Cas system [14–17]. However, the vectorization of these tools into viral vectors to optimize their use *ex vivo* or *in vivo* raises several problems. Indeed, ZFNs function as dimers and generally require cotransduction of three vectors (one for each dimer and one for the recombinating substrate) [18–20]. Moreover ZFNs may induce

cellular toxicity due to off target activity [21–23]. TALENs have an important size with repeat domains hampering their vectorization [24]. CRISPR-Cas is a very recent tool and its vectorization has not yet been described. One may however expect its vectorization into viral vectors will be challenging as the system is based on the concomitant use of a chimeric DNA displaying hairpin structures [25] and of Caspase 9 which induces apoptosis when over-expressed [26]. These features will undoubtedly represent challenges for the vectorization of CRISPR-Cas into viral vectors for targeted integration.

Site-specific recombinases such as Cre [27–34] or FLP [35–38] of the tyrosine recombinases family are other genome editing tools more easily vectorizable and widely used for the purpose of site specific integration. However, the use of these recombinases is limited by the absence of endogenous recognition site in mammalian cells and by the bidirectionality of the recombination reaction they mediate. Within the superfamily of site-specific recombinases, phage integrases catalyse unidirectional recombination events [39]. Among these the PhiC31 phage integrase (phiC31-int), of the large serine recombinases family, is the most commonly used site-specific integrase for gene transfer purposes [40,41]. In its natural context, phiC31-int mediates efficient recombination between the phage attachment site (*attP*) and the bacterial attachment site (*attB*). The recombination of these two sites results in the unidirectional and site-specific integration of the phage genome into the bacterial chromosome (reviewed in [39]) leading to an integrated phage genome flanked by the recombinant *attL* (left) and *attR* (right) sites (Figure 1). When used for gene transfer into eukaryotic cells, phiC31-int can catalyse integration of a plasmid containing an *attB* sequence into endogenous pseudo *attP* sites (p*attP*) displaying a high degree of homology with the wild type *attP* site [42]. Hence, associated with transfection techniques, phiC31-int has been successfully exploited to stably modify the genome into particular genomic sites of many types of eukaryotic cells *in vitro* [43–50] for transgenesis [48,51–55] and gene therapy applications [56–65]. The use of phiC31-int presents several advantages. First, the recombinase can be used to generate conservative recombination between *attB* and pseudo *attP* sites [42]. Second, Chalberg et al demonstrated that the majority of phiC31-int mediated recombination events in the human genome occur in intergenic regions [66]. However, in most of these studies the vectorization of phiC31-int relied on cotransfection (or nucleofection) of plasmids for both the delivery of phiC31-int and of the transgene, thus limiting this technology to *in vitro* or *ex vivo* applications.

A strategy to increase the efficiency of DNA delivery *in vivo* is to use viral-derived vectors. As the expression of the genome editing tool must be transitory to avoid genotoxicity, it could therefore be delivered by a transient viral vector [67]. Even though phiC31-int has already been delivered by an adenoviral vector system [68,69] the use of such vectors is limited by their cell toxicity and immunogenicity [70–72]. In contrast, lentiviral vectors (LV) have the advantage to be non-immunogenic [73,74] and offers the possibility to be pseudotyped by different envelops, allowing a high degree of flexibility regarding the tropism of the particles and the

type of cell(s) they transduce (for review see Cronin J. et al [75]). Most importantly, it was shown that LV integrase can be modified to obtain non integrating lentiviral vectors (NILVs) [76–78], which act as episomal vectors. Hence, NILVs are vectors of choice to deliver genome editing tools and have been successfully used to deliver transposases [79,80], FLP [81] or ZFNs [82,83]. However, the use of NILVs has never been described for the vectorization of a serine recombinase.

In the present study, we combine the unidirectional site-specific recombination capability of phiC31-int with the efficiency of NILVs for gene transfer. For targeted integration, two different NILVs, one delivering the DNA sequence to be integrated and containing the natural *attB* site, the other expressing the phiC31-int are used. Through a step by step approach, we demonstrate for the first time that phiC31-int can be vectorized in NILV and we provide clues to further improve the system. However, analysis of integration events reveals that significant DNA damages can result from phiC31-int mediated recombination. In conclusion, the vectorization of serine recombinases in NILVs is feasible and constitutes a promising tool for basic research; however, one should remain cautious about the chromosomal aberrations that can be induced by these recombinases, particularly for clinical uses.

Results and Discussion

NILV genomes can be used as a substrate for site-specific integration mediated by phiC31-int into human genome

We first assessed the ability of a NILV DNA genome to be used as a substrate for phiC31-int. We therefore generated a Hela cell line constitutively expressing phiC31-int thanks to an integrative lentiviral vector expressing the phiC31-int under the control of the CMV promoter (LV CMV-phiC31-int) to. Hela cells were transduced and clonal populations were isolated and analyzed by RT-PCR to estimate the phiC31-int expression level (Figure 2A). The clone Hi16 was selected for its robust constitutive expression of phiC31-int.

The ability of the constitutively expressed phiC31-int to mediate recombination between a NILV bearing an *attB* site with genomic p*attP* sites was then tested. Hi16 cells were transduced with a non-integrative lentiviral vector expressing the Neomycine (Neo) resistance gene under the control of the CMV promoter and containing an *attB* site (NILV *attB*-CMV-Neo). After two weeks of G418 selection we obtained four cell clones which genomes were analyzed. Theoretically Neo integration is expected to arise from 3 distinct mechanisms: (i) phiC31-int-**specific** integration, (ii) **residual** integration mediated by a residual activity of the mutant HIV integrase (review about this field [84]) or (iii) **illegitimate** integration due to recombination of the episomal DNA molecule with the cellular genome by host cell mechanisms (Figure 3A). If the analysis is realized with clonal populations, these three different mechanisms should be discriminated by performing 2 PCR assays, one amplifying the LTR and the other amplifying the *attB* site. A positive LTR PCR reveals the presence of a LTR-LTR junction, indicating that the integration has occurred through a LTR-independent mechanism, either involving *attB* site-specific recombination or illegitimate recombination. In contrast, a positive *attB* PCR reveals that the *attB* site is intact, indicating an *attB*-independent integration, either involving LTR dependent (residual) integration or illegitimate recombination. In summary, cells are analyzed without knowing whether their genome contains one or several integration of NILV. A positive result for LTR PCR only or for *attB* PCR only allows to identify the mechanism of integration without ambiguities (ie respectively

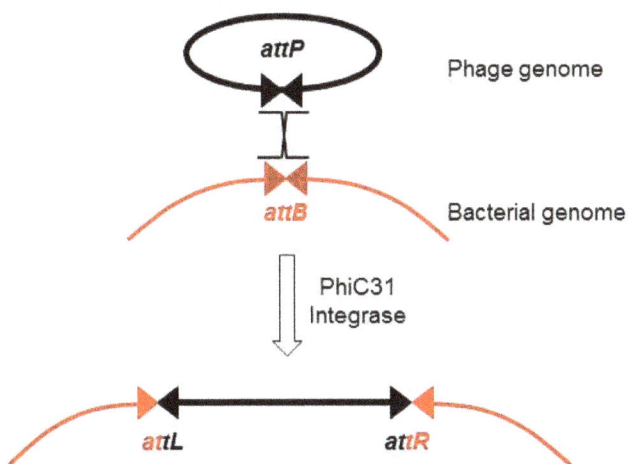

Figure 1. Scheme of phiC31-int mediated recombination in bacterial host. PhiC31 integrase performs precise recombination between an *attB* site located in the *Streptomyces* genome and an *attP* site located on the phiC31 phage genome. The outcome is integration of the phage into the host genome.

A

B

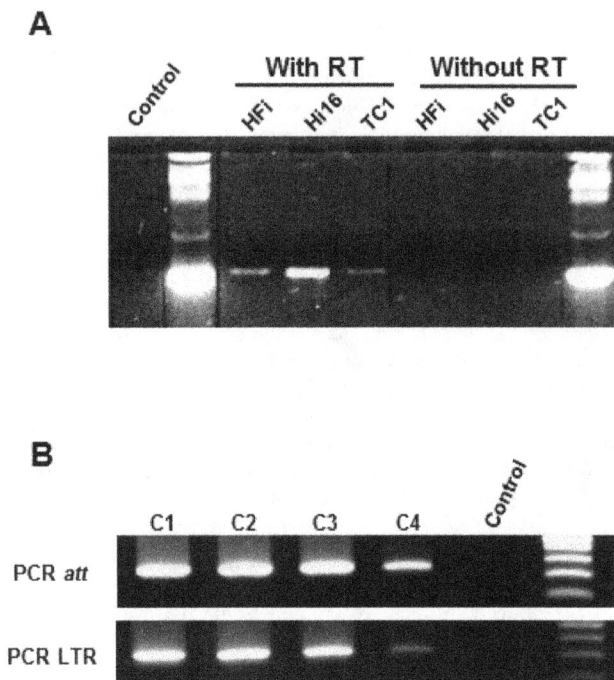

Figure 2. Analysis of cell lines which constitutively expressed phiC31-int. A) PhiC31 RT-PCR on three different cell lines. HFi and Hi16 are derived from Hela cell line and TC1 from NIH-3T3 cell line. Control condition lane lacks RNA. B) PCR which detects LTR junctions or intact *attB* sites after transduction with a NILV *attB*-CMV-Neo.

phic31-int specific integration or residual integration). As the LTR PCR does not amplify the 1-LTR region generated when the linear genome circularized through homologous recombination of both LTR regions, recombination involving 1-LTR circles would always result in LTR− in the PCR test. In cells where a single integration of a 1-LTR circle event arose, attB+/LTR− or attB−/LTR− profiles may be detected, respectively when the recombination is independent or dependent on phiC31-int. In all our experiments, we never observed attB−/LTR− clones. On the contrary, if both PCR are positive integration could result either from illegitimate recombination or from double independent recombination events (ie: phiC31-int site specific recombination and HIV integrase mediated residual integration). We collected the four clones and checked integration pattern of NILV with PCR. The PCR analysis revealed that all clones were positive for both LTR and *attB* PCRs (Figure 2B). Thus this result does not allow to conclude about the nature of the integration events.

To clear up this ambiguity, genomic DNA of isolated clones was further analyzed by inverse PCR (iPCR) to isolate potential specific integration events. *attB* based primers (P1/P1′) were used to amplify junctions of recombination. To avoid the PCR background generated by the non-recombined *attB* site, an enzymatic cocktail including an enzyme which cuts only once in the vector sequence, in 5′ to the *attB* site (Figure 3B) was used. In this way, we were able to isolate pseudo *attR* junctions with iPCR, (but not the pseudo *attL* junction). Using this adapted strategy, a lentiviral genome integrated by phiC31-int at the human locus Xq22.1 was detected (Figure 4). Interestingly this locus had already been described as a preferential p*attP* site in the human genome [66]. Moreover, the core sequence is exactly in the same position, suggesting that the recombination event occurred very precisely at this locus.

Although NILV integration also occurred through other means than phiC31-int recombination, our results clearly demonstrate that a NILV is a suitable substrate for phiC31-int mediated recombination in human cells. We therefore investigated whether phiC31-int could function when vectorized in a NILV.

NILVs allow adequate expression of phiC31-int to mediate recombination into a reporter system containing an *attP* site artificially introduced in human cells

The ability of a NILV to express phiC31-int and mediate recombination between another NILV genome carrying an *attB* site and a wild-type *attP* site artificially introduced in a human cell genome was further tested. First, a clonal Hela cell line containing the wild-type *attP* site inserted in its genome (HDsred line) was generated using an integrative LV (CMV-*attP*-DsRED2). HDsred cells were then transduced with two NILVs, one allowing expression of phiC31-int (NILV CMV-phiC31-int), and the other expressing Neo and containing the *attB* site (NILV *attB*-CMV-Neo). After cotransduction cells were grown with G418 to select integration events. The genomic DNA extracted from Neo resistant clones was analyzed by PCR to detect *attL* recombination junction (Figure 5A). Results showed only background signal generated by non-recombined *attP* site (figure 5B). To prevent this amplification, the genomic DNA was digested by a restriction enzyme that cuts both *attP* and *attR* sites but not *attL* site (Figure 5A) prior to PCR amplification. PCR results obtained after the enzymatic treatment revealed an *attL* junction in the population transduced with the highest dose of phiC31-int vector (Figure 5C). These results have been further confirmed by nested PCR (Figure 5D) and PCR product sequencing. Taken together, these results demonstrate that a NILV can deliver a functional phiC31-int capable to integrate an episomal lentiviral substrate containing an *attB* site into an *attP* site artificially introduced into the human genome.

Although the used PCR strategy does not allow to estimate the targeting efficiency of the *attP* site with the double vector system, the need of a nuclease digestion before PCR to reveal specific integration events suggests that the efficiency of the two NILVs system is low. As this assay involved the two natural recombination sites of PhiC31-int, further reduced efficiency would be expected when targeting endogenous p*attP* sites. Consequently, we next focused on the improvement of the two NILVs system efficiency.

Modification of the phiC31-int sequence to improve the efficiency of the NILV phiC31-int to allow target integration in pseudosites *attP*

It was previously shown that C-terminal addition of a nuclear localization system (NLS) to phiC31-Int improves its efficiency in eukaryotic cells [85]. We therefore tested this improved integrase in NILV vectorization strategy. To compare the two versions of phiC31-int, Hela cells were cotransduced with the following NILVs vectors: CMV-Neo with or without *attB* site and CMV-phiC31-int with or without NLS. Cells were grown with G418 and the resistant clones were quantified. The results from experiments in which CMV-phiC31-int was cotransduced with either CMV-Neo or *attB*-CMV-Neo are presented in Figure 6A. The CMV-Neo and the *attB*-CMV-Neo conditions did not display significant differences, indicating that no significant PhiC31-int recombinase activity occurred. In contrast, when the cells are transduced with CMV-phiC31-int-NLS instead of CMV-phiC31-Int (Figure 6B) the presence of the NLS sequence induced significant differences between the CMV-Neo and the *attB*-CMV-Neo conditions. These

A

B

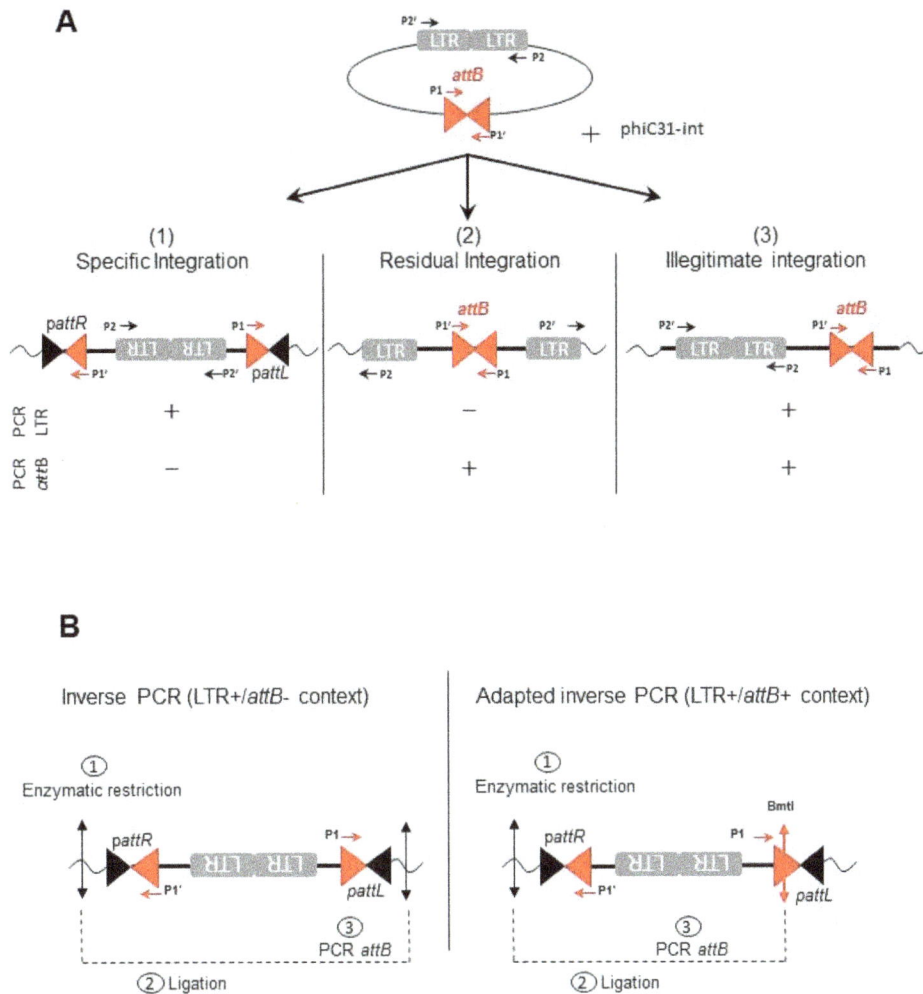

Figure 3. Analysis strategies to detect the specific integrations mediated by phiC31-int. A) Illustration of the three mechanisms of the phiC31-int mediated integration of a NILV containing an *attB* sequence. According to the type of integration, the PCR results in three different profiles: - PCRs LTR+/*attB*− : integration type (1), specific integration. - PCRs LTR−/*attB*+: integration type (2), residual integration. - PCRs LTR+/*attB*+: integration type (3), illegitimate integration. P1/P1′ are the primers used for *attB* PCR and P2/P2′ are the primers used for LTR PCR. B) Schematic representations of the inverse PCR and the adapted inverse PCR strategies used to characterize phiC31-int integration sites.

results show that addition of a C-terminal NLS to phiC31-int significantly increased its recombination efficiency with p*attP* sites. Indeed the phiC31-int-NLS mediated integration was 2 to 2.5 fold above the background level produced by NILV residual integration (Figure 6B).

The use of the phiC31-int-NLS vectorized in a NILV allows to significantly increase the efficiency of recombination. We therefore further tested this hybrid lentivirus phiC31-int-NLS vector to target genomic p*attP* site into murine and human cells.

The two NILVs system allows site specific recombination in murine and human cells but induced aberrant chromosomal rearrangements

To target p*attP* site in the murine and human genome with the two vectors system, Hela cells and NIH-3T3 cells were cotransduced with the NILV *attB*-CMV-Neo and the hybrid lentivirus phiC31-int-NLS vector. After two weeks of selection, Neo resistant cells were collected and several clonal populations were isolated to facilitate interpretation of PCR analyses. The clones were analyzed with the LTR and *attB* PCRs assay (Figure 3A) to

determine the proportion of specific integration events compared to NILV residual integration and illegitimate recombination events. We categorized clones in the 3 following groups: LTR+/ *attB*− clones (group I, ie phiC31-int-NLS integration), LTR−/ *attB*+ clones (group II, ie residual integration) and LTR+/*attB*+ clones (group III, ie illegitimate integration or mixed integration profile). The results are presented in Table 1.

We obtained 108 clones for murine cells and 28 for human cells. 7.5% of murine clones are in group I and 20.3% in group III (Table 1A). Consequently the integration mediated by the hybrid lentivirus phiC31-int-NLS vector is comprised between 7.5% and 27.8%. As no human clone corresponded to group III, the proportion of the hybrid lentivirus phiC31-int-NLS vector specific integration corresponds to the proportion of group I, ie 46.4% (Table 1B).

To determine which integration sites were targeted and confirm the type of integration of clonal groups I and III, we performed an analysis by iPCR as previously described (Figure 3B). The sequencing of iPCR products demonstrates that all human and murine clones tested contain in their genome an integrated vector with a recombinant pattern at the *attB* site. Therefore, the two

A

attB site	TGC CAG GGC GTG CCC **TTG** GGC TCC CCG GGC GCG
attP site	CAA CTG GGG TAA CCT **TTG** AGT TCT CTC AGT TGG
attR site	CAA CTG GGG TAA CCT **TTG** GGC TCC CCG GGC GCG
attL site	TGC CAG GGC GTG CCC **TTG** AGT TCT CTC AGT TGG

B

attB site	TGC CAG GGC GTG CCC **TTG** GGC TCC CCG GGC GCG
Locus Xq22.1	TGC CTC ATT TAA TCT **AT** AGG TTC TCC TTG TTC
Pseudo attR site	TGC CTC ATT TAA TCT **TG** GGC TCC CCG GGC GCG

Figure 4. DNA sequence of *att* and p*attP* sites. A) Wild type *attP* and *attB* sites. After recombination two hybrids sites are formed: *attL* and *attR*. B) Recombination between *attB* site and the human locus Xq22.1 This recombination generates a p*attR* which has been isolated by inverse PCR. Xq22.1 had been described previously as a human p*attP* by MP Calos et al., who isolated the same p*attR*.

vectors system that we developed allows targeting p*attP* sites in the murine and human genomes (Table 2).

Twelve murine integration sites were isolated, three from group I with the two junctions (p*attL* and p*attR*) and nine from group III with only the p*attR* junction (Table 2). For two of the three group I clones both flanking regions were sequenced but the isolated junctions were too short to allow identification of the integration locus. Surprisingly, these two clones of group I present abnormal p*attL* and p*attR* junctions where several bases were missing, probably due to a deletion mechanism. Similarly, the sequence analysis of group III clones shows that recombination between the *attB* site and p*attP* site is not as precise as expected. Indeed, 6 out of 9 clones have missing bases in the p*attR* junctions. However, because we isolated only one flanking region for clones of group III, we cannot conclude about deletion events, as the missing bases could result from a gap of the *attB* core region involved in recombination.

Five human integration sites were isolated, all from group I. Both junctions were isolated for three sites and only the p*attR* junction for two sites (Table 2). As for the murine integration site analysis, missing bases in the recombined pseudo-sites were detected, including in clones for which both L and R junction could be determined. This further confirms that missing bases indeed reveal a deletion mechanism, probably occurring during the phiC31-int mediated recombination between a natural *attB* site and a p*attP* site. Furthermore, the integration site of the three sites for which the two junctions were isolated could be localized exactly (Table 2). Interestingly, for these three sites chromosomal gaps were observed between the p*attR* site and the p*attL* site. The size gaps are 13 bp, 795 pb and 4.8 kbp. The two first gaps could result from a mechanism of deletion which could occur through NHEJ pathway. Nevertheless, these hypotheses cannot explain the

gap of 4.8 kbp. Indeed, in this case, the two flanking regions isolated during iPCR are in the same chromosomal orientation, which is not the case when normal recombination occurs. We hypothesize that the observed aberrant recombination results from two successive recombination events involving two p*attP* sites located 4.8 kb from each other that led to an inversion of the 4.8 kbp sequence (figure 7). This mechanistic model was previously proposed to explain chromosomal rearrangements in mammalian cells resulting from aberrant recombinations mediated by phiC31-int [86].

In conclusion the hybrid lentivirus-phiC31-int-NLS vector that we developed allows targeted integration in p*attP* sites in murine and human genomes but seem to induce frequent deletions of base pairs in the *attB* site present into the vector or into the endogenous p*attP* site. Moreover, it may induce chromosomal deletions and translocations.

Conclusions

Our work establishes the ability of the hybrid lentivirus-phiC31-int-NLS to integrate a NILV substrate into a*ttP* or p*attP* sites in murine and human cell lines. Although the number of integration sites isolated in this study does not allow to determine preferential genome recognition sites for integration mediated by phiC31-int vectorized in NILV, a human p*attP* site already described could be isolated [66]. Most importantly, we demonstrated that the use of hybrid lentiviral phiC31-int-NLS vector can induce DNA damages, probably due to the activity of the recombinase. Indeed, other reports already described similar results using non-viral transfection of PhiC31-int. Anja Ehrhardt et al have shown that 15% of transgenes integrated by phiC31-int were flanked by chromosomal DNA sequence from different chromosomes [86]

A

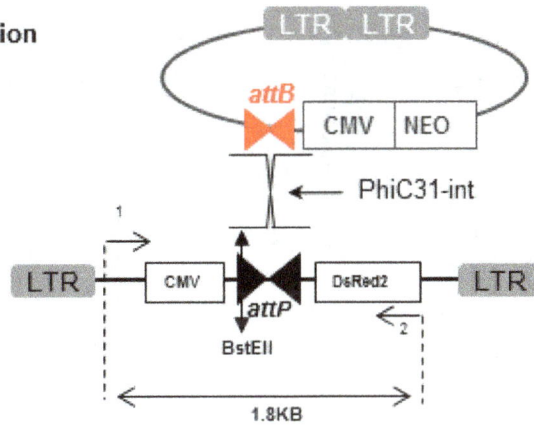

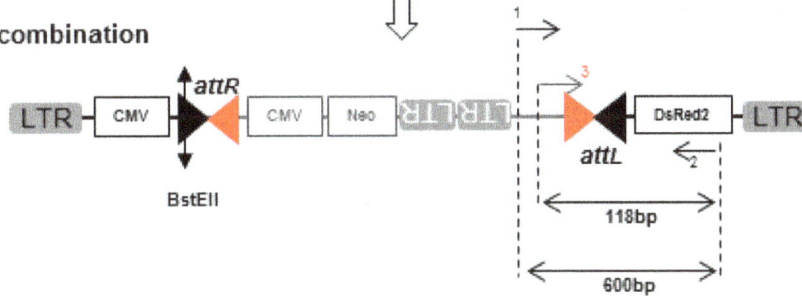

B

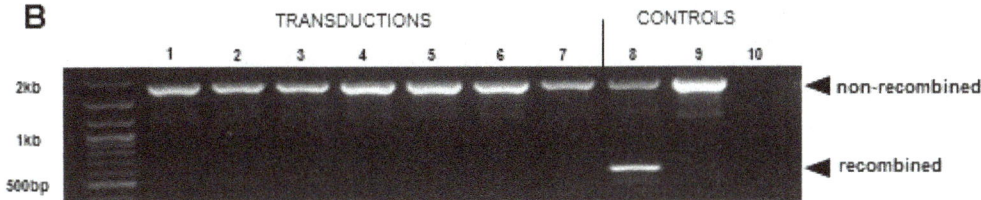

C

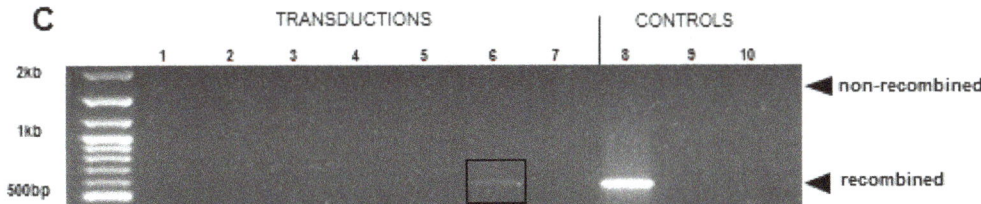

D

Figure 5. Detection of recombination mediated by phiC31-int between an *attB* site contained into a NILV and a genomic *attP* site. A) Scheme of the DsRed2 PCR before and after the enzymatic restriction treatment. B) PCR DsRed2 results without restriction enzyme treatment. Lanes 1 to 3: cotransduction with CMV-Neo and CMV-PhiC31 increasing vector input of 50–150–300 ng of p24. Lanes 4 to 6: cotransduction with *attB*-CMV-Neo and CMV-PhiC31 increasing vector input of 50–150–300 ng of p24. Lane 7: *attB*-CMV-Neo. Lane 8: positive control generated by triple-transfection (CMV-phiC31-int, *attB*-CMV-Neo and CMV-*attP*-DsRed2). Lane 9: negative control without vector. Lane 10: negative control of PCR. C) PCR DsRed2 results after restriction enzyme treatment. Lanes are similar to figure B. D) Nested PCR from the product isolated from lane 6 to confirm the specificity of PCR DsRed2 amplification.

A

B

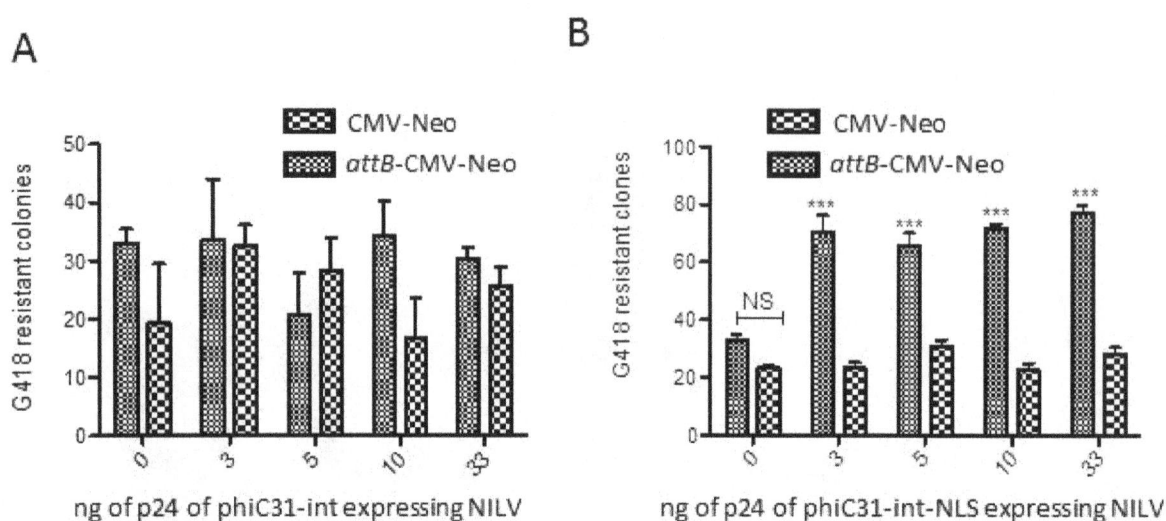

Figure 6. Effect of NLS sequence on phiC31-int activity in NILV context. A) Cotransduction of NILVs CMV-PhiC31 and CMV-Neo or *attB*-CMV-Neo. Four p24 doses of PhiC31 vector were used (D1: 3 ng, D2: 5 ng, D3: 10 ng, D4: 33 ng). B) Cotransduction of NILVs CMV-PhiC31-NLS and CMV-Neo or *attB*-CMV-Neo. Four p24 doses of PhiC31 vector were used (D1: 3 ng, D2: 5 ng, D3: 10 ng, D4: 33 ng). No significant differences are observed between sample with or without *attB* sequence in the vector pTRIP-CMV-Neo. Satistics: two ways ANOVA with Bonferroni posttest (Prism 5).

and Ji Liu et al have shown that phiC31-int induces DNA damages and chromosomal rearrangements in primary and adult human fibroblasts [87,88]. In addition, Ehrard et al. have shown that phiC31-int is competent to integrate linear DNA fragments. They did not characterize the mechanisms and the consequences of this type of event are unknown, but one may hypothesize that such event would induce chromosome break. Considering that the cycle of NILVs involves linear intermediates, it would be of particular interest to further investigate the ability of phiC31-int to integrate these linear forms and the consequences of such events. Taken together, these observations limit the use of the hybrid lentiviral phiC31-int-NLS vector for clinical applications, it remains useful in transgenesis contexts where non aberrant recombination events can be selected. Alternatively, the hybrid lentiviral system may be used to vectorize other recombinases or genome editing tools with higher safety features. Indeed other serine recombinases have been shown to function in human cells [89–91] and could be valuable candidates for vectorization in NILVs. Moreover, the modification of recombinases and other genome editing tools by directed evolution techniques [92,93] could be used to render them hyper-specific and hyper-efficient in order to improve their safety features. For instance, directed evolution has proven efficient to modify the efficiency and/or specify of a variety of molecular tools,

including Sleeping Beauty [94] Cre recombinase [95], FLP recombinase [96], ZFNs [97] or phiC31-int [98].

Methods

Plasmids

The encapsidation plasmid expressing a functional integrase (p8.91 IN$_{WT}$) has been described previously [77]. The encapsidation plasmid expressing a deficient integrase (p8.91 IN$_{D64V}$) was derived from the plasmid p8.91 IN$_{WT}$ and the plasmid pCMVΔR(int-)8.2 previously described [74] and kindly provided by D.B. Kohn (UCLA, Los Angeles (CA), USA). This plasmid contains a point mutation in the coding region of the integrase catalytic domain, creating a D64V change in the amino acid sequence. The plasmid p8.91 IN$_{D64V}$ was generated by replacing the BclI-AflII fragment of p8.91 IN$_{WT}$ by the corresponding fragment (containing the substitution) from pCMVΔR(int-)8.2.

The envelope expression plasmid pMDG(VSV) was used to express the VSV-G from the human CMV immediate early promoter [74].

The vector plasmid pTrip-CMV-phic31-int-WPRE was derived from the plasmid pTrip-CMV-GFP-WPRE previously described [77] and the plasmid pCMV-phic31-int previously described [41] and kindly provided by M.P. Calos (Stanford University, Stanford

Table 1. Repartition in 3 groups of human and murine clones according to the *attB* and LTR PCR results.

		Group I (LTR+) *Specific integration*	Group II (attB+) *Residual integration*	Group III (LTR+/attB+) *Unknow integration*
Mouse				
	Clone number	8	78	22
	%	7,5%	72,2%	20,3%
Human				
	Clone number	13	15	0
	%	46,4%	53,6%	0,0%

Table 2. Mapping and description of *pattP* sites isolated by iPCR on human and murine cell lines.

Chromosome	Number of junction isolated	Deletions: The half of att site present into the vector	Deletions: Chromosome	Genomic location: Pseudo site	Genomic location: Context	If intronic: gene name	If intergenic, flanking gene names: 5'side gene	Distance (kb)	3'side gene	Distance (kb)
Mouse										
Groupe I										
1	2	*pattR*: 2 bp / *pattL*: 42 bp	ND	1E4	repeat sequence					
2	2	*pattR*: No / *pattL*: 13 bp	ND	2H3-H4	repeat sequence					
7	2 (only *pattR* exploitable)	*pattR*: 4 bp	ND	7F2 108099955	Exonic	MOR204-16				
Groupe III										
1	1	*pattR*: 8 bp	ND	1H1 159228123	Intergenic		PAPP-A2	277	AI316802	49
5	1	*pattR*: 16 bp	ND	5B1 29812070	Exonic	4632413E21Rik				
5	1	*pattR* : 21 bp	ND	5C31 58057809	Intronic	Pcdh7				
6	1	*pattR*: No	ND	6G1 135251898	Intergenic		Gsg1	7	1700023A18Rik	51
7	1	*pattR*: 4 bp	ND	7A1 7104238	Intergenic		AIE1	90	5730403M16Rik	10
9	1	*pattR*: 12 bp	ND	9A1 3001529	Intronic	AC131780.5-201				
13	1	*pattR*: No	ND	13D1 102691720	Intergenic		AA414921	984	F630107B15	1.5
17	1	*pattR*: No	ND	17A3.3 23698186	Intergenic		EG622645	305	4.1B	53
X	1	*pattR*: 10 bp	ND	XA7.3 75331158	Intronic	Cf-8				
Human										
Groupe I										
1	2	*pattR*: No / *pattL*: 20 bp	13 bp	1q32.1 205953621/634	Intergenic		SLC26A9	15	RAB7	28
9	1	*pattR*: 9 bp / *pattL*: ND	ND	9p21.1 33615936	Intergenic		bA255A11.3	43	TCRBV20S2	1.5
17	2	*pattR*: No / *pattL*: 12 bp	Inversion of 4,8 kb	17q11.2 26164583/59669	Intronic	HSD24				
17	1	*pattR*: No / *pattR*: ND	ND	17q21.32 45406611	Intronic	C17orf57				
17	2	*pattR*: 10 bp / *pattL*: 34 bp	795 pb	17q25.3 82068091/886	Intergenic		DUS1L	2559	FASN	11080

Step 1 : PhiC31-int mediated NILV integration between *attB* site and p*att1* site

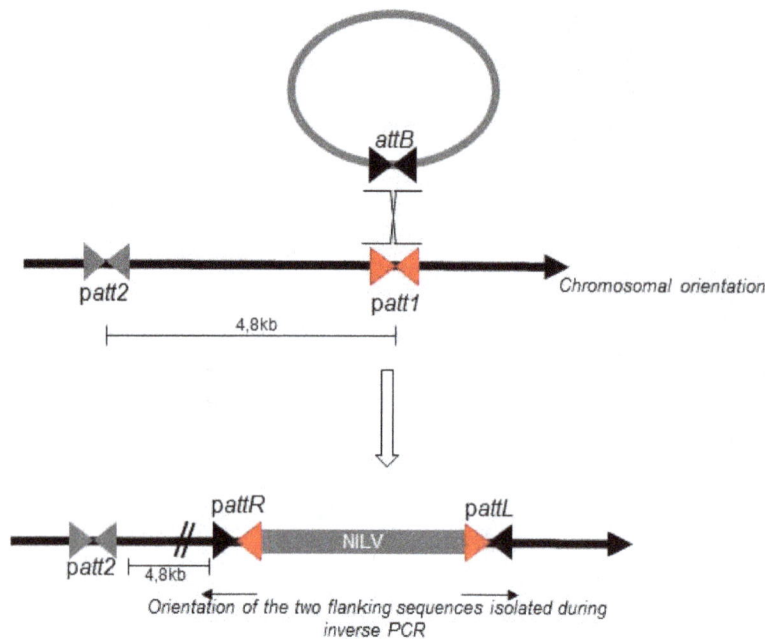

Step 2 : Intramolecular recombination mediated by phiC31-int between the NILV p*attL* site and p*att2* site

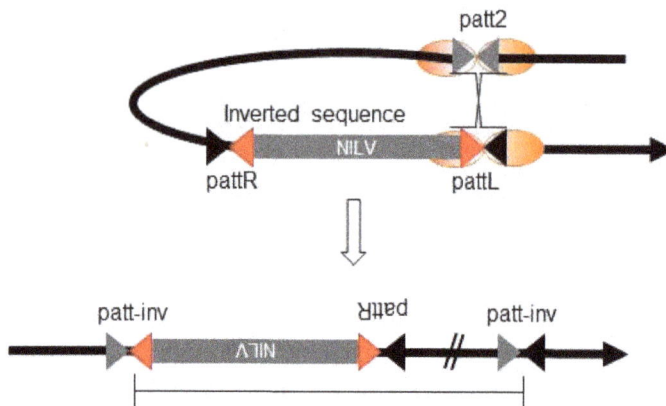

Figure 7. Hypothetical model to explain the inversion of 4.8 Kb detected by iPCR. Step 1: Integration of a NILV mediated by phiC31-int into a p*attP* site. Step2: Recombination mediated by phiC31-int between the p*attL* generated during step 1 and another p*att* site located at 4 kb.

(CA), USA). The plasmid pTrip-CMV-phic31-int-WPRE was generated by replacing the BamHI-SnabI fragment (CMV-GFP) of the pTrip-CMV-GFP-WPRE by the CMV-phic31-int fragment from the plasmid pCMV-phic31-int. The vector plasmid pTrip-CMV-phic31-int-NLS-WPRE was derived from the plasmid pTrip-CMV-phic31-int-WPRE and the plasmid pCMV-phic31-int. The C-terminal region of phic31-int was amplified with a primer containing the SV40 NLS sequence (5'-CCCGTTGGCAGGA-AGCACTTCCGG-3'/5'- ATTCGCGGATCCGCTAAACCT-TCCTCTTCTTCTTAGGCGCCGCTACGTCTTCCGTGCC-GTCC-3') from the plasmid pCMV-phic31-int. This PCR product was subcloned into the plasmid pCMV-phic31-int in place of the C-terminal region of phic31-int by using Eco47III and BamHI

restriction enzymes to generate the plasmid pCMV-phic31-int-NLS. The plasmid pTrip-CMV-phic31-int-NLS-WPRE was finally generated by replacing the SpeI-BamHI fragment (GFP) of the pTrip-CMV-phic31-int-WPRE by the corresponding fragment (phic31-int-NLS) from pCMV-phic31-int-NLS. The vector plasmid pTrip-*attB*-CMV-Neo-WPRE was derived from the plasmid pTrip-CMV-Neo-WPRE previously described [77] and the plasmid p*attB* previously described [41] and kindly provided by MP Calos. The plasmid pTrip-a*ttB*-CMV-Neo-WPRE was generated by inserting the SalI fragment of the p*attB* plasmid (*attB* sequence) into the linearized plasmid pTrip-CMV-Neo-WPRE.

The plasmid pTrip-CMV-a*ttP*-DsRed2 was derived from the plasmid pTrip-CMV-DsRed2 kindly provided by P. Ravassard.

Hybridization of 2 single strands DNA fragments (5'- CCCCAA-CTGGGGTAACCTTTGAGTTCTCTCAGTTGGGGG-3'/5'-C-CCCCAACTGAGAGAACTCAAAGGTTACCCCAGTTGGGG-3') generated a double stranded DNA fragment corresponding to the *attP* sequence flanked by BamHI cohesive ends. This fragment was inserted into the linearized plasmid pTrip-CMV-*attP*-DsRed2 to generate the plasmid pTrip-CMV-*attP*-DsRed2.

Lentiviral production

Lentiviral vectors were generated by the transient transfection of 293T cells by using the calcium phosphate precipitation method previously described [77]. Briefly, cells were cotransfected with the vector plasmid (pTrip-CMV-phic31-int-WPRE, pTrip-CMV-phic31-int-NLS-WPRE, pTrip-*attB*-CMV-Neo-WPRE or pTrip-CMV-*attP*-DSred2-WPRE), the transcomplementation plasmid (p8.91 IN_{WT} for integrative vectors or p8.91 IN_{64} for non-integrative vectors), and the plasmid encoding the vesicular stomatitis virus envelope glycoprotein (pMD-G). Supernatant was collected 48 hours after transfection, treated with DNaseI (Roche) and filtered (0.45 μm). Viral particles were then concentrated by ultracentrifugation (90 min, 22,000 rpm, rotor SW28) and resuspended in 0.1M PBS. The HIV p24 Gag antigen was quantified for each stock by ELISA (HIV-1 P24 antigen assay; Beckman Coulter, Fullerton, CA) according to manufacturer's instructions.

Cell culture

Human epithelial HeLa, Hi16, HeLa-DsRED2 and 293T cells and murine NIH 3T3 cells were grown in Dulbecco's modified medium (Invitrogen) supplemented with antibiotics (100 U/mL penicillin and 100 mg/mL streptomycin) and 10% heat inactivated fetal calf serum (Eurobio). The cells were plated and cultured in a humidified incubator at 37°C in a 5% $CO2$ and 90% air atmosphere.

Genomic DNA extraction

Genomic DNA extractions were performed with a lysis buffer composed of TrisHCl 10 mM (pH 7.5), EDTA 10 mM, SDS 0.6%, RNase A (Qiagen) 100 μg/mL and proteinase K (Eurobio) 100 μg/mL. The lysates were purified by phenol/chloroform and precipitated using sodium acetate and ethanol.

Generation of a reporter cell line HeLa-DsRED2

The HeLa-DsRED2 cell line containing an *attP* site and expressing DsRED2 fluorescent protein was generated using the lentiviral vector CMV-*attP*-DsRED2. HeLa cells were transduced with LV CMV-*attP*-DsRED2 (unconcentrated supernatant). Cells were grown for 3 days and seeded at clonal density in 96-well plates (0.3 cell per well) to generate single cell derived colonies. Clones were analyzed for DsRED2 expression by flow cytometry and PCR amplification of the vector genome.

Transduction

Cell suspensions were incubated for 3 hours with required vectors in medium supplemented with 1 μM DEAE-dextran. After 3 hours of incubation, cells were seed at desired density in fresh medium and grown for the purpose of the experiment.

Hi16 cells (2.10^6 cells/mL) were transduced with NILV-phic31-int (300 ng of p24) and NILV-Neo (100 ng of p24). Cells were seeded in 10 cm plates and grown in medium supplemented with 1 mg/mL of G418 for 12 days. Cells were then seeded in 96 well plates at low density (0.3 cell per well) to generate single cell derived colonies.

HeLa DsRED2 cells (2.10^6 cells/mL) were transduced with NILV CMV-phic31-int (50, 150, 300 ng of p24) and NILV CMV-Neo (100 ng of p24). Cells were seeded in 10 cm plates and grown in medium supplemented with 1 mg/mL of G418 (renewed every 3 days) for 12 days before extraction and analysis of genomic DNA.

Hela and NIH-3T3 cells (5.10^5 cells/mL) were transduced with NILV CMV-phic31-int (300 ng of p24) and NILV CMV-Neo (100 ng of p24). Cells were seeded in 10 cm plates and grown in medium supplemented with 1 mg/mL of G418 for 12 days. Cells were then seeded in 96 well plates at low density (0.3 cell per well) to generate single cell derived colonies.

Evaluation of Recombination Frequency

HeLa were directly transduced in suspension (8.10^4 cells/mL) with NILV CMV-phic31-int (3, 6, 18 and 36 ng of p24) and NILV CMV-Neo (12 ng of p24) during 3 hours in 150 μL of medium supplemented with 1 μM DEAE-dextran. Cells were then seeded in 6-wells plates with 2 mL of fresh medium. The day after, medium was removed and replaced with fresh medium supplemented with 1 mg/mL of G418. The medium was replaced every 3 days. Cells were grown 12 days, until clones developed and were then fixed with PFA 4% and stained with trypan blue. Clones on each well were counted. We transduced cells in three replicate tubes for each condition, and results are expressed as the mean of three measurements.

PCR reactions

RT-PCR. To analyze phic31-int expression by RT-PCR, total RNAs were extracted from HeLa cells using the RNeasy minikit (Qiagen), according to the manufacturer's instructions. Then, RNAs were reverse transcribed using the Superscript First Strand Synthesis kit (Invitrogen), according to the manufacturer's instructions. Phic31-int cDNA was amplified using the primers 5'-GCGAAGATTCTCGACACG-3' and 5'-TCGCAGTACAGC-TTGTCC-3' at the concentration of 10 μM.

PCR on genomic DNA. PCR performed on genomic DNA used 500 ng of DNA, 1.5 mM of MgCl2 and 10 μM of each primer. The primers were as follows: amplification of *attB* region: 5'-CAATTTGCTGAGGGCTATTGAG-3' and 5'- CTGTCC-CTGTAATAAACCCG-3'; amplification of the LTRs region: 5'-CTCAATAAAGCTTGCCTTGAGTGC-3' and 5'-TCAGAT-CTGGTCTAACCAGAGAGACC-3'; amplification of the Ds Red2 region: 5'-AGGCCAGACAATTATTGTCTGG-3' and 5'-ATGGTCTTCTTCTGCATCACG-3'; amplification of DsR ED2 (nested primers) 5'- AAGAATCCTGGCTGTGGAAAG-3' and 5'-AACTCGGTGATGACGTTCTCG-3'

Inverse PCR. Genomic DNA (10 μg) was primarily submitted to enzymatic restriction. The enzymatic cocktail used was: XbaI (100 U), StuI (100 U) BsrGI (30 U) and BstXI (30 U). After 16 hours incubation, the enzymes were heat inactived at 65°C for 30 minutes. Cohesive ends were filled in with Klenow (15 U) and dNTPs (2 mM) at 25°C for 20 minutes. Klenow was inactived with 1 mM EDTA. The products of digestion were purified by phenol/chloroform and precipitated using sodium acetate and ethanol. The ligation of the digestion products was performed by ligase (1.000 U, NEB) within ligation buffer supplemented with ATP (1 mM). The products of ligation were purified by phenol/chloroform and precipitated using sodium acetate and ethanol. PCR was performed with 100 ng of DNA with primers allowing the amplification of the *attB* region.

Adapted inverse PCR was performed using the same protocol with the addition of BmtI (30 U) in the enzymatic restriction cocktail.

The products of inverse PCR were visualized on 0.8% agarose gel with ethidium bromide staining. These products were extracted and purified using the Wizard SV Gel and PCR Clean-up System (Promega) according to the manufacturer's instructions, then cloned in a plasmid using pGEM-T Easy Vector System I according to the manufacturer's instructions. Next, inverse PCR products were sequenced using T7 and/or Sp6 primers.

Sequence analysis

All sequencing was performed by Eurofins genomics. Sequences were aligned with vector and genomic DNA, and recombination junctions were identified by sequence matching to *attB*. Human and murine blasts were performed using the *NCBI* and *ensembl genome* databases. The chromosomal localization of pseudosites attP has been performed using the Genebank GRCm38.p2

C57BL/6J assembly in mouse and the Primary Assembly GRCh38 for Human.

Acknowledgments

We thank Professor Michele P. Calos for providing us the pCMV-phiC31 (pCMVint). We also would like to thank Dr. Marie-José Lecomte for critical reading of the manuscript, Delphine Muller and Aurore Berthier for the technical support.

Author Contributions

Conceived and designed the experiments: NG C. Sarkis. Performed the experiments: NG DA SU HP AL. Analyzed the data: NG C. Sarkis MCS C. Serguera SP. Contributed reagents/materials/analysis tools: NG AL JM. Wrote the paper: NG SP C. Sarkis C. Serguera.

References

1. Hacein-Bey-Abina S, Von Kalle C, Schmidt M, McCormack MP, Wulffraat N, et al. (2003) LMO2-associated clonal T cell proliferation in two patients after gene therapy for SCID-X1. Science 302: 415–419. doi:10.1126/science.1088547

2. Urnov FD, Miller JC, Lee Y-L, Beausejour CM, Rock JM, et al. (2005) Highly efficient endogenous human gene correction using designed zinc-finger nucleases. Nature 435: 646–651. doi:10.1038/nature03556

3. Geurts AM, Cost GJ, Freyvert Y, Zeitler B, Miller JC, et al. (2009) Knockout rats via embryo microinjection of zinc-finger nucleases. Science 325: 433. doi:10.1126/science.1172447

4. Morton J, Davis MW, Jorgensen EM, Carroll D (2006) Induction and repair of zinc-finger nuclease-targeted double-strand breaks in Caenorhabditis elegans somatic cells. Proc Natl Acad Sci U S A 103: 16370–16375. doi:10.1073/pnas.0605633103

5. Bozas A, Beumer KJ, Trautman JK, Carroll D (2009) Genetic analysis of zinc-finger nuclease-induced gene targeting in Drosophila. Genetics 182: 641–651. doi:10.1534/genetics.109.101329

6. Hockemeyer D, Soldner F, Beard C, Gao Q, Mitalipova M, et al. (2009) Efficient targeting of expressed and silent genes in human ESCs and iPSCs using zinc-finger nucleases. Nat Biotechnol 27: 851–857. doi:10.1038/nbt.1562

7. Gaj T, Guo J, Kato Y, Sirk SJ, Barbas CF 3rd (2012) Targeted gene knockout by direct delivery of zinc-finger nuclease proteins. Nat Methods 9: 805–807. doi:10.1038/nmeth.2030

8. Boch J, Scholze H, Schornack S, Landgraf A, Hahn S, et al. (2009) Breaking the code of DNA binding specificity of TAL-type III effectors. Science 326: 1509–1512. doi:10.1126/science.1178811

9. Moscou MJ, Bogdanove AJ (2009) A simple cipher governs DNA recognition by TAL effectors. Science 326: 1501. doi:10.1126/science.1178817

10. Tesson L, Usal C, Ménoret S, Leung E, Niles BJ, et al. (2011) Knockout rats generated by embryo microinjection of TALENs. Nat Biotechnol 29: 695–696. doi:10.1038/nbt.1940

11. Carlson DF, Tan W, Lillico SG, Stverakova D, Proudfoot C, et al. (2012) Efficient TALEN-mediated gene knockout in livestock. Proc Natl Acad Sci U S A 109: 17382–17387. doi:10.1073/pnas.1211446109

12. Sander JD, Cade L, Khayter C, Reyon D, Peterson RT, et al. (2011) Targeted gene disruption in somatic zebrafish cells using engineered TALENs. Nat Biotechnol 29: 697–698. doi:10.1038/nbt.1934

13. Hockemeyer D, Wang H, Kiani S, Lai CS, Gao Q, et al. (2011) Genetic engineering of human pluripotent cells using TALE nucleases. Nat Biotechnol 29: 731–734. doi:10.1038/nbt.1927

14. Cho SW, Kim S, Kim JM, Kim J-S (2013) Targeted genome engineering in human cells with the Cas9 RNA-guided endonuclease. Nat Biotechnol 31: 230–232. doi:10.1038/nbt.2507

15. Mali P, Yang L, Esvelt KM, Aach J, Guell M, et al. (2013) RNA-guided human genome engineering via Cas9. Science 339: 823–826. doi:10.1126/science.1232033

16. Friedland AE, Tzur YB, Esvelt KM, Colaiácovo MP, Church GM, et al. (2013) Heritable genome editing in C. elegans via a CRISPR-Cas9 system. Nat Methods. doi:10.1038/nmeth.2532

17. Hwang WY, Fu Y, Reyon D, Maeder ML, Tsai SQ, et al. (2013) Efficient genome editing in zebrafish using a CRISPR-Cas system. Nat Biotechnol 31: 227–229. doi:10.1038/nbt.2501

18. Bitinaite J, Wah DA, Aggarwal AK, Schildkraut I (1998) FokI dimerization is required for DNA cleavage. Proc Natl Acad Sci U S A 95: 10570–10575.

19. Wah DA, Bitinaite J, Schildkraut I, Aggarwal AK (1998) Structure of FokI has implications for DNA cleavage. Proc Natl Acad Sci U S A 95: 10564–10569.

20. Smith J, Bibikova M, Whitby FG, Reddy AR, Chandrasegaran S, et al. (2000) Requirements for double-strand cleavage by chimeric restriction enzymes with zinc finger DNA-recognition domains. Nucleic Acids Res 28: 3361–3369.

21. Pruett-Miller SM, Reading DW, Porter SN, Porteus MH (2009) Attenuation of zinc finger nuclease toxicity by small-molecule regulation of protein levels. PLoS Genet 5: e1000376. doi:10.1371/journal.pgen.1000376

22. Cornu TI, Cathomen T (2010) Quantification of zinc finger nuclease-associated toxicity. Methods Mol Biol Clifton NJ 649: 237–245. doi:10.1007/978-1-60761-753-2_14

23. Ramalingam S, Kandavelou K, Rajenderan R, Chandrasegaran S (2011) Creating designed zinc-finger nucleases with minimal cytotoxicity. J Mol Biol 405: 630–641. doi:10.1016/j.jmb.2010.10.043

24. Holkers M, Maggio I, Liu J, Janssen JM, Miselli F, et al. (2013) Differential integrity of TALE nuclease genes following adenoviral and lentiviral vector gene transfer into human cells. Nucleic Acids Res 41: e63. doi:10.1093/nar/gks1446

25. Jore MM, Lundgren M, van Duijn E, Bultema JB, Westra ER, et al. (2011) Structural basis for CRISPR RNA-guided DNA recognition by Cascade. Nat Struct Mol Biol 18: 529–536. doi:10.1038/nsmb.2019

26. Druskovic M, Suput D, Milisav I (2006) Overexpression of caspase-9 triggers its activation and apoptosis in vitro. Croat Med J 47: 832–840.

27. Orban PC, Chui D, Marth JD (1992) Tissue- and site-specific DNA recombination in transgenic mice. Proc Natl Acad Sci U S A 89: 6861–6865.

28. Smith AJ, De Sousa MA, Kwabi-Addo B, Heppell-Parton A, Impey H, et al. (1995) A site-directed chromosomal translocation induced in embryonic stem cells by Cre-loxP recombination. Nat Genet 9: 376–385. doi:10.1038/ng0495-376

29. Tsien JZ, Chen DF, Gerber D, Tom C, Mercer EH, et al. (1996) Subregion- and cell type-restricted gene knockout in mouse brain. Cell 87: 1317–1326.

30. Rohlmann A, Gotthardt M, Willnow TE, Hammer RE, Herz J (1996) Sustained somatic gene inactivation by viral transfer of Cre recombinase. Nat Biotechnol 14: 1562–1565. doi:10.1038/nbt1196-1562

31. Pfeifer A, Brandon EP, Kootstra N, Gage FH, Verma IM (2001) Delivery of the Cre recombinase by a self-deleting lentiviral vector: efficient gene targeting in vivo. Proc Natl Acad Sci U S A 98: 11450–11455. doi:10.1073/pnas.201415498

32. Marumoto T, Tashiro A, Friedmann-Morvinski D, Scadeng M, Soda Y, et al. (2009) Development of a novel mouse glioma model using lentiviral vectors. Nat Med 15: 110–116. doi:10.1038/nm.1863

33. DuPage M, Dooley AL, Jacks T (2009) Conditional mouse lung cancer models using adenoviral or lentiviral delivery of Cre recombinase. Nat Protoc 4: 1064–1072. doi:10.1038/nprot.2009.95

34. Anton M, Graham FL (1995) Site-specific recombination mediated by an adenovirus vector expressing the Cre recombinase protein: a molecular switch for control of gene expression. J Virol 69: 4600–4606.

35. O'Gorman S, Fox DT, Wahl GM (1991) Recombinase-mediated gene activation and site-specific integration in mammalian cells. Science 251: 1351–1355.

36. Moldt B, Staunstrup NH, Jakobsen M, Yáñez-Muñoz RJ, Mikkelsen JG (2008) Genomic insertion of lentiviral DNA circles directed by the yeast Flp recombinase. BMC Biotechnol 8: 60. doi:10.1186/1472-6750-8-60

37. Nakano M, Odaka K, Ishimura M, Kondo S, Tachikawa N, et al. (2001) Efficient gene activation in cultured mammalian cells mediated by FLP recombinase-expressing recombinant adenovirus. Nucleic Acids Res 29: E40.

38. Kondo S, Takata Y, Nakano M, Saito I, Kanegae Y (2009) Activities of various FLP recombinases expressed by adenovirus vectors in mammalian cells. J Mol Biol 390: 221–230. doi:10.1016/j.jmb.2009.04.057

39. Smith MCM, Brown WRA, McEwan AR, Rowley PA (2010) Site-specific recombination by phiC31 integrase and other large serine recombinases. Biochem Soc Trans 38: 388–394. doi:10.1042/BST0380388

40. Thorpe HM, Smith MC (1998) In vitro site-specific integration of bacteriophage DNA catalyzed by a recombinase of the resolvase/invertase family. Proc Natl Acad Sci U S A 95: 5505–5510.

41. Groth AC, Olivares EC, Thyagarajan B, Calos MP (2000) A phage integrase directs efficient site-specific integration in human cells. Proc Natl Acad Sci U S A 97: 5995–6000. doi:10.1073/pnas.090527097

42. Thyagarajan B, Olivares EC, Hollis RP, Ginsburg DS, Calos MP (2001) Site-specific genomic integration in mammalian cells mediated by phage phiC31 integrase. Mol Cell Biol 21: 3926–3934. doi:10.1128/MCB.21.12.3926-3934.2001

43. Belteki G, Gertsenstein M, Ow DW, Nagy A (2003) Site-specific cassette exchange and germline transmission with mouse ES cells expressing phiC31 integrase. Nat Biotechnol 21: 321–324. doi:10.1038/nbt787

44. Nakayama G, Kawaguchi Y, Koga K, Kusakabe T (2006) Site-specific gene integration in cultured silkworm cells mediated by phiC31 integrase. Mol Genet Genomics MGG 275: 1–8. doi:10.1007/s00438-005-0026-3

45. Ishikawa Y, Tanaka N, Murakami K, Uchiyama T, Kumaki S, et al. (2006) Phage phiC31 integrase-mediated genomic integration of the common cytokine receptor gamma chain in human T-cell lines. J Gene Med 8: 646–653. doi:10.1002/jgm.891

46. Ma Q, Sheng H, Yan J, Cheng S, Huang Y, et al. (2006) Identification of pseudo attP sites for phage phiC31 integrase in bovine genome. Biochem Biophys Res Commun 345: 984–988. doi:10.1016/j.bbrc.2006.04.145

47. Thyagarajan B, Liu Y, Shin S, Lakshmipathy U, Scheyhing K, et al. (2008) Creation of engineered human embryonic stem cell lines using phiC31 integrase. Stem Cells Dayt Ohio 26: 119–126. doi:10.1634/stemcells.2007-0283

48. Lister JA (2011) Use of phage φC31 integrase as a tool for zebrafish genome manipulation. Methods Cell Biol 104: 195–208. doi:10.1016/B978-0-12-374814-0.00011-2

49. Groth AC, Fish M, Nusse R, Calos MP (2004) Construction of transgenic Drosophila by using the site-specific integrase from phage phiC31. Genetics 166: 1775–1782.

50. Allen BG, Weeks DL (2005) Transgenic Xenopus laevis embryos can be generated using phiC31 integrase. Nat Methods 2: 975–979. doi:10.1038/nmeth814.

51. Allen BG, Weeks DL (2006) Using phiC31 integrase to make transgenic Xenopus laevis embryos. Nat Protoc 1: 1248–1257. doi:10.1038/nprot.2006.183.

52. Meredith JM, Underhill A, McArthur CC, Eggleston P (2013) Next-generation site-directed transgenesis in the malaria vector mosquito Anopheles gambiae: self-docking strains expressing germline-specific phiC31 integrase. PLoS One 8: e59264. doi:10.1371/journal.pone.0059264

53. Fish MP, Groth AC, Calos MP, Nusse R (2007) Creating transgenic Drosophila by microinjecting the site-specific phiC31 integrase mRNA and a transgene-containing donor plasmid. Nat Protoc 2: 2325–2331. doi:10.1038/nprot.2007.328

54. Bischof J, Maeda RK, Hediger M, Karch F, Basler K (2007) An optimized transgenesis system for Drosophila using germ-line-specific phiC31 integrases. Proc Natl Acad Sci U S A 104: 3312–3317. doi:10.1073/pnas.0611511104

55. Hollis RP, Stoll SM, Sclimenti CR, Lin J, Chen-Tsai Y, et al. (2003) Phage integrases for the construction and manipulation of transgenic mammals. Reprod Biol Endocrinol RBE 1: 79. doi:10.1186/1477-7827-1-79

56. Bertoni C, Jarrahian S, Wheeler TM, Li Y, Olivares EC, et al. (2006) Enhancement of plasmid-mediated gene therapy for muscular dystrophy by directed plasmid integration. Proc Natl Acad Sci U S A 103: 419–424. doi:10.1073/pnas.0504505102

57. Ortiz-Urda S, Thyagarajan B, Keene DR, Lin Q, Calos MP, et al. (2003) PhiC31 integrase-mediated nonviral genetic correction of junctional epidermolysis bullosa. Hum Gene Ther 14: 923–928. doi:10.1089/104303403765701204

58. Olivares EC, Hollis RP, Chalberg TW, Meuse L, Kay MA, et al. (2002) Site-specific genomic integration produces therapeutic Factor IX levels in mice. Nat Biotechnol 20: 1124–1128. doi:10.1038/nbt753

59. Bauer JW, Laimer M (2004) Gene therapy of epidermolysis bullosa. Expert Opin Biol Ther 4: 1435–1443. doi:10.1517/14712598.4.9.1435

60. Held PK, Olivares EC, Aguilar CP, Finegold M, Calos MP, et al. (2005) In vivo correction of murine hereditary tyrosinemia type I by phiC31 integrase-mediated gene delivery. Mol Ther J Am Soc Gene Ther 11: 399–408. doi:10.1016/j.ymthe.2004.11.001

61. Chalberg TW, Genise HL, Vollrath D, Calos MP (2005) phiC31 integrase confers genomic integration and long-term transgene expression in rat retina. Invest Ophthalmol Vis Sci 46: 2140–2146. doi:10.1167/iovs.04-1252

62. Chavez CL, Keravala A, Chu JN, Farruggio AP, Cuéllar VE, et al. (2012) Long-term expression of human coagulation factor VIII in a tolerant mouse model using the φC31 integrase system. Hum Gene Ther 23: 390–398. doi:10.1089/hum.2011.110

63. Keravala A, Portlock JL, Nash JA, Vitrant DG, Robbins PD, et al. (2006) PhiC31 integrase mediates integration in cultured synovial cells and enhances gene expression in rabbit joints. J Gene Med 8: 1008–1017. doi:10.1002/jgm.928

64. Portlock JL, Keravala A, Bertoni C, Lee S, Rando TA, et al. (2006) Long-term increase in mVEGF164 in mouse hindlimb muscle mediated by phage phiC31 integrase after nonviral DNA delivery. Hum Gene Ther 17: 871–876. doi:10.1089/hum.2006.17.871

65. Keravala A, Chavez CL, Hu G, Woodard LE, Monahan PE, et al. (2011) Long-term phenotypic correction in factor IX knockout mice by using ΦC31 integrase-mediated gene therapy. Gene Ther 18: 842–848. doi:10.1038/gt.2011.31

66. Chalberg TW, Portlock JL, Olivares EC, Thyagarajan B, Kirby PJ, et al. (2006) Integration specificity of phage phiC31 integrase in the human genome. J Mol Biol 357: 28–48. doi:10.1016/j.jmb.2005.11.098

67. Wilson JH (2003) Pointing fingers at the limiting step in gene targeting. Nat Biotechnol 21: 759–760. doi:10.1038/nbt0703-759

68. Ehrhardt A, Yant SR, Giering JC, Xu H, Engler JA, et al. (2007) Somatic integration from an adenoviral hybrid vector into a hot spot in mouse liver results in persistent transgene expression levels in vivo. Mol Ther J Am Soc Gene Ther 15: 146–156. doi:10.1038/sj.mt.6300011

69. Robert M-A, Zeng Y, Raymond B, Desfossé L, Mairey E, et al. (2012) Efficacy and site-specificity of adenoviral vector integration mediated by the phage φC31 integrase. Hum Gene Ther Methods 23: 393–407. doi:10.1089/hgtb.2012.122

70. Raper SE, Chirmule N, Lee FS, Wivel NA, Bagg A, et al. (2003) Fatal systemic inflammatory response syndrome in a ornithine transcarbamylase deficient patient following adenoviral gene transfer. Mol Genet Metab 80: 148–158.

71. Schnell MA, Zhang Y, Tazelaar J, Gao GP, Yu QC, et al. (2001) Activation of innate immunity in nonhuman primates following intraportal administration of adenoviral vectors. Mol Ther J Am Soc Gene Ther 3: 708–722. doi:10.1006/mthe.2001.0330

72. Byrnes AP, Rusby JE, Wood MJ, Charlton HM (1995) Adenovirus gene transfer causes inflammation in the brain. Neuroscience 66: 1015–1024.

73. Kafri T, Blömer U, Peterson DA, Gage FH, Verma IM (1997) Sustained expression of genes delivered directly into liver and muscle by lentiviral vectors. Nat Genet 17: 314–317. doi:10.1038/ng1197-314

74. Naldini L, Blömer U, Gallay P, Ory D, Mulligan R, et al. (1996) In vivo gene delivery and stable transduction of nondividing cells by a lentiviral vector. Science 272: 263–267.

75. Cronin J, Zhang X-Y, Reiser J (2005) Altering the tropism of lentiviral vectors through pseudotyping. Curr Gene Ther 5: 387–398.

76. Yáñez-Muñoz RJ, Balaggan KS, MacNeil A, Howe SJ, Schmidt M, et al. (2006) Effective gene therapy with nonintegrating lentiviral vectors. Nat Med 12: 348–353. doi:10.1038/nm1365

77. Philippe S, Sarkis C, Barkats M, Mammeri H, Ladroue C, et al. (2006) Lentiviral vectors with a defective integrase allow efficient and sustained transgene expression in vitro and in vivo. Proc Natl Acad Sci U S A 103: 17684–17689. doi:10.1073/pnas.0606197103

78. Grandchamp N, Henriot D, Philippe S, Amar L, Ursulet S, et al. (2011) Influence of insulators on transgene expression from integrating and non-integrating lentiviral vectors. Genet Vaccines Ther 9: 1. doi:10.1186/1479-0556-9-1

79. Vink CA, Gaspar HB, Gabriel R, Schmidt M, McIvor RS, et al. (2009) Sleeping beauty transposition from nonintegrating lentivirus. Mol Ther J Am Soc Gene Ther 17: 1197–1204. doi:10.1038/mt.2009.94

80. Staunstrup NH, Moldt B, Mátés L, Villesen P, Jakobsen M, et al. (2009) Hybrid lentivirus-transposon vectors with a random integration profile in human cells. Mol Ther J Am Soc Gene Ther 17: 1205–1214. doi:10.1038/mt.2009.10

81. Moldt B, Staunstrup NH, Jakobsen M, Yáñez-Muñoz RJ, Mikkelsen JG (2008) Genomic insertion of lentiviral DNA circles directed by the yeast Flp recombinase. BMC Biotechnol 8: 60. doi:10.1186/1472-6750-8-60

82. Cornu TI, Cathomen T (2007) Targeted genome modifications using integrase-deficient lentiviral vectors. Mol Ther J Am Soc Gene Ther 15: 2107–2113. doi:10.1038/sj.mt.6300345

83. Lombardo A, Genovese P, Beausejour CM, Colleoni S, Lee Y-L, et al. (2007) Gene editing in human stem cells using zinc finger nucleases and integrase-defective lentiviral vector delivery. Nat Biotechnol 25: 1298–1306. doi:10.1038/nbt1353

84. Sarkis C, Philippe S, Mallet J, Serguera C (2008) Non-integrating lentiviral vectors. Curr Gene Ther 8: 430–437.

85. Andreas S, Schwenk F, Küter-Luks B, Faust N, Kühn R (2002) Enhanced efficiency through nuclear localization signal fusion on phage PhiC31-integrase: activity comparison with Cre and FLPe recombinase in mammalian cells. Nucleic Acids Res 30: 2299–2306.

86. Ehrhardt A, Engler JA, Xu H, Cherry AM, Kay MA (2006) Molecular analysis of chromosomal rearrangements in mammalian cells after phiC31-mediated integration. Hum Gene Ther 17: 1077–1094. doi:10.1089/hum.2006.17.1077

87. Liu J, Skjørringe T, Gjetting T, Jensen TG (2009) PhiC31 integrase induces a DNA damage response and chromosomal rearrangements in human adult fibroblasts. BMC Biotechnol 9: 31. doi:10.1186/1472-6750-9-31

88. Liu J, Jeppesen I, Nielsen K, Jensen TG (2006) Phi c31 integrase induces chromosomal aberrations in primary human fibroblasts. Gene Ther 13: 1188–1190. doi:10.1038/sj.gt.3302789

89. Gregory MA, Till R, Smith MCM (2003) Integration site for Streptomyces phage phiBT1 and development of site-specific integrating vectors. J Bacteriol 185: 5320–5323.

90. Kolot M, Meroz A, Yagil E (2003) Site-specific recombination in human cells catalyzed by the wild-type integrase protein of coliphage HK022. Biotechnol Bioeng 84: 56–60. doi:10.1002/bit.10747

91. Olivares EC, Hollis RP, Calos MP (2001) Phage R4 integrase mediates site-specific integration in human cells. Gene 278: 167–176.

92. Miller OJ, Bernath K, Agresti JJ, Amitai G, Kelly BT, et al. (2006) Directed evolution by in vitro compartmentalization. Nat Methods 3: 561–570. doi:10.1038/nmeth897

93. Tay Y, Ho C, Droge P, Ghadessy FJ (2010) Selection of bacteriophage lambda integrases with altered recombination specificity by in vitro compartmentalization. Nucleic Acids Res 38: e25. doi:10.1093/nar/gkp1089

94. Zayed H, Izsvák Z, Walisko O, Ivics Z (2004) Development of hyperactive sleeping beauty transposon vectors by mutational analysis. Mol Ther J Am Soc Gene Ther 9: 292–304. doi:10.1016/j.ymthe.2003.11.024

95. Santoro SW, Schultz PG (2002) Directed evolution of the site specificity of Cre recombinase. Proc Natl Acad Sci U S A 99: 4185–4190. doi:10.1073/pnas.022039799

96. Bolusani S, Ma C-H, Paek A, Konieczka JH, Jayaram M, et al. (2006) Evolution of variants of yeast site-specific recombinase Flp that utilize native genomic sequences as recombination target sites. Nucleic Acids Res 34: 5259–5269. doi:10.1093/nar/gkl548

97. Guo J, Gaj T, Barbas III CF (2010) Directed Evolution of an Enhanced and Highly Efficient FokI Cleavage Domain for Zinc Finger Nucleases. J Mol Biol 400: 96–107. doi:10.1016/j.jmb.2010.04.060

98. Sclimenti CR, Thyagarajan B, Calos MP (2001) Directed evolution of a recombinase for improved genomic integration at a native human sequence. Nucleic Acids Res 29: 5044–5051.

9

Comprehensive Characterization of Human Genome Variation by High Coverage Whole-Genome Sequencing of Forty Four Caucasians

Hui Shen[1,2], Jian Li[1,2], Jigang Zhang[1,2], Chao Xu[1,2,3], Yan Jiang[1,2,3], Zikai Wu[1,2,3], Fuping Zhao[1,2], Li Liao[1,2], Jun Chen[1], Yong Lin[3], Qing Tian[1,2], Christopher J. Papasian[2], Hong-Wen Deng[1,2,3]*

1 Center for Bioinformatics and Genomics, Department of Biostatistics and Bioinformatics, School of Public Health and Tropical Medicine, Tulane University, New Orleans, Louisiana, United States of America, 2 School of Medicine, University of Missouri-Kansas City, Kansas City, Missouri, United States of America, 3 Center of System Biomedical Sciences, University of Shanghai for Science and Technology, Shanghai, P. R. China

Abstract

Whole genome sequencing studies are essential to obtain a comprehensive understanding of the vast pattern of human genomic variations. Here we report the results of a high-coverage whole genome sequencing study for 44 unrelated healthy Caucasian adults, each sequenced to over 50-fold coverage (averaging $65.8\times$). We identified approximately 11 million single nucleotide polymorphisms (SNPs), 2.8 million short insertions and deletions, and over 500,000 block substitutions. We showed that, although previous studies, including the 1000 Genomes Project Phase 1 study, have catalogued the vast majority of common SNPs, many of the low-frequency and rare variants remain undiscovered. For instance, approximately 1.4 million SNPs and 1.3 million short indels that we found were novel to both the dbSNP and the 1000 Genomes Project Phase 1 data sets, and the majority of which (~96%) have a minor allele frequency less than 5%. On average, each individual genome carried ~3.3 million SNPs and ~492,000 indels/block substitutions, including approximately 179 variants that were predicted to cause loss of function of the gene products. Moreover, each individual genome carried an average of 44 such loss-of-function variants in a homozygous state, which would completely "knock out" the corresponding genes. Across all the 44 genomes, a total of 182 genes were "knocked-out" in at least one individual genome, among which 46 genes were "knocked out" in over 30% of our samples, suggesting that a number of genes are commonly "knocked-out" in general populations. Gene ontology analysis suggested that these commonly "knocked-out" genes are enriched in biological process related to antigen processing and immune response. Our results contribute towards a comprehensive characterization of human genomic variation, especially for less-common and rare variants, and provide an invaluable resource for future genetic studies of human variation and diseases.

Editor: Philip Awadalla, University of Montreal, Canada

Funding: The investigators of this work were partially supported by grants from the National Institutes of Health (P50AR055081, R01AG026564, R01AR050496, R01AR057049, and R03TW008221), the Franklin D. Dickson/Missouri Endowment from University of Missouri-Kansas City, and the Edward G. Schlieder Endowment from Tulane University. The work was also benefited by Shanghai Leading Academic Discipline Project (S30501). The funders had no role in study design, data collection and analysis, decision to publish, or preparation of the manuscript.

Competing Interests: The authors have declared that no competing interests exist.

* E-mail: hdeng2@tulane.edu

Introduction

Genome-wide association studies (GWAS) have identified a large number of genetic variants that are associated with a variety of human complex diseases/traits [1,2]. A major challenge in this post-GWAS era is to pinpoint the functional variants underlying the observed associations and to identify the missing heritability [3–6], which requires a comprehensive identification and characterization of genetic variants in the human genome. The rapidly evolving massively-parallel DNA sequencing technology enables efficient and cost-effective whole genome sequencing [7–10], and is revolutionizing our understanding of the human genome architecture and variation, human evolution, and the genomics of common and rare disorders [11–26]. The 1000 Genomes Project Consortium recently reported results for the Phase 1 of the project [12,20]. By performing mainly low-coverage whole genome sequencing and exon-targeted sequencing, the consortium

identified approximately 15 million single nucleotide polymorphisms (SNPs), 1 million short insertions and deletions (indels), and over 20,000 structural variants [12,20]. More recently, a few high-coverage sequencing studies have been carried out at whole genome level [14] or at target genes [27,28] and discovered a large number of previously unidentified variants, suggesting that a considerable number of human genetic variants, particularly rare variants, remain to be discovered beyond those currently archived in the dbSNP and the 1000 Genomes Project.

These initial studies attest to the necessity for performing whole genome sequencing analysis on additional human samples, particularly at high-coverage, in order to gain a comprehensive understanding of the human genomic variation. Here we report the results of a whole genome sequencing study for 44 Caucasian subjects from a single population in Midwest USA, all of whom were sequenced at high coverage. This study represents the first few high-coverage analyses of multiple genomes for healthy human subjects in a single population of the same ethnicity. Our results contribute towards a more comprehensive characterization of human genomic variation, especially for less-common and rare variants, and provide a valuable resource for future genetic studies of human variation and diseases.

Results

Sequence data generation and mapping

Genomic DNA samples from 44 healthy, self-reported US Caucasian adults from Midwest USA (in Kansas City and its vicinities), including 22 males and 22 females, were sequenced at Complete Genomics Inc. (Mountain View, CA). All participants signed an informed-consent document before entering the study.

For each individual, 147.7–229.6 gigabases (Gb) of sequence were generated and mapped to the NCBI human reference genome (build 37.1, GRCh37/hg19), resulting in an average of 65.8-fold (range: 51.7–80.3×) genomic coverage (Table 1 and Table S1), which is significantly higher than other reported population-based whole genome sequencing studies. Diploid calls were confidently made for an average 96.0% of the autosomal bases in the reference genome, with a range of 93.8% to 96.9% across the 44 genomes (Table 1 and Table S1).

SNP and indel identification and characterization

In total, 10,871,465 distinct SNPs were identified in the 44 genomes, with an average of 3.3 million SNPs per genome (Table 1). The average SNP transition to transversion ratio was 2.13, and the average SNP heterozygote to homozygote ratio was 1.56 (Table S1), both consistent with previous reports [7,9,21]. In addition, we identified a total of 1,350,484 distinct short insertions (range 1–76 bp) and 1,464,731 distinct short deletions (range 1–192 bp) in our sample, with an average of 207,000 short insertions and 214,000 short deletions in an individual genome (Table 2, Table S1). The estimated heterozygote to homozygote ratios were 1.4–1.7 for short indels (Table 2, Table S1), also comparable to those calculated from a previous study [11]. Multiple SNPs and/or indels that are in close proximity (i.e., less than two reference bases between two variant sequences) were grouped together as a single variant locus, termed block substitutions. In total, we identified 394,517 distinct block substitutions in our sample, with each individual genome carrying approximately 71,398 such block substitutions (Table 2, Table S1). These block substitutions can be length-conversing (the sample's allele and reference are the same length) or length-changing, and the change in sequence length caused by these substitutions ranges 0–183 bp (Figure S3).

Of all the SNPs (n = 10,862,507) mapped to autosomes and X chromosome, 7,658,805 (70.5%) were already present in dbSNP (v131), and 9,134,659 (84.1%) were in the 1000 Genomes Project

Phase 1 data set (released on 5/21/2011, http://www.1000genomes.org/), whereas the rest 1,390,686 (12.8%) SNPs were in neither dataset and thus considered to be novel (Fig. 1a). In contrast, a large proportion (46.2%) of the identified short indels was novel relative to dbSNP (v131) and the 1000 Genomes Project Phase 1 data set (Fig. 1b). Interestingly, short indels reported in any of the three data sets were largely unique to that data set, showing only ~20–40% overlap with any other data set (Fig. 1b). The limited overlap of indels identified between different data sets or between different individual genomes has been previously recognized [11,12,15,25]. This observation may reflect the substantial challenges in making accurate calls of indels. For instance, the estimated false discovery rates for indels were fairly high by both the Complete Genomics approach (~3.0–6.5%) [8] and the 1000 Genomes project approach (~3.7–8.1%) [29]. Moreover, 18% of indel sites identified by the 1000 Genomes project were found to be inconsistent or ambiguous [29].These results highlighted the need for developing advanced experimental and analytic methods for identification and characterization of indels.

Consistent with previous reports [12,13], high densities of variations were observed at the human leukocyte antigen (HLA) region, and a number of telomeric and sub-telomeric regions (Fig. S1). In general, we observed strong correlations in genomic distributions between distinct forms of variants (Fig. S2). Specifically, short insertions and deletions showed very strong correlation throughout the entire genome, whereas block substitutions were relatively poorly correlated with other forms of variants (Fig. S2).

The genome-wide distribution of novel SNPs (Fig. 1c) exhibited similar patterns as that of the total SNPs (Fig. S1). As expected, the vast majority (~96%) of novel SNPs represented low-frequency (minor allele frequency, MAF, ≤5%) variants (Fig. 1d), as most common SNPs in human populations were already included in dbSNP. Intriguingly, several regions contained relatively high proportions of novel SNPs (Fig. S1). For instance, 75–83% of all SNPs identified at centromeric regions on chromosomes 3, 7, 16 and X, were not in the dbSNP (Table S2). In addition, similarly high fractions of novel SNPs were also observed in these regions, when comparing our data with the 1000 Genomes Project Phase 1 data set (novel rate: 80–97%) and when comparing the 1000 Genomes Project Phase 1 data set with the dbSNP (novel rate: 84–91%) (Table S2). The high discrepancy between SNPs from different data sets that are mapped to these regions may suggest that it is difficult to accurately sequence these regions with currently available sequencing techniques.

To evaluate the accuracy of our data, we first estimated the SNP genotyping concordance between whole genome sequencing and SNP arrays for 20 subjects that had been genotyped with the

Table 1. Summary information of population-based whole-genome sequencing studies.

	Aligned Bases (Gb)	Coverage depth	Genome Covered	SNPs (Total)	Indels/Subs (Total)
Current study (N = 44)	188.08	65.8×	96.0%	3,307,678 (10,871,465)	492,486 (3,209,732)
Korean WGS (N = 10)	74.09	26.1×	NA	3,602,372 (8,367,302)	332,561 (1,191,599)
Duke WGS[1] (N = 20)	NA	31.1×	97.5%	3,473,639 (10,530,094)	609,795 (2,736,907)
1000 G_LC[2] (N = 179)	10.51	3.56×	86.0%	3,019,919 (14,894,361)	361,669 (1,330,158)
1000 G_HCT[3] (N = 6)	118.5	41.6×	79.0%	3,001,156 (5,907,699)	352,474 (682,148)

Numbers shown are average numbers per individual except where indicated otherwise.
[1]For autosomes only; [2] 1000 G_LC: Low-coverage samples from the 1000 Genomes Project; [3] 1000 G_HCT: High-coverage trios from the 1000 Genomes Project.

Table 2. Summary of identified SNPs and indels/block substitutions.

Variant Type	Total No. of Variants	Average No. of Variants in individual genomes
SNP	10,871,465	3,307,678
Intergenic	6,674,155	2,054,900
Intragenic	4,197,310	1,252,778
Intron	3,473,672	1,043,427
UTR	614,753	181,267
Splicing acceptor site	6,644	1,898
Splicing donor site	1,547	398
Coding domain	79,762	18,796
Synonymous	36,538	9,612
Non-synonymous	43,224	9,184
Missense	42,549	9,082
Nonsense	616	87
Nonstop	59	15
Indels	2,815,215	421,088
Coding domain	3,606	381
Frameshift	2,344	217
Frameshift-preserving	1,262	164
Block substitutions	394,517	71,398
Coding domain	1,870	274
Synonymous	30	20
Frameshift	264	17
Missense	1,546	234
Nonsense	29	3
Nonstop	1	<1

Affymetrix Genome-Wide Human SNP Array 6.0. Out of the 906,600 SNPs called on the SNP arrays, 801,557 (98.5%) were successfully called by the present sequencing project and the genotype concordance rates were 99.83% and 99.45% for homozygous and heterozygous SNPs, respectively. By comparing with the recently reported 10 Korean genomes [14], the 69 genomes data set from Complete Genomics Inc (http://www.completegenomics.com/sequence-data/download-data/), and an exome sequencing data of 100 Han Chinese subjects (Lin et al., unpublished), we *in silico* validated 177,554 SNPs and 411,658 indels that were not archived in the dbSNP (v131) or the 1000 Genomes Project Phase 1 data sets, supporting that these novel variants are likely to be genuine polymorphisms and not just unique to our specific population. To further evaluate the accuracy of the novel variants, we tested 25 novel nonsynonymous SNPs by traditional Sanger sequencing. Twenty-four SNPs out of the 25 were validated, highlighting the high quality and reproducibility of our data.

Gene-based characterization of SNPs and indels

Of all the identified SNPs, 79,762 were mapped to CDS (coding sequences) of RefSeq transcripts, including 36,538 synonymous, 42,549 missense, 616 nonsense, and 59 nonstop (causing the loss of a stop codon) SNPs (Table 2). Consistent with previous reports [12,13], we observed a significantly higher fraction of low-frequency variants (MAF≤0.05) in non-synonymous SNPs, when compared to synonymous or noncoding SNPs, presumably reflecting purifying selection against deleterious mutations

(Fig. S3). Of all the identified indels, only a very small fraction (0.13%) is mapped to coding sequence regions and most of the coding indels were predicted to cause frameshift changes (Table 2), which is consistent with previous reports [12,13]. We observed a noticeable preference for 3n-bp indels in coding regions compared to indels in non-coding genic regions, such as in introns and UTRs (Fig. S3). This preference has also been observed in other studies [9,13,30] and may reflect the selection pressure for minimally impacting variants in coding regions. In addition, we detected 1,870 block substitutions that are mapped to coding sequence regions, and almost all of the coding block substitutions were predicted to cause non-synonymous changes (Table 2). This is expected, as block substitutions involve multiple variant bases and thus are more likely to result in amino acid changes. Similar bias towards non-synonymous variants were also reported in a previous study by Rosenfeld et al [31], who focused on a specific type of block substitution, namely, double nucleotide polymorphisms (DNPs) and triple nucleotide polymorphisms (TNPs), which by definition are length-conserving block substitutions involving two or three consecutive nucleotides.

We performed GO (Gene Ontology) analyses to determine whether genes containing high density of amino acid changing variants were enriched in specific biological processes. The GO terms related to sensory perception, such as 'sensory perception of chemical stimulus' ($p = 3.47 \times 10^{-8}$) and 'sensory perception of smell' ($p = 7.21 \times 10^{-7}$) showed significant enrichment for genes with high density of amino acid changing variants (Table 3). In contrast, amino acid changing variants were significantly under-

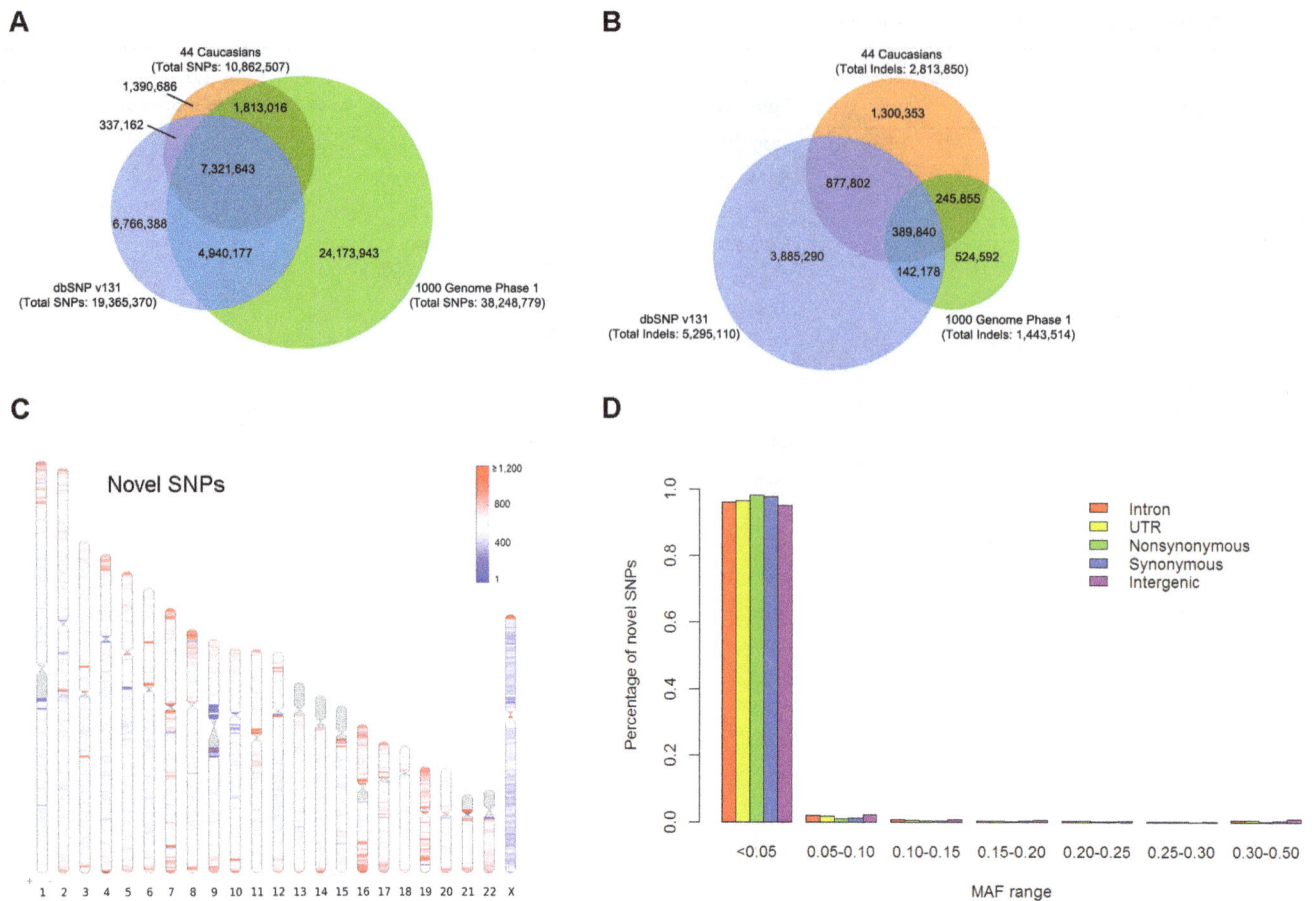

Figure 1. Summary characterizations of the identified variants. A–B, Venn diagram showing SNPs and indels identified in the present study overlapping with those archived in the dbSNP (v131) and the 1000 Genomes Project Phase 1 data sets (released on 5/21/2011). To account for differences in placement of many indels between different data sets, indels were considered to match if they were within 25 bp distance and of the same size. Only SNPs and indels mapped to autosomes and X chromosome were analyzed. **C,** Genome-wide distribution of novel SNPs. Total number of novel SNPs (compared to dbSNP v131 and the 1000 Genomes Project pilot phase) were calculated in non-overlap 1-megabases (Mb) windows across the human genome and plotted in ideograms using *Idiographica* [41]. The diversities were illustrated by colors, with red indicating higher numbers or proportions and blue indicating lower numbers or proportions. Genomic regions in which no SNPs were identified or no reference sequences could be determined are shown in grey. **D,** Allele frequency spectrum of novel SNPs.

represented in multiple biological processes related to 'cellular macromolecule metabolic process' ($p = 1.38 \times 10^{-19}$) and 'RNA metabolic process' ($p = 1.52 \times 10^{-17}$) (Table 3). These results were not unexpected because these metabolic and signal transduction related biological processes are critical for cell survival and functioning, consequently, are expected to be under strong purifying selection against deleterious mutations. On the other hand, sensory perception related genes are known to be highly associated with environmental adaptation. Thus, variants may have accumulated that differ in frequencies among human populations, suggesting balanced selection, the possible relaxation of purifying selection, and/or an increased mutation rate [9,15].

Using the Variant Annotation Tool (version 2.0.1.) [32] with the GENCODE v7 annotation [33], we examined SNPs and indels for variants predicted to result in the complete loss-of-function (LoF). LoF variants were defined as SNPs predicted to create or disrupt a stop codon, frameshift indels, and variants predicted to disrupt a splice site. On average, each person carried approximately 179 LoF variants with ~44 of these variants occurred in a homozygous state, which would result in complete "knock-out" of the annotated genes. Altogether, we found 182 unique "knocked-out" genes in our samples. While about 40% of these genes were

"knocked out" in just one or two individuals, 46 genes were "knocked out" in over 30% of our samples (Fig. 2), suggesting that a number of genes are commonly "knocked-out" in general populations. GO analysis suggested that these "knocked-out" genes were significantly enriched in several biological processes, particularly those related to antigen processing and immune responses (Table 3). Intriguingly, 8 LoF variants located in 8 genes (Table S3) were found to be homozygous in all the 44 genomes, as well as in other independently sequenced human genomes [26,30], suggesting that the reference sequences at these sites may represent rare alleles or sequencing errors.

Disease-associated SNPs and indels

Using *Trait-o-matic* (https://github.com/xwu/trait-o-matic), we found that each individual genome carried an average of 90 non-synonymous SNPs that were known to be associated with OMIM diseases/traits. Among the 44 samples, 29 subjects carried 14 heterozygous variants causing autosomal dominant diseases/traits, and 4 subjects were homozygous for 4 variants causing autosomal recessive diseases/traits (Table S4). However, the 44 subjects were all apparently healthy with no recognizable or self-reported

Table 3. Top 10 GO terms significantly enriched or depleted for deleterious coding variants, and enriched for "knocked-out" genes.

GO Accession #	Biological Process	P-value
Enriched for deleterious coding variants		
GO:0050907	detection of chemical stimulus involved in sensory perception	2.97E–09
GO:0007606	sensory perception of chemical stimulus	3.47E–08
GO:0050911	detection of chemical stimulus involved in sensory perception of smell	1.03E–07
GO:0009593	detection of chemical stimulus	3.22E–07
GO:0007608	sensory perception of smell	7.21E–07
Depleted for amino acid changing variants		
GO:0044260	cellular macromolecule metabolic process	1.38E–19
GO:0009987	cellular process	1.31E–18
GO:0016070	RNA metabolic process	1.52E–17
GO:0043170	macromolecule metabolic process	1.96E–15
GO:0006139	nucleobase-containing compound metabolic process	2.04E–15
Enriched for "knocked-out" gene		
GO:0002474	antigen processing and presentation of peptide antigen via MHC class I	1.79E–23
GO:0019882	antigen processing and presentation	1.18E–21
GO:0048002	antigen processing and presentation of peptide antigen	1.65E–18
GO:0006611	protein export from nucleus	2.95E–12
GO:0006955	immune responses	3.56E–08

P-values were computed for significance of enrichment by *Gorilla* (http://cbl-gorilla.cs.technion.ac.il/).

diseases at the time of enrollment. This might be explained by the fact that the majority of these variants were known to have reduced/age-dependent penetrance (e.g., R468H mutation in *MFN2* gene for Charcot-Marie-Tooth disease-2A2) and/or did not cause severe health-threatening symptoms (e.g., wet/dry ear wax). This perspective is supported by the fact that 8 of the 18 variants were present in low to intermediate frequency in general populations (Table S4). Alternatively, some of these variants might have been erroneously assigned as disease mutations. Each

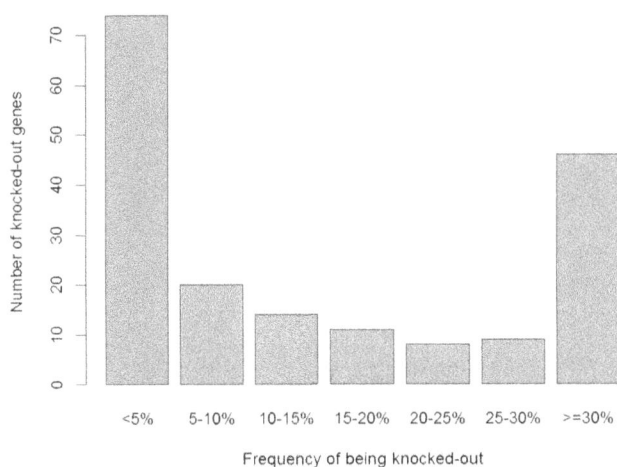

Figure 2. Identification of "knocked-out" genes. A, Frequency spectrum of observed "knocked-out" genes. Genes containing homozygous LoF variants were expected to be silent or knocked-out. Numbers of "knocked-out" genes were counted with respect to the frequency of "knock-out" occurrence in the 44 genomes.

genome also contained approximately 276 additional non-synonymous SNPs, including about 34 potential nonsense/nonstop SNPs and 11 frameshift indels and block substitutions in genes associated with OMIM diseases/traits.

Mitochondrial and Y chromosome analyses

By comparing to the revised Cambridge Reference Sequence (rCRS) of the Human Mitochondrial DNA [34], we identified a total of 306 mitochondrial variants, including 285 SNPs, 13 short indels, and 8 block substitutions (Table S5). Approximately 54% of these SNPs, and almost all the indels and block substitutions, were novel relative to dbSNP (v131). In the 37 mitochondrial genes, we identified 54 non-synonymous SNPs (53 missense and 1 nonstop), and one frameshift block substitution.

A total of 8,673 SNPs, 1,352 indels, and 998 block substitutions were identified on the Y chromosome (Table S5). Similar to the mitochondrial genome, approximately 54% of these SNPs and the majority (~65%) of indels/block substitutions on the Y chromosome were novel relative to dbSNP (Table S5).

Copy number variants (CNVs)

Through the analysis of read depth in 2-kb sliding windows, we discovered an average of 3,749 CNVs (size range of 2–154 kb) totaling about 12 Mb in each genome. Approximately 45% of the CNVs detected in our study overlapped (≥50% of sequence overlap) with CNVs currently annotated in the Database of Genomic Variants (DGV) [35]. Similar to previous reports [9,20], we observed a steadily decreasing number of events with increasing CNV length (Fig. S4). Out of the 493 CNVs identified in the 20 subjects with the Affymetrix Genome-Wide Human SNP Array 6.0, 255 overlapped with CNVs called by the sequencing data.

Rates of novel variants discovery

To evaluate how the rate of novel SNP/indel identification changed as the number of sequenced genomes increased, we adopted the permutation method used by Pelak et al. [21]. Briefly, we permuated the order of the 44 personal genomes 1000 times and determined the mean number of "new" variants added by each additional personal genome, simultaneously considering variants archived in dbSNP/1000 Genome Project Phase 1 data sets as well as those discovered in the previously considered genomes. As shown in Figure 3, one randomly selected individual genome contained an average of 44,500 novel SNPs and ~107,000 novel indels (not in dbSNP/1000 Genome Project Phase 1 data sets). As additional individuals were sequenced, the number of additional novel SNPs and indels (not in dbSNP/1000 Genome Project Phase 1 data sets and the previously sequenced genomes) added per genome gradually declined. After about 40 genomes were sequenced, the discovery rate of additional novel variants appeared to plateau at approximately 29,500 SNPs and 14,300 indels per genome.

Discussion

Over the past several years, GWAS assaying several hundred thousand to a few million SNPs in thousands of individuals have successfully identified numerous genetic variants that were significantly associated with over two hundred human complex diseases/traits [1,2]. However, most of these variants are likely to be genetic 'markers', rather than actual causative variants, and the vast majority of these variants, individually or in combination, can explain only a small proportion (general <10%) of the heritability for these diseases/traits [5,6]. The missing heritability is partially attributable to the imprecise estimation of genetic effects based on the disease/trait-associated markers, and rare/low-frequency variants, especially variants other than SNPs, which are poorly covered by current GWAS approaches [5,6,36]. In this context, performing an exhaustive inspection of all genetic variants located in the associated regions is essential to pinpoint causative variants and to generate deeper insights into genetic and functional mechanisms underlying the observed associations. Exhaustive inspection of all genetic variants requires a more comprehensive

description and understanding of genetic variants in the human genome. Recent advances in DNA sequencing technology have greatly facilitated sequencing of individual genomes. In the present study, we carried out high-coverage (>50×), whole genome sequencing for 44 apparently healthy Caucasians, and identified approximately 11 million SNPs and 3.3 million short indels and block substitutions, including ~1.4 million SNPs and ~1.3 million indels that were not cataloged in the dbSNP (v131) and the 1000 Genome Project Phase 1 data set.

The recently reported pilot phase of the 1000 Genomes Project identified a massive number of SNPs and estimated that approximately 95% of all accessible common SNPs were catalogued in their data set [12,20]. By comparing the SNPs identified in the current study with the 1000 Genomes Project Phase 1 and the dbSNP (v131) data sets, we showed that an average of 95.6% of the SNPs in any individual genome was already reported (Table S1). Our study also identified 1,390,686 novel SNPs and the majority of which (~96%) were with low and rare frequencies (MAF ≤5%), highlighting the power of rare variant identification by our high-coverage genome sequencing strategy. Based on our evaluation of the discovery rate of novel variants, we can obtain a reasonable estimation of the novel variants contained in a single individual genome. For instance, a single individual genome in our sample contained an average of ~44,500 SNPs and 107,000 indels that were not archived in the dbSNP (v131) or the 1000 Genome Project Phase 1 data, many of which were expected to be at low frequencies. This number should be interpreted cautiously, particularly for the novel indels, considering the relatively high false discovery rate (3.0–6.5%) [8] in indel identification. On the other hand, given the fact that potential deleterious genetic variants were enriched in the low and rare frequency spectrum, our data suggested that a considerable number of genuine variants that are related to disease susceptibility may still be undiscovered. Consequently, it is critical that additional high-coverage sequencing studies be conducted in additional subjects in order to identify these variants.

Interestingly, we showed that each individual genome from the general population carries approximately 179 potential LoF variants with ~44 genes expected to be completely disrupted ("knocked-out") due to LoF variants in homozygous state. Though

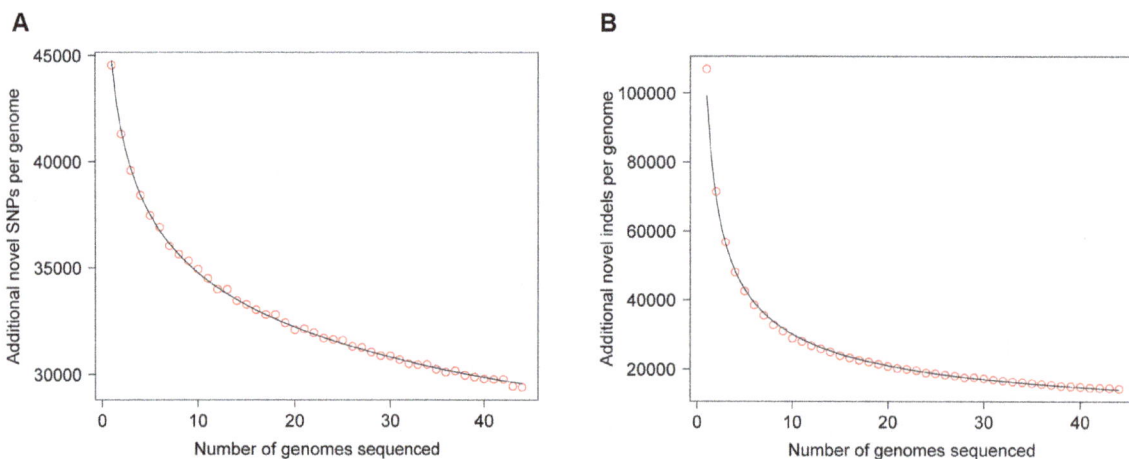

Figure 3. The number of novel SNPs and indels discovered as the number of sequenced genomes increased. We evaluated how many additional "new" **A**) SNPs and **B**) indels, respectively, were identified per genome as the number of sequenced genomes increased, considering both variants archived in databases (dbSNP v131 and the 1000 Genome Project Phase 1 data) and variants "discovered" in previously considered genomes. The 44 genomes were added into the analyses in a random order. With 1000 permutations, the average numbers of novel variants added per genome are shown, along with the best fitting trendline for each plot.

several earlier studies suggested individual genome may carry 250–300 [12,18] and perhaps even more [21] potential LoF variants, these numbers should be interpreted cautiously because LoF variants were expected to be highly enriched for false positives [37]. A recent study [38] suggested that many of these previously identified putative LoF variants may represent a variety of sequencing and annotation errors. After applying a series of stringent filters to the putative LoF variants, it was estimated that human genome typically contain ~100 genuine LoF variants with ~20 genes completed inactivated [38]. These high-confidence LoF variants have important implications for the interpretation of clinical sequencing studies. On the other hand, these stringent filters were also expected to remove a significant number of true positive LoF variants as well and thus the number of these high-confidence LoF variants may be regarded as a lower bound for the number of LoF variants carried by individual genome [38]. Therefore, our results and similar numbers (~150 LoF variants per individual genome) recently reported by the 100 Genome Project Consortium [29] may represent a reasonable estimation of the true number of LoF variants and "knocked-out" genes per individual genome. In total, we identified 182 "knocked-out" genes in our samples. The GO term analysis suggested that these "knocked-out" events were more frequently occurred in genes participating in antigen processing and presentation and immune responses, probably reflecting the fact that many genes in the immune systems have redundant biological functions [35]. As more individual genomes being sequenced in the future, we expect to slowly identify many more human genes that are not essential for survival and can be "knocked-out" in the general population. Detailed clinical phenotyping and biological characterization at tissue/cell-level in these individuals will shed light on the functional mechanisms and impacts of these specific "nonessential" genes.

Complete identification and characterization of human DNA variants is essential to further decipher the genetic basis of human evolution and disease susceptibility, and for a comprehensive understanding of human biology. This study, along with previously reported whole genome sequencing projects represents initial, but significant, steps towards a better understanding of the vast human genomic variation.

Materials and Methods

Study population

Forty four unrelated healthy Caucasian adults, including twenty two females and twenty two males, were recruited through the Kansas City Osteoporosis Study (KCOS), a genetic repertoire of ~6,000 subjects collected for genomic studies of complex diseases/traits. All subjects were living in Kansas City, Missouri and its surrounding areas and were self-identified as being of European origin. The study was approved by the University of Missouri-Kansas City (UMKC) Institutional Review Board, and each participant signed an informed-consent document before entering the study.

Genomic DNA preparation and sequencing

Genomic DNA was isolated from peripheral blood using the Gentra Puregene Blood kit (Qiagen, Valencia, CA) according to the recommended protocol. DNA concentration was measured by Nanodrop 1000 (Thermo Scientific, Wilmington, DE) and Quant-iT^{TM} Pico Green dsDNA kit (Invitrogen, Carlsbad, CA). Whole genome DNA sequencing was performed by Complete Genomics, Inc (Mountain View, CA), using its paired end library preparation

and sequencing-by-ligation methodology as established previously [8,16].

The resulting mate-paired reads were initially mapped to the NCBI reference genome (build 37.1, GRCh37/hg19) using a fast algorithm, and these initial mappings were both expanded and refined by local de novo assembly, which was applied to all regions of the genome that appeared to contain variations based on initial mappings [8]. Mapped reads were assembled into a best-fit diploid sequence with two separate resultant sequences for each locus in diploid regions (exceptions: mitochondria were assembled as haploid, and for males the non-pseudo-autosomal regions were assembled as haploid) by using a custom software suite that implements both Bayesian and de Bruijn graph techniques [8]. Variants were called by independently comparing each of the diploid assemblies to the reference. Data for each genome were delivered as lists of sequence variants (SNPs, short indels, and block substitutions) relative to the NCBI reference genome accompanied with variant quality scores. Block substitutions were called where a series of nearby reference bases had been replaced with a different series of bases in an allele. Block substitutions can be length-conserving (the same number of bases as the corresponding reference sequence region) or length-altering. For sites with multiple variant bases, if at least two reference bases on both alleles were called between two variant sequences, then the sites were broken into smaller variant events. The sequence variant data from this study are available at http://tulane.edu/publichealth/bio/genetics-and-genomics-study-of-osteoporosis.cfm.

CNV analysis

CNVs were identified by using *CNV-seq* (http://tiger.dbs.nus.edu.sg/cnv-seq/) [39], which compares the numbers of mapped reads in a sliding window between two individuals and assesses the hypothesis of no copy number variation based on a probabilistic model. For this analysis, one of the sequenced genomes was randomly picked as the reference genome and the remaining 43 genomes were compared with this reference genome for numbers of mapped reads in 2 kb sliding windows.

Variant annotation

Based on their locations mapped to the RefSeq transcripts, the identified variants were classified into different groups, including intergenic, CDS, intron, splicing donor and acceptor sites, and 5'-/3'- untranslated regions (UTRs). Since variants may be fall into multiple categories depending on the alternative transcript isoforms examined, a hierarchy CDS>UTRs>splicing sites>intron>intergenic was applied, such that a variant was only counted once at its highest level in the hierarchy. SNPs and indels mapped to CDS and splicing sites were further assessed for their potential effects on the gene products to identify potential LoF variants that are predicted to create or disrupt a stop codon, shift the normal reading frame, and disrupt a splice site. The identification of LoF variants was performed by using the Variant Annotation Tool (version 2.0.1.) [32] with the GENCODE v7 annotation reference [33].

The program *Gorilla* (http://cbl-gorilla.cs.technion.ac.il/) [40] was employed to test whether genes containing high or low density of amino acid changing variants (non-synonymous SNPs and frameshift indels/block substitutions) were enriched in certain GO terms (biological processes). The density of deleterious coding variants for each Refseq gene was calculated by dividing the total number of deleterious variants in the corresponding Refseq transcripts by the sum of CDS lengths of the transcripts. A ranked list of genes was generated by sorting the Refseq genes in three steps, first based on the density of deleterious coding variants, then by the total number of deleterious variants, and finally by the sum of CDS lengths. The ranked list of genes was then uploaded into *Gorilla* for GO enrichment analysis.

Non-synonymous SNPs identified in these personal genomes were screened for known associations with inherited disorders/traits by using *Trait-o-matic* (https://github.com/xwu/trait-o-matic), which finds and cross-references the non-synonymous SNPs with records in the Online Mendelian Inheritance in Man (OMIM) database (http://www.omim.org).

Variant correlation

The human genome was divided into consecutive 300-kb windows and the numbers of different types of variants (SNP, insertion, deletion, and block substitution) were counted within each window. Pairwise Pearson's correlation coefficients between different types of variants were computed using variant counts from 100 consecutive windows through the *cor* function in R.

Genome-wide SNP genotyping

Among the 44 subjects, 20 have been genotyped using the Affymetrix Genome-Wide Human SNP Array 6.0 (Affymetrix, Santa Clara, CA, USA) through our recent genome-wide association study for BMD (unpublished data). Genotyping was performed following the manufacturer's recommended protocol. SNPs were identified using Birdsuite (version 1.5.2, http://www.broad.mit.edu/mpg/birdsuite/analysis.html).

Supporting Information

Figure S1 Genome-wide distribution of SNPs, indels, block substitutions, and proportion of novel SNPs.

Figure S2 Genome-wide correlations between different types of variants.

Figure S3 SNP allele frequency and size distributions of indels and block substitutions.

Figure S4 Size distribution of identified CNVs.

Table S1 Characteristics of study subjects and genome sequencing data.

Table S2 Top 5 regions showing highest proportion of novel SNPs.

Table S3 Fifteen loss-of-function variants that were presented as homozygous form in all 44 genomes.

Table S4. Heterozygous autosomal dominant and homozygous autosomal recessive variants identified in the 44 genomes.

Table S5 Summary of variants identified on mitochondrial and Y chromosome.

Acknowledgments

We thank Aaron Solomon, Erica Beilharz, and Zuoming Deng at Complete Genomics Inc. for their assistance in setting up and performing the whole genome sequencing, and providing consultation for data analysis.

Author Contributions

Conceived and designed the experiments: HS JL HWD. Analyzed the data: JL JZ CX YJ ZW FZ LL JC YL QT. Wrote the paper: HS CJP HWD.

References

1. Feero WG, Guttmacher AE, Collins FS (2010) Genomic medicine – an updated primer. N Engl J Med 362: 2001–2011.
2. Manolio TA, Brooks LD, Collins FS (2008) A HapMap harvest of insights into the genetics of common disease. J Clin Invest 118: 1590–1605.
3. Freedman ML, Monteiro AN, Gayther SA, Coetzee GA, Risch A, et al. (2011) Principles for the post-GWAS functional characterization of cancer risk loci. Nat Genet 43: 513–518.
4. Kingsley CB (2011) Identification of causal sequence variants of disease in the next generation sequencing era. Methods Mol Biol 700: 37–46.
5. Eichler EE, Flint J, Gibson G, Kong A, Leal SM, et al. (2010) Missing heritability and strategies for finding the underlying causes of complex disease. Nat Rev Genet 11: 446–450.
6. Manolio TA, Collins FS, Cox NJ, Goldstein DB, Hindorff LA, et al. (2009) Finding the missing heritability of complex diseases. Nature 461: 747–753.
7. Bentley DR, Balasubramanian S, Swerdlow HP, Smith GP, Milton J, et al. (2008) Accurate whole human genome sequencing using reversible terminator chemistry. Nature 456: 53–59.
8. Drmanac R, Sparks AB, Callow MJ, Halpern AL, Burns NL, et al. (2010) Human genome sequencing using unchained base reads on self-assembling DNA nanoarrays. Science 327: 78–81.
9. McKernan KJ, Peckham HE, Costa GL, McLaughlin SF, Fu Y, et al. (2009) Sequence and structural variation in a human genome uncovered by short-read, massively parallel ligation sequencing using two-base encoding. Genome Res 19: 1527–1541.
10. Pushkarev D, Neff NF, Quake SR (2009) Single-molecule sequencing of an individual human genome. Nat Biotechnol 27: 847–850.
11. Ahn SM, Kim TH, Lee S, Kim D, Ghang H, et al. (2009) The first Korean genome sequence and analysis: full genome sequencing for a socio-ethnic group. Genome Res 19: 1622–1629.
12. Durbin RM, Abecasis GR, Altshuler DL, Auton A, Brooks LD, et al. (2010) A map of human genome variation from population-scale sequencing. Nature 467: 1061–1073.
13. Fujimoto A, Nakagawa H, Hosono N, Nakano K, Abe T, et al. (2010) Whole-genome sequencing and comprehensive variant analysis of a Japanese individual using massively parallel sequencing. Nat Genet 42: 931–936.
14. Ju YS, Kim JI, Kim S, Hong D, Park H, et al. (2011) Extensive genomic and transcriptional diversity identified through massively parallel DNA and RNA sequencing of eighteen Korean individuals. Nat Genet 43: 745–752.
15. Kim JI, Ju YS, Park H, Kim S, Lee S, et al. (2009) A highly annotated whole-genome sequence of a Korean individual. Nature 460: 1011–1015.
16. Lee W, Jiang Z, Liu J, Haverty PM, Guan Y, et al. (2010) The mutation spectrum revealed by paired genome sequences from a lung cancer patient. Nature 465: 473–477.
17. Ley TJ, Mardis ER, Ding L, Fulton B, McLellan MD, et al. (2008) DNA sequencing of a cytogenetically normal acute myeloid leukaemia genome. Nature 456: 66–72.
18. Lupski JR, Reid JG, Gonzaga-Jauregui C, Rio Deiros D, Chen DC, et al. (2010) Whole-genome sequencing in a patient with Charcot-Marie-Tooth neuropathy. N Engl J Med 362: 1181–1191.
19. Mardis ER, Ding L, Dooling DJ, Larson DE, McLellan MD, et al. (2009) Recurring mutations found by sequencing an acute myeloid leukemia genome. N Engl J Med 361: 1058–1066.
20. Mills RE, Walter K, Stewart C, Handsaker RE, Chen K, et al. (2011) Mapping copy number variation by population-scale genome sequencing. Nature 470: 59–65.
21. Pelak K, Shianna KV, Ge D, Maia JM, Zhu M, et al. (2010) The characterization of twenty sequenced human genomes. PLoS Genet 6.
22. Roach JC, Glusman G, Smit AF, Huff CD, Hubley R, et al. (2010) Analysis of genetic inheritance in a family quartet by whole-genome sequencing. Science 328: 636–639.
23. Sobreira NL, Cirulli ET, Avramopoulos D, Wohler E, Oswald GL, et al. (2010) Whole-genome sequencing of a single proband together with linkage analysis identifies a Mendelian disease gene. PLoS Genet 6: e1000991.
24. Tong P, Prendergast JG, Lohan AJ, Farrington SM, Cronin S, et al. (2010) Sequencing and analysis of an Irish human genome. Genome Biol 11: R91.

25. Wang J, Wang W, Li R, Li Y, Tian G, et al. (2008) The diploid genome sequence of an Asian individual. Nature 456: 60–65.

26. Wheeler DA, Srinivasan M, Egholm M, Shen Y, Chen L, et al. (2008) The complete genome of an individual by massively parallel DNA sequencing. Nature 452: 872–876.

27. Tennessen JA, Bigham AW, O'Connor TD, Fu W, Kenny EE, et al. (2012) Evolution and functional impact of rare coding variation from deep sequencing of human exomes. Science 337: 64–69.

28. Nelson MR, Wegmann D, Ehm MG, Kessner D, St Jean P, et al. (2012) An abundance of rare functional variants in 202 drug target genes sequenced in 14,002 people. Science 337: 100–104.

29. Genomes Project C, Abecasis GR, Auton A, Brooks LD, DePristo MA, et al. (2012) An integrated map of genetic variation from 1,092 human genomes. Nature 491: 56–65.

30. Levy S, Sutton G, Ng PC, Feuk L, Halpern AL, et al. (2007) The diploid genome sequence of an individual human. PLoS Biol 5: e254.

31. Rosenfeld JA, Malhotra AK, Lencz T (2010) Novel multi-nucleotide polymorphisms in the human genome characterized by whole genome and exome sequencing. Nucleic Acids Res 38: 6102–6111.

32. Habegger L, Balasubramanian S, Chen DZ, Khurana E, Sboner A, et al. (2012) VAT: a computational framework to functionally annotate variants in personal genomes within a cloud-computing environment. Bioinformatics 28: 2267–2269.

33. Harrow J, Frankish A, Gonzalez JM, Tapanari E, Diekhans M, et al. (2012) GENCODE: the reference human genome annotation for The ENCODE Project. Genome Res 22: 1760–1774.

34. Andrews RM, Kubacka I, Chinnery PF, Lightowlers RN, Turnbull DM, et al. (1999) Reanalysis and revision of the Cambridge reference sequence for human mitochondrial DNA. Nat Genet 23: 147.

35. Zhang J, Feuk L, Duggan GE, Khaja R, Scherer SW (2006) Development of bioinformatics resources for display and analysis of copy number and other structural variants in the human genome. Cytogenet Genome Res 115: 205–214.

36. Hindorff LA, Sethupathy P, Junkins HA, Ramos EM, Mehta JP, et al. (2009) Potential etiologic and functional implications of genome-wide association loci for human diseases and traits. Proc Natl Acad Sci U S A 106: 9362–9367.

37. MacArthur DG, Tyler-Smith C (2010) Loss-of-function variants in the genomes of healthy humans. Hum Mol Genet 19: R125–130.

38. MacArthur DG, Balasubramanian S, Frankish A, Huang N, Morris J, et al. (2012) A systematic survey of loss-of-function variants in human protein-coding genes. Science 335: 823–828.

39. Xie C, Tammi MT (2009) CNV-seq, a new method to detect copy number variation using high-throughput sequencing. BMC Bioinformatics 10: 80.

40. Eden E, Navon R, Steinfeld I, Lipson D, Yakhini Z (2009) GOrilla: a tool for discovery and visualization of enriched GO terms in ranked gene lists. BMC Bioinformatics 10: 48.

41. Kin T, Ono Y (2007) Idiographica: a general-purpose web application to build idiograms on-demand for human, mouse and rat. Bioinformatics 23: 2945–2946.

10

CNV Analysis in Tourette Syndrome Implicates Large Genomic Rearrangements in *COL8A1* and *NRXN1*

Abhishek Nag[1], Elena G. Bochukova[2], Barbara Kremeyer[1], Desmond D. Campbell[1], Heike Muller[1], Ana V. Valencia-Duarte[3,4], Julio Cardona[5], Isabel C. Rivas[5], Sandra C. Mesa[5], Mauricio Cuartas[3], Jharley Garcia[3], Gabriel Bedoya[3], William Cornejo[4,5], Luis D. Herrera[6], Roxana Romero[6], Eduardo Fournier[6], Victor I. Reus[7], Thomas L Lowe[7], I. Sadaf Farooqi[2], the Tourette Syndrome Association International Consortium for Genetics, Carol A. Mathews[7], Lauren M. McGrath[8,9], Dongmei Yu[9], Ed Cook[10], Kai Wang[11], Jeremiah M. Scharf[8,9,12], David L. Pauls[8,9], Nelson B. Freimer[13], Vincent Plagnol[1], Andrés Ruiz-Linares[1]*

1 UCL Genetics Institute, Department of Genetics, Evolution and Environment, University College London, London, United Kingdom, 2 University of Cambridge Metabolic Research Laboratories, Institute of Metabolic Science, Addenbrooke's Hospital, Cambridge, United Kingdom, 3 Laboratorio de Genética Molecular, SIU, Universidad de Antioquia, Medellín, Colombia, 4 Escuela de Ciencias de la Salud, Universidad Pontificia Bolivariana, Medellín, Colombia, 5 Departamento de Pediatría, Facultad de Medicina, Universidad de Antioquia, Medellín, Colombia, 6 Hospital Nacional de Niños, San José, Costa Rica, 7 Department of Psychiatry, University of California San Francisco, San Francisco, California, United States of America, 8 Psychiatric and Neurodevelopmental Genetics Unit, Center for Human Genetics Research, Boston, Massachusetts, United States of America, 9 Department of Psychiatry, Massachusetts General Hospital, Boston, Massachusetts, United States of America, 10 University of Illinois, Chicago, Illinois, United States of America, 11 University of Southern California, Los Angeles, California, United States of America, 12 Department of Neurology, Massachusetts General Hospital, Boston, Massachusetts, United States of America, 13 Center for Neurobehavioral Genetics, University of California Los Angeles, Los Angeles, California, United States of America

Abstract

Tourette syndrome (TS) is a neuropsychiatric disorder with a strong genetic component. However, the genetic architecture of TS remains uncertain. Copy number variation (CNV) has been shown to contribute to the genetic make-up of several neurodevelopmental conditions, including schizophrenia and autism. Here we describe CNV calls using SNP chip genotype data from an initial sample of 210 TS cases and 285 controls ascertained in two Latin American populations. After extensive quality control, we found that cases (N = 179) have a significant excess (P = 0.006) of large CNV (>500 kb) calls compared to controls (N = 234). Amongst 24 large CNVs seen only in the cases, we observed four duplications of the *COL8A1* gene region. We also found two cases with ~400kb deletions involving *NRXN1*, a gene previously implicated in neurodevelopmental disorders, including TS. Follow-up using multiplex ligation-dependent probe amplification (and including 53 more TS cases) validated the CNV calls and identified additional patients with rearrangements in *COL8A1* and *NRXN1*, but none in controls. Examination of available parents indicates that two out of three *NRXN1* deletions detected in the TS cases are *de-novo* mutations. Our results are consistent with the proposal that rare CNVs play a role in TS aetiology and suggest a possible role for rearrangements in the *COL8A1* and *NRXN1* gene regions.

Editor: Ge Zhang, Cincinnati Children's Hospital Medical Center, United States of America

Funding: This work was funded by National Institutes of Health (http://www.nih.gov/) grants NS043538 (to ARL), NS037484 (to NBF), MH085057 (to JMS), NS40024 (to DLP and JMS) and NS16648 (the Tourette Syndrome Association International Consortium for Genetics), a CODI grant (http://www.udea.edu.co/portal/page/portal/Programas/GruposInvestigacion/h.Convocatorias/CODI/estrategiaSostenibilidad) (Sostenibilidad, Universidad de Antioquia), a grant from the Judah Foundation (http://www.judahfoundation.org/), and the Tourette Syndrome Association (http://tsa-usa.org/) (fellowship to BK and a grant to DLP). VP is supported by a grant from the United Kingdom Medical Research Council (http://www.mrc.ac.uk) (G1001158) and by the NIHR Moorfields Biomedical Research Council (http://www.brcophthalmology.org/). The funders had no role in study design, data collection and analysis, decision to publish, or preparation of the manuscript.

Competing Interests: The authors have declared that no competing interests exist.

* E-mail: a.ruizlin@ucl.ac.uk

Introduction

TS is a childhood onset neuropsychiatric illness characterised by the occurrence of multiple, motor and vocal tics and is often associated with obsessive-compulsive disorder (OCD) and attention-deficit hyperactivity disorder (ADHD) [1–5]. Twin studies have estimated a sibling relative risk ratio for TS of about 6–8 [2], one of the highest amongst neuropsychiatric disorders. However, identification of genetic variants underlying TS has proven difficult

[5–7]. Genome-wide linkage and candidate gene association studies have failed to provide robust evidence implicating specific loci, and a recent GWAS has not identified common variants associated with TS at genome-wide significance thresholds [8]. The observation of chromosomal abnormalities in TS families [9–11] has suggested the possibility that genomic rearrangements could play an important role in this disorder, but prior studies have provided conflicting evidence regarding the involvement of copy number variants (CNVs) in TS [12,13]. To further evaluate the role of CNVs in TS, we performed a genomewide study of CNVs in a case/control sample from two well-studied, closely related Latin American population isolates.

Ethics Statement

This research was approved by the BioEthics Committee of Universidad de Antioquia (Colombia) and the NHS National Research Ethics Service, Central London Committee REC 4 (UK). Written consent was obtained from all subjects. In the case of minors, written consent was obtained from a parent or legal guardian.

Patients and Methods

We studied CNVs in a sample of 210 unrelated TS cases ascertained in two closely related Latin American population isolates and 285 unrelated population controls. The populations of Antioquia, Colombia, and of the Central Valley of Costa Rica (CVCR) have similar and partly shared demographic histories and are genetically closely related [14,15]. They are therefore expected to show an enrichment for shared predisposing factors for complex genetic conditions, including TS [14–17]. Of the cases, 81 were recruited at the Neuropaediatrics Clinic of Hospital Universitario San Vicente de Paúl (Antioquia, Colombia) and 129 were recruited at Hospital Nacional de Niños (San José, Costa Rica). Diagnosis was based on DSM-IV criteria, focusing on narrowly defined moderate to severe TS. The mean age of cases was 13 years, with a mean age for the start of symptoms at 6.4 years. In addition to TS, 48% of the cases have a diagnosis of ADHD and 53% have OCD. An additional set of 53 TS cases used for MLPA-based follow-up (see below) was also recruited through the Neuropaediatrics Clinic of Hospital Universitario San Vicente de Paúl (Antioquia, Colombia), following the same diagnostic procedures. Population controls were obtained in Antioquia as part of on-going genetic diversity studies in the region [18]. For both, cases and controls, genealogical enquiries confirmed local ancestry in at least 6/8 great-grandparents. Because matched population controls from the CVCR were unavailable, and based on the close genetic relatedness of Antioquia and the CVCR, Antioquian controls were contrasted with Antioquian and Costa Rican cases accounting for stratification (see below). All samples were genotyped using Illumina Human660 arrays as part of the TSAICG genome-wide association study of TS [8].

We obtained CNV calls from the raw hybridization intensities using PennCNV [19]. We excluded from this analysis samples that were outliers based on either the variability of the raw intensity data (using the standard deviation of the logR ratio), or on the total number of CNVs called (see Methods S1 and Figure S2). This resulted in 413 samples being retained for further analysis (179 cases and 234 controls). To make the final CNV calls, we used the following criteria: (i) we merged neighbouring CNVs when the

distance separating them was less than half of the total distance from the start of the first CNV to the end of the second CNV, (ii) we only called CNVs containing at least 10 SNPs, and (iii) we ignored CNVs located in centromeric and telomeric regions.

The CNV burden for each sample was determined by counting all CNVs and stratifying them by size into four categories: <10 kb, 10–100 kb, 100–500 kb and >500 kb. All calls for CNVs >500 kb ("large CNVs") were confirmed individually by plotting the LogR ratio and B allele frequency for the SNPs in the region (Figure S4). The CNV burden was then contrasted between cases and controls using Fisher's exact test.

Principal component analysis (PCA) of the genotype data was performed using EIGENSTRAT [20], as implemented in the EIGENSOFT package (http://genepath.med.harvard.edu/reich/Software.htm).

Results

Overall, in the final dataset we made an average of 3.5 CNV calls per subject with a median CNV length of 76.4 kb. Of these, 60% correspond to deletions and 40% to duplications (Figure S3).

We contrasted the total CNV burden between TS cases and controls, stratified by size into four categories: <10 kb, 10–100 kb, 100–500 kb and >500 kb (Table 1). We found a statistically significant increase in the frequency of CNVs >500 kb in cases (27 or 0.15 per individual) compared to controls (15 or 0.06 per individual; $p = 0.006$). In total, 25 cases (14%) versus 15 controls (6.4%) were found to carry large CNVs, representing an excess of ~7.6% (95% C.I. = 1.6–13.6%, one-sided Fisher's exact test $p = 0.006$). Of the 27 large CNVs found in cases, 24 occurred in regions free of CNVs in controls. Two of the TS cases had two large CNVs each, while no control carried more than one large CNV. Since no controls were available for the CVCR samples, we evaluated the effect of population stratification by testing the correlation of CNV burden with ancestry of the samples, evaluated using PCA. The presence of large CNVs was not correlated with ancestry ($p > 0.05$ for PCs 1 to 4). We also verified that OR estimates for large CNVs are consistent whether the CVCR cases are included (95% ci: 1.27–4.96) or not (95% ci: 1.08–5.95), but as expected from a reduction in sample size, when the burden analysis is restricted to Antioquia the significance decreases (one-sided Fisher's exact test $p = 0.16$). Because cases and controls were genotyped in two batches (one batch of CVCR cases and one batch of Antioquia cases and controls), we also tested for correlation of genotyping batch with the presence of large CNVs, but found no significant effect.

We next explored the potential involvement in TS of CNVs at specific genome regions, stratifying by size. We first examined the 24 (out of 27) regions with CNVs >500 Kb that were detected only in the cases. Of these, 4 did not include exons of any

Table 1. CNV burden in TS cases and controls.

CNV size (kb)	Count in cases	Frequency per case	Count in controls	Frequency per control	p-value
<10	10	0.06	22	0.09	NS
10–100	382	2.13	498	2.13	NS
100–500	194	1.08	300	1.28	NS
>500	27	0.15	15	0.06	0.006
Total	613	3.42	835	3.56	NS

NS = Not significant.

Table 2. Chromosomal regions harbouring large (>500 kb) CNVs overlapping annotated gene exons in at least two TS cases and not in controls.

Location	CNV Type[a]	Start position[b]	End position	Size	# of markers	Gene(s)	Figure
2p22.3	Dup	32,487,194	33,186,442	699,249	145	*BIRC6,TTC27,LTBP1*	S4–4
	Dup	32,487,194	33,174,461	687,268	134	*BIRC6,TTC27,LTBP1*	S4–5
3q12.1	Dup	100,269,291	100,876,782	607,492	105	*COL8A1*	S4–9
	Dup	100,269,291	100,886,715	617,425	113	*COL8A1*	S4–10
	Dup	100,269,291	100,886,715	617,425	108	*COL8A1*	S4–11
	Dup	100,249,016	100,886,715	637,700	105	*COL8A1*	S4–12
5q21.1	Dup	101,503,405	102,033,686	530,282	66	*SLCO4C1,SLCO6A1*	S4–21
	Dup	101,532,676	102,033,686	501,011	70	*SLCO4C1,SLCO6A1*	S4–22

[a]Dup = duplication;
[b]Based on build 36 of the human genome.

annotated gene. The remaining 20 mapped to 15 different genomic regions. Two of these contain genes for uncharacterized proteins with no known functions (*LOC284749* and *FLJ46357*). The remaining 18 large CNVs were located in 13 gene regions (Table S1). Of these regions, 10 presented rearrangements in a single case and some of these regions could be of potential relevance for TS (such a region on 22q11 overlapping DiGeorge's syndrome critical region (Figure S4–43) which has been implicated in rare unusual TS cases [21,22] and has also been found to be associated with schizophrenia [23–25]). Three regions showed rearrangements in more than one TS case. A ~600 Kb region on 3q12.1 (overlapping the *COL8A1* gene) was duplicated in four cases (Table 2). Two other regions on 2p22.3 and 5q21.1 (overlapping the *BIRC6/TTC27/LTBP1* and the *SLCO4C1/ SLCO6A1* genes, respectively) were duplicated in two cases each (Table 2). We also examined genome regions with CNVs <500 kb but focusing solely on those encompassing exons of the same gene in at least two TS cases but not in controls. We identified four such regions, each carrying a CNV in two patients (Table 3). The largest rearrangements (two ~400 kb deletions) encompass exons 1–3 of the *Neurexin1* (*NRXN1*) gene on 2p16.3 (Figures S4–6 and S4–7).

We followed up the *COL8A1* and *NRXN1* findings using multiplex ligation-dependent probe amplification (MLPA; Meth-

ods S1) targeting exons 1 and 2 of *COL8A1* and exons 1 to 4 of *NRXN1* (with two additional probes 3′ and 5′ of this gene) (Table S2). We carried out MLPA in the Antioquian samples included in the SNP-based analysis for which DNA was available (92 cases and 142 controls). We validated the five SNP-based CNV calls (four on *COL8A1* and one on *NRXN1*) made on these samples (Figure S5-1). MLPA identified an additional three *COL8A1* deletions and two *NRXN1* deletions not detected in the SNP-based CNV calls (Figures S5-2 and S5-3). No CNVs in *COL8A1* or *NRXN1* were detected by MLPA in the controls. We also applied the *COL8A1* and *NRXN1* MLPA assay to an additional set of 53 TS cases from Antioquia but did not detect further rearrangements in these individuals. Aggregating the results of the SNP-based CNV calls and MLPA (Table 4), in a total of 232 cases examined we found 7 with rearrangements in *COL8A1* (all from Antioquia) and 4 in *NRXN1* (3 from Antioquia and 1 from the CVCR). None of the 234 Antioquian controls showed rearrangements in these two gene regions in the SNP-based calls or MLPA. To further support the notion that the CNVs observed here are not simply population polymorphisms, we checked the Database of Genomic Variants (DGV; http://dgvbeta.tcag.ca/dgv/app/home), a curated catalogue of human structural variation, for CNVs in the *NRXN1* and *COL8A1* gene regions. While there is a considerable number of CNVs in both regions, all of the CNVs that lie within the

Table 3. Regions harbouring smaller CNVs (<500 kb) overlapping gene exons in at least two TS cases but not in controls.

Location	CNV type[a]	Start position[b]	End position	Size	# of markers	Gene(s)[b]
2p16.3	Del	50,817,046	51,203,727	386,682	103	*NRXN1*[c]
	Del	51,022,554	51,422,546	399,993	86	*NRXN1*[d]
10q23.33	Del	97,352,018	97,391,986	39,969	16	*ALDH18A1*
	Del	97,353,334	97,391,986	38,653	15	*ALDH18A1*
12q24.33	Dup	131,674,763	131,772,074	97,312	20	*P2RX2,POLE*
	Dup	131,665,952	131,772,074	106,123	24	*P2RX2,POLE*
21q22.12	Dup	36,412,525	36,502,751	90,227	23	*CBR3,DOPEY2*
	Dup	36,412,525	36,479,912	67,388	15	*CBR3,DOPEY2*

[a]Dup = duplication; Del = deletion;
[b]Based on build 36 of the human genome;
[c]Figure S4–6;
[d]Figure S4–7.

Table 4. Number of TS cases and controls with CNVs affecting *COL8A1* and *NRXN1* detected using SNP-based calls, MLPA or both.

Gene	SNP-based calls		MLPA[b]				Additional cases N=53[c]	Aggregated totals		p-value[a]
	Cases N=179	Controls N=234	Cases N=92 (of 179)			Controls N=142 (of 234)		Cases N=232	Controls N=234	
			SNP-based calls	Validated	Additional CNVs detected					
COL8A1	4	0	4	4	3	0	0	7	0	0.004
NRXN1	2	0	1	1	2	0	0	4	0	0.03

[a] One-tailed Fisher's exact test.
[b] MLPA was applied to a subset of the samples examined in the initial SNP-based calls.
[c] This set of 53 follow-up samples was not included in the initial SNP-based calls but was examined only with MLPA.

respective gene itself are between a few hundred bp and ~100 kb long, and therefore significantly shorter than the variants described here. More importantly, the majority of these variants do not affect any of the exons of the respective genes, the only exception being a 100 kb deletion affecting *NRXN1* exons 7-9 (DGV Variation_2383). This variant affects a different region from the variants observed here; in addition, it was found only in one out of 540 chromosomes and is therefore also not likely to represent a common population polymorphism. Overall, the size and position of the variants identified here, both in *NRXN1* and *COL8A1*, do not show any overlap with common population polymorphism.

To evaluate the possibility that the *COL8A1* and *NRXN1* rearrangements detected in TS cases could represent *de-novo* mutations, we applied the MLPA assay to the parents of TS cases with rearrangements in these two gene regions. We considered only the patients for which DNA from both parents was available and confirmed relatedness in each trio. This included two cases with *COL8A1* duplications and three cases with *NRXN1* deletions (all from Antioquia). The same duplication was found in a parent in each of the two cases with *COL8A1* duplications examined, indicating that this variant was inherited. This and the observation of similar boundaries for the *COL8A1* duplications in the SNP-based CNV calls (Table 2) suggest that this variant is segregating in the Antioquian population. Deletion of *NRXN1* 5′ exons was found in the father of one of the cases with a *NRXN1* deletion (GT64.1) but not in the parents of the two other cases with this deletion, indicating a *de novo* mutation in these two trios. The father of case GT64.1 has a diagnosis of OCD, a condition that shows significant co-morbidity and may share common predisposing factors with TS (interestingly, the paternal grand-father is reported to have suffered from OCD; however, his CNV type is unknown). One of the two *de novo* *NRXN1* deletions identified occurred in a proband that had no family history of TS (case GT5.1, Figure S5-2a). The second case with a *de novo* *NRXN1* deletion (GT34.1, Figure S5-2b) had a history of TS/OCD on the paternal side of his family.

Discussion

Our results provide statistically significant evidence of a high burden of large CNVs (>500kb) in TS, thereby supporting the proposal for an involvement of rare CNVs in various neurodevelopmental disorders, including TS, and their possible aetiological overlap [12,13,26–28]. We also find suggestive evidence for the involvement of rearrangements specifically affecting the *NRXN1* and *COL8A1* genes. In the aggregated data (Table 4) we find a nominally significant association of *COL8A1* and *NRXN1* rearrangements with TS (p-values of 0.004 and 0.03 respectively). Due to the limited sample size, these p-values would not reach significance accounting for multiple testing. Data from the Database of Genomic Variants further supported the notion that the variants observed here are not part of the spectrum of common population polymorphisms. When considering the trio data, the lack of a straightforward co-segregation between the structural variants observed in our study and the TS phenotype implies the involvement of further predisposing loci in the aetiology of TS; however, this is not unexpected for such a phenotypically and genetically complex condition and does not conflict with a role for *NRXN1* or *COL8A1* in TS predisposition. Overall, our results strongly warrant further investigation of these two genes in TS.

The importance of *NRXN1* in mediating cell-cell interactions in the central nervous system, as well as its confirmed involvement in other neurodevelopmental disorders, make this gene an excellent candidate gene for TS [12,29,30]. Our results are consistent with those of a previous study reporting deletions affecting *NRXN1* exons 1–3 in TS, the same exons found to be deleted in our study [12]. The fact that two of the three *NRXN1* rearrangements, for which inheritance status could be confirmed, were found to be *de novo* events, is in line with recent findings stressing a role for *de novo* mutations in neurodevelopmental disease. The potential involvement of *COL8A1* in TS is intriguing. A growing body of evidence suggests that collagen subunits are involved in neural development, influencing processes such as axonal guidance, synaptogenesis and Schwann cell differentiation [31,32]. *COL8A1* has also been found to be up-regulated during repair processes in the mouse brain [32]. Interestingly, the top signal in the recent GWAS of TS [8] also implicated a collagen gene (*COL27A1*).

In conclusion, our results are consistent with the view that TS is genetically a highly heterogeneous disorder, in which rare variants, including *de-novo* mutations, could underlie a substantial fraction of cases. Recently, Cooper et al (2011) conducted a large-scale study to investigate the role of CNVs in ~15,000 children with intellectual disability and estimated that ~14.2% are due to CNVs >400 kb. Similarly, the 7.6% excess of large CNVs in TS patients observed here could be taken as a rough estimate of the proportion of cases that might be caused by CNVs. The analysis of larger TS study samples should enable a more definite assessment of the role of large rearrangements at specific gene regions in this disorder. More extensive surveys of parent-TS offspring trios are also required to estimate the proportion of cases that could be due to highly penetrant *de-novo* mutations. Finally, sequencing studies should allow a full assessment of the role of rare variants in the aetiology of TS.

Supporting Information

Figure S1   No significant correlation was observed between PCs 1–4 and presence of large CNVs. Left panel: PCA1 versus PCA2. Right panel: PCA3 versus PCA4.

Figure S2   Samples with NumCNV>30 or LRR_SD>0.24 were excluded from subsequent analyses.

Figure S3   The 413 DNA samples that passed QC yielded an average of 14.47 CNV calls per subject. On applying call-level filtering criteria to these calls, an average of 3.50 CNV calls per subject (spanning 10 to 522 SNPs) were obtained. Deletions (865/1448) were more frequently observed compared to duplications (583/1448). Deletions were observed more frequently in the small CNV category while duplications were observed more frequently in the large CNV category (Figure S3).

Figure S4   (1–44): Sample ID, population origin and case/control status are shown as figure heading. LogR ratio and B allele frequency are shown in the top and bottom panels, respectively. CNV boundaries are indicated by red dotted lines. Human RefSeq genes are shown below each panel (vertical lines indicating exons). Genomic position (in Mb) based on the hg18 human genome sequence.

Figure S5   CNV calls using Multiplex Ligation-dependent Probe Amplification (MLPA). Figure S5-1: Validation of the SNP-based CNV calls in *COL8A1* and *NRXN1* by MLPA. Top panel: heterozygous duplication in *COL8A1* (exons 1 and 2). Representative MLPA data and MLPA target probes for *COL8A1* are shown. Bottom panel: Detection of a heterozygous deletion in *NRNX1* (exons 1, 2, 3). MLPA target probes for *NRXN1* are shown, the unlabelled target regions are probes located either on chromosome 2 but outside the deleted region or on other chromosomes (Table S2). Patient MLPA traces are in red, overlaid upon the normal control MLPA traces in black. Arrows point to the deleted/duplicated probes. Figure S5-2: Detection of *de novo* deletions in *NRNX1* (exons 2 and 3) in TS cases. A, trio 5. B, trio 34. Patient MLPA traces are in red overlaid upon the normal control MLPA traces in black. The parents' traces are in blue, overlaid upon normal controls in black. Arrows point to the MLPA probes in *NRXN1*. Figure S5-3: Two additional TS cases (GT5.1 and GT34.1) with deletions involving either exon 1, 2 or 3 of *NRXN1* detected by MLPA. Representative MLPA data are shown. Patient traces are in red, overlaid upon the control traces in black. Arrows point to the MLPA probes in *NRXN1*. Figure S5-4: Three additional TS cases (GT7.1, GT29.1 and GT114.1) with deletion of exon 2 of *COL8A1* detected by MLPA. Representative MLPA data are shown. Patient traces are in red, overlaid upon the control traces in black. Arrows point to the MLPA probes in *COL8A1*.

Figure S6   (1 to 5): Sample ID, population origin and case/control status are shown as figure heading. LogR ratio and B allele frequency are shown in the top and bottom panels, respectively. CNV boundaries are indicated by red dotted lines. The structure of *NRXN1* (Figures S6-1 and S6-2) or *COL8A1* (Figures S6-3 to S6-5) is shown below each panel with exons shown as vertical lines. Genomic position (in Mb) provided make use of the hg18 human genome sequence as reference.

Table S1   Chromosomal regions harbouring large (>500 kb) CNVs overlapping annotated gene exons in TS cases but not in controls. [a]Dup = duplication; [b] According to build 36 of the human genome.

Table S2   Target probes used in the MLPA assay.

Methods S1   CNV Quality Control and CNV validation by Multiplex ligation-dependent probe amplification (MLPA).

Acknowledgments

The authors would like to thank all volunteers who participated in the study.

Author Contributions

Conceived and designed the experiments: VP ARL. Performed the experiments: AN EB BK HM AVVD VP. Analyzed the data: AN EB BK DDC VP ARL. Contributed reagents/materials/analysis tools: JC ICR SCM MC JG GB WC LDH RR EF VIR TLL ISF CAM LMM DY EC KW JMS DLP NBF. Wrote the paper: AN BK VP ARL.

References

1. American Psychiatric Association (2000) Diagnostic and Statistical Manual of Mental Disorders. Washington, DC: American Psychiatric Association.

2. Price RA, Kidd KK, Cohen DJ, Pauls DL, Leckman JF (1985) A twin study of Tourette syndrome. ArchGenPsychiatry 42: 815–820.

3. Saccomani L, Fabiana V, Manuela B, Giambattista R (2005) Tourette syndrome and chronic tics in a sample of children and adolescents. Brain Dev 27: 349–352.

4. Stewart SE, Illmann C, Geller DA, Leckman JF, King R, et al. (2006) A controlled family study of attention-deficit/hyperactivity disorder and Tourette's disorder. J Am Acad Child Adolesc Psychiatry 45: 1354–1362.

5. O'Rourke JA, Scharf JM, Yu D, Pauls DL (2009) The genetics of Tourette syndrome: a review. J Psychosom Res 67: 533–545.

6. State MW (2010) The genetics of child psychiatric disorders: focus on autism and Tourette syndrome. Neuron 68: 254–269.

7. State MW (2011) The genetics of Tourette disorder. Curr Opin Genet Dev 21: 302–309.

8. Scharf JM, Yu D, Mathews CA, Neale BM, Stewart SE, et al. (2012) Genome-wide association study of Tourette's syndrome. Mol Psychiatry.

9. Verkerk AJ, Mathews CA, Joosse M, Eussen BH, Heutink P, et al. (2003) CNTNAP2 is disrupted in a family with Gilles de la Tourette syndrome and obsessive compulsive disorder. Genomics 82: 1–9.

10. Abelson JF, Kwan KY, O'Roak BJ, Baek DY, Stillman AA, et al. (2005) Sequence variants in SLITRK1 are associated with Tourette's syndrome. Science 310: 317–320.

11. Ercan-Sencicek AG, Stillman AA, Ghosh AK, Bilguvar K, O'Roak BJ, et al. (2010) L-histidine decarboxylase and Tourette's syndrome. N Engl J Med 362: 1901–1908.

12. Sundaram SK, Huq AM, Wilson BJ, Chugani HT (2010) Tourette syndrome is associated with recurrent exonic copy number variants. Neurology 74: 1583–1590.

13. Fernandez TV, Sanders SJ, Yurkiewicz IR, Ercan-Sencicek AG, Kim YS, et al. (2012) Rare copy number variants in tourette syndrome disrupt genes in histaminergic pathways and overlap with autism. Biol Psychiatry 71: 392–402.

14. Carvajal-Carmona LG, Ophoff R, Service S, Hartiala J, Molina J, et al. (2003) Genetic demography of Antioquia (Colombia) and the Central Valley of Costa Rica. Hum Genet 112: 534–541.

15. Service S, Deyoung J, Karayiorgou M, Roos JL, Pretorious H, et al. (2006) Magnitude and distribution of linkage disequilibrium in population isolates and implications for genome-wide association studies. Nat Genet 38: 556–560.

16. Herzberg I, Valencia-Duarte AV, Kay VA, White DJ, Muller H, et al. (2010) Association of DRD2 variants and Gilles de la Tourette syndrome in a family-based sample from a South American population isolate. Psychiatr Genet 20: 179–183.

17. Herzberg I, Jasinska A, Garcia J, Jawaheer D, Service S, et al. (2006) Convergent linkage evidence from two Latin American population isolates supports the presence of a susceptibility locus for bipolar disorder in 5q31–34. Hum Mol Genet 15: 3146–3153.

18. Wang S, Ray N, Rojas W, Parra MV, Bedoya G, et al. (2008) Geographic patterns of genome admixture in Latin American Mestizos. PLoS Genet 4: e1000037.

19. Wang K, Li M, Hadley D, Liu R, Glessner J, et al. (2007) PennCNV: an integrated hidden Markov model designed for high-resolution copy number variation detection in whole-genome SNP genotyping data. Genome Res 17: 1665–1674.

20. Price AL, Patterson NJ, Plenge RM, Weinblatt ME, Shadick NA, et al. (2006) Principal components analysis corrects for stratification in genome-wide association studies. NatGenet.

21. Robertson MM, Shelley BP, Dalwai S, Brewer C, Critchley HD (2006) A patient with both Gilles de la Tourette's syndrome and chromosome 22q11 deletion syndrome: clue to the genetics of Gilles de la Tourette's syndrome? J Psychosom Res 61: 365–368.

22. Clarke RA, Fang ZM, Diwan AD, Gilbert DL (2009) Tourette syndrome and klippel-feil anomaly in a child with chromosome 22q11 duplication. Case Report Med 2009: 361518.

23. Karayiorgou M, Morris MA, Morrow B, Shprintzen RJ, Goldberg R, et al. (1995) Schizophrenia susceptibility associated with interstitial deletions of chromosome 22q11. Proc Natl Acad Sci U S A 92: 7612–7616.

24. Liu H, Abecasis GR, Heath SC, Knowles A, Demars S, et al. (2002) Genetic variation in the 22q11 locus and susceptibility to schizophrenia. Proc Natl Acad Sci U S A 99: 16859–16864.

25. Xu B, Roos JL, Levy S, van Rensburg EJ, Gogos JA, et al. (2008) Strong association of de novo copy number mutations with sporadic schizophrenia. Nat Genet 40: 880–885.

26. Scharf JM, Mathews CA (2010) Copy number variation in Tourette syndrome: another case of neurodevelopmental generalist genes? Neurology 74: 1564–1565.

27. Merikangas AK, Corvin AP, Gallagher L (2009) Copy-number variants in neurodevelopmental disorders: promises and challenges. Trends Genet 25: 536–544.

28. Itsara A, Cooper GM, Baker C, Girirajan S, Li J, et al. (2009) Population analysis of large copy number variants and hotspots of human genetic disease. Am J Hum Genet 84: 148–161.

29. Vrijenhoek T, Buizer-Voskamp JE, van der Stelt I, Strengman E, Sabatti C, et al. (2008) Recurrent CNVs disrupt three candidate genes in schizophrenia patients. Am J Hum Genet 83: 504–510.

30. Glessner JT, Wang K, Cai G, Korvatska O, Kim CE, et al. (2009) Autism genome-wide copy number variation reveals ubiquitin and neuronal genes. Nature 459: 569–573.

31. Fox MA (2008) Novel roles for collagens in wiring the vertebrate nervous system. Curr Opin Cell Biol 20: 508–513.

32. Hubert T, Grimal S, Carroll P, Fichard-Carroll A (2009) Collagens in the developing and diseased nervous system. Cell Mol Life Sci 66: 1223–1238.

11

Substrate-Dependent Evolution of Cytochrome P450: Rapid Turnover of the Detoxification-Type and Conservation of the Biosynthesis-Type

Ayaka Kawashima, Yoko Satta*

Department of Evolutionary Studies of Biosystems, The Graduate University for Advanced Studies (Sokendai), Shonan Village, Hayama, Kanagawa, Japan

Abstract

Members of the cytochrome P450 family are important metabolic enzymes that are present in all metazoans. Genes encoding cytochrome P450s form a multi-gene family, and the number of genes varies widely among species. The enzymes are classified as either biosynthesis- or detoxification-type, depending on their substrates, but their origin and evolution have not been fully understood. In order to elucidate the birth and death process of cytochrome *P450* genes, we performed a phylogenetic analysis of 710 sequences from 14 vertebrate genomes and 543 sequences from 6 invertebrate genomes. Our results showed that vertebrate detoxification-type genes have independently emerged three times from biosynthesis-type genes and that invertebrate detoxification-type genes differ from vertebrates in their origins. Biosynthetic-type genes exhibit more conserved evolutionary processes than do detoxification-type genes, with regard to the rate of gene duplication, pseudogenization, and amino acid substitution. The differences in the evolutionary mode between biosynthesis- and detoxification-type genes may reflect differences in their respective substrates. The phylogenetic tree also revealed 11 clans comprising an upper category to families in the cytochrome P450 nomenclature. Here, we report novel clan-specific amino acids that may be used for the qualitative definition of clans.

Editor: Vladimir N. Uversky, University of South Florida College of Medicine, United States of America

Funding: This work was supported by Japan Society for the Promotion of Science (JSPS) Grants-in-Aid for Scientific Research (B) Grant Number 21370106 and for JSPS Fellows. This work was also supported in part by the Center for the Promotion of Integrated Sciences (CPIS) of Sokendai. The funders had no role in study design, data collection and analysis, decision to publish, or preparation of the manuscript.

Competing Interests: The authors have declared that no competing interests exist.

* Email: satta@soken.ac.jp

Introduction

Enzymes in the cytochrome P450 (CYP) family are heme-binding proteins. The first CYP protein was discovered in rat liver microsomes [1], and it was later functionally characterized as a monooxygenase [2]. Monooxygenases incorporate one of the two atoms of molecular oxygen into the substrate, which results in hydroxylation in the most cases. *CYP* genes form a multi-gene family and encode proteins with amino-acid sequence identities higher than 40%. Each family comprises subfamilies with amino-acid sequence identities higher than 55%. In the classification of CYPs, a clan is defined as a higher-order category of CYP families [3]. Although clans can be useful for defining the relationships among *CYP* genes in different phyla within each kingdom [4], the definition of "clan" is rather arbitrary compared with the definitions of "family" and "subfamily."

CYP genes are present in vertebrates, invertebrates, plants, fungi, and even some prokaryotes [5]. The number of known *CYP* genes in metazoan, plant, and fungus genomes is moderately large. For example, there are 115 *CYP* genes in the human genome, 97 in the sea squirt (*Ciona intestinalis*) (Cytochrome P450 Homepage: http://drnelson.uthsc.edu/cytochromep450.html [6]), 120 in the sea urchin (*Strongylocentrotus purpuratus*) [7], 457 in rice (*Oryza sativa*) [6], 272 in *Arabidopsis thaliana* [8], and 159 in *Aspergillus oryzae* [9]. In contrast, there are relatively a small number of *CYP* genes in eubacteria or Archaea, ranging from none in *Escherichia coli* to 33

in *Streptomyces avermitilis* [9]. Among metazoan *CYP* genes, *CYP51* is particularly conserved and participates in the synthesis of cholesterol, which is an essential component of the eukaryotic cell membrane. A possible prokaryotic homolog (*CYP51B1*) to the metazoan *CYP51* is reported in the genome of *Mycobacterium tuberculosis* [10]. It is therefore thought that *CYP51* is the most ancient *CYP* gene. Although the functional role of CYPs in prokaryotes is not well defined [11,12,13], the presence of eukaryotic *CYP* genes in prokaryotes indicates that the emergence of *CYP*s preceded the origin of eukaryotes [14]. However, it has also been suggested that bacterial *CYP51* arose through lateral transfer from plants [15]. Indeed, the absence of *CYP* genes in some bacteria, such as *E. coli*, suggests that they are not essential in prokaryotes.

CYPs are classified into two types, the detoxification type (D-type) and the biosynthesis type (B-type), on the basis of their substrates [4,16,17]. In humans, the D-type detoxifies xenobiotics such as plant alkaloids, aromatic compounds, fatty acids, and especially drugs. On the other hand, the B-type is involved in the biosynthesis of physiologically active chemicals such as steroids, cholesterols, vitamin D3, and bile acids. In general, however, only a few CYPs have well-defined substrates, and even in humans, the substrates of some CYPs remain unidentified. For this reason, the nomenclature of vertebrate CYPs (family or subfamily) is largely determined by significant phylogenetic clustering with known functional sequences in humans. We also adopt this phylogenetic method to classify each vertebrate gene as either B- or D-type.

Fission yeasts have two B-type *CYP* genes (*CYP51F1* and *CYP61A3*) but no D-type genes. D-type genes are therefore thought to have emerged from a B-type gene in eukaryotes [4,14,16]. Nelson *et al.* [9] performed a phylogenetic analysis of 1,572 metazoan *CYP*s and reported the presence of 11 clans. They further investigated the origin and evolutionary processes of these clans. However, the length of intermodal branches connecting the different clans were too short to unambiguously discern the phylogenetic relationships between clans, making it difficult to fully characterize early *CYP* evolution.

In vertebrates, the number of *CYP* genes per genome varies greatly between species; for example, humans have 115, mice have 185, and zebrafish have 81 (Cytochrome P450 Homepage). Among the 115 human *CYP*s, 57 are functional and 58 are pseudogenes, suggesting rapid gene turnover. The 115 genes constitute 18 families, and the number of subfamilies within each family ranges from 1 to 13. These *CYP* genes are distributed on all chromosomes except chromosomes 5, 16, and 17. Five clusters of closely related genes are located on chromosomes 1, 7, and 10 (one cluster each) and chromosome 19 (two clusters) [18,19]. One such cluster on chromosome 19 has been studied in primates and rodents as well from an evolutionary viewpoint, indicating that an initial tandem duplication occurred in an early mammalian ancestor and that gene duplications and/or rearrangements frequently occurred in a lineage-specific manner [20].

In this paper, we aim to elucidate the birth and death processes of vertebrate *CYP* genes. In particular, we compare and contrast the origin and evolution of B- and D-types, and present an evolutionary model of vertebrate *CYP* genes.

Materials and Methods

Sequence datasets and identification of B- and D-type genes in vertebrates and invertebrates

The nucleotide sequences of 115 *CYP* genes in the human genome were obtained from the Cytochrome P450 Homepage. Using these sequences as queries, we performed a basic local alignment search tool (BLAST) search by using BLASTn and downloaded coding sequences (CDS) of homologous nucleotide sequences from 14 vertebrate species (*Pan troglodytes*: CHIMP2.1.4, *Macaca mulatta*: MMUL_1, *Callithrix jacchus*: C_jacchus3.2.1, *Bos taurus*: UMD 3.1, *Canis lupus familiaris*: CanFam3.1, *Mus musculus*: GRCm38.p2, *Rattus norvegicus*: Rnor_5.0, *Monodelphis domestica*: BROADO5, *Gallus gallus*: Galgal4, *Taeniopygia guttata*: 3.2.4, *Anolis carolinensis*: AnoCar2.0, *Xenopus tropicalis*: JGI 4.2, *Oryzias latipes*: MEDAKA1.70, and *Danio rerio*: Zv9) from NCBI (http://www. ncbi.nlm.nih.gov/) or ENSEMBL databases (http://www. ensembl.org/index.html). In the BLAST search, the top two hits and the top five hits were retrieved when B- and D-type genes were used as queries, respectively. The nucleotide sequences of refseq from NCBI were obtained, and sequences from ENSEMBL were filtered by length (>1000 bp) and their identity with human genes. The extent of sequence identity was dependent on the divergence time between each vertebrate species and humans. For example, in fish, we filtered out sequences with identity >60%. Orthology was confirmed by the presence of a syntenic region and the presence of adjacent loci, if any.

The following invertebrate species were included in the analysis: amphioxus (*Branchiostoma floridae*), sea squirt (*C. intestinalis*), sea urchin (*S. purpuratus*), sea anemone (*Nematostella vectensis*), water flea (*Daphnia pulex*), and fruit fly (*Drosophila melanogaster*). Protein sequences obtained from the Cytochrome P450 Homepage were used for the analysis of invertebrate CYPs. Only protein sequences >350 amino acids in length were included in the phylogenetic analysis. Because of the too extensive sequence divergence between vertebrate and invertebrate *CYP* genes, BLAST searches of the NCBI and ENSEMBL databases were not performed.

Molecular evolutionary analysis

Vertebrate nucleotide sequences and invertebrate amino acid sequences in *CYP* coding regions were aligned separately using ClustalW [21] implemented in MEGA5 [22], and each alignment was further edited by hand. In the alignment of the vertebrate nucleotide sequences, we first translated them into the amino acid sequences and after checked by eye, reconverted them to the nucleotide sequences. We excluded sites at which >20% of the operating taxonomic units (OTUs) showed gaps. As a result, 28.7% of the aligned sites showed >60% identity, 48.5% showed >50% identity, and 71.9% showed >30% identity (data not shown). We then constructed Neighbor-joining (NJ) trees [23] using either nucleotide differences per site (p-distance) [24] or amino acid distances (JTT distance) [25]. We performed missing data treatment under both the pairwise deletion and complete deletion options. The maximum likelihood (ML) [26] method was used to test the tree topology. All methods for tree construction were implemented in MEGA5 [22].

Pseudogenization or deletion of genes

The nucleotide sequences of the *CYP* pseudogenes in the human genome were obtained from the Cytochrome P450 Homepage. We selected genes containing >1000 bp out of the 1500 bp CDS. We retrieved orthologous genes from other vertebrate genomes by performing BLAST searches, using the human sequences as queries. The orthologous sequences were aligned with their human counterparts by ClustalW. Based on this alignment, we searched other vertebrates for nonsense or frame-shift mutations found in humans. To estimate the time of pseudogenization, we calculated the ratio of non-synonymous substitutions to synonymous substitutions, always per site, for pairs comprising a pseudogene and an orthologous functional gene. Using this ratio, we estimated the pseudogenization time for all *CYP* pseudogenes from the formula in Sawai *et al.* [27]. We used the TimeTree (http://www.timetree.org/index.php [28]) as a reference for calibrating species divergence time. When an orthologous gene is absent (deletion) in any non-human vertebrate, we searched for the syntenic region in the genome in order to confirm the deletion.

Estimation of functional constraint

In order to compare the amino-acid substitution rate for each functional *CYP* gene in primates, we normalized the rate with the synonymous substitution rate. This normalization for each gene measured the degree of "functional constraint". To be complete, we assumed that the gene tree is the same as the species tree for four primates (humans, chimpanzees, rhesus macaques, and marmosets) and placed the numbers of synonymous and non-synonymous substitutions on each branch by the least squares method [29]. The degree of functional constraint $1 - f$ is obtained from the ratio (f) of the sum of non-synonymous substitutions to that of synonymous substitutions of all branches in each tree. Finally, we compared the degree of functional constraint or directly the f value between B-type and D-type genes by using the Mann–Whitney U test [30].

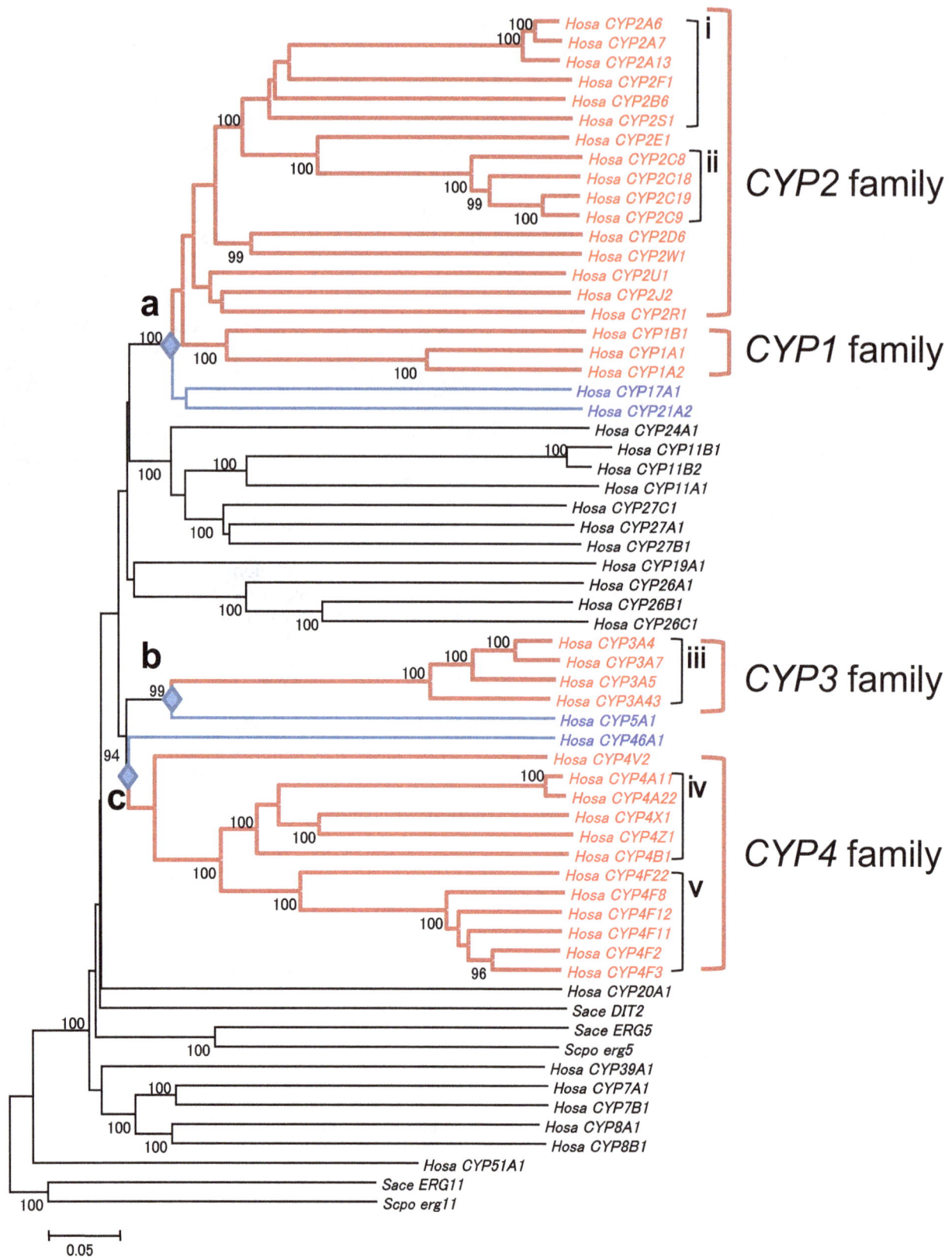

Figure 1. Phylogenetic tree of *cytochrome P450* genes in humans. The tree includes all functional *CYP* genes in humans (*Hosa*) and all yeasts (*Sace*, *Saccharomyces cerevisiae*; *Scpo*, *Schizosaccharomyces pombe*). The tree was constructed using the NJ method for nucleotide differences between the CDS and rooted with yeast *CYP51* gene sequences (*Sace ERG11* and *Scpo erg11*). Red text indicates D-type *CYP* genes, and black and blue text indicate B-type *CYP* genes. The *CYP1–4* families are indicated by a red bracket on the right side of the tree. Three diamonds (*a*, *b*, and *c*) indicate duplications of B- and D-type genes. The B-type genes that were the ancestors of D-type genes are indicated with a blue line and character. Black

brackets and roman numerals (i–v) at the tips of the tree show five clusters of D-type genes: i, the *CYP2* family on chromosome 19q; ii, the *CYP2C* subfamily on chromosome 10q; iii, the *CYP3A* subfamily on chromosome 7q; iv, the *CYP4* family on chromosome 1p; v, the *CYP4F* subfamily on chromosome 19p. The number near each node indicates the bootstrap value (>94%) supporting the node.

Results

Vertebrate D-type genes emerged independently three times from B-type genes

Among the 57 functional *CYP* genes in the human genome, 35 are D-type genes and 22 are B-type genes. This classification is based on the description of the enzyme substrate [31], if any, and subfamily or family classification [6]. D-type genes constitute four *CYP* families: *CYP1* (3 genes), *CYP2* (16 genes), *CYP3* (4 genes), and *CYP4* (12 genes). B-type genes are grouped into 14 families: *CYP5* (1 gene), *CYP7* (2 genes), *CYP8* (2 genes), *CYP11* (3 genes), *CYP17* (1 gene), *CYP19* (1 gene), *CYP20* (1 gene), *CYP21* (1 gene), *CYP24*

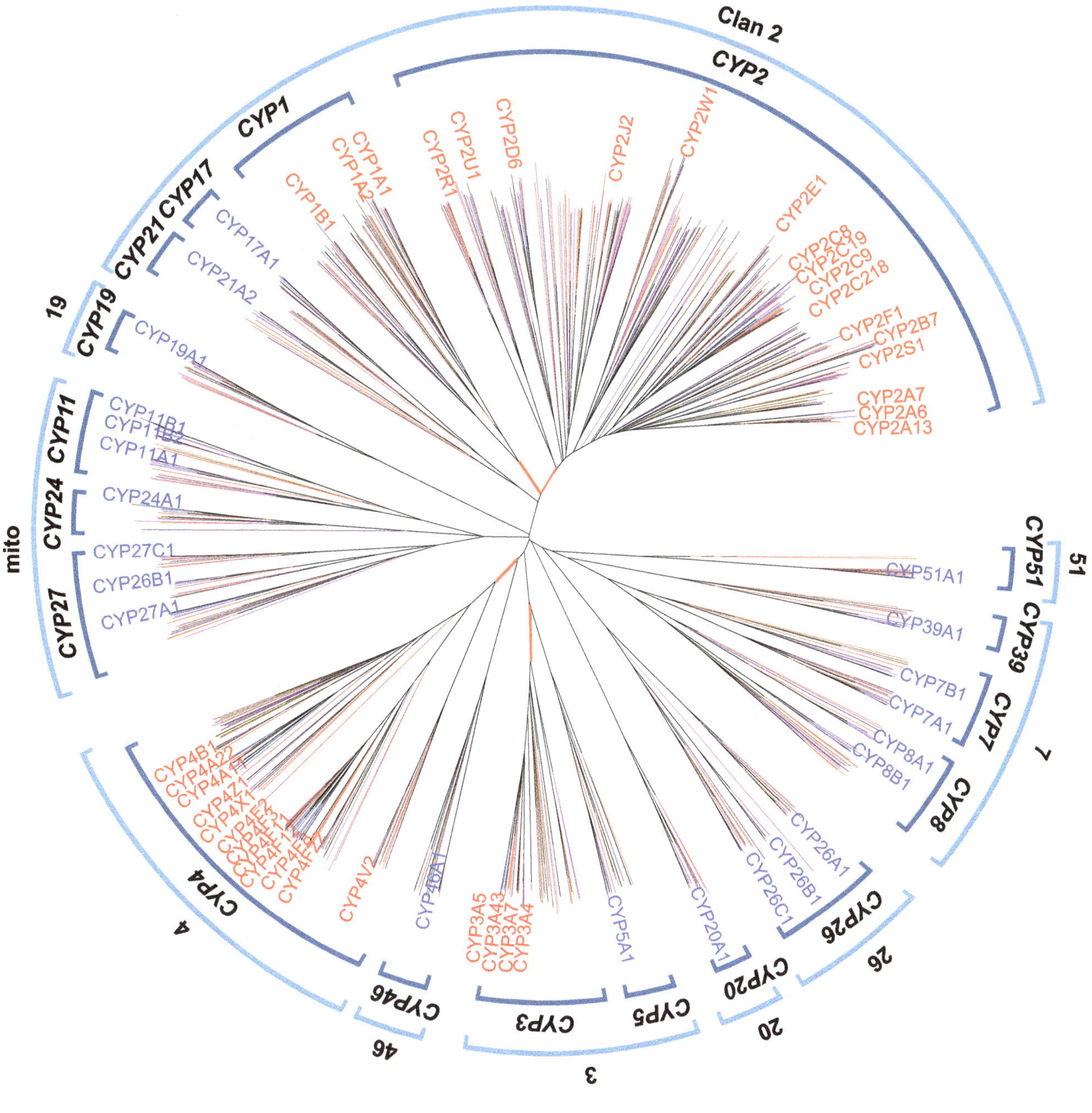

Figure 2. The phylogenetic tree of B- and D-type *CYP* genes in vertebrates. An internal bracket at the tips of the tree indicates the *CYP* family in vertebrates, and an external bracket indicates clusters for a clan. D-type *CYP* genes in humans are shown in red, and B-type *CYP* genes are shown in blue. Red-shaded branches indicate the divergence of D-type from B-type. Text colors indicate the following: red for D- and blue for B-type in humans; dark brown for *Bos taurus*; light blue for *Canis lupus familiaris*; pink for *Mus musculus*; aqua for *Rattus norvegicus*; dark red for *Monodelphis domestica*; dark orange for *Gallus gallus*; purple for *Taeniopygia guttata*; brown for *Anolis carolinensis*; blue-purple for *Xenopus tropicalis*; red-purple for *Oryzias latipes*; orange for *Danio rerio*.

Table 1. Presence or absence of vertebrate orthologs to human *CYP* genes.

CYP family	genes	Species name													
		Patr	Mamu	Caja	Bota	Cafa	Mumu	Rano	Modo	Anca	Gaga	Tagu	Xetr	Orla	Dare
1	A1	1	1	1	1	1	1	1							
	A2	1	1	1	1	1	1	1							
	A1 or A2								2	3	1	1	1	1	1
	B1	1	1	1	1	1	1	1	1	1	0	1	1	1	1
	Others	0	0	0	1	0	0	0	1	2	1	0	3	2	3
2	A6	0	1	0	0	0									
	A7	1	2	0	0	1									
	A13	1	2	1	1	1									
	Other A*						4	2	1	0	0	0	0	0	0
	B6	1	1	1	1	1	4	4	2	0	0	0	0	0	0
	C8	1	1	1											
	C9	1	1	1											
	C18	1	1	1											
	C19	1	1	1											
	Other C*				7	2	9	6	8	0	0	0	0	0	0
	D6	1	1	1	2	0	5	5	1	1	1	1	5	0	0
	E1	1	1	1	1	1	1	1	1	0	0	0	0	0	0
2	F1	1	1	1	1	3	1	1	1	1	0	0	0	0	0
	J2	1	1	1	5	1	6	3	6	1	4	2	1	0	0
	R1	1	1	1	1	1	1	0	1	1	1	1	1	1	1
	S1	1	1	1	1	1	2	1	1	0	0	0	0	0	0
	U1	1	1	1	1	1	1	1	1	0	0	1	1	1	1
	W1	1	1	1	1	1	1	1	1	2	1	1	0	0	0
	Others	1	1	1	1	2	1	2	1	15	7	7	18	10	20
3	A4	1	1	1											
	A5	1	1	1											
	A7	1	1	1											
	A43	1	1	1											
	Other A*				3	4	8	4	4	3	2	1	5	1	1
	Others	0	0	0	0	0	0	0	0	0	0	0	0	2	5
4	A11	1	1	1	2	1									
	A22	1	1	1	2	1									
	A11 or 22						5	1	0	0	0	0	0	0	0
	OtherA*	0	0	0	0	3									
	B1	1	1	1	1	1	2	1	1	6	2	2	4	2	3
	F2	1	1	1											
	F3	1	1	1											
	F8	1	1	1											
	F11	1	1	1											
	F12	1	1	1											
	F22	1	1	1											
	Other F*				6	3	9	8	6	1	0	1	3	1	1
	V2	1	1	1	1	1	1	1	0	1	1	0	2	1	2
	X1	1	1	1	1	2	1	1	2	0	0	0	0	0	0
	Z1	1	1	1	0	0	0	0	0	0	0	0	0	0	0
	5A1	1	1	1	1	1	1	1	1	1	1	1	1	1	1
7	A1	1	1	1	1	1	1	1	1	1	1	1	1	1	1
	B1	1	1	1	1	1	1	1	1	1	1	1	0	0	1

Table 1. Cont.

CYP family	genes	Patr	Mamu	Caja	Bota	Cafa	Mumu	Rano	Modo	Anca	Gaga	Tagu	Xetr	Orla	Dare
Species name															
8	A1	1	1	1	1	1	1	1	1	0	0	0	1	1	1
	B1	1	1	1	1	1	1	1	2	1	1	1	2	1	3
11	A1	1	1	1	1	1	1	1	1	0	1	1	1	1	1
	B1	1	1	1	1	0	1	1	1	1	0	0	0	0	1*
	B2	1	1	1	0	1	1*	1*	0	0	0	0	0	0	0
	17A1	1	1	1	2	1	1	1	0	1	1	1	1	1	2
	19A1	1	1	1	1	1	1	1	1	1	1	1	1	2	2
	20A1	1	1	1	1	1	1	1	1	1	1	1	1	1	1
	21A2	1	1	1	1	1	1	1	1	1	1	1	1	2	2
	24A1	1	1	1	1	1	1	1	1	1	0	2	2	1	1
26	A1	1	1	1	1	1	1	1	1	1	1	1	1	1	1
	B1	1	1	1	1	1	1	1	0	0	1	1	1	1	1
	C1	1	0	1	1	1	1	1	1	1	1	1	1	1	1
27	A1	1	1	1	1	1	1	1	1	1	0	2	1	2	2
	B1	1	1	1	1	1	1	1	1	1	0	0	1	1	1
	C1	1	1	1	1	1	0	0	1	1	1	1	1	1	1
	39A1	1	1	1	1	1	1	1	1	1	1	1	1	0	1
	46A1	1	1	1	1	1	1	1	0	1	1	1	2	2	2
	51A1	1	1	1	1	1	1	1	1	1	1	1	1	1	1

Others: *CYP* genes are not orthologs to the human *CYP* genes listed here but are subfamily members belonging to each family.
*: Genes are included in the subfamily, but the subfamily number differs from that in humans.

(1 gene), *CYP26* (3 genes), *CYP27* (3 genes), *CYP39* (1 gene), *CYP46* (1 gene), and *CYP51* (1 gene) (Table S1). Using the definition proposed by Nelson [32], the 57 CYPs can be classified into 10 clans: clans 2, 3, 4, mito, 7, 19, 20, 26, 46, and 51. Clan "mito" contains genes encoding enzymes that operate in mitochondria. Of the 10 clans, 6 (2, 3, 4, mito, 7, and 26) contain more than two families, whereas 4 (19, 20, 46, and 51) contain only one single family. The amino acid alignment of the 57 functional *CYP* genes showed that four amino aid sites are conserved. Two of these (310F and 316C) sare located near the heme-binding region (Figure S1). The latter site (316C) is known to be structurally close to the iron ion in the heme-binding region and to operate as an active center of the enzyme [33]. This conserved cysteine is said as the proximal Cys [33]. The other two sites (242E and 245R) are located about 80 amino acids upstream from the proximal Cys. Although it is unknown whether these amino acids are involved in any specific function, their conservation suggests some evolutionary or functional importance. Furthermore, several clan-specific amino acids were found in the 57 functional human *CYP*s (Figure S2). Some of them were conserved not only in vertebrates but also in metazoans, although the number of conserved sites correlates with the number of genes in each clan (data not shown).

To characterize the phylogenetic relationships among the 57 functional *CYP* genes in the human genome, an NJ tree was constructed based on the total nucleotide differences (p-distances) between the CDSs (Figure 1). In the resulting tree, members of each family formed monophyletic groups with respect to other families, and each monophyletic group was supported by a relatively high bootstrap value. The phylogeny showed that D-type genes emerged independently from B-type genes at least three times: first, an ancestral gene of *CYP17A1* and *CYP21A1* was

duplicated, generating the ancestor of the *CYP1* and *CYP2* families (node *a* in the tree, Figure 1). Second, the *CYP3A* subfamily arose from the common ancestor of *CYP3* and *CYP5* (node *b* in the tree). Third, an ancestor of *CYP46A1* was duplicated, generating the ancestor of the *CYP4* family (node *c* in the tree). All nodes (*a*, *b*, and *c*) were supported by high bootstrap values (94~100% in Figure 1). In addition to these bootstrap values, amino acids that could distinguish B- from D-type genes were also identified (Figure S3). For example, an amino acid site in the middle of the sequence supported node *a*. In the D-type genes, F was shared by all members of the *CYP1* family whereas V was shared by all members of the *CYP2* family. In contrast, the B-type *CYP*s, *CYP17A1*, and *CYP21A1*, shared T at that site. Similarly, several other amino acid changes that support nodes *a*, *b*, and *c* were observed (Figure S3).

To investigate the duplicaion times of three major D-types from their ancestral B-types, orthologs and paralogs of human B-type and D-type *CYP* genes were retrieved from 14 vertebrate genomes. This resulted in a total of 710 *CYP* nucleotide sequences so that we examined twice as many vertebrate sequences as in the previous study (388) [9]. The presence or absence of vertebrate orthologs to the 57 functional human genes is summarized in Table 1, showing that almost all 14 genomes contain orthologs of B-type genes. We used the pairwise deletion option and constructed a phylogenetic tree (Figure 2); its topology readily confirmed the orthologous relationship between human and other vertebrate B-type genes. However, it was difficult to identify orthologous relationships between D-type genes from humans and other vertebrates, especially in the *2A*, *2C*, *3A*, and *4F* subfamilies, owing to frequent species-specific duplications. Nevertheless, monophyletic relationships within each D-type family (*CYP1–4*) were observed with

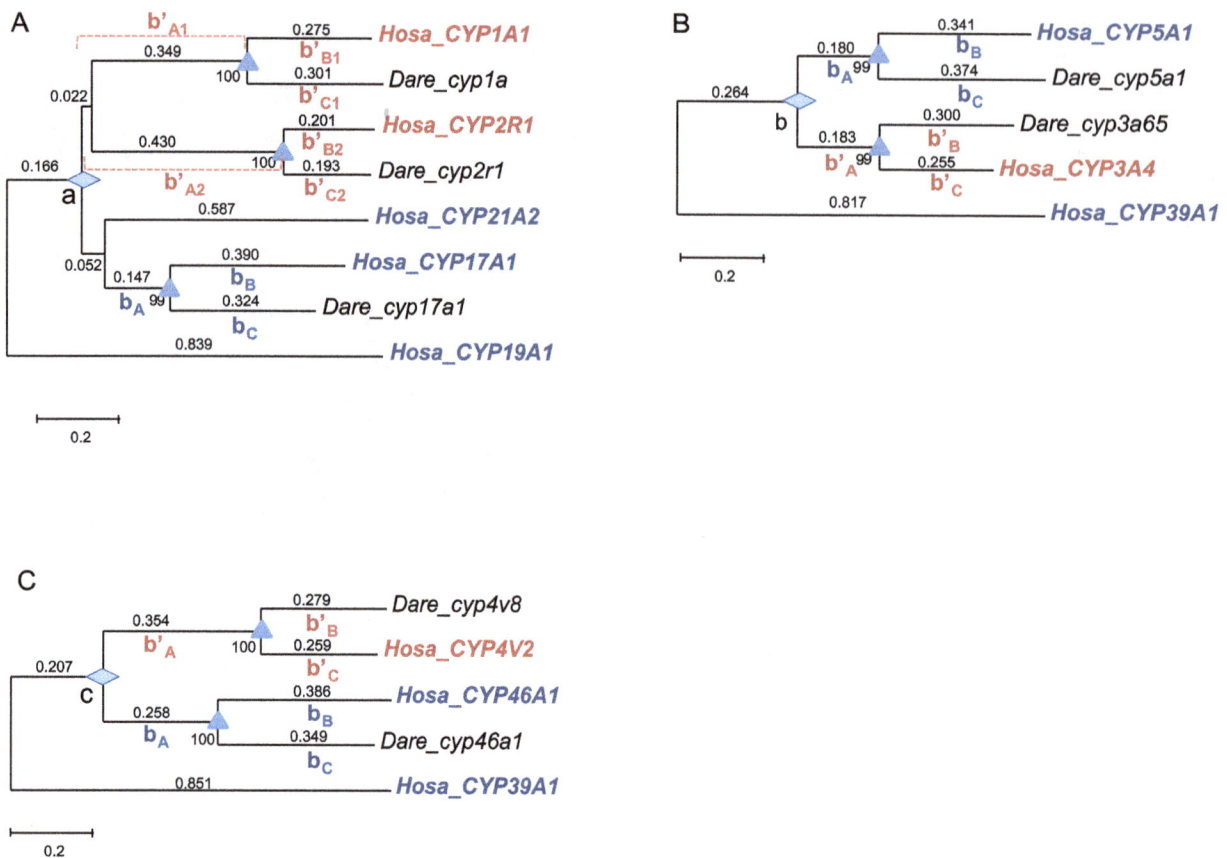

Figure 3. Duplication time of B-type and D-type genes. A) The divergence between *CYP1/2* and *17A1/21A2*. B) The divergence between *CYP3A* and *5A1*. C) The divergence between *CYP4* and *46A1*. The divergence between humans and zebrafish was used as a calibration time (=400 mya), and is shown as a triangle in each tree. The duplication event is shown as a diamond. b_A, b_B, and b_C represent the branch length between the duplication event and species divergence. The branches after species divergence in B-type genes and the branches b'_A, b'_B, and b'_C represent the length of D-type genes. The number near each branch shows the branch length.

relatively high bootstrap values (>80%), so that vertebrate genes in each monophyletic group are classified as the D-type. The phylogenetic analysis revealed that human D- and B-type genes had already emerged when vertebrates diverged, and that three duplication events occurred in the B-type genes from which the D-type genes were originated.

Assuming a molecular clock and that zebrafish and humans diverged 400 million years ago (mya) (TimeTree; http://www. timetree.org/), we calculated the total branch lengths leading to both B- and D-type genes (branch b_A, b_B, and b_C to B-type and b'_A, b'_B, and b'_C to D-type in Figure 3A–C) to estimate the timing of the emergence of the *CYP1–4* families (nodes *a*, *b*, and *c* in Figure 1). Since b_B, b_C, b'_B, and b'_C correspond to 400 million years (myr), each ratio of (b_A+b_B) to b_B, (b_A+b_C) to b_C, ($b'_A+b'_B$) to b'_B, and ($b'_A+b'_C$) to b'_C yielded an estimate of the duplication time. The estimates varied from 623–1316 mya for *a*, 601–664 mya for *b*, and 681–926 mya for *c*. To be conservative, we used the youngest estimate for each node: 623±35 mya for *a*, 601±34 mya for *b*, and 681±37 mya for *c*. As anticipated, these estimates preceded the emergence of vertebrates (608 mya, TimeTree) but occurred after the divergence of vertebrates and chordates (774 mya, TimeTree). This finding suggests that invertebrates do not possess orthologs to vertebrate D-type genes, despite the presence of D-type *CYPs* in insects, which function in insecticide resistance and detoxification of plant alkaloids [34].

Evolutionary relationship between invertebrate and vertebrate *CYPs*

To further examine the relationships between vertebrate and invertebrate D-type *CYP* genes, we searched for homologs of human D-type *CYPs* in six invertebrate species (amphioxus: *B. floridae*, sea squirt: *C. intestinalis*, sea urchin: *S. purpuratus*, sea anemone: *N. vectensis*, water flea: *D. pulex*, and fruit fly: *D. melanogaster*). A total of 543 *CYP* amino acid sequences were retrieved from the Cytochrome P450 Homepage. A preliminary search to determine the phylogenetic position of vertebrate *CYPs* in the tree that included both vertebrate and invertebrate *CYPs* revealed that each vertebrate *CYP* family formed a monophyletic group. To simplify the phylogenetic analysis, amino acid sequences from these invertebrates were aligned only with sequences from humans, as a representative vertebrate, and the tree was constructed on the basis of amino acid distances (Figure 4).

The amino-acid distance tree shows that 10 clans (clans 2, 3, 4, mito, 7, 19, 20, 26, 46, and 51) are common to vertebrates; the tree also reveals one *Drosophila*-specific clan. A previous study of 1,572 *CYP* sequences also identified 11 clans in metazoans, but with inclusion of clan 74, which was present only in lancelets, sea anemones, and Trichoplax, but absent in vertebrates [9]. In the present analysis, despite the inclusion of both lancelet and sea anemone, clan 74 was not detected. However, a further phylogenetic analysis that included only yeasts, humans, lancelets,

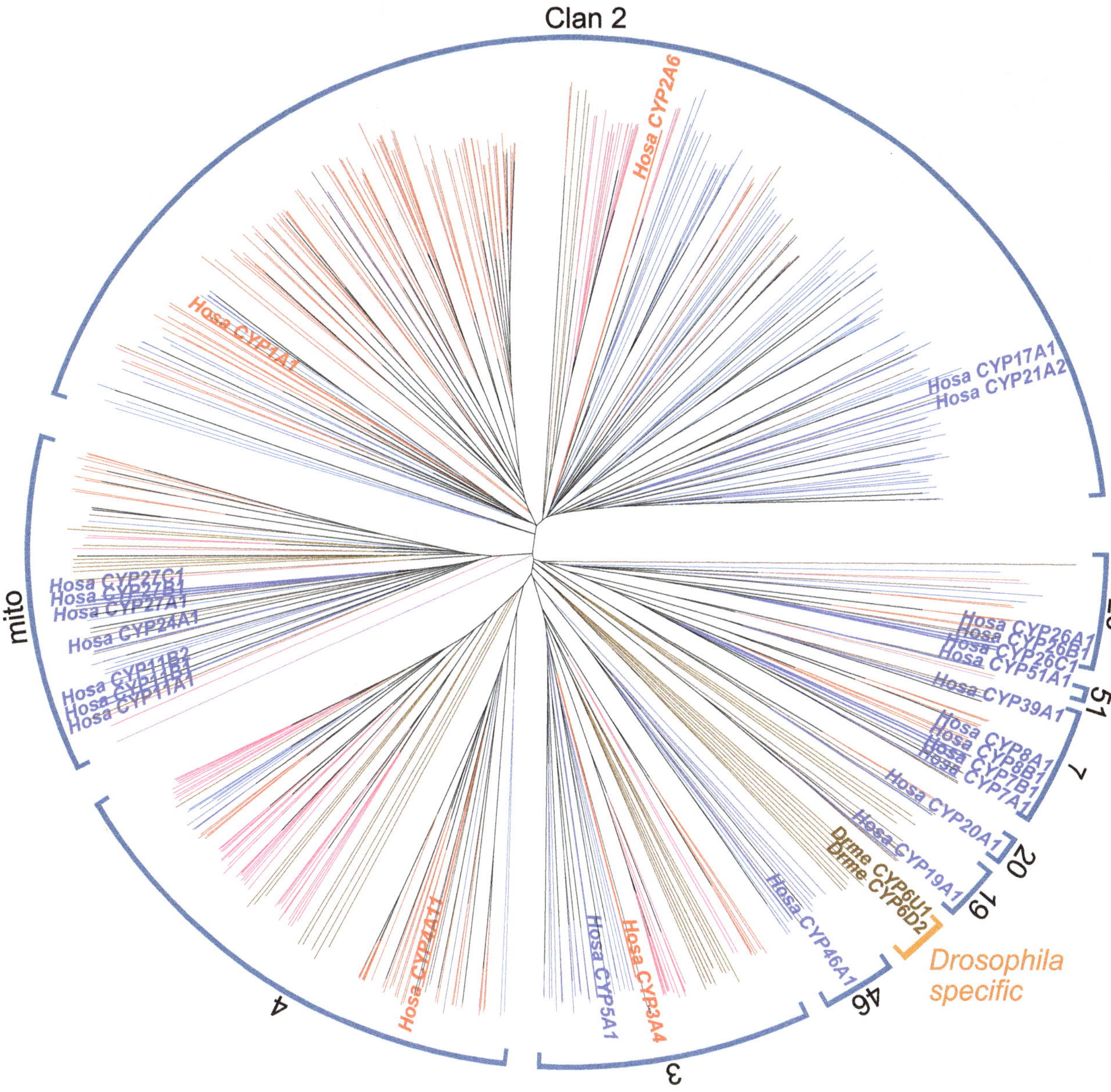

Figure 4. NJ tree of all invertebrate *CYP* genes. D- and B-type *CYP* genes in humans are shown in red and blue text, respectively. The numbers near the brackets indicate clans. Orange character indicates the *Drosophila*-specific clan. Abbreviations and their color (in parentheses) are defined as follows: *Hosa*, *Homo sapiens* (red for D- and blue for B-type); *Brfl*, *Branchiostoma floridae* (dark brown); *Neve*, *Nematostella vectensis* (light blue); *Dapu*, *Daphnia pulex* (pink); *Stpu*, *Strongylocentrotus purpuratus* (aqua); *Ciin*, *Ciona intestinalis* (dark red); *Drme*, *Drosophila melanogaster* (brown); *Sace*, *Saccharomyces cerevisiae*; *Scpo*, *Schizosaccharomyces pombe* (purple). Gene names of *CYP6D2* and *6U1* in *D. melanogaster* are shown in clan 19. Bootstrap values supporting nodes of clusters mentioned in the text are shown.

and sea anemones identified clan 74, although it was supported by a relatively low bootstrap value (55%). In addition, the genes that comprised the *Drosophila*-specific clan (*CYP6D2*, *6U1*, *28A5*, *28C1*, *28D1*, *308A1*, *309A1*, *350A1*, and *317A1*) were all included in clan 3 [9]. This holds true when we draw trees with different methods (maximum likelihood), although the bootstrap value for this clan is too low (<20%) to confirm this inclusion. We also observed some other differences from the previous study [9]: clan 51 did not include any sea-urchin gene, and clan 20 included neither sea urchin nor sea-anemone gene (Figure 4). The absence of a sea-urchin *CYP51* ortholog can be explained by the incompleteness of the database used here. In fact, a blast search of the NCBI

database using human *CYP51* as a query identified a *CYP51* gene (Accession number: NM_001001906) in the recently published sea urchin genome. However, clan 20-like genes were absent from the sea urchin and the sea-anemone genomes in the database. In addition, clan 19 in the present tree appeared to include the *Drosophila* genes (*313A1*, *313B1*, *316A1*, and *318A1*) that were included in clan 4 in the previous study. In fact, the *Drosophila*-specific genes in clan 19 shared 16 of 433 amino acids with human *CYP19* (Figure S4), and these 16 amino acids were conserved among vertebrate CYPs. However, an ML tree supported the presence of the *Drosophila* sequences in clan 4, with very low bootstrap support (6%).

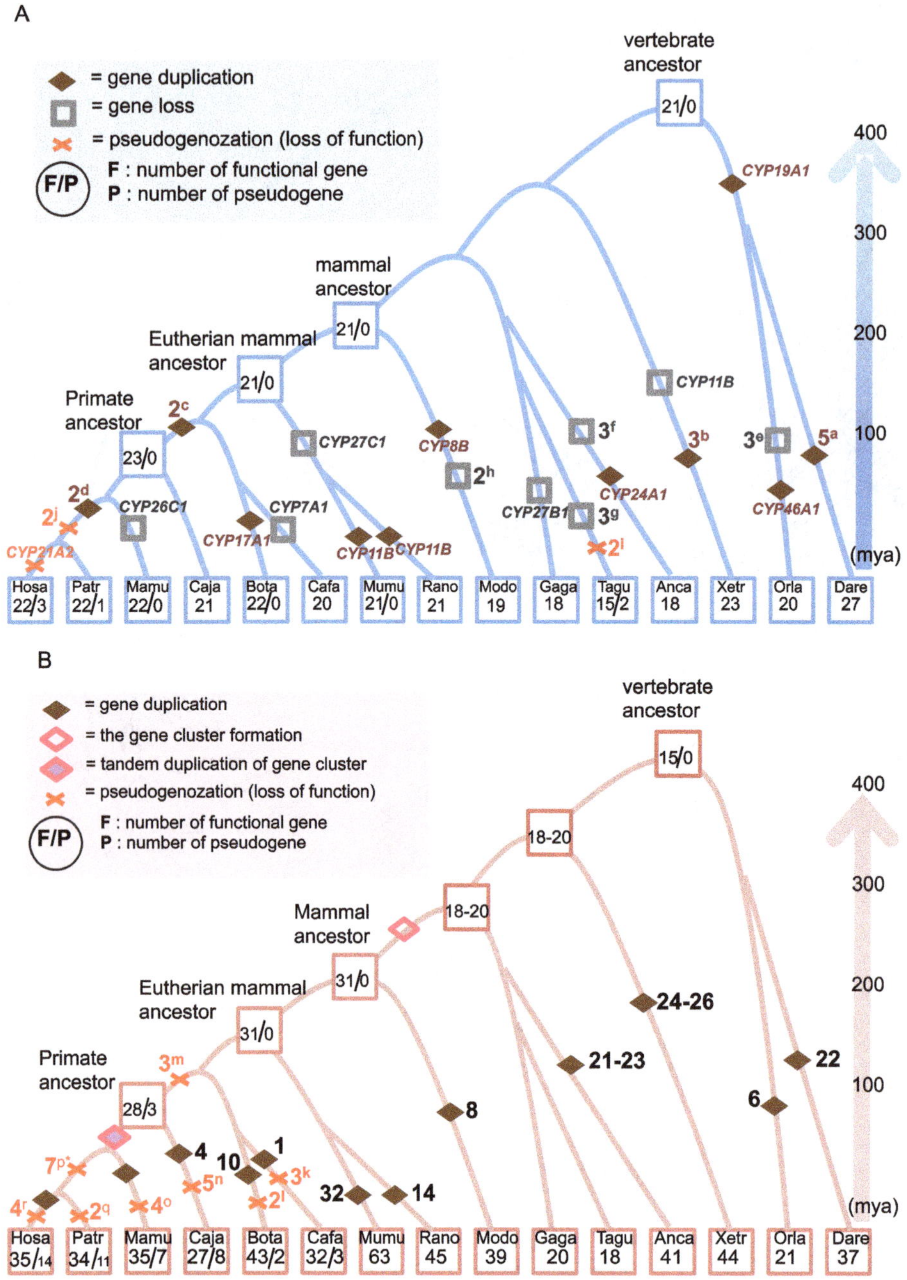

Figure 5. The birth and death processes of *CYP* genes in vertebrates. A) B-type *CYP* genes and B) D-type *CYP* genes. In both figures, numbers inside squares represent the number of functional genes and pseudogenes in each species and its ancestors. Diamonds, crosses, and rectangles indicate gene duplication, pseudogenization, and deletion events, respectively. The number adjacent to each symbol represents the number of events. The letter adjacent to the number indicates the list of *CYP* genes, as follows. a: *CYP8B2*, *CYP8B3*, *CYP17A1*, *CYP27A1*, and *CYP46A1*, b: *CYP8B*, *CYP27A1*, and *CYP46A1*, c: *CYP11B* and *CYP21A2*, d: *CYP51* (two genes), e: *CYP7B*, *CYP11B*, and *CYP39A1*, f: *CYP11A*, *CYP21A2*, and *CYP26*, g: *CYP11B*, *CYP21A2*, and *CYP27* (A or B), h: *CYP17A1* and *CYP26B1*, i: *CYP24A1* and *CYP27A1*, j: *CYP21A1P* and *CYP51* (two genes), k: *CYP4F9P*, *CYP4F23P*, and *CYP4F24P*, l: *CYP4A11* and *CYP2F1P*, m: *CYP2T2P*, *3P*, and *CYP2G1P*, n: *CYP4A11*, *CYP4B1*, *CYP4F22*-like (two genes), o: *CYP2A7P1*, *CYP2A13*, *CYP2B6P*, and *CYP4F11*, p: *CYP2B7P1*, *CYP2D8P1*, *CYP2F1P*, *CYP4F9P*, and *CYP4F23P*, and *CYP4F24P*, *CYP1D1P* were found in *Pan paniscus*, but were absent from *Pan troglodytes*, q: *CYP2B6* and *CYP2C18*, r: *CYP2A7P1*, *CYP2G2P*, and *CYP4Z2P*.

Clans including invertebrate *CYP* genes were supported by low bootstrap values, and clan definitions were dependent on the methods used for tree construction. Thus, the notion of clan becomes ambiguous and ill-defined for distantly related metazoan *CYP* genes.

The origin of D-type genes in vertebrates and invertebrates

Clan 2 included human *CYP17A1* and *CYP21A2* (B-type) as well as members of the *CYP1* and *CYP2* families (D-type). Similarly, clan 3 included both types of *CYP* genes: *CYP5A1* (B-type) and *CYP3A* subfamilies (D-type). These two cases indicate that the emergence of D-type from B-type genes occurred after the

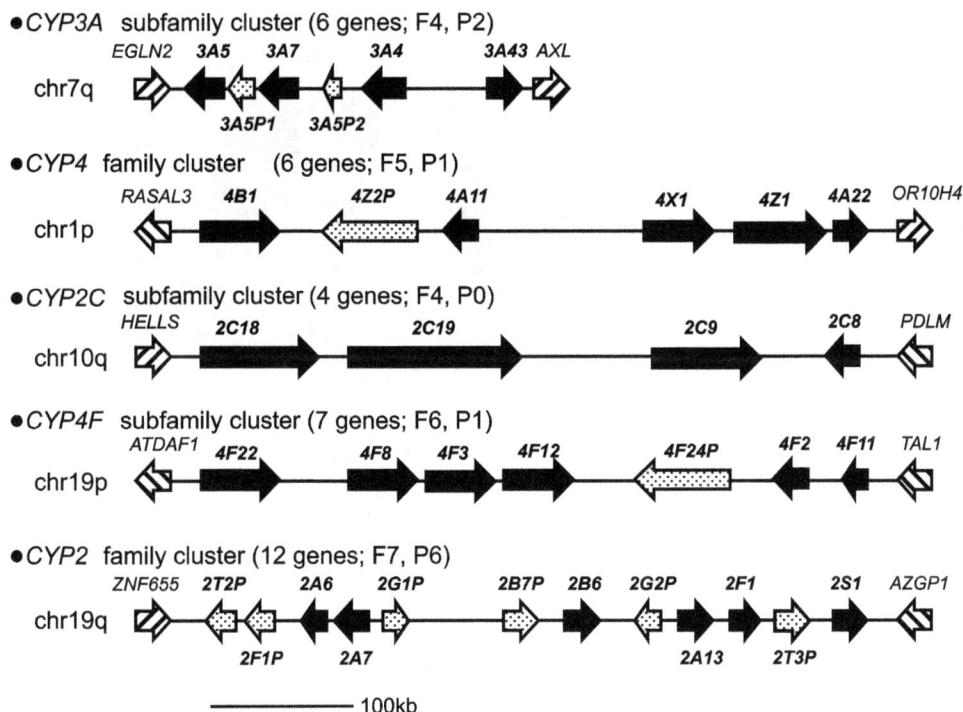

Figure 6. CYP gene clusters in the human genome. A striped arrow represents an anchor gene in a syntenic region of each cluster. Black arrows and dotted arrows represent functional CYP genes and pseudogenes, respectively. The length of each gene cluster is approximately 500 kb, except the CYP3A cluster (250 kb). The number of total genes, functional genes, and pseudogenes in each cluster are shown after the cluster name.

emergence of the clan. However, clan 4 included only the CYP4 family from humans but not CYP46A1, an ancestor of the CYP4 family. This is the only case where the emergence of the D-type predates clan emergence. In addition, clan 4 included both vertebrate and invertebrate genes. Vertebrate CYP4 likely acquired its detoxification function in the stem lineage of vertebrates when invertebrate sequences were B-type; alternatively, the ancestor of clan 4 may have already possessed D-type functions when invertebrate genes in clan 4 encoded D-type enzymes.

Fruit flies are known to possess two D-type CYP genes, CYP6D2 and CYP6U1, which function in insecticide metabolism. In the tree generated in this study, these CYP genes were distantly related to human D-type genes, suggesting that D-type genes in fruit flies emerged independently from those in vertebrates.

Gene duplications and losses in the B- and D-type lineages during vertebrate evolution

Nearly all of the 14 vertebrate genomes examined here contained 21 orthologs to the 22 functional human B-type genes. On the basis of the presence or absence of CYP genes in each vertebrate genome, we parsimoniously estimated the number of genes in each ancestor of amniotes, mammals, eutherian mammals, primates, catarrhini, and hominoids, as well as the number of gains of genes in each taxonomic lineage. The number of ancestral genes remained stable throughout the evolution of vertebrates: the number of genes in each vertebrate ancestor did not change over the course of evolution until the emergence of a primate ancestor. A gene-duplication event occurred in the primate ancestor, generating CYP11B2. In the ancestor of hominoids, emergence of new genes occurred twice, generating the ancestors of the CYP51P1 and CYP51P2 genes (Figure 5A).

In contrast to the rather stable mode of evolution observed in the stem, lineage-specific gains and losses of genes occurred relatively frequently. For instance, a shared duplication of CYP19A1 occurred in the lineage leading to the common ancestor of zebrafish and medaka. In addition, lineage-specific gene duplications occurred in the zebrafish (CYP8B2, CYP8B3, CYP17A1, CYP27A1, and CYP46A1), medaka (CYP46A1), frog (CYP8B1, CYP27A1, and CYP46A1), green anole (CYP24A1), and opossum (CYP8B1) lineages. Interestingly, gene duplications of CYP8B, CYP46A1, and CYP27A1 occurred independently several times in a species-specific manner. Similarly, lineage-specific gene losses (deletions) were observed; for instance, deletions occurred in a lineage leading to the medaka (CYP7B1, CYP11B1, and CYP39A1), frog (CYP11B1), green anole (CYP11A1, CYP21A2, CYP26A1), chicken (CYP27B1), zebra finch (CYP11B1, CYP21A2 and CYP27), and opossum (CYP17A1 and CYP26B1). Several deletions affecting the same genes (CYP11B1 and CYP21A2) occurred independently in medaka, frog, green anole, and zebrafinch.

Although only a limited number of genomic sequences are available, we identified 19 gene gains and 16 losses among the 15 available genomes, including the human genome. Assuming that the total branch length in the vertebrate tree is 2,685 myr (for individual species divergence times, see Figure S5), we estimated the rate of gene gains and losses to be 0.7 and 0.6 per 100 myr, respectively.

Using a similar analysis, we also examined the 14 vertebrate genomes for the presence of paralogs and orthologs of 35 human D-type genes. This analysis revealed that the number of genes varies from 15 to 31 in ancestral species, and from 18 to 63 in extant species (Figure 5B). In contrast to the relatively stable evolutionary mode of B-type genes, D-type genes underwent more frequent gene duplications and pseudogenization (Table 1).

A

B

C

D

Figure 7. Phylogenetic tree of the D-type family. A) *CYP1* family, B) *CYP2* family, C) *CYP3* family, and D) *CYP4* family. Each NJ tree was based on the total nucleotide substitutions among members. The origin of each of the five clusters (corresponding to i–v in Figure 1) is indicated with a diamond in Figure B–D. Each subfamily is indicated by a bracket. In Figure B, the *CYP2T* subfamily is not shown because no functional gene belonging to this subfamily is present in the human genome. In Figures C and D, the red dashed rectangle outlines a specific clade.

One important difference between D- and B-type genes is that D-type genes cluster on chromosomes, and these clusters are composed of closely related genes. This difference is reflected in the phylogeny, which shows that the genes in each cluster are monophyletic (Figure 1). In the human genome, five clusters have been identified: the *CYP2* family clusters on chromosome 19q [18,19], the *CYP2C* subfamily clusters on chromosome 10q, the *CYP3A* subfamily clusters on chromosome 7q, the *CYP4* family clusters on chromosome 1p, and the *CYP4F* subfamily clusters on chromosome 19p. Each cluster region occupies approximately 500 kb, with the exception of *CYP3A*, which occupies 250 kb. Each cluster included the following number of genes: 12 for *CYP2*, four for *CYP2C*, six for *CYP3A* and *CYP4*; and seven for *CYP4F* (Figure 6).

Using the phylogenic analyses of each *CYP1–4* family in vertebrates, we identified several species-specific gene duplications. The topology of the tree for the *CYP1* family revealed four subfamilies (*1A*, *1B*, *1C*, and *1D*), and showed that these subfamilies diverged in the ancestor of vertebrates. The *CYP1A* and *1B* subfamilies were conserved from fish to humans, whereas primates lacked *CYP1D*, and mammals lacked *CYP1C* (Figure 7A). The *CYP2* family was shown to be composed of 16 subfamilies (*CYP2A*, *2B*, *2C*, *2D*, *2E*, *2F*, *2G*, *2H*, *2J*, *2K*, *2R*, *2S*, *2T*, *2U*, *2W*, and *2AC*), three of which (*CYP2B*, *2E* and *2S*) were specific to mammals, while the *2A/G* and *F* subfamilies were present only in

mammals and reptiles. These five subfamilies (except the *CYP2E* subfamily) diverged successively to form the *CYP2* cluster in an ancestor of mammals (Figure 7B). However, *CYP2U* and *2R* were shown to be common to all vertebrates. The *CYP3* family tree contained only two subfamilies, *CYP3A* and the fish-specific *3C* family (Figure 7C). *CYP3A* comprised amphibian-, bird-, and mammal-specific clades. In each taxonomic group, members of the *CYP3A* subfamily appear to have been duplicated independently. The tree constructed for the *CYP4* family included six subfamilies (*4A*, *4B*, *4F*, *4V*, *4X*, and *4Z*) (Figure 7D). *CYP4A* and *4X/Z* were specific to mammals, whereas the other three subfamilies (*4B*, *F*, and *V*) were common to all vertebrates. In particular, the members of the *4F* subfamily formed several species-specific clusters, except *CYP4F22*. It is unclear, however, whether these species-specific clusters resulted from gene conversion or from recent duplication of the subfamily in each species. The evolution of D-type genes has involved frequent species-specific gene duplications, compared to B-type genes (Figure 5B). In D-type genes, it is unclear how many gene duplications occurred before eutherian divergence. We estimated the rate of duplication subsequent to the eutherian radiation, which revealed 53 duplications in 432 myr, or the rate of 12.7 duplications per 100 myr. No gene loss was observed. These results are in contrast to the results for B-type genes.

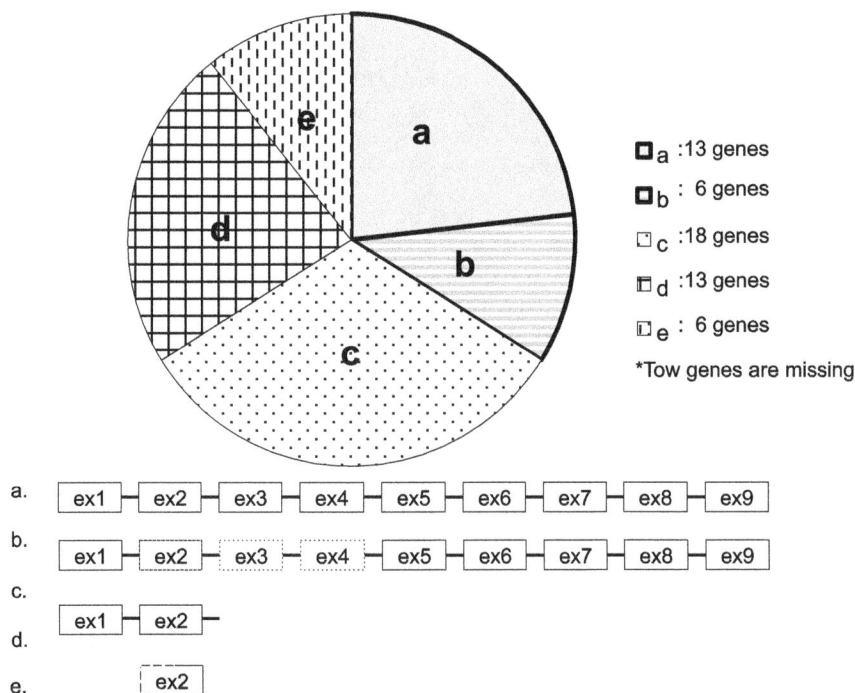

Figure 8. Categorization of the 58 human *CYP* pseudogenes. Among the 58 pseudogenes, paralogs were detected by a BLAST search. a. The number of exons and introns is the same as in the paralogous genes (13 genes), b. They contain greater than half the number of exons and introns of their paralogs (6 genes), c. One or two exons or introns remained (18 genes), d. A portion of an exon remained (13 genes), e. The BLAST search returned no hits (6 genes). *Two pseudogenes were absent from the human genome databases. Approximately one-third (a and b) of the human *CYP* pseudogenes were used for phylogenetic analysis.

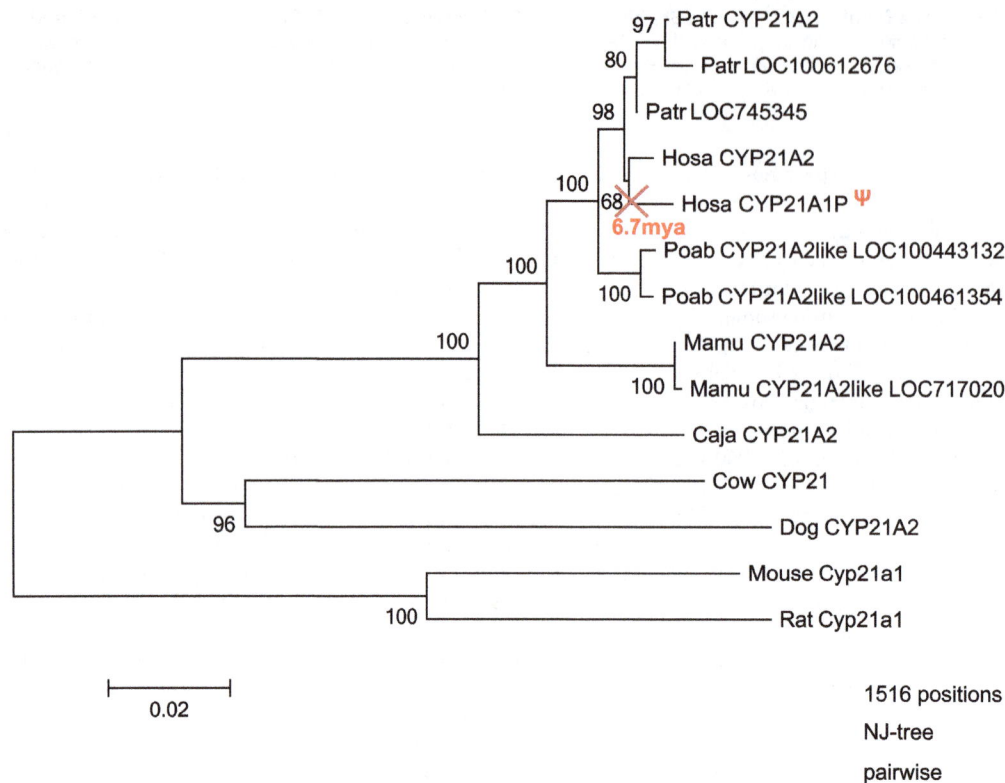

Figure 9. Time of pseudogenization of *CYP21A1P* in humans. The phylogenetic tree was obtained using the NJ method, using the CDS. The pseudogene is represented by Ψ. The cross shows the time at which function was lost.

B- and D-type *CYP* pseudogenes

The evolutionary modes of D- and B-type *CYP* genes differed also in pseudogenization, which is defined as a loss of gene function. Among the 58 pseudogenes present in the human genome, more than half (41 of 58) are fragmented, with few exons and introns remaining. The total length of such pseudogenes is less than one-tenth of that of a functional *CYP* gene, which prevented identification of several of the original genes (Figure 8). We identified the original functional genes for 17 pseudogenes, among which 3 were B-type (*CYP21A1P*, *CYP51P1*, and *CYP51P2*) and 14 were D-type (*CYP1D1P*, *CYP2A7P1*, *CYP2B7P1*, *CYP2D7P1*, *CYP2D8P1*, *CYP2F1P*, *CYP2G1P*, *CYP2G2P*, *CYP2T2P*, *CYP2T3P*, *CYP4F9P*, *CYP4F23P*, *CYP4F24P*, and *CYP4Z2P*) (Table S1). Of the 3 B-type pseudogenes, *CYP51P1* and *CYP51P2* are processed pseudogenes, and the biological causes of their pseudogenization are not related to a relaxation of functional constraints. In this sense, *CYP21A1P* is only a pseudogene due to relaxation of functional constrains. Rhesus macaques, orangutans, and humans have two copies of *CYP21A*, and chimpanzees have three (Figure 9). However, a pseudogene for *CYP21A* is present only in humans, and the time of pseudogenization was estimated to be 6.7 mya, around the divergence of humans from chimpanzees. The presence of this pseudogene is clinically significant in humans: partial gene conversion from a pseudogene to a functional gene causes 21-hydroxylase deficiency; furthermore, copy number variation has been observed in the region containing *CYP21A* and the neighboring *C4A* in the *HLA* region of human chromosome 6 [35].

In contrast, among the 14 D-type pseudogenes, four (*CYP2G1P*, *2G2P*, *2T2P*, and *2T3P*) have been reported to be human-specific, on the basis of a comparison between humans and mice [36]. We searched for orthologs to the human pseudogenes in other primate genomes and found that all but *CYP2G2P* are pseudogenized in other primates as well, but are functional in non-primate vertebrates (Figure S6). Our findings showed that *CYP2G1P*, *2T2P*, and *2T3P* are primate-specific pseudogenes, whereas *CYP2G2P* is a human-specific pseudogene. Using an accelerated non-synonymous substitution rate in pseudogenes [27], we calculated that *CYP2G2P* emerged 2.6 mya. In addition to *CYP2G2P*, further analysis revealed a single human-specific pseudogene, *4Z2P*, with a pseudogenization time of 6.4 mya. On the basis of the results of our analysis, *CYP2D7P1* also appeared to be a human-specific pseudogene. Interestingly, however, pseudogenization of this ortholog has also been found in orangutans, but the cause is different from that for humans [37]. It appears that this gene lost its function in humans and orangutans independently. In D-type genes, in addition to the 14 pseudogenes present in the human genome, 7 pseudogenes were identified in chimpanzees, macaques, marmosets, dogs, and cows. Among the seven, six were specie-specific, one (*2C18*) to chimpanzees, two (*2A13* and *4F11*) to macaques, and three (*4B1* and two *4F22*-like genes) to marmosets. The remaining *CYP2B6P* was pseudogenized independently in chimpanzees and macaques. Among the 11 pseudogenes, with the exception of the three human-specific pseudogenes, *CYP2A7P1* was pseudogenized in macaques and humans independently, at 28.4 mya and 5.9 mya, respectively. Pseudogenization of the remaining 10 genes occurred in the primate or hominoid stem lineage.

It is unclear how many times pseudogenization occurred in D-type genes before eutherian divergence. We estimated the rate subsequent to the eutherian radiation at 30 pseudogenizations over 432 myr, yielding a rate of 6.9 per 100 myr. In contrast, the

A

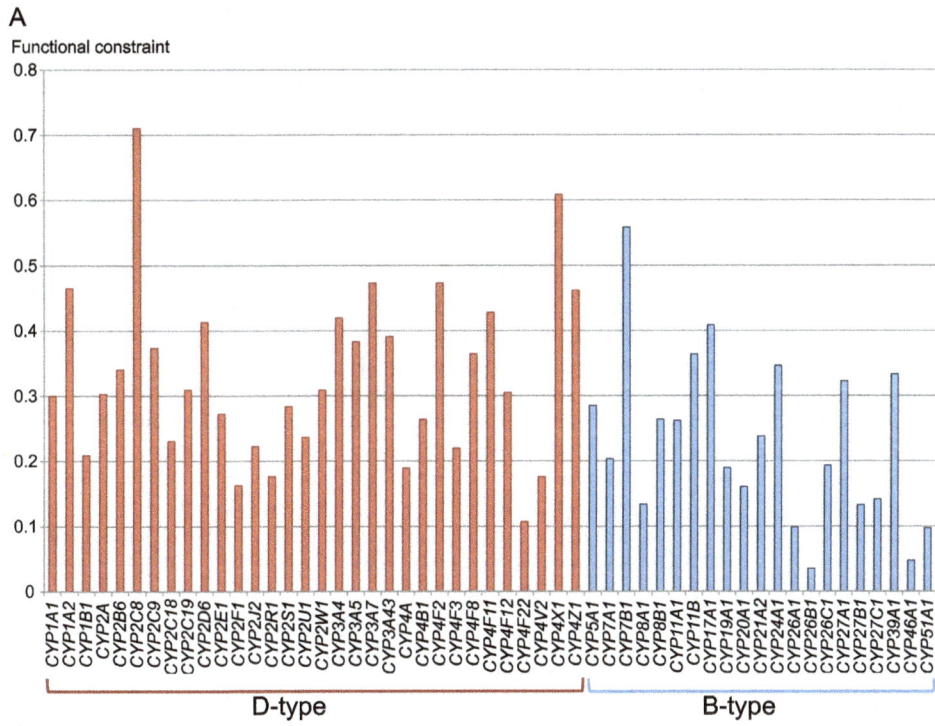

B

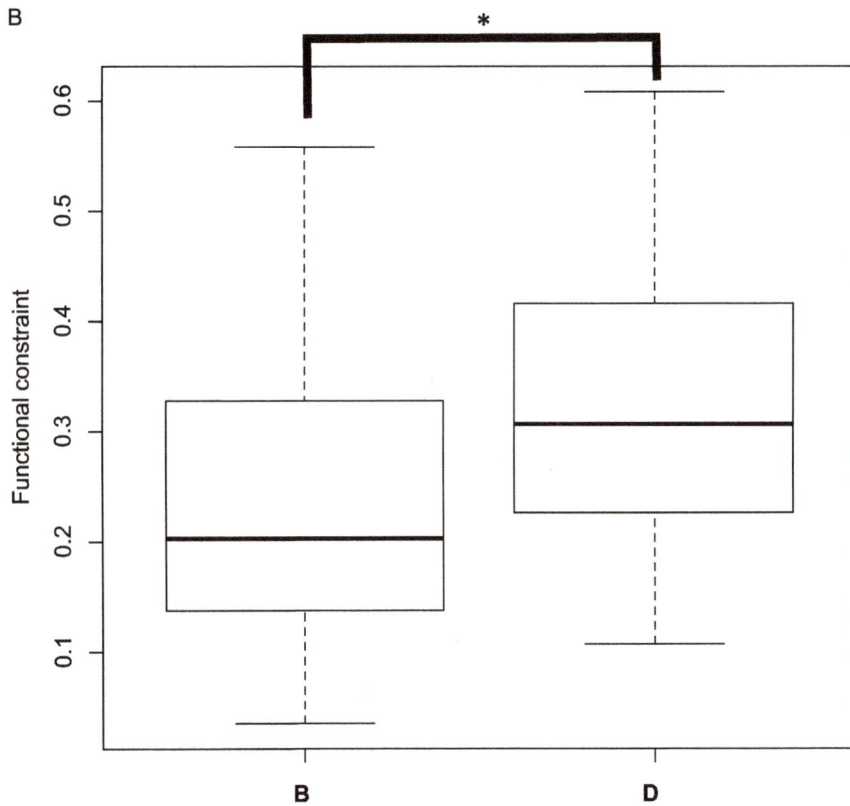

Figure 10. Functional constraint of *CYP* genes. A) Functional constraint was estimated for each *CYP* gene. The y-axis shows the functional constraint obtained via the ratio of per-site non-synonymous substitutions to synonymous substitutions (D_N/D_S). Red bars indicate D-type genes, and blue bars indicate B-type genes. B) Comparison of median values for functional constraints between primate B- and D-type genes. "B" indicates functional constraint of B-type genes, and "D" indicates functional constraint of D-type genes. The *P*-value was 0.01282 (significance was defined as $P<0.05$). The *P*-value was calculated using the Mann–Whitney *U* test.

number of pseudogenization events in B-type genes was estimated to be only five over 2,685 myr, yielding a rate of 0.19 per 100 myr.

Evolutionary rate of B- and D-type genes

Our results revealed that births and deaths of genes were more frequent in D-type genes than in B-type genes. As such, it was important to compare the evolutionary rate of B- and D-type genes. For this comparison, the non-synonymous substitution rate was normalized to the synonymous rate, and the ratio (*f*) for each B- and D-type gene was calculated in primates (see Materials and Method, Figure 10A). The average *f*-values for B- and D-type genes were calculated to be 0.24±0.14 and 0.33±0.13, respectively, and the median values for B- and D-type genes were determined to be 0.23 and 0.31, respectively (Figure 10B). The average and median values for D-type genes were significantly greater than those for B-type genes (Wilcoxon's test, *P*-value = 0.0173), suggesting that the degree of functional constraint (1-*f*) is stronger in B-type than in D-type genes. These results are consistent with the rapid birth and death process of D-type genes.

Discussion

The origin of D-type *CYP* genes

The origin of B-type genes is assumed to be single and ancient, because fission yeast possesses B-type genes and because a possible ortholog to the B-type gene *CYP51* is present even in prokaryotic genomes. However, D-type genes are thought to have different origins. The present phylogenetic analysis demonstrated that four D-type families are conserved among all vertebrates, and that the D-type families were derived from three gene-duplication events of B-type genes in the stem lineage of vertebrates. Based on the molecular clock hypothesis, B- to D-type gene duplications occurred approximately 600–700 mya, consistent with the phylogenetic analysis. However, the D-type *CYPs* impart resistance to insecticides in invertebrates; in fruit flies, two such enzymes are CYP6U1 and CYP6D2. The phylogenetic analysis of both human and fruit fly *CYP* genes indicated an independent emergence of D-type genes. Moreover, other invertebrate genomes contain human D-type-like genes, but orthology has not been confirmed. It appears that D-type genes in vertebrates and insects evolved independently from different origins, which is consistent with the idea of a rapid turnover of D-type genes.

In this paper, we focused on an early stage of *CYP* gene diversification in vertebrates and showed the emergence of D-type from B-type genes. However, some exceptions should be noted. For example, *CYP2R1* is categorized as a B-type gene on the basis of its function, but the nucleotide sequence showed that it is closely related to other D-type *CYP2* genes. From this observation, it appears that *CYP2R1* has been converted from a D-type to a B-type *CYP* gene. This is supported by the observation that the amino acid sequence of CYP2R1 is highly conserved in all vertebrates, reflecting the high degree of functional constraint against the gene.

The evolution of *CYP* genes is driven by substrate specificity

The birth and death (pseudogenization) rates of B- and D-type genes differed in magnitude: the rates in B-type genes were 0.7 and

0.2 per 100 myr, respectively, whereas those in D-type genes were 12.7 and 6.9 per 100 myr, respectively. Compared with D-type genes, the evolution of B-type genes was highly conserved with regard to their mode of birth and death processes as well as amino acid substitutions. The substrates of B-type enzymes are chemicals that play important roles in metabolism of vitamin D, steroids, and cholesterol. In contrast, the substrates of D-type enzymes are xenobiotics such as plant alkaloids. In light of this substrate specificity, we hypothesize that the conserved evolutionary pattern observed in B-type enzymes reflects the importance and conservation of their substrates, whereas the rapid evolution of D-type enzymes indicates that their substrates are flexible and highly dependent on environmental factors. Future studies of the evolution of the substrate-recognition sites will be required to confirm this hypothesis.

Supporting Information

Figure S1 Conserved amino acids at a position. The *x*-axis indicates amino-acid positions in an alignment of 710 vertebrate *CYP* genes (after excluding gaps), and the y-axis indicates the ratio of conserved amino acids at each position. Red bars indicate highly (>95%) conserved positions. The chart below the bar graph reports the approximate position of six substrate recognition sites (SRS): SRS1–6. The bracket represents the heme-binding region (~10 amino acids) of the *CYP* gene.

Figure S2 Conserved amino acids within each clan of vertebrate CYP genes. Conserved amino acids within each clan are shown in colored columns: the red column is specific for clan 2; aqua, mito; orange, 4; blue, 3; yellow, 26, and brown, 7. The four sites in purple (E, R, G, and C) are highly conserved (>95%) sites among all vertebrate species. This figure shows only human *CYP* genes: positions are 275–285 and 350–360.

Figure S3 Amino acids distinguishing B-type from D-type genes within each clan. The yellow column indicates the amino acids in B-type genes that differ from D-type genes. The other colors are the same as in Figure S2. A, Clan 2; B, Clan 3; C, Clan 4 and 46.

Figure S4 Conserved amino acids between human and *Drosophila* genes in a *CYP19* clan. Human (*CYP19A1*) and Drosophila (*CYP313A1, 313B1, 316A1,* and *318A1*) genes share the same amino acids at 16 of 423 aligned sites. The shared sites are shown in red.

Figure S5 Phylogenetic tree and species-divergence time. The phylogenetic tree was constructed on the basis of the divergence time of 15 species (time tree). The species names at the tip of the tree are abbreviated as in Table 1. *Hosa, Homo sapiens.* The number at each node represents species divergence time in mya. The scale under the tree indicates time in myr.

Figure S6 The cause of functional loss in human-specific *CYP* pseudogenes. There are four human-specific *CYP* psuedogenes (*CYP2G1P, 2P, CYP2T2P,* and *3P*). Possible causal mutations, premature stop codons (red) and frame-shift mutations (blue), were identified in human and other primate CYP nucleotide and amino acid alignments. The row labeled "exon" for *CYP2G1P* and *2P* shows the number of exons in which mutations were found, and the row labeled "bp" indicates the nucleotide position of the CDS in functional genes from rat and mouse.

Table S1 The number of *CYP* gene in Human. After exclusion of truncated pseudogeens each category includes genes as below. **a**: *CYP1A1, CYP1A2, CYP1B1, CYP2A6, CYP2A7, CYP2B6, CYP2C8, CYP2C9, CYP2C18, CYP2C19, CYP2D6, CYP2E1, CYP2F1, CYP2J2, CYP2R1, CYP2S1, CYP2U1, CYP2W1, CYP3A4, CYP3A5, CYP3A7, CYP3A43, CYP4A11, CYP4A20, CYP4A22, CYP4B1, CYP4F2, CYP4F3, CYP4F8, CYP4F11, CYP4F12, CYP4F22, CYP4V2, CYP4X1,* **b**: *CYP1D1P, CYP2A7P1, CYP2B7P1, CYP2D7P1, CYP2D8P1, CYP2F1P, CYP2G1P,* *CYP2G2P, CYP2T2P, CYP2T3P, CYP4F9P, CYP4F23P, CYP4F24P, CYP4Z2P,* **c**: *CYP5A1, CYP7A1, CYP7B1, CYP8A1, CYP8B1, CYP11A1, CYP11B1, CYP11B2, CYP17A1, CYP19A1, CYP20A1, CYP21A2, CYP24A1, CYP26A1, CYP26B1, CYP26C1, CYP27A1, CYP27B1, CYP27C1, CYP39A1, CYP46A1, CYP51A1,* **d**: *CYP21A1P, CYP51P1, CYP51P2*.

Acknowledgments

We thank Dr. Takahata for his critical discussion of this manuscript.

Author Contributions

Conceived and designed the experiments: AK YS. Performed the experiments: AK YS. Analyzed the data: AK YS. Contributed reagents/materials/analysis tools: AK YS. Wrote the paper: AK YS.

References

1. Klingenberg M (1958) Pigments of rat liver microsomes. Arch Biochem Biophys 75: 376–386.
2. Omura T, Sato R (1962) A new cytochrome in liver microsomes. J Biol Chem 237: PC1375–PC1376.
3. Nelson DR (1998) Metazoan cytochrome P450 evolution. Comp Biochemi and Phys Part C: Pharm Toxic Endocr 121: 15–22.
4. Gotoh O (2012) Evolution of cytochrome p450 genes from the viewpoint of genome informatics. Biol Pharm Bull 35: 812–817.
5. Munro AW, Lindsay G (1996) Bacterial cytochromes P-450. Mol Microbiol 20: 1115–1125.
6. Nelson DR (2009) The cytochrome p450 homepage. Human Genomics 1:59–65.
7. Sea Urchin Genome Sequencing Consortium (2006) The genome of the sea urchin *Strongylocentrotus purpuratus*. Science 314: 941–952.
8. Mao G, Seebeck T, Schrenker D, Yu O (2013) CYP709B3, a cytochrome P450 monooxygenase gene involved in salt tolerance in *Arabidopsis thaliana*. BMC Plant Biol 13: 169.
9. Nelson DR, Goldstone JV, Stegeman JJ (2013) The cytochrome P450 genesis locus: the origin and evolution of animal cytochrome P450s. Phil Trans R Soc B 368: 20120474.
10. Quaderer R, Omura S, Ikeda H, Cane DE (2006) Pentalenolactone biosynthesis. Molecular cloning and assignment of biochemical function to PtlI, a cytochrome P450 of Streptomyces avermitilis. J Am Chem Soc 128: 13036–13037.
11. Aoyama Y, Horiuchi T, Gotoh O, Noshiro M, Yoshida Y (1998) CYP51-like gene of *Mycobacterium tuberculosis* actually encodes a P450 similar to eukaryotic CYP51. J Biochem 124: 694–696.
12. Yoshida Y, Aoyama Y, Noshiro M, Gotoh O (2000) Sterol 14-demethylase P450 (CYP51) provides a breakthrough for the discussion on the evolution of cytochrome P450 gene superfamily. Biochem Biophys Res Comm 273: 799–804.
13. Debeljak N, Fink M, Rozman D (2003) Many facets of mammalian lanosterol 14α-demethylase from the evolutionarily conserved cytochrome P450 family CYP51. Arch Biochem Biophys 409: 159–171.
14. Qi X, Bakht S, Qin B, Leggett M, Hemmings A, et al. (2006) A different function for a member of an ancient and highly conserved cytochrome P450 family: From essential sterols to plant defense. Proc Natl Acd Sci USA 103: 18848–18853.
15. Nelson DR (1999) Cytochrome P450 and the individuality of species. Arch Biochem Biophys 369: 1–10.
16. Rezen T, Debeljak N, Kordis D, Rozman D (2004) New aspects of lanosterol 14a-demethlase and cytochrome P450 evolution: Lanosterol/cycloartenol diversification and lateral transfer. J Mol Evol 59: 51–58.
17. Nebert DW, Dalton TP (2006) The role of cytochrome P450 enzymes in endogenous signaling pathways and environmental carcinogenesis. Nature Rev Cancer 6: 947–960.
18. Hoffman SMG, Hu S (2006) Dynamic evolution of the CYP2ABFGST gene cluster in primates. Mutation Res 616: 133–138.
19. Hu S, Wang H, Knisely AA, Raddy S, Kovacevic D, et al. (2008) Evolution of the CYP2ABFGST gene cluster in rat, and a fine-scale comparison among rodent and primate species. Genetica 133: 215–226.
20. Thomas JH (2007) Rapid birth-death evolution specific to xenobiotic cytochrome P450 genes in vertebrates. PLoS Genetics 3: e67
21. Larkin MA, Blackshields G, Brown NP, Chenna R, McGettigan PA, et al. (2007) Clustal W and Clustal X version 2.0. Bioinformatics 23: 2947–2948.
22. Tamura K, Peterson D, Peterson N, Stecher G, Nei M, et al. (2011) MEGA5: Molecular evolutionary genetics analysis using maximum likelihood, evolutionary distance, and maximum parsimony methods. Mol Biol Evol 28: 2731–2739.
23. Saitou N, Nei M (1987) The neighbor-joining method: a new method for reconstructing phylogenetic trees. Mol Biol Evol 4: 406–425.
24. Nei M and Kumar S (2000). Molecular Evolution and Phylogenetics. Oxford University Press, New York.
25. Jones D, Taylor WR, Thronton JM (1992) The rapid generation of mutation data matrices from protein sequences. Comput Appl Biosci 8: 275–282.
26. Felsenstein J (1981) Evolutionary trees from DNA sequences: a maximum likelihood approach. J Mol Evol 17: 368–76.
27. Sawai H, Go Y, Satta Y (2008) Biological implication for loss of function at major histocompatibility complex loci. Immunogenetics 60: 295–302.
28. Hedges SB, Dudley, Kumar S (2006) Time Tree: a public knowledge-base of divergence times among organisms. Bioinformatics 23: 2971–2.
29. Rzhetsky A, Nei M (1992) Statistical properties of the ordinary least-squares, generalized least-squares, and minimum-evolution methods of phylogenetic inference. J Mol Evol 35: 367–75.
30. Mann HB, Whitney DR (1947) On a test of whether one of two random variables is stochastically larger than the other. Ann Math Statist 18: 50–60.
31. Ohmura T, Ishimura Y, Fujii Y (2009) Molecular biology of P450. 2nd ed.
32. Nelson DR, Koymans L, Kamataki T, Stegeman JJ, Feyereisen R, et al. (1996) P450 superfamily: update on new sequences, gene mapping, accession numbers and nomenclature. Pharmacogenetics, 6(1): 1–42
33. Meunier B, de Visser SP, Shaik S (2004) Mechanism of oxidation reactions catalyzed by cytochrome P450 enzymes. Chem Rev 104: 3947–3980.
34. Feyereisen R (2011) Arthropod CYPomes illustrate the tempo and mode in P450 evolution. Bioch et Biophys Acta 1814: 19–28.
35. Urabe K, Kimura A, Harada F, Iwanaga T, Sasazuki T (1990) Gene conversion in steroid 21-hydroxylase genes. Am J Hum Genet 46: 1178–1186.
36. Nelson DR, Zeldin DC, Hoffman SM, Maltais LJ, Wain HM, et al. (2004) Comparison of cytochrome P450 (CYP) genes from the mouse and human genomes, including nomenclature recommendations for genes, pseudogenes and alternative-splice variants. Pharmacogenetics 14(1): 1–18.
37. Yoshiki Yasukochi, Yoko Satta (2011) Evolution of the CYP2D gene cluster in humans and four non-human primates. Genes Genet Syst 86: 109–116.

12

Whole Genome Sequence of a Turkish Individual

Haluk Dogan, Handan Can, Hasan H. Otu*

Department of Genetics and Bioengineering, Istanbul Bilgi University, Istanbul, Turkey

Abstract

Although whole human genome sequencing can be done with readily available technical and financial resources, the need for detailed analyses of genomes of certain populations still exists. Here we present, for the first time, sequencing and analysis of a Turkish human genome. We have performed 35x coverage using paired-end sequencing, where over 95% of sequencing reads are mapped to the reference genome covering more than 99% of the bases. The assembly of unmapped reads rendered 11,654 contigs, 2,168 of which did not reveal any homology to known sequences, resulting in ~1 Mbp of unmapped sequence. Single nucleotide polymorphism (SNP) discovery resulted in 3,537,794 SNP calls with 29,184 SNPs identified in coding regions, where 106 were nonsense and 259 were categorized as having a high-impact effect. The homo/hetero zygosity (1,415,123:2,122,671 or 1:1.5) and transition/transversion ratios (2,383,204:1,154,590 or 2.06:1) were within expected limits. Of the identified SNPs, 480,396 were potentially novel with 2,925 in coding regions, including 48 nonsense and 95 high-impact SNPs. Functional analysis of novel high-impact SNPs revealed various interaction networks, notably involving hereditary and neurological disorders or diseases. Assembly results indicated 713,640 indels (1:1.09 insertion/deletion ratio), ranging from −52 bp to 34 bp in length and causing about 180 codon insertion/deletions and 246 frame shifts. Using paired-end- and read-depth-based methods, we discovered 9,109 structural variants and compared our variant findings with other populations. Our results suggest that whole genome sequencing is a valuable tool for understanding variations in the human genome across different populations. Detailed analyses of genomes of diverse origins greatly benefits research in genetics and medicine and should be conducted on a larger scale.

Editor: Huiping Zhang, Yale University, United States of America

Funding: This work is funded by Istanbul Bilgi University Research Fund. The funders had no role in study design, data collection and analysis, decision to publish, or preparation of the manuscript.

Competing Interests: The authors have declared that no competing interests exist.

* E-mail: hasan.otu@bilgi.edu.tr

Introduction

Following the publication of two draft sequences [1,2], a highly accurate and nearly complete assembly of the human genome was published in 2004 [3]. In parallel with the low-cost/high-throughput advances in DNA sequencing technology, human whole genome sequencing (WGS) is being performed worldwide at an increasing pace. Individual WGS began to surface with Venter's and Watson's genomes [4,5], and this approach was quickly adapted to individuals from diverse ethnic backgrounds [6]. Understanding DNA sequence variation sheds light on the relationship between genotype and phenotype, and WGS has proven to be a powerful tool. The 1000 Genomes Project, for example, has performed 185 human WGSs from four populations and discovered about 20,000 novel structural variants in its Pilot Phase [7]. In Phase I of the project, the number of sequenced individuals increased to 1,092 covering 14 populations and identifying 38M single nucleotide polymorphisms (SNPs), 1.4 M indels and over 14 K larger deletions [8].

Efforts (other than WGS) that target discovery of human genome variations also exist, such as the HapMap project [9]. Latest HapMap results cover 1,184 individuals from 11 populations and involve genotyping of common SNPs and sequencing of relatively small regions (~100 Kbp). Overall, HapMap and similar consortiums have catalogued over 10 million SNPs, 3 million indels, and associated linkage-disequilibrium patterns. This ongoing process of identifying genomic variants has paved the way for genome-wide association studies over the past few years, and

disease susceptibility has been found to be associated with these variants for over a thousand regions so far. This accumulated knowledge in the post-genomic era is opening new frontiers in medicine and public health using a personalized approach, and WGS is becoming the method of choice with its ability to construct a nearly complete picture of identifying structural variations.

Despite the increasing use of human WGS for both research and clinical purposes, there remain two areas that require further attention: i) there are populations for which WGS or SNP discovery efforts have not been done; ii) very few of the human WGS performed so far provide high-coverage sequencing results with detailed analysis. Out of the 185 individuals for whom WGS has been performed in the 1000 Genomes Project's Pilot Phase, only six were subjected to high-coverage sequencing (~42x) while the remaining individuals were subjected to low-coverage sequencing (2–6x). All of the individuals studied in Phase I of the project were analyzed using low-coverage sequencing (2–6x).

There have been various efforts to perform high coverage WGS of different populations with detailed analysis [10,11,12,13,14,15]. In order to provide a better and more complete picture of human genome variations, we believe more individuals from diverse populations need to be sequenced and analyzed at a sufficiently detailed level. Therefore, in this paper, we present a high-coverage WGS of a Turkish individual and the results of the associated analysis.

Turkey, the most populous well-defined region inhabited by Turks, is an interesting geographical region as it lies at the crossroads between Europe and Asia. Historically, migration from Central Asia and ancestral contribution to regions surrounding Anatolia, such as the Balkans, Middle East, Caucasian and Caspian regions, has positioned the Turkish population as an interesting genetic resource that requires further detailed analysis. Certain important diseases, such as hemoglobinopathies (e.g. sickle-cell disease), thalassemias, and Behcet's disease, are highly prevalent in the Turkish population; and diseases exist where the Turkish population does not exhibit the variant believed to be the cause [16,17,18]. The current study provides a baseline for high-throughput/wide-spectrum analysis of genome variations in the Turkish population, which may lead to a better understanding of the relationship between the genotype and the phenotype through comparative analysis.

Materials and Methods

Ethics Statement

This study is approved by the Committee on Ethics in Research on Humans of Istanbul Bilgi University.

Individual Selection, DNA Isolation, and Genotyping

The genomic DNA (gDNA) used in this study came from a healthy male individual, who was anonymous and was reported to come from Turkish ethnicity for at least three generations. Informed consent was obtained prior to the collection of the blood sample from which gDNA was isolated using a QIAamp DNA blood kit (Qiagen®). The individual gave written consent to the publication of his genome sequence. A quality control inspection and rough quantitation of the gDNA sample was performed by agarose gel electrophoresis and UV-induced ethidium bromide fluorescence (Figure S1). The quality of the sample on the gel was visually compared to New England BioLabs 2-Log DNA Ladder molecular weight size marker. The sample was of acceptable quality for continued processing. The individual was genotyped using Illumina Human CytoSNP-12 V2.1 (Illumina®) SNP chip following the manufacturer's instructions.

Library Preparation and Sequencing

The genomic DNA sample was used to generate a paired-end library suitable for the HiSeq sequencing platform (Illumina®) prepared using the TrueSeq DNA Sample Preparation kit, following the manufacturer's instructions. Quality control analysis of the library using an Agilent 2100 Bioanalyzer indicated that the library was of acceptable quality, containing the expected fragment size and yield, for continued sample processing (Figure S2). The library generated was used in the cBot System for cluster generation in three flow cell lanes. The flow cell containing amplified clusters was sequenced using 2×101 base pair paired-end sequencing on a Hi-Seq 2000. Bad quality reads were eliminated from the final output of the sequencing machine. In brief, for each cluster, a "chastity" score was calculated, which is the Highest_Intensity/(Highest_Intensity + Next_Highest_Intensity) for a base call in the first 25 cycles. A cluster was retained if it contained at most one base call instance where the chastity parameter was less than the threshold. Remaining reads were further trimmed and filtered using Trimmomatic [19] where reads with a high quality (average Q≥20) score and a minimum length of 36 (after trimming) were kept.

Sanger sequencing was used for validation of 20 small indels identified by computational analysis of the whole genome sequence data. DNA was isolated using a Dual DNA isolation kit (GenedireX Inc., Taiwan). Twenty regions were amplified from 50 ng genomic DNA with 10 pmol of forward and reverse primer pairs. Polymerase chain reaction (PCR) was performed using the following cycling profile: initial denaturation at 95°C for 5 min. followed by 10 cycles of 95°C for 30 s, 63°C for 30 s, and 72°C for 30 s; 25 cycles of 95°C for 30 s, 56°C for 30 s, and 72°C for 30 s; and a final extension step at 12°C for 5 min. Amplicons were purified using GenedireX PCR Clean-Up kit (GenedireX Inc., Taiwan) and quantified. ABI 3100 DNA analyzer and ABI BigDye Terminator cycle sequencing was used for Sanger sequencing.

Mapping and *De Novo* Assembly

The reference genome used in this study was NCBI human reference genome build 37.1 (GRCh37/hg19 assembly). We adopted two workflows for read mapping. First, we used the vendor supplied Eland and Casava (v1.8) pipeline using recommended settings with a variants covariance cut-off value of 3. Alternatively, we used the Burrows-Wheeler Alignment tool (BWA v0.6.2) for mapping reads to the reference genome [20]. The vendor-supplied method of mapping was used only for SNP calling in the respective pipeline. The mapping results provided in the manuscript are based on the BWA approach. In the application of the BWA tool, we used a minimum seed length of 20 (with a maximum seed distance of 2), an output alignment score cut-off of 30, a maximum edit distance of 0.04, and a maximum insert size of 500. Specifically, we used the *bwasw* algorithm to index the database and generated suffix array indices for two ends in a paired-end read separately. We then combined the two results with the *sampe* algorithm to produce the final sequence alignment/ map (SAM) file. BWA analysis results were investigated with SAMStat v1.08 [21] to determine the quality and statistics associated with the mapping step. Unmapped reads were assembled using the iterative De Bruijn Graph De Novo Assembler (IDBA v1.1.0) where the minimum seed length for overlapping nucleotides was set to be 25 [22]. We required at least five pair-end connections to join two contigs and a minimum contig length of 100. The resulting contigs were analyzed using BLAST v2.2.26 [23] on the NCBI's RefSeq genomic database using an E-value cut-off of 10^{-10}.

SNP/Indel and CNV/SV Identification

SNP identification was done using the vendor-supplied Eland-Casava pipeline with the recommended settings and The Genome Analysis Toolkit (GATK v2.2) [24] applied on the BWA output. GATK has also been used in indel identification. Prior to variant discovery, reads were subjected to local realignment, coordinate sort, quality recalibration, and duplicate removal. In the GATK analysis, we used a minimum confidence score threshold of Q30 with default parameters. Annotation of the discovered SNPs/ indels and their potential effects were analyzed using snpEff v3.1 [25]. During the SNP/indel discovery and analysis phases, we adopted NCBI's dbSNP build 135. CNV/SV events were discovered using read-depth-based CNVnator and paired-end-based CLEVER algorithms [26,27]. In the CLEVER approach, we used a maximum insert length of 50,000 and a maximum allowed coverage of 200. In the CNVnator analysis, we used a bin (window) size of 100 with default parameters. CNV/SV calls using the SNP chip data were obtained by QuantiSNP with default parameters [28].

Functional Analysis

We used Ingenuity Software Knowledge Base (IKB), (Redwood City, CA) for the functional interaction analysis of the genes affected by high-impact, novel SNPs. IKB uses interactions between genes and/or gene products based on manual curation of scientific literature providing a robust interaction database.

Once a gene list of interest is identified, IKB uses known interactions between these genes to build an interaction network. The final network includes genes that are not in the input list but are highly connected to the genes in the input list. This feature enables the investigation of new modules of interactions that are not covered by existing canonical pathways. The results are also analyzed in terms of drugs, small metabolites, functions, and diseases that are overlaid on the resulting network. To this end, an annotation of the resulting interaction network is achieved where functional entities involved in the network are underlined.

Results

Individual Selection, DNA Isolation, and Genotyping

Prior to the sequencing step, we compared the microarray genotyping results of the individual used in this project, who was an anonymous, healthy male claiming to have come from Turkish ancestry for at least three generations, with those obtained from the HapMap project [9] and a recent genome-wide association study targeting Behcet's disease [29]. The latter has utilized 1,215 cases and 1,278 controls from Turkey, genotyped on Illumina's HumanCNV370-Quad v3.0 1 (Illumina®) chip. HapMap populations represent African ancestry in the southwestern USA (ASW); Utah, USA inhabitants with ancestry from northern and western Europe (CEU); Han Chinese in Beijing, China (CHB); Chinese in metropolitan Denver, Colorado, USA (CHD); Gujarati Indians in Houston, Texas, USA (GIH); Japanese in Tokyo, Japan (JPT); Luhya in Webuye, Kenya (LWK); Maasai in Kinyawa, Kenya (MKK); Mexican ancestry in Los Angeles, California, USA (MXL); Tuscans in Italy (TSI); and Yoruba in Ibadan, Nigeria (YRI). We compared the SNP calls coming from the three data sets using Eigenstrat v4.2 [30] to investigate the clustering of individuals based on a principal components analysis (PCA). We then performed PCA analysis only on the Turkish samples used in the Behcet study and the individual used in the current study. In both PCA analyses shown in Figure 1, we utilized only the healthy controls from the Behcet study. Our results show that the individual chosen for WGS represents a typical member of the Turkish population, which differs from the populations used in the HapMap project.

Trimming, Mapping, and Assembly of the Reads

DNA sequencing generated 1,238,722,496 paired-end reads corresponding to ~125,111 M bases of data yielding ~35x coverage. Quality and length-based trimming and filtering dropped 4.44% of the reads eliminating a total 5.03% of total base pairs. The remaining ~1.18 billion reads (accounting for ~116,720 M bases) proved to be of high quality (mode of the average read quality Q is ~38), with sufficient length (mode is ~100 bp), and included no Ns (Figure S3). Of the high quality reads, 95.28% (~1.13 billion) were successfully mapped to the reference genome (GRCh37/hg19) covering 99.6% of the bases in the reference genome using the BWA mapping approach. Approximately 50 million unmapped reads (accounting for 4,946 M bases) were assembled using IDBA, which generated 11,654 contigs with lengths ranging between 100 – 43,190 base pairs amounting to ~10 Mbp of potentially novel sequence. Mean contig length was 856 bp with an N50 of 1,378 bp and an N80 of 497 bp. Of the contigs, 9,486 (~81%) received a hit in the RefSeq database. Most of the contigs that received a hit were found to be homologues to alternate, reference, or other human sequences (~97%), while the remaining 313 contigs were found to be homologous to nonhuman primates and other sequences. The 2,168 contigs that were not found to be homologous to any

sequences in the RefSeq database represented a total of 927,213 base pairs of assembly with a mean contig length of 427 bp and an N50 of 469 bp. These results are summarized in Table 1.

SNP Identification

Casava and GATK workflows identified 3,642,449 and 4,301,769 SNPs, respectively. In order to increase the reliability of our findings, all downstream analysis was performed with SNPs identified by both methods, which resulted in 3,537,794 variants. Of these concordant SNP calls, 97.8% were in agreement with the SNPs called by the genotyping performed on the array, showing high reproducibility. The transitions (2,383,204) transversions (1,154,590), Ts/Tv, ratio was 2.06, and the homozygosity (1,415,123) and heterozygosity (2,122,671) proportions were 40% and 60%, respectively, with both ratios and percentages resembling expected figures in similar studies [14]. Of the SNPs, 47% were in an intronic region; and 43% of the SNPs were in an intergenic region. About 7% of the SNPs were in upstream or downstream regions of a gene and an additional 1.2% of the SNPs were in an untranslated 3′ or 5′ region. Of these SNPs, 29,184 were identified in coding regions with 15,876 synonymous, 13,202 nonsynonymous, and 106 nonsense SNPs. In Figure 2, we show the distribution of the SNPs based on the region in which they were found.

SNPs, where (i) the variant hits a splice acceptor/donor site, (ii) a start codon is changed into a nonstart codon, or (iii) a stop codon is gained or lost due to the variant are categorized as having a "high-impact" effect. There were a total of 259 SNPs with a high-impact effect and 167 of these were found to be on gene sequences. Out of the ~3.5 million SNPs identified, there were 480,396 potentially novel SNPs that did not exist in dbSNP. Of these potentially novel SNPs, 49% and 41% were in intergenic and intronic regions, respectively. Over 8% were upstream or downstream of a gene, and 2,925 (or 0.5%) of the SNPs were found to be in coding regions. There were 48 nonsense SNPs and a total of 95 SNPs with a high impact. These high-impact SNPs affected 47 genes (45 well characterized). In Table S1, we list these genes along with their effects and annotations. In Table 2, we show the 23 well-characterized genes that were affected by a novel nonsense SNP. Gene annotations were obtained using the GeneALaCart tool of the GeneCards suite (www.genecards.org) [31].

Indel Detection

We identified 713,640 indels, which consisted of 341,382 insertions and 372,258 deletions. Of these indels, 159,593 (or 22%) were found to be novel. The length distribution ranged from -52 bp to 34 bp (see Figure S4 for the histogram of indel lengths) where the average ± standard deviation values were 17.03±9.72 bp for insertions and -26.50±15.15 bp for deletions. Of the indels, 40.8% and 49.5% were in an intergenic and intronic region, respectively. An additional 8.3% were equally divided between upstream and downstream regions of genes; and about 1% were in 3′ and 5′ untranslated regions, the majority (~95%) being in the 3′ UTR. Only 50 and 53 indels were in a splice site acceptor and donor regions, respectively; and 1,934 (or 0.2%) affected a coding region. Of these, 246 indels caused a frame shift, while 104 resulted in a codon deletion and 75 resulted in a codon insertion. When we imposed a window-based filtering such that no two indels co-occurred within 20 bp of each other, we identified 655,195 indels, out of which 123,478 (or 19%) were novel.

We performed Sanger sequencing on 20 regions containing 20 predicted indels and validated 18 of these indels. We randomly selected the indels for validation with the constraints that they

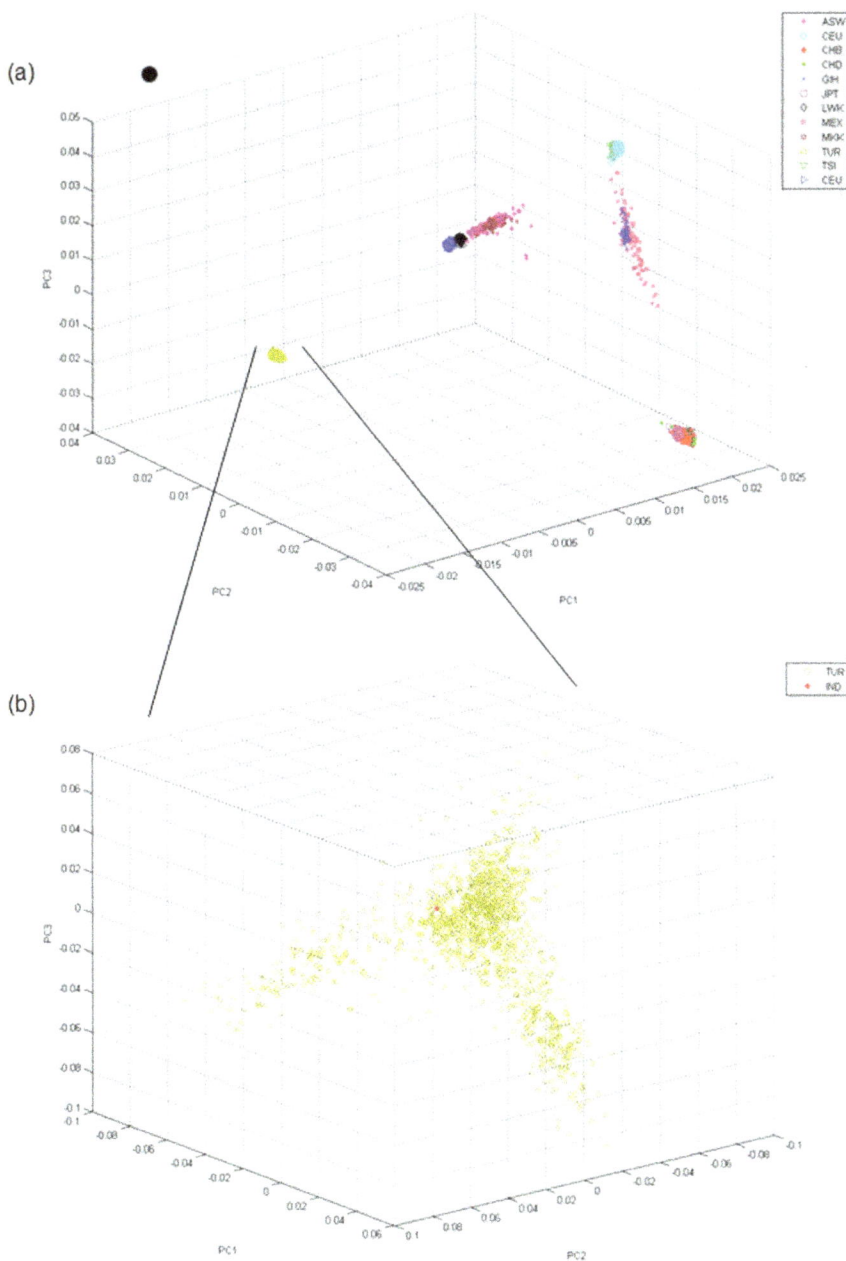

Figure 1. Position of the Turkish population and the sequenced individual based on Principal Components Analysis of genotyping data. (a) PCA of the genotyping data from the populations in the HapMap project and GWAS targeting the Turkish population (TUR); (b) PCA of the genotyping data from the GWAS targeting the Turkish population and the individual used for sequencing in this project (IND).

overlapped with coding sequences in possibly detrimental ways and showed different homo/heterozygosity status in a ratio of ~2:1, as observed in previous studies [4]. Seventeen of the 20 indels used for validation were overlapping with known genes, and 3 were overlapping with predicted gene regions. Fourteen indels were homozygous, and 6 were heterozygous. The 20 indels represented 3 frame-shift deletions, 5 frame-shift insertions, 3 nonframe-shift deletions, 8 nonframe-shift insertions and 1 stop-gain SNV all overlapping with coding sequencing in potentially detrimental ways. We list the details of the 20 indels used for validation and utilized forward and reverse primers in Table S2 and Table S3.

Structural Variant Discovery

Employed read-depth and paired-end CNV/SV discovery methods [26,27] identified 9,109 such events including 7302 deletions, 1663 duplications, and 144 insertions. Length normalized distribution of these calls followed a uniform distribution across chromosomes. On average, we observed 3.11 CNV/SV events per chromosome per million base pairs with a standard deviation of 0.77 (Figure S5). Of the predicted CNV/SV calls, 58.5% overlapped with the structural variants identified as part of the 1000 Genomes Project. The length distribution of the total and novel CNV/SV events revealed that 3,820 out of 9,109 total (or 42%) and 1,786 out of 3,780 novel (or 48%) events were less than 1 Kbp (Figure S6). When we compared the CNV/SV events

Table 1. Read Sequencing and Analysis Statistics.

Trimming and Filtering

No. of Reads (Raw)	Total Base Pairs (Raw)	No. of Reads (Trimmed and Filtered)	Total Base Pairs (Trimmed and Filtered)
~1.24×10^9	~125×10^9	~1.18×10^9	~117×10^9

Mapping

No. of Mapped High Quality Reads	Total Base Pairs Mapped	No. of Unmapped High Quality Reads	Total Base Pairs Unmapped
~1.13×10^9	~112×10^9	~50×10^6	~5×10^9

Assembly of Unmapped Reads

No. of Contigs	Total Length of the Assembly (bp)	Min.–Max.–Mean Contig Length (bp)	N50 (bp)
11,654	9,987,256	100–43,190–856	1,378

Homology Search

Contigs Without a Hit	Total Length of Unhit Contigs (bp)	Min.–Max.–Mean Unhit Contig Length	N50 of Unhit Contigs (bp)
2,168 (19%)	927,213	100–9,345–427	469

Contigs With a Hit	Reference Genome Alternate Assemblies	Other Human Sequences	Non-human primates	Other
9,486 (81%)	983 (8.5%)	7,814 (67.0%)	376 (3.2%)	218 (1.9%)	95 (0.8%)

called by two different algorithms, we identified 1,629 concordant, high-confidence calls. Of these high confidence calls, 1,223 (or 75%) overlapped with CNV/SVs identified as part of the 1000 Genomes Project. We also verified the CNV/SV calls with the results of the SNP chip data and found 394 concordant calls.

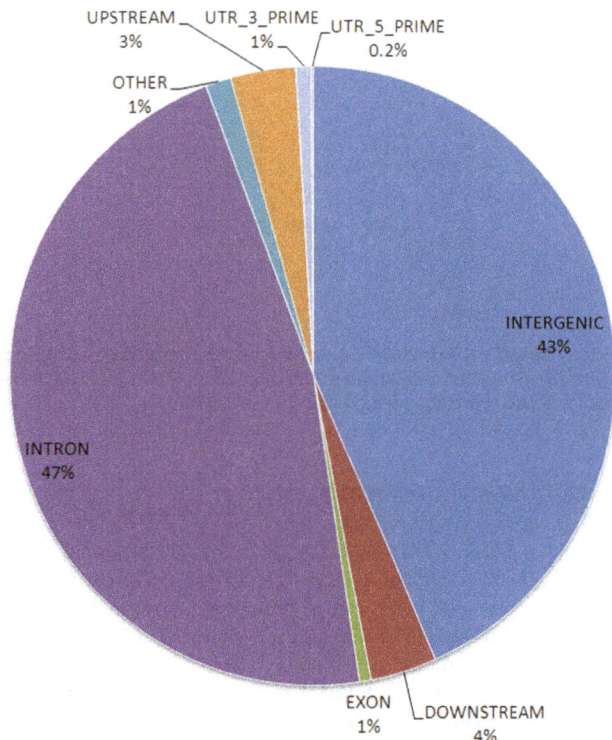

Figure 2. Distribution of the 3,537,794 identified SNPs based on their genomic location.

Discussion

In this paper, we present high-depth coverage (~35×) and detailed analysis of the whole genome sequence of a Turkish individual. Although whole genome human sequencing is almost routinely done, very few of these efforts provide high coverage and analysis; and various populations are not included in large consortium efforts. Therefore, we believe the current study provides a reference data set in understanding human genome variation on a large scale and a population-dependent context and is an initial step in exploring Turkish genomic features. Despite its importance as the first Turkish whole genome analysis, we acknowledge that further studies are required to validate and generalize our findings to the population scale. Nevertheless, we believe the data presented here will provide a cornerstone for such studies and enrich the analysis of human genomic variation across diverse populations.

Our genotyping results from chip and sequence data, and the comparative analysis of these with HapMap and other chip genotyping results suggest that the sequenced individual represents a typical member of the Turkish population. The sequencing analysis proves that the data generated is of high quality as only 5% of the total sequence is filtered out, and the remaining reads almost completely cover the reference human genome. High N50 values indicate successful assembly of unmapped reads, which also resulted in ~1 Mbp of unmapped human genome sequence that did not reveal any homology in the RefSeq database. SNP identification rendered high reproducibility with ~98% consistency between the sequencing and microarray genotyping results and validated about 86% of identified SNPs in the dbSNP database. We believe the remaining 480,396 potentially novel SNPs contribute to understanding the human genomic variation and the relationship between genotypes and phenotypes.

The length distribution of the 9,109 identified CNV/SV events showed peaks at the 300 bp and 6,000 bp marks potentially representing short and long interspersed elements, respectively. A similar trend is seen in the 1,629 high confidence calls (Figure S6). The relatively high number of CNV/SV calls are due to the inclusion of short variants (variants smaller than 1 Kbp), which constitute 42% of total calls and 48% of novel calls. We validated

Table 2. Annotation for the 23 genes that were affected by a novel nonsense SNP.

Symbol	Descriptions	Chr	Disorder/Disease	Function	Pathway
ABCA9	ATP-binding cassette A9	17	Pseudoxanthoma elasticum	Monocyte differentiation; Lipid homeostasis	ABC transporters
ADCK3	aarF domain containing kinase 3	1	Spinocerebellar ataxia	Protein serine/threonine kinase activity	
ANKRD35	Ankyrin repeat domain-containing protein 35	1		Protein binding	
CAD	CAD trifunctional protein	2	Fibrosarcoma	Aspartate carbamoyltransferase activity	Pyrimidine metabolism; Transcription/Ligand-dependent activation of the ESR1/SP pathway
CDC27	cell division cycle 27	17		Cell cycle checkpoint	Cell cycle_Regulation of G1/S transition
DPRX	Divergent-paired related homeobox	19		Sequence-specific DNA binding TF activity	
FRG2C	FSHD region gene 2 family, member C	3			
GIMAP6	GTPase, IMAP family member 6	7		GTP binding	
HSPBAP1	27 kDa heat shock protein-associated protein 1	3	Intractable epilepsy; Renal carcinoma	Cellular stress response	
HTR2C	5-hydroxytryptamine receptor 1C	X	Schizophrenia; Migraine; Prader-Willi syndrome; Attention deficit hyperactivity disease	Phosphatidylinositol phospholipase C activity	Calcium signaling pathway; Neuroactive ligand-receptor interaction
KBTBD3	BTB and kelch domain-containing protein 3	11		Protein binding	
KRTAP2-2	Keratin-associated protein 2.2	17		Keratin filament	
MLL3	Myeloid/lymphoid leukemia3	7	Leukemia	Methyltransferase activity	Lysine degradation
MYT1	Myelin transcription factor I	20	Dysembryoplastic neuroepithelial tumor; Periventricular leukomalacia	Oligodendrocyte lineage development	
PCNT	Pericentrin	21	Seckel syndrome; Microcephaly	M transition of mitotic cell cycle	Centrosome maturation
PPP2R2B	Protein phosphatase 2, regulatory subunit B	5	Spinocerebellar ataxia	Apoptotic process	mRNA surveillance pathway; Tight junction; Reg'n. of CFTR activity
PROSER1	Proline and serine rich 1	13			
TBCK	TBC1 domain containing kinase	4			
TCP10L2	T-complex 10 like prtn. 2	6	Spina bifida	Cytosol	
TECTA	Tectorin alpha	11	Nonsyndromic deafness; Scotoma; Sensorineural hearing loss	Cell-matrix adhesion	
TFAP2B	Transcription factor AP-2 beta	6	Patent ductus arteriosus; Skeletal muscle neoplasm	Cellular ammonia/urea/creatinine homeostasis	
XIAP	X-linked inhibitor of apoptosis protein	X	Leukemia; Lymphoma	Caspases, apoptosis regulation; inflammation	Ubiquitin mediated proteolysis; SMAC-mediated apoptosis
ZNF778	Zinc finger protein 778	16	KBG syndrome; Learning disability	Zinc ion binding	

CNV/SV events that fall within a size range which can be validated with the SNP chip. We have been able to verify 394 CNV/SV calls where the majority of the variants were greater than 10 kbp in length. We found 3,780 CNV/SV calls that were not identified in the 1000 Genomes Project and may be variants potentially specific to the Turkish population.

In order to position our results in a better population genetics context, we compared the SNPs identified in the sequenced Turkish individual to the SNPs found in Utah, USA inhabitants with ancestry from Europe (CEU) and to the SNPs found in Han Chinese in Beijing, China (CHB). Both CEU and CHB populations were included in the HapMap project, which

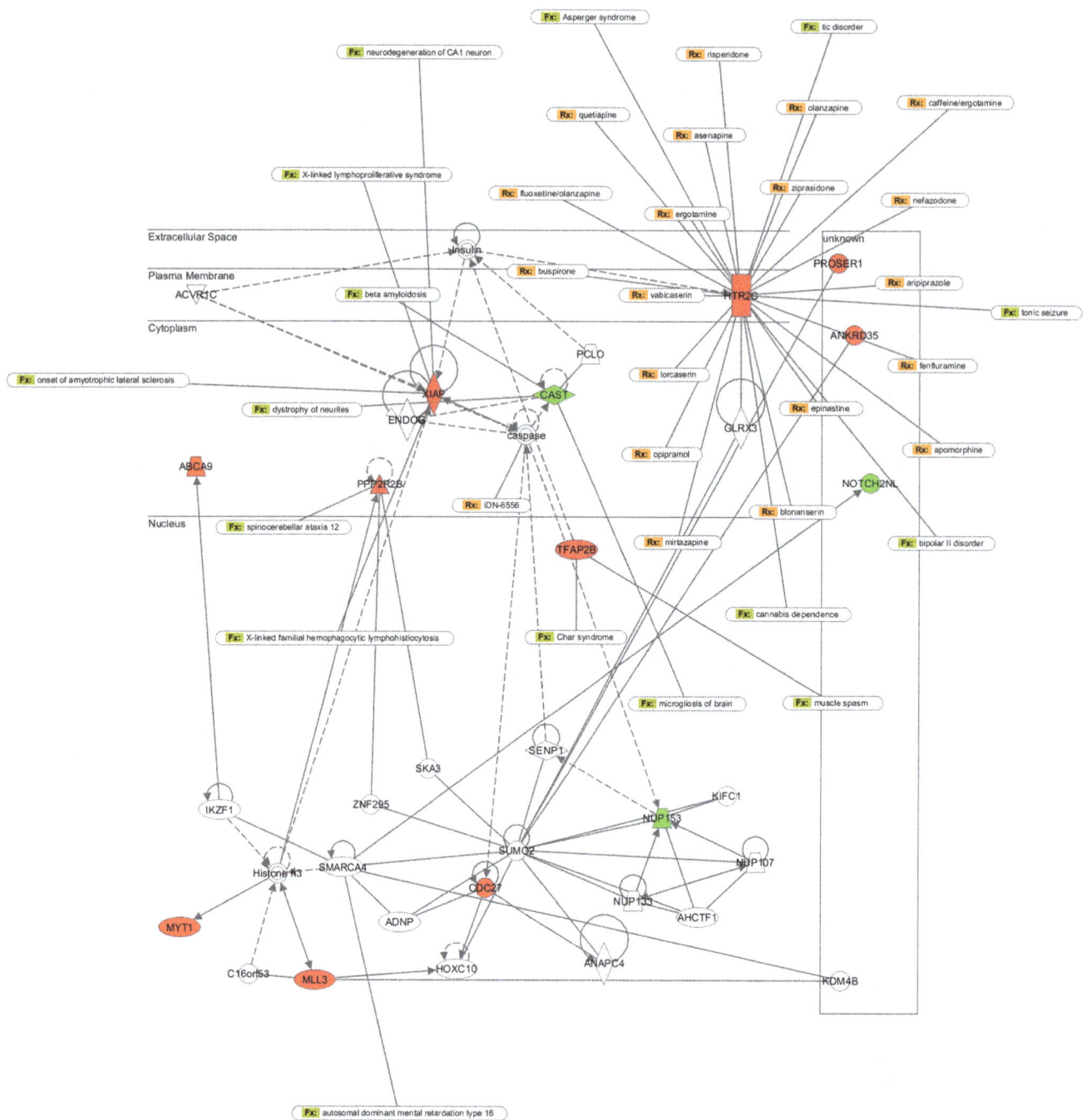

Figure 3. Ingenuity Network analysis of 45 genes affected by a high impact novel SNP. Genes indicated by red are affected by a nonsense SNP and genes indicated by green are affected by an SNP targeting a splice site donor/acceptor region. Drug targets, hereditary and neurological disorders/diseases are indicated where applicable.

constitute the source of the identified SNPs in these populations used in our analysis. From a historical perspective, precursors of the Turks originated in Central Asia and Turks are known to have been inhabitants of regions that are parts of modern day China. Turks' migration toward the West ended up mostly in Anatolia with minor settlements in Europe. Turks' presence in Europe was later expanded throughout the Ottoman era. We, therefore, performed our comparative analysis with CEU and CHB, which

are the two most closely related populations to the Turkish population with available large-scale genetic data. We found 665,032 SNPs commonly shared by the three populations, which corresponds to 47% of all of the CEU SNPs and 50% of all the CHB SNPs (Figure S7). SNPs exclusively shared by the Turkish and CEU populations were 3% of the total CEU SNPs (43,592 out of 1,412,090 SNPs) while SNPs exclusively shared by the Turkish and CHB populations were 1% of the total CHB SNPs (15,657 out of 1,328,223 SNPs). Although more evidence is needed to make conclusive remarks, our results may suggest that the Turkish population is almost equidistant to the CEU and CHB populations, being slightly closer to the CEU.

Novel SNPs predicted by our results have the potential to explain genetic features specific to the Turkish population. The 23 well-characterized genes that were affected by the novel nonsense SNPs identified in this study were found to affect 25 different phenotypes listed in OMIM, potentially leading to additional genetic disease mechanisms (see Table 2). We used Ingenuity Software Knowledge Base (IKB), (Redwood City, CA) to further identify networks that explain underlying interactions for the 47 genes that were affected by a high-impact novel SNP. The biological functions identified by IKB are grouped in 66 categories. Out of these categories, two of the most significant ones were hereditary disorders and neurological diseases. The former included 14 disorders, 3 of which were X-linked via the gene XIAP; and the latter involved 27 diseases, most notably due to HTR2C. These results are summarized in Table S4.

IKB analysis revealed three interaction networks that included 15, 13, and 3 of the 47 genes. In Figure 3 and Figure S8, we show the first two networks where corresponding drug and hereditary and neurological disorders/diseases are overlaid on the networks. In this representation, we show the subcellular layout; and for the network shown in Figure 3, HTR2C and XIAP seem to play central roles at the membrane and cytoplasm levels. We observe a downstream effect of HTR2C on SMT3 suppressor of mif two 3 homolog 2 (SUMO2) through glutaredoxin 3 (GLRX3). It has been shown that 5-hydroxytryptamine receptors are associated with protein networks involved in synaptic localization along with multidomain proteins such as GLRX3 [32]. GLRX3 may participate in the inhibition of apoptosis and play a role in cellular growth, regulating the function of the thioredoxin system; and contains subunits that may serve as a redox sensor [33]. SUMO2, which binds with GLRX3 [34], is known to be involved in a number of processes like apoptosis, protein stability, and transcriptional regulation. On the other hand, SUMO2 is active in acetylation of histone H3 [35] through SMARCA4 [36]; and XIAP is involved in methylation of histone H3 [37]. This cascade of events may suggest a mechanism in which two central genes affected by novel nonsense SNPs identified in this study, XIAP and HTR2C, are involved in various neurological diseases such as amyotrophic lateral sclerosis [38]. In the other interaction network identified by IKB, we see a key role played by ubiquitin C (UBC), which is linked to the genes affected by novel high impact SNPs. The suggested outcomes involve various neurological disorders in addition to tumorigenesis and blastomas (Figure S8). Overall, we believe that functional and systems level analysis of population-dependent genomic variations may shed light on disease mechanisms as demonstrated here.

Data Accession

The whole genome sequencing reads have been deposited in the NCBI Short Read Archive (SRA) database under the accession number SRA056442. SNPs and indels have been deposited in the NCBI dbSNP database with the handle ID "BILGI_BIOE". CNV/SV calls have been deposited in the NCBI dbVar database with project ID PRJNA171612.

Supporting Information

Figure S1   QC gel image (0.8% agarose) of gDNA sample (1) compared to molecular weight marker (M). Sample quality was considered to be acceptable if the gDNA supplied a single visible band while lacking any significant degradation products (degraded DNA seen as smear of small fragments).

Figure S2   Library quality. The electropherogram for the generated library displaying expected yield and size.

Figure S3   Quality statistics for the forward reads only (almost identical results are obtained for the reverse reads) following trimming and filtering. a) Average base quality with respect to the position of the base in the read; b) Histogram of the sequence lengths; c) histogram of the average quality scores of the reads; d) Ns seen in the reads with respect to the position of the base in the read.

Figure S4   Length distribution of the identified 713, 640 indels.

Figure S5   Distribution of 9,109 identified CNV/SV calls across chromosomes. A) Number of CNV/SV events; B) Length normalized (per million base pairs) CNV/SV events.

Figure S6   Length distribution of 9,109 identified, 3870 novel, and 1629 high confidence CNV/SV calls. Note that the bin size for SVs less than 10 Kbp is 200 bp while the bin size for SVs more than 10 Kbp is 2 Kbp. The last bar in the graphs on the right-hand column represents SVs more than 250 Kbp.

Figure S7   Overlap of SNPs identified in the Turkish individual used in the manuscript (TUR); Utah, USA inhabitants with ancestry from Europe (CEU); and Han Chinese in Beijing, China (CHB).

Figure S8   IKB Network analysis of 45 genes affected by a high impact novel SNP. Genes indicated by red are affected by a nonsense SNP and genes indicated by green are affected by a SNP targeting a splice site donor/acceptor region. Hereditary and Neurological Disorders/Diseases are indicated where applicable.

Table S1   45 well characterized genes that were affected by a high-impact SNP. Effect Types: 1: Stop gained; 2: Splice site acceptor; 3: Splice site donor; 4: Stop lost; 5: Start lost.

Table S2   20 predicted indels used for validation by Sanger Sequencing (V: Validated NV: Not Validated).

Table S3   Forward and reverse primers used in Sanger Sequencing.

Table S4   Biological Function categories known to involve 45 well characterized genes that were affected by a high-impact SNP.

Author Contributions

Conceived and designed the experiments: HHO HC. Performed the experiments: HC HD. Analyzed the data: HHO HD. Contributed reagents/materials/analysis tools: HC HHO. Wrote the paper: HHO HC. Interpreted the results: HHO HC HD.

References

1. Lander ES, Linton LM, Birren B, Nusbaum C, Zody MC, et al. (2001) Initial sequencing and analysis of the human genome. Nature 409: 860–921.

2. Venter JC, Adams MD, Myers EW, Li PW, Mural RJ, et al. (2001) The sequence of the human genome. Science 291: 1304–1351.

3. Consortium IHGS (2004) Finishing the euchromatic sequence of the human genome. Nature 431: 931–945.

4. Levy S, Sutton G, Ng PC, Feuk L, Halpern AL, et al. (2007) The diploid genome sequence of an individual human. PLoS Biol 5: e254.

5. Wheeler DA, Srinivasan M, Egholm M, Shen Y, Chen L, et al. (2008) The complete genome of an individual by massively parallel DNA sequencing. Nature 452: 872–876.

6. Pritchard JK (2011) Whole-genome sequencing data offer insights into human demography. Nat Genet 43: 923–925.

7. Abecasis GR, Altshuler D, Auton A, Brooks LD, Durbin RM, et al. (2010) A map of human genome variation from population-scale sequencing. Nature 467: 1061–1073.

8. Abecasis GR, Auton A, Brooks LD, DePristo MA, Durbin RM, et al. (2012) An integrated map of genetic variation from 1,092 human genomes. Nature 491: 56–65.

9. Altshuler DM, Gibbs RA, Peltonen L, Dermitzakis E, Schaffner SF, et al. (2010) Integrating common and rare genetic variation in diverse human populations. Nature 467: 52–58.

10. Ahn SM, Kim TH, Lee S, Kim D, Ghang H, et al. (2009) The first Korean genome sequence and analysis: full genome sequencing for a socio-ethnic group. Genome Res 19: 1622–1629.

11. Fujimoto A, Nakagawa H, Hosono N, Nakano K, Abe T, et al. (2010) Whole-genome sequencing and comprehensive variant analysis of a Japanese individual using massively parallel sequencing. Nat Genet 42: 931–936.

12. Ju YS, Kim JI, Kim S, Hong D, Park H, et al. (2011) Extensive genomic and transcriptional diversity identified through massively parallel DNA and RNA sequencing of eighteen Korean individuals. Nat Genet 43: 745–752.

13. Kim JI, Ju YS, Park H, Kim S, Lee S, et al. (2009) A highly annotated whole-genome sequence of a Korean individual. Nature 460: 1011–1015.

14. Tong P, Prendergast JG, Lohan AJ, Farrington SM, Cronin S, et al. (2010) Sequencing and analysis of an Irish human genome. Genome Biol 11: R91.

15. Bentley DR, Balasubramanian S, Swerdlow HP, Smith GP, Milton J, et al. (2008) Accurate whole human genome sequencing using reversible terminator chemistry. Nature 456: 53–59.

16. Akin H, Onay H, Turker E, Cogulu O, Ozkinay F (2010) MEFV mutations in patients with Familial Mediterranean Fever from the Aegean region of Turkey. Mol Biol Rep 37: 93–98.

17. Erbilgin Y, Sayitoglu M, Hatirnaz O, Dogru O, Akcay A, et al. (2010) Prognostic significance of NOTCH1 and FBXW7 mutations in pediatric T-ALL. Dis Markers 28: 353–360.

18. Yazici Y, Yurdakul S, Yazici H (2010) Behcet's syndrome. Curr Rheumatol Rep 12: 429–435.

19. Krueger F, Kreck B, Franke A, Andrews SR (2012) DNA methylome analysis using short bisulfite sequencing data. Nat Methods 9: 145–151.

20. Li H, Durbin R (2009) Fast and accurate short read alignment with Burrows-Wheeler transform. Bioinformatics 25: 1754–1760.

21. Lassmann T, Hayashizaki Y, Daub CO (2011) SAMStat: monitoring biases in next generation sequencing data. Bioinformatics 27: 130–131.

22. Peng Y, Leung H, Yiu S, Chin F. IDBA–A Practical Iterative de Bruijn Graph De Novo Assembler; 2010. Springer. pp. 426–440.

23. Altschul SF, Gish W, Miller W, Myers EW, Lipman DJ (1990) Basic local alignment search tool. J Mol Biol 215: 403–410.

24. DePristo MA, Banks E, Poplin R, Garimella KV, Maguire JR, et al. (2011) A framework for variation discovery and genotyping using next-generation DNA sequencing data. Nat Genet 43: 491–498.

25. Cingolani P, Platts A, Wang le L, Coon M, Nguyen T, et al. (2012) A program for annotating and predicting the effects of single nucleotide polymorphisms, SnpEff: SNPs in the genome of Drosophila melanogaster strain w1118; iso-2; iso-3. Fly (Austin) 6: 80–92.

26. Abyzov A, Urban AE, Snyder M, Gerstein M (2011) CNVnator: an approach to discover, genotype, and characterize typical and atypical CNVs from family and population genome sequencing. Genome Res 21: 974–984.

27. Marschall T, Costa IG, Canzar S, Bauer M, Klau GW, et al. (2012) CLEVER: clique-enumerating variant finder. Bioinformatics 28: 2875–2882.

28. Colella S, Yau C, Taylor JM, Mirza G, Butler H, et al. (2007) QuantiSNP: an Objective Bayes Hidden-Markov Model to detect and accurately map copy number variation using SNP genotyping data. Nucleic Acids Res 35: 2013–2025.

29. Remmers EF, Cosan F, Kirino Y, Ombrello MJ, Abaci N, et al. (2010) Genome-wide association study identifies variants in the MHC class I, IL10, and IL23R-IL12RB2 regions associated with Behcet's disease. Nat Genet 42: 698–702.

30. Price AL, Patterson NJ, Plenge RM, Weinblatt ME, Shadick NA, et al. (2006) Principal components analysis corrects for stratification in genome-wide association studies. Nat Genet 38: 904–909.

31. Safran M, Dalah I, Alexander J, Rosen N, Iny Stein T, et al. (2010) GeneCards Version 3: the human gene integrator. Database (Oxford) 2010: baq020.

32. Becamel C, Alonso G, Galeotti N, Demey E, Jouin P, et al. (2002) Synaptic multiprotein complexes associated with 5-HT(2C) receptors: a proteomic approach. EMBO J 21: 2332–2342.

33. Cheng NH, Zhang W, Chen WQ, Jin J, Cui X, et al. (2011) A mammalian monothiol glutaredoxin, Grx3, is critical for cell cycle progression during embryogenesis. FEBS J 278: 2525–2539.

34. Golebiowski F, Matic I, Tatham MH, Cole C, Yin Y, et al. (2009) System-wide changes to SUMO modifications in response to heat shock. Sci Signal 2: ra24.

35. Ni Z, Karaskov E, Yu T, Callaghan SM, Der S, et al. (2005) Apical role for BRG1 in cytokine-induced promoter assembly. Proc Natl Acad Sci U S A 102: 14611–14616.

36. Bruderer R, Tatham MH, Plechanovova A, Matic I, Garg AK, et al. (2011) Purification and identification of endogenous polySUMO conjugates. EMBO Rep 12: 142–148.

37. Ougolkov AV, Bone ND, Fernandez-Zapico ME, Kay NE, Billadeau DD (2007) Inhibition of glycogen synthase kinase-3 activity leads to epigenetic silencing of nuclear factor kappaB target genes and induction of apoptosis in chronic lymphocytic leukemia B cells. Blood 110: 735–742.

38. Inoue H, Tsukita K, Iwasato T, Suzuki Y, Tomioka M, et al. (2003) The crucial role of caspase-9 in the disease progression of a transgenic ALS mouse model. EMBO J 22: 6665–6674.

13

Human-Specific HERV-K Insertion Causes Genomic Variations in the Human Genome

Wonseok Shin[1], Jungnam Lee[1], Seung-Yeol Son[2], Kung Ahn[3], Heui-Soo Kim[3], Kyudong Han[1]*

1 Department of Nanobiomedical Science and WCU Research Center, Dankook University, Cheonan, Republic of Korea, 2 Department of Microbiology, College of Advanced Science, Dankook University, Cheonan, Republic of Korea, 3 Department of Biological Sciences, College of Natural Sciences, Pusan National University, Busan, Republic of Korea

Abstract

Human endogenous retroviruses (HERV) sequences account for about 8% of the human genome. Through comparative genomics and literature mining, we identified a total of 29 human-specific HERV-K insertions. We characterized them focusing on their structure and flanking sequence. The results showed that four of the human-specific HERV-K insertions deleted human genomic sequences via non-classical insertion mechanisms. Interestingly, two of the human-specific HERV-K insertion loci contained two HERV-K internals and three LTR elements, a pattern which could be explained by LTR-LTR ectopic recombination or template switching. In addition, we conducted a polymorphic test and observed that twelve out of the 29 elements are polymorphic in the human population. In conclusion, human-specific HERV-K elements have inserted into human genome since the divergence of human and chimpanzee, causing human genomic changes. Thus, we believe that human-specific HERV-K activity has contributed to the genomic divergence between humans and chimpanzees, as well as within the human population.

Editor: Richard Cordaux, University of Poitiers, France

Funding: This research was supported by the World Class University (R31-10069) and by the Basic Science Research (2011-0009080) program, through the National Research Foundation of Korea (NRF, http://www.nrf.re.kr/nrf_eng_cms/), funded by the Ministry of Education, Science, and Technology. The funders had no role in study design, data collection and analysis, decision to publish, or preparation of the manuscript.

Competing Interests: The authors have declared that no competing interests exist.

* E-mail: jim97@dankook.ac.kr

Introduction

Repetitive mobile elements are responsible for half of the human genome. Among them, human endogenous retroviruses (HERVs) and related sequences account for ~8% of the human genome [1]. It is thought that HERVs are derived from exogenous retrovirus infections early in the evolution of primates because they have a similar structure to the provirus of an infectious virus [2]. A full-length HERV element is approximately 9.5 kb in length and consists of an internal region of four essential viral genes (*gag*, *pro*, *pol*, and *env*) and two long terminal repeats (LTRs); *gag* stands for group-specific antigen which is the retroviral capsid protein, *pro* encodes for a protease, and *pol* contains a reverse transcriptase domain [3,4]. HERVs are distinguished from other LTR retrotransposons by the presence of the envelope (*env*) gene, which codes for viral membrane proteins [5]. The LTRs contain many regulatory elements such as promoters, enhancers, and polyadenylation signals required for retroviral gene expression [6,7].

Since the initial infection of HERV into its host genome, the elements have lost their ability to synthesize mature retroviral particles by accumulating mutations preventing them from infecting other cells [8]. Nonetheless, they have successfully propagated within genomes via retrotransposition and vertical inheritance, reaching ~203,000 copies in the human genome [1]. HERVs fall into three different classes (I-III) based on sequence similarity to different genera of infectious retroviruses, and each class comprises many families with independent origins [1,3]. There are 31 HERV families in the human genome and they are

named according to the specificity of the tRNA primer-binding site [3,9]. It was reported that most HERV families underwent radiations in their host genomes after the divergence of Old and New World monkeys [8]. Among the three HERV classes, class II HERVs exist in the lowest frequency in the human genome, but they include the HERV-K family, which is the youngest family and is known to have actively mobilized since the divergence of humans and chimpanzees [1,10]. The HERV-K subfamily could be integrated and endogenized into the human genome by germ-line infection, which was supported by the evidence of purifying selection on the *env* gene of HERV-K elements [11].

It has been suggested that the HERV-K family is the most biologically active family because it retains the ability to encode functional retroviral proteins and produce retrovirus-like particles [12,13,14]. Due to this, the HERV-K family has been the subject of many studies but to date no functional provirus capable of producing infectious particles has been detected [10]. Although the HERV-K family emerged in the catarrhine lineage prior to the divergence of hominoids and Old World monkeys, some of its members inserted into the human genome after the divergence of humans and chimpanzees [8]. Thus, the HERV-K family may have contributed to the genomic differences between humans and chimpanzees through species-specific insertion and subsequent related genomic rearrangements. In this study, we identified 29 human-specific HERV-K elements in the human genome and examined the human genomic changes caused by these insertions. Our analyses focused on the mechanisms through which the HERV-K insertions caused the observed changes. In addition, we conducted a polymorphism test of the HERV-K insertions in human populations, the result of which indicates that HERV-K elements may also be contributing to genomic variations within the human species.

Results and Discussion

Identification of Human-specific HERV-K Insertions

To identify human-specific HERV-K elements, we first extracted 2,618 HERV-K elements from the human genome. However, some of these elements contained other internal non-HERV repeat element insertions or internal sequence deletions. In these cases, each HERV-K fragment was counted as a separate element by the tool we used to extract them, rather than counting the un-fragmented element only once. Thus, we manually inspected the HERV-K candidate loci and reassembled all fragmentary elements, resulting in a revised total of 1,390 loci (Table 1). To detect human-specific insertion loci in these 1,390 HERV-K elements, we examined the orthologous loci of each human-derived HERV-K element in the chimpanzee, orangutan, and rhesus macaque genomes. In this way, we identified 26 human-specific HERV-K loci in the human genome. Four previous studies have attempted to identify human-specific HERV-K loci [15,16,17,18]. A comparison of our results showed that our strategy recovered five human-specific HERV-K loci that these previous studies missed. However, three of the human-specific HERV-K loci previously reported in the literature (HERV-K103, 113, and 134) were missing from our dataset. We examined these three loci in detail. Two were solitary LTRs in the human reference genome sequence and since we did not include solitary LTRs in our dataset of human-specific HERV-K loci, it is unsurprising these two loci were missed by our strategy. Close examination of the third missing locus revealed this locus to be polymorphic in human populations. In other words, we were unable to detect the locus because the HERV-K element is absent in the human reference genome sequence. Given this, we assert that our strategy to identify human-specific HERV-K elements in the human reference genome is robust. Thus, as shown in the Figure S1, at least 29 human-specific HERV-K elements are existed in the human genome.

We characterized the human-specific HERV-K elements focusing on their size. A full-length HERV-K element consists of ~7.5 kb of internal region and two LTRs, each of which is ~1 kb. However, most of the HERV-K elements in the human genome contain internal deletions of variable sizes. In this study, we considered the element whose internal region is >7 kb to be a full-length element. The size of HERV-K internal regions ranged from 97 to 7546 bp, and 17 out of the 29 human-specific HERV-K elements were full-length elements according to our criterion. HERV-K elements have been grouped into two types, type I and type II, according to the presence/absence of a 292 bp sequence

at the *pol-env* boundary of the elements [2]. Only type II elements contain the 292 bp sequence. We further examined the full-length human-specific HERV-K elements. As shown in Table 2, eight and nine elements are identified as type I and type II, respectively, including three previously studied insertions [16,18].

Additionally, we found two interesting human-specific HERV-K loci of non-standard sequence architecture. Each of these consists of two HERV-K internals and three LTRs. One of the two loci, HERV-K108, may have resulted from ectopic homologous recombination between two different LTRs, the mechanism for which was introduced in another study on HERV-K and is depicted in Figure 1A [19]. The three LTRs of HERV-K108 showed a high degree of sequence similarity and were closely related in the phylogenetic tree in Figure 2. The other locus, HERV-K124 also contains three different LTRs. However, it was unclear what mechanism may be responsible for the observed sequence architecture of this locus. If LTR-LTR recombination were to explain this locus, we would expect the three LTRs to have a high degree of sequence similarity to one another, but the 3′ and internal HERV-K124 LTRs are truncated and inverted relative to 5′ HERV-K124 LTR. We therefore speculate that HERV-K124 was generated in two steps: LTR inversion and template switching, as shown in Figure 1B. Although the LTR inversion is a rare event, a possible mechanism responsible for the LTR inversion was suggested in one of previous studies on HERV-K [20].

Genomic Environment of Human-specific HERV-K Insertions

We aligned the human-specific HERV-K elements based on their LTR sequences except for eight loci because those elements contained LTRs that were too short (23–257 bp) resulting in ambiguity in the alignment. Next, we reconstructed the phylogenetic relationships between these LTRs. It is known that the two LTRs of an HERV element tend to have a high sequence identity to one another. As shown in Figure 2, this expected within-element sequence identity was found in all of our loci except HERV-K115. We suspect that gene conversion may have led to the differences observed between the two LTR sequences of the HERV-K115 [21].

To examine the genomic environment of the human-specific HERV-K insertions, we analyzed the GC content and gene density of genomic regions flanking the elements (Table S1). GC content was calculated for the 20 kb of flanking genomic sequence on each side of each locus. The GC content of these flanking regions averaged 41.6%. This is only slightly higher than the human reference genomic average GC content of 41% [1]. In addition, we analyzed the gene density of the 1 Mb of flanking genomic sequence to each side of the human-specific HERV-K elements and the results are described in Table S1. The gene density of these insertions averaged about 17 genes per Mb, which is substantially higher than the ~10 genes per Mb average reported for the human genome [1]. It has been previously reported that HERV-K elements are preferentially integrated into GC-rich regions, and thus gene-rich regions [22], and our findings are consistent to with this assertion.

Polymorphic Distribution of Human-specific HERV-K Insertions

The HERV-K family has been shown to be actively mobilizing in the human genome since the divergence of human and chimpanzee, and thus some of these elements are likely to be polymorphic in the human population. To evaluate the polymorphism levels associated with human-specific HERV-K loci, we

Table 1. Summary of human-specific HERV-K insertions.

Classification	No. of loci
Computationally predicted HERV-K loci	1390
Number of human-specific HERV-K insertion events	29
Full-length human-specific HERV-K insertion	17
Truncated human-specific HERV-K insertion	8
Non-classical insertion of HERV-K	4

Table 2. The structural characterization of human-specific full-length HERV-K.

Type	HERV	Chromosomal position (hg19)	Length (bp) (5'/3'LTR/Internal)	Comment	Stop codon/Region
I	K101	chr 22: 18926187-18935361	968/964/7243	In frame *pol* broken	TGA/*pro*
	K102	chr 1: 155596457-15605636	968/968/7244	In frame *pol-env* fusion	TGA/*gag*
	K103	chr 10: 27182399-27183366	968/968/7245	In frame *pol-env* fusion/*env* broken	–
	K106	chr 3: 112743124-112752282	960/960/7239	In frame *env* broken	–
	K107	chr 5: 156084717-156093896	968/968/7244	In frame *pol-env* fusion	–
	K117	chr 3: 18528336-185289515	968/968/7244	In frame *pol-env* fusion/*env* broken	TAG/*env*
	K133	chr 21: 19933659-19941962	966/257/7081	In frame *pol-env* fusion/*env* broken	TAG, TGA/*gag, pro, pol, env*
	K134	chr 12: 55727215-55728183	969/968/7243	In frame pol broken	TGA/*pol*
II	K104	chr 5: 30486760-30496205	951/960/7535	–	TGA/*gag, pol, env*
	K108a	chr 7: 4622057-4631528	968/968/7535	Dual internal sequences, triple LTRs	TAG/*gag, env*
	K108b	chr 7: 4630561-4640031	968/968/7535		TAG/*gag*
	K109	chr 6: 78426662-78436083	960/960/7502	–	TAG, TGA/*pol*
	K113	chr 19: 21841536-21841541	968/968/7536	–	–
	K115	chr 8: 7355397-7364859	960/968/7535	–	–
	K118	chr 11: 101565794-101575259	968/968/7530	–	TGA/*gag, env*
	K119	chr 12: 58721242-58730698	968/968/7521	–	–
	K121	chr 3: 125609302-125618439	804/804/7530	–	TAG, TGA/*gag, pro, pol, env*
	K132	chr 19: 21841536-21841541	23/995/7869	*Alu* insertion within internal/*pol* broken	–

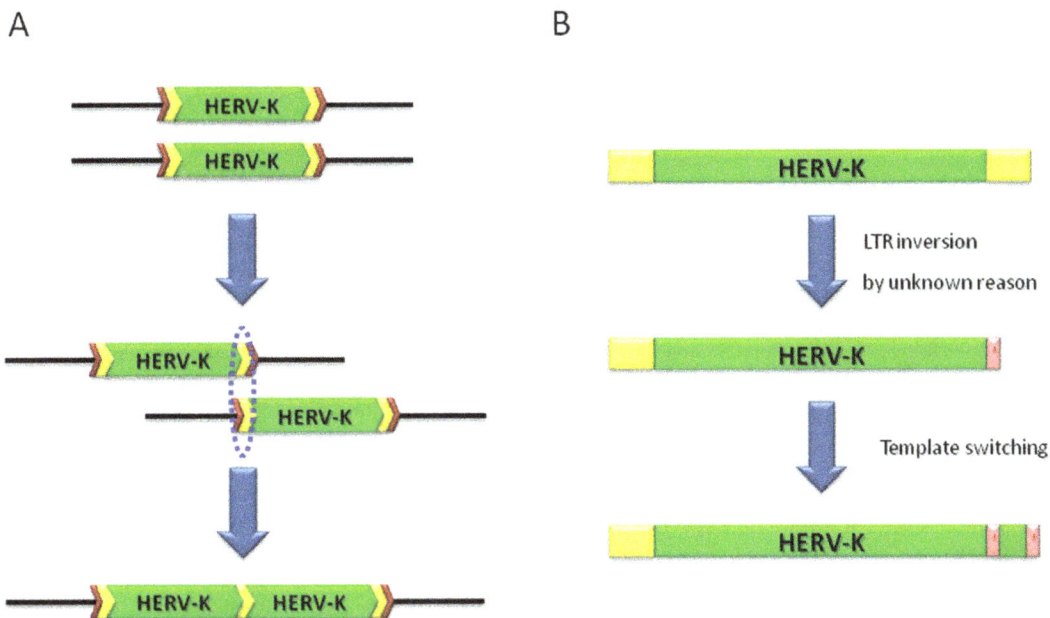

Figure 1. Comparison of human-specific HERV-K108 and HERV-K124 elements. Both of HERV-K108 and HERV-K124 have two HERV-K internal regions (green). However, their sequence architecture is the result of different mechanisms. (A) HERV-K108. After the insertion of the HERV-K element, non-allelic homologous recombination between two different LTRs (yellow chevrons) of the HERV-K element occurred. This resulted in a locus containing two HERV-K internal regions and three LTRs. This locus retains the original TSDs (red chevrons) created upon its initial insertion. (B) HERV-K124. Compared to the HERV-K108, which has two intact internal regions and three intact LTRs, the second internal region of HERV-K124 has largely deleted and its internal and 3' LTRs inverted and partially deleted. The mechanism(s) responsible for this element's sequence architecture is not clearly resolved, but we depict here a potential mechanism capable of generating this element. Yellow boxes indicate standard LTRs, pink boxes indicate inverted partial LTRs, and green boxes indicate HERV-K internal regions.

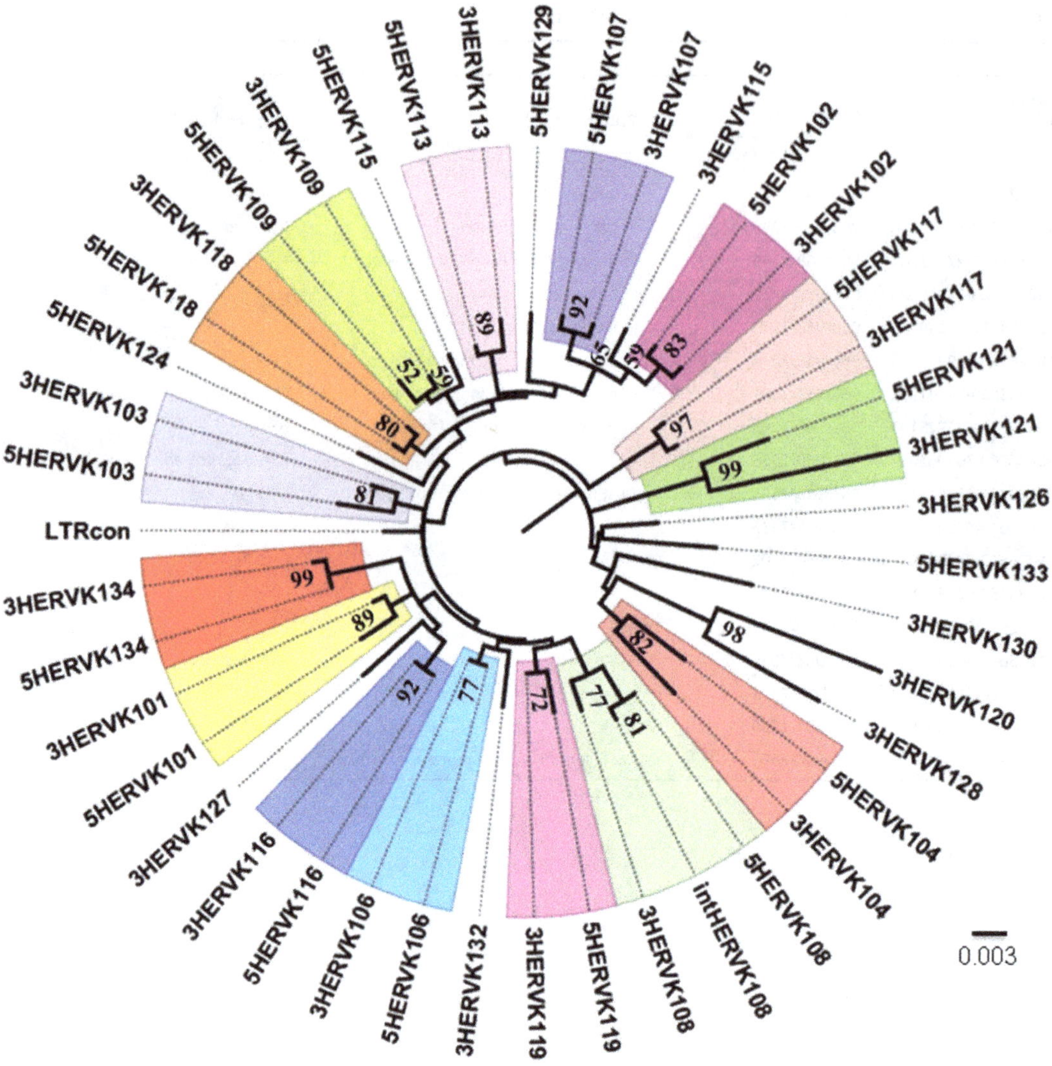

Figure 2. The phylogenetic tree of human-specific HERV-K LTRs. This is a maximum likelihood tree reconstructed using Kimura-2-parameter distance model. Most HERV-K elements contain an LTR at their 5′ and 3′ ends. In cases where the two LTR sequences are similar to one another, they are shown in the same colour. LTRs from the same element but having divergent sequences are not clustered in the same colour. Short LTRs causing ambiguity on this tree were excluded from this analysis. Bootstrap values for nodes (% of 1000 replicates) scoring higher than 50% are reported.

genotyped 25 loci in 80 humans (20 from Asian, 20 from South American, 20 from European and 20 from African American) whose DNAs were purchased from the Coriell Institute for Medical Research. We were not able to amplify the remaining four loci because they reside either in regions of segmental duplication or in centromeric regions. As shown in Figure 3B, there are three possible states for each sister chromatid at a human-specific HERV-K insertion locus: absence of the HERV-K element, presence of the element and presence of a solitary LTR. Among the human-specific HERV-K elements, three loci, HERV-K 109, 118, and 134, exhibit all the three forms in the human populations tested. The polymorphism test found that the polymorphism level of the human-specific HERV-K elements is about 48% (12/25) which is higher than levels reported for other human-specific retrotransposons [23,24,25]. We examined the recombination rate of the genomic regions where the human-specific HERV-K elements reside because a high recombination rate could contribute to the observed increase their polymorphism level. As shown in Table 3, the recombination rates in the genomic

regions flanking human-specific HERV-K elements averaged ~1.2 cM per Mb on both long and short arms. We compared the result with the genome-wide average recombination rates, ~1 cM and ~2 cM per Mb on the long and short arms, respectively [1]. Based on the result, we conclude that recombination rate is not a major factor responsible for the higher polymorphism levels observed in human-specific HERV-K elements.

Through the polymorphism test, we found that both type I and II full-length human-specific HERV-K elements are polymorphic in the 80 human individuals. This indicates that both types were capable of retrotransposition after the divergence of human and chimpanzee and increases likelihood that members of these groups are currently able to retrotranspose in the human genome.

Structural Analysis of Human-specific Full-length HERV-K

The majority of HERVs in the human genome exist in truncated form and are characterized by multiple stop codons, insertions, and deletions [26,27]. It is suspected that a smaller

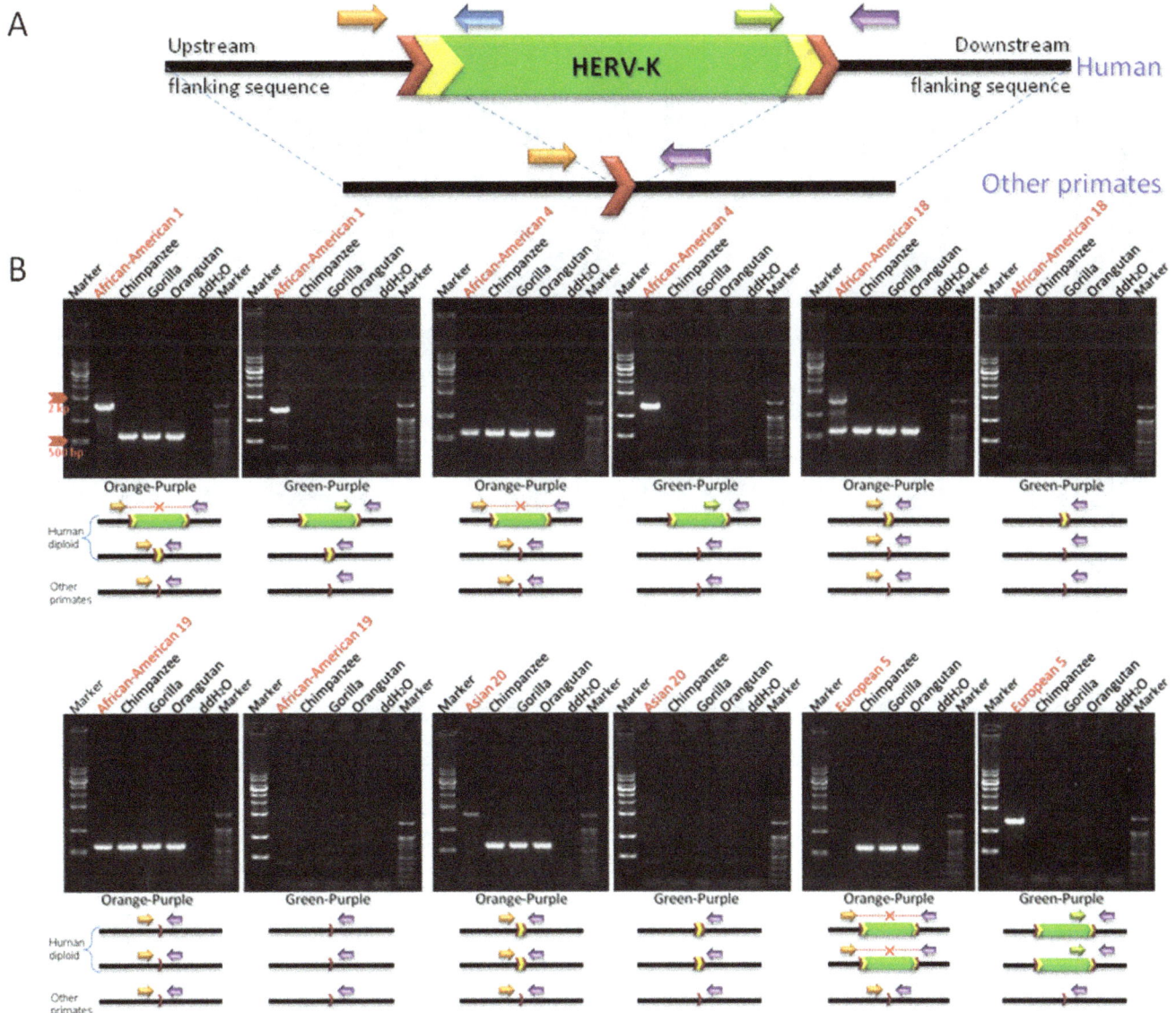

Figure 3. Variable polymorphic patterns of a HERV-K118 in human diploid genomes. Human-specific HERV-K118 insertion locus was amplified by PCR using the genomic DNAs of human population and other primates as template. (A) A typical primate HERV-K element. The ~7.5 kb structure of the HERV-K internal region is shown in green. Yellow chevrons are LTRs (~1 kb) and red chevrons are target site duplications (TSDs). (B) Gel chromatographs show PCR products of targeted human-specific HERV-K loci on a panel containing human three non-human primates. High bands indicate the presence of an insertion, while low bands indicate its absence. Orange and purple arrows indicate primers designed in the conserved flanking regions of all species. Green arrows indicate internal primers designed within the human-specific HERV-K. As shown in the gel pictures, human-specific HERV-K insertion loci exhibit a variety polymorphic patterns in human diploid genomes.

subset of human-specific HERV-K elements are capable of retrotransposition and thus contain intact open reading frames (ORFs) because their proteins and particles have been detected in the human genome [28]. We therefore examined whether any of the identified human-specific full-length HERV-Ks contain intact ORFs. As shown in Table 2, five human-specific type I HERV-Ks, HERV-K102, 103, 107, 117, and 133, exhibit fused *pol* and *env* genes in the same frame. A search for stop codons in the gene components of the human-specific type I HERV-Ks revealed that HERV-K101, 102, 117, and 134 have stop codons in their *pro*, *gag*, *env*, and *pol* genes, respectively, and HERV-K133 contains stop codons in all of these genes (Figure 4). In sum, a total of three HERV-K elements have retained intact ORFs in the human

genome, indicating that they have a potential to produce the viral particles [29].

As mentioned above, the type of HERV-K element is determined according to the presence/absence of a 292 bp 'deletion' at the *pol-env* boundary. It has been reported that the ancestral precursor of the type I HERV-K lacked the 292 bp sequence and that this deletion must not have been directly related to the precursor's ability to retrotranspose in the human genome. This is because the human genome contains at least eight type I full-length HERV-K elements which must be offspring of the precursor [30]. However, we could not rule out other possible origins for Type I insertions. For example, they could result from the recombination between competent Type II viruses and transcripts of preexisting Type I.

Table 3. Characteristic of human-specific HERV-K insertions.

No.	HERV	Genomic location	Features[c]	Rec. rate (cM/Mb; avg)	size (bp) 5' LTR	internal	3' LTR	internal	3' LTR	Reference
1	K116	chr1:75842771-75849143	Inserted into human-specific L1PA2	0.7	968	4437	968			[40]
2	K102[a]	chr1:155596457-155605636	-	1.3	968	7244	968			[2]
3	K120	chr2:130719538-130722650	Inserted into SD region	1.1	23	2129	961			[18]
4	K106[a]	chr3:112743124-112752282	Polymorphic	0.4	960	7239	960			[2]
5	K121[a]	chr3:125609302-125618439	Inserted into SD region	0.8	804	7530	804			[41]
6	K122	chr3:148281441-148285419	13 bp L1 sequence in 3' end of ERV, polymorphic	2.1	23	3920	23			[18]
7	K123	chr3:170955654-170955804	Non-classical insertion, HERV-K9 subfamily	1.5	0	143	0			This Study
8	K117[a]	chr3:185280336-185289515	Inserted into SD region, polymorphic	2	968	7244	968			[40]
9	K124	chr4:161579938-161582439	Second HERV-K internal to 206 bp in first HERV-K internal of 3' end	0.9	968	1171	78[b]	206	78[b]	This Study
10	K104[a]	chr5:30486760-30496205	-	1.8	951	7535	960			[2]
11	K107[a]	chr5:156084717-156093896	Polymorphic	0.6	968	7244	968			[42]
12	K125	chr6:74042982-74043123	Non-classical insertion	0.6	0	142	0			This Study
13	K109[a]	chr6:78426662-78436083	Polymorphic	0.7	960	7502	960			[2]
14	K108[a]	chr7:4622057-4640031	LTR-LTR homologous recombination, polymorphic	1.6	968	7535	968	7536	968	[2]
15	K126	chr7:104388369-104393269	13bp L1 sequence in 3' end of ERV	1.1	0	3921	967[b]			[18]
16	K115[a]	chr8:7355397-7364859	Inserted into SD region, polymorphic	0.9	960	7535	968			[30]
17	K127	chr8:140472149-140475259	-	2.7	23	2120	968			[18]
18	K128	chr10:101580569-101587739	-	0.1	23	6162	968			[16]
19	K118[a]	chr11:101565794-101575259	Polymorphic	0.6	968	7530	968			[43]
20	K119[a]	chr12:58721242-58730698	Polymorphic	0.3	968	7521	968			[43]
21	K129	chr12:111007843-111009348	-	0.6	968	515	23			[44]
22	K130	chr16:34231397-34234142	Non-classical insertion	0.1	0	1788	958			This Study
23	K131	chr17:6078917-6079053	Non-classical insertion	3.5	0	96	41			This Study
24	K132[a]	chr19:28128498-28137384	Inserted into satellite DNA region of centromere	0.4	23	7546	995			[45]
25	K133[a]	chr21:19933659-19941962	TSD contains partial LTR50, MIRb, and AT_rich	3	966	7081	257			[46]
26	K101[a]	chr22:18926187-18935361	Inserted into SD region	3.3	968	7243	964			[2]
27	K103[a]	chr10:27182399-27183366	Solitary LTR in hg19, inserted into SD region, polymorphic	0.9	968	7245	968			[2]
28	K113[a]	chr19:21841536-21841541	Absence in hg19, inserted into SD region, polymorphic	0.1	968	7536	968			[30]
29	K134[a]	chr12:55727215-55728183	Solitary LTR in hg19, polymorphic	1.1	969	7243	968			[17]

[a]Full-length human-specific HERV-K locus.
[b]Sequence is reversed.
[c]TSD, Target Site Duplication; SD, Segmental Duplication.

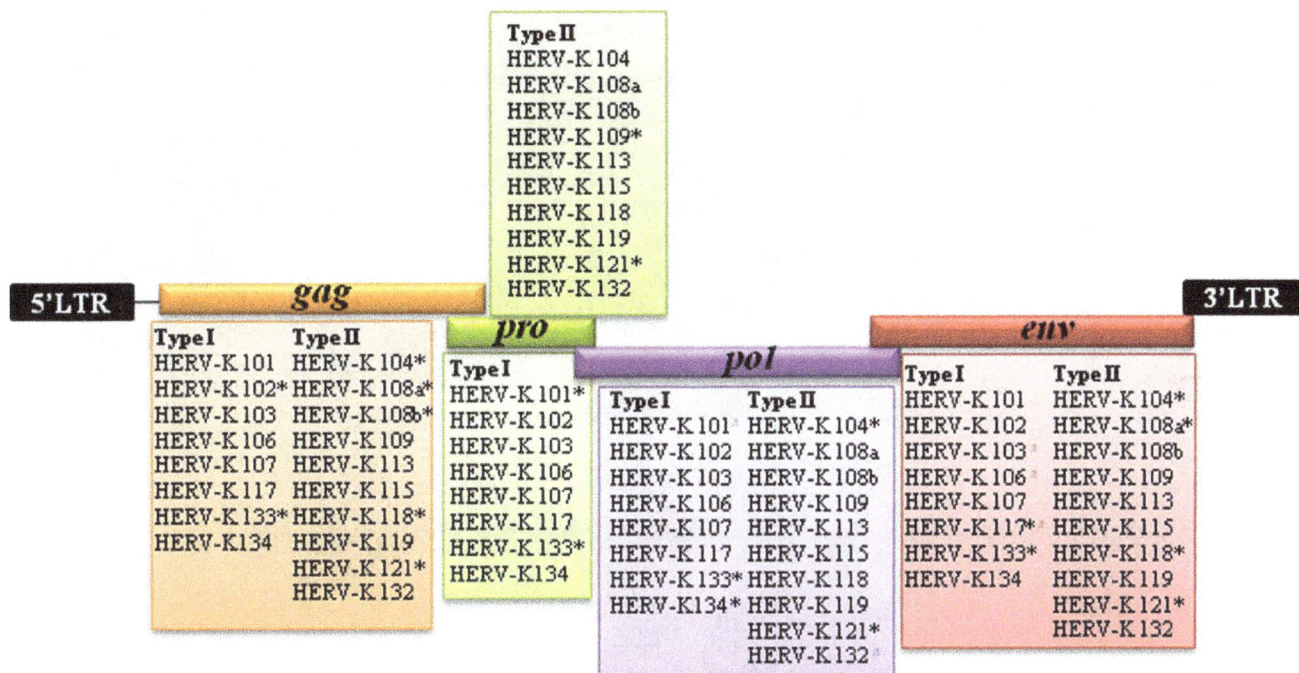

Figure 4. Diagram of a human-specific full-length HERV-K element. The ORFs of *gag*, *pro*, *pol*, and *env* are depicted as colored boxes. HERV-K members that contain versions of *gag*, *pro*, *pol*, and *env* are listed under each HERV genes (* and # indicate that the HERV-K locus contains stop codon or broken frame, respectively).

Among the human-specific type II HERV-Ks, HERV-K104, 108, 118, and 121 contained stop codons in multiple genes while HERV-K132 had an *Alu* insertion within its *pol* gene. These five HERV-Ks are therefore not functionally and structurally intact in the human genome. However, HERV-K113, 115, and 119 possess intact gene components, which indicates that they have the potential to encode the functional proteins required for their mobilization. HERV-K113 and 115 were previously identified to be full-length and polymorphic (HERV-K presence/absence) in human populations [30]. The result of our polymorphic test on HERV-K119 showed that this element is also polymorphic in the 80 human individuals, but its pattern of polymorphism is different from that of the other elements; the polymorphism at the HERV-K113 and 115 loci takes the form of an absence or presence of the HERV-K element between individuals, but the HERV-K119 locus exists as either a full-length HERV-K or a solitary LTR. We speculate that this architecture is the product of a homologous recombination event between the two LTRs of a full-length HERV-K element. Given this, we suspect that the HERV-K119 element is relatively older than the other two elements (HERV-K113 and 115). These intact full-length HERV-K elements could play a role in human disease. This possibility has been suggested by several reports describing HERV-encoded transcripts and proteins in tumors [31,32] and tissue from patients with autoimmune diseases [27,33,34].

Human-specific HERV-K Insertion-associated Genetic Variations

We found four non-classical HERV-K insertion loci in our dataset (Table 1). Figure 5B depicts one possible mechanism responsible for the non-classical insertion. These elements are 5′ and 3′ truncated, meaning that they also do not have classical TSDs. Additionally, they are involved in target site deletions in the

human genome. Through a comparison of the human-specific HERV-K flanking sequence and its corresponding chimpanzee pre-insertion sequence, we calculated the deletion size. However, the chimpanzee orthologous sequence of HERV-K130 insertion contained two unsequenced regions. We amplified one of the regions for sequencing (accession number: JQ811903) and the primer sequences are described in Table S2. We estimated the size of the other region using the orangutan reference genome sequence. The deletion sizes of the target sites of the non-classical HERV-Ks range from 6 bp to 10,207 bp. We further examined their genomic environments and found that three of them occurred in intergenic regions and one occurred in an intronic region. It has been reported that non-classical insertions are associated with double-strand break (DSB) repair, a mechanism proposed to aid in stability of fragile sites in the host genome [35]. Also, it has been suggested that DSBs can be repaired through homologous recombination (HR) or non-homologous end joining (NHEJ) to ensure the maintenance of genome integrity in eukaryotic organisms [36]. As for the four non-classical HERV-K insertions, we examined the microhomology between each HERV-K element and its pre-insertion sequence from the chimpanzee genome. Microhomology, if present, could mediate the insertion of the HERV-K between DSB ends via a NHEJ-associated process. We identified 5′ and 3′ microhomologies for three out of the four loci but were not able to detect microhomology for the HERV-K130 locus, as shown in Table S3. We conclude, therefore, in the cases where microhomology exists at both ends of the HERV-K insertion, the likelihood of DSB repair through NHEJ is increased.

In this study, we identified 29 human-specific HERV-K insertions including previously reported three loci (HERV-K103, 113, and 134) that have integrated into the human genome since the divergence of humans and chimpanzees. During this time,

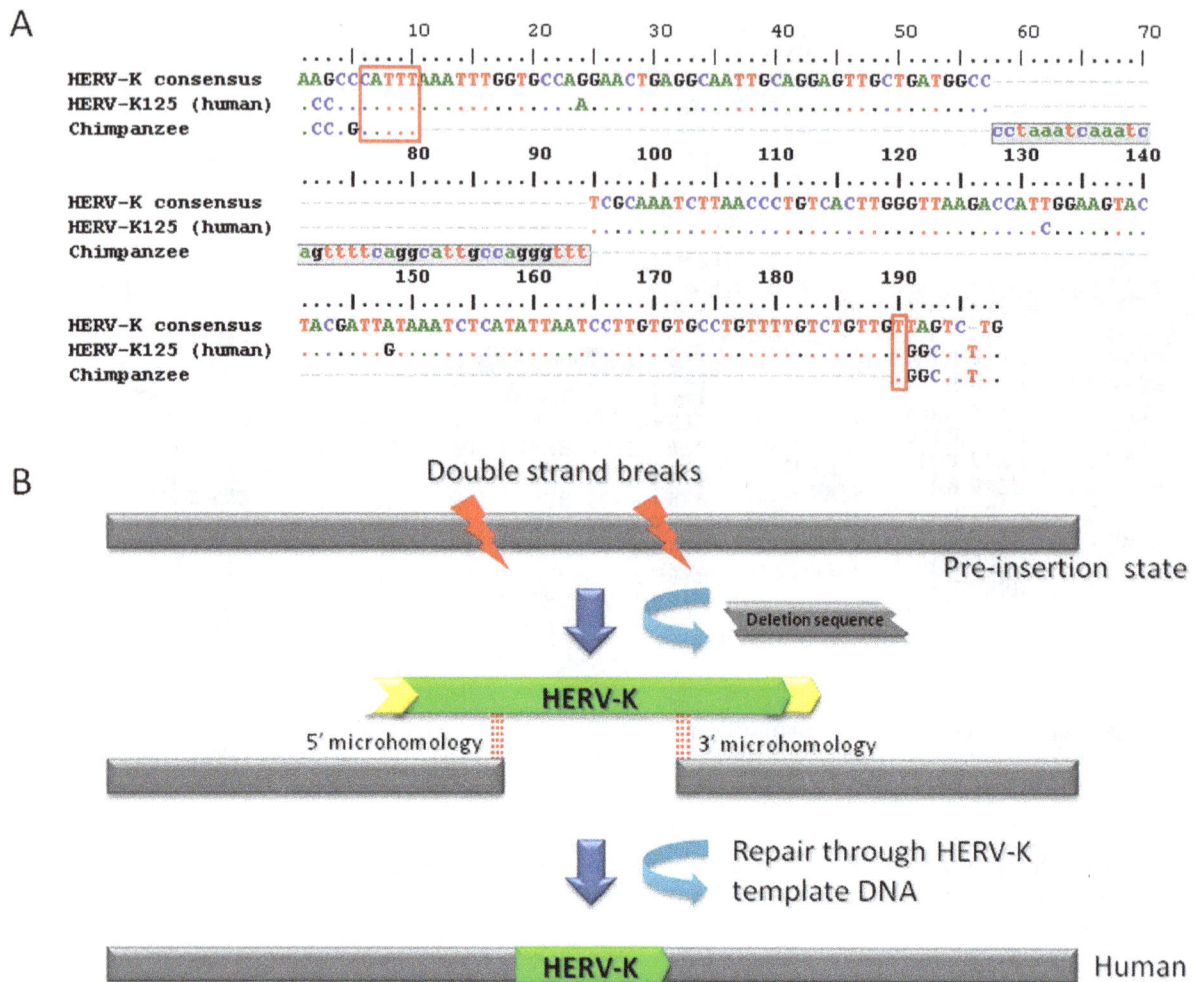

Figure 5. Non-classical insertion of human-specific HERV-K element in the human genome. Four non-classical insertions of human-specific HERV-K were observed in the human genome. The human-specific locus, HERV-K125, is depicted here. (A) An alignment of the non-classical insertion of human-specific HERV-K125 element, and its pre-insertion site to the HERV-K consensus sequence. This alignment reveals a 37 bp deletion of the pre-insertion site in the human genome (gray region in the chimpanzee sequence). Red boxes indicate microhomology at either end of the non-classical insertion, which suggests the involvement of an NHEJ mechanism. (B) A schematic diagram that describes the non-classical insertion of an HERV-K element (green box) and the deleted-region of genomic sequence (broken gray box).

HERV-K activity contributed to genomic variation between the two species. Through a polymorphism test, we found that the polymorphic rate of these elements is 48%. This indicates that the activity of the HERV-K family has resulted in genomic variations between and within human populations. It is currently unknown whether there are any retrotranspoitionally competent copies of HERV-K element in the human genome. However, based on the results of this study, we assert that HERV-K element activity is a cause of genomic differences between the human and chimpanzee genomes as well as genomic diversity within the human population.

Materials and Methods

Computational Data Mining and Manual Inspection of Human-specific HERV-K Loci

To computationally screen the human genome (hg19; February 2009 freeze) for potential human-specific HERV-K loci, we first extracted all HERV-K loci from the human genome by using UCSC Table Browser utility (http://genome.ucsc.edu/cgi-bin/

hgTables?org = Human&db = hg19&hgsid = 226995881&hgta_doMainPage = 1). For each HERV-K locus, we next extracted 2 kb flanking sequences, up and down stream. This human sequence was then used as a query against other primate genome sequences (panTro3; October. 2010 freeze, ponAbe2; July 2007 freeze, rheMac2; January 2006 freeze), using UCSC's BLAT utility (http://genome.ucsc.edu/cgi-bin/hgBlat). For each hit in the BLAT search, we retrieved the human, chimpanzee, orangutan and rhesus macaque sequences. Repeat elements existing in these nonhuman sequences were annotated using the RepeatMasker (http://www.repeatmasker.org/cgi-bin/WEBRepeatMasker) tool. Based on these repeat element annotations, we confirmed whether each HERV-K locus was specific to the human genome or not.

PCR Amplification and DNA Sequence Analysis

To experimentally verify the human-specific HERV-K insertion candidates, we conducted PCR analysis with four different DNA templates: *Homo sapiens* (human; NA10851, Coriell Cell Repository, Camden, NJ), *Pan troglodytes* (common chimpanzee), *Gorilla*

gorilla (gorilla), and *Pongo pygmaeus* (Bornean orangutan). Genomic DNA for three apes was kindly provided by Dr. Takenaka (Primate Research Institute, Kyoto University). Oligonucleotide primers for the PCR amplification of human-specific HERV-K insertion candidates were designed, using the Primer3 utility (http://biotools.umassmed.edu/bioapps/primer3_www.cgi) (Table S4). PCR amplification of each locus was performed in 20 μl reaction using 20–30 ng template DNA, 200 nM of each oligonucleotide primer, and 10 μl of EmeraldAmp GT PCR Master Mix (TaKaRa, Ohtsu, Japan). Each sample was subjected to an initial denaturation step of 5 min at 95°C, followed by 35 cycles of PCR at 1 min of denaturation at 95°C, 1 min at the annealing temperature, and 1 to 2 min of extension at 72°C depending on the PCR product size, followed by a final extension step of 10 min at 72°C. The PCR products were loaded on 1–2% agarose gels, stained with ethidium bromide, and visualized using UV fluorescence. For the loci whose expected product size was >2 kb, we used Ex TaqTM polymerase (TaKaRa Japan), 2X EF-Taq Pre mix 2 (SolGent, Korea), and KOD FX (Toyobo, Japan) to carry out PCR following the manufacturer's instructions.

If needed, we purified PCR products from the agarose gel using the Wizard® SV gel and PCR Clean-up system (Promega) and cloned them into vectors using the pGME®-T Easy Vector system (Promega, http://www.promega.com) according to the manufacturer's instructions. The sequencing of the PCR product was performed on an ABI 3730xl DNA analyzer (Applied Biobiosystems, www.appliedbiosystems.com) at the oligonucletides synthesis and sequencing facility, MACROGEN (http://dna.macrogen.com/eng). The resulting DNA sequences were analyzed using the BioEdit v.7.0.5.3 sequence alignment software package and have been deposited in Genbank under accession numbers JQ966584-JQ966591 and JQ999963-JQ999964.

Data Analyses

We downloaded the HERV-K consensus sequence, including LTRs, from the RepeatMasker utility (http://www.repeatmasker.org/cgi-bin/WEBRepeatMasker) and aligned human-specific HERV-K elements with this consensus sequence using the software BioEdit v.7.0.5.3 [37]. To reconstruct the phylogenetic relationships among the human-specific HERV-K elements, we used the software MEGA 5.03 [38]. A maximum likelihood tree based on the observed number of nucleotide differences and a Kimura-2-parameter distance model was built. Each node of the tree was evaluated based on 1000 bootstrap replicates and the percentage of replicates in which each node in the final tree was reconstructed is reported in Figure 2. To examine the GC content of the flanking sequences of the human-specific HERV-K elements, we extracted 20 kb of flanking sequence up and down stream of each element using the Human BLAT search Tool server (http://genome.ucsc.edu/cgi-bin/hgBlat?commend=start). The percentage of GC nucleotides in the flanking sequence was then calculated using the EMBOSS GeeCee server (http://bioweb.pasteur.fr/seqanal/interfaces/

geecee.html). For the gene density analysis, we counted the number of genes within a 2 Mb window of flanking sequence centered on each human-specific HERV-K element using the National Center for Biotechnology Information Map Viewer utility (http://www.ncbi.nlm.nih.gov/projects/mapview/map_search.cgi?taxid = 9606).

RetroTector10 Program Application

To determine the genomic structure of human-specific full-length HERV-Ks located on a specific locus, we used the RetroTector10 program (http://www.kvir.uu.se/RetroTector/RetroTectorProject.html) [39]. It contains three basic modules: first, the recognition of LTR candidates; second, the detection of chains of conserved retroviral motifs fulfilling the distance constraints; and third, the attempted reconstruction of the original retroviral protein sequences, combination of the alignment, and properties of the protein ends.

Supporting Information

Figure S1   The 29 human-specific HERV-K insertion loci in the human genome. Blue and green circles indicate the chromosomal locations of full-length and truncated human-specific HERV-K elements, respectively. Among them, 12 loci were polymorphic and 4 loci were non-classical insertions. The karyotype images were created using the idiographica webtool (http://www.ncrna.org/idiographica/).

Table S1   GC content and gene density in flanking regions of human-specific HERV-K loci.

Table S2   PCR primers for the sequences deleted by HERV-K130 insertion.

Table S3   Additional information on human-specific HERV-K insertions.

Table S4   PCR primers for human-specific HERV-K loci.

Acknowledgments

We would like to thank Dr. Thomas J. Meyer for thoughtful comments on the manuscript.

Author Contributions

Performed the computational analysis: WS JL KA. Conceived and designed the experiments: WS KH. Performed the experiments: WS JL. Analyzed the data: WS JL SYS KA HSK KH. Contributed reagents/materials/analysis tools: KH. Wrote the paper: WS JL KA KH.

References

1. Lander ES, Linton LM, Birren B, Nusbaum C, Zody MC, et al. (2001) Initial sequencing and analysis of the human genome. Nature 409: 860–921.
2. Barbulescu M, Turner G, Seaman MI, Deinard AS, Kidd KK, et al. (1999) Many human endogenous retrovirus K (HERV-K) proviruses are unique to humans. Curr Biol 9: 861–868.
3. Griffiths DJ (2001) Endogenous retroviruses in the human genome sequence. Genome Biol 2: REVIEWS1017.
4. Khodosevich K, Lebedev Y, Sverdlov E (2002) Endogenous retroviruses and human evolution. Comp Funct Genomics 3: 494–498.
5. Balada E, Ordi-Ros J, Vilardell-Tarres M (2009) Molecular mechanisms mediated by human endogenous retroviruses (HERVs) in autoimmunity. Rev Med Virol 19: 273–286.
6. Buzdin A, Ustyugova S, Khodosevich K, Mamedov I, Lebedev Y, et al. (2003) Human-specific subfamilies of HERV-K (HML-2) long terminal repeats: three master genes were active simultaneously during branching of hominoid lineages. Genomics 81: 149–156.
7. Dunn CA, van de Lagemaat LN, Baillie GJ, Mager DL (2005) Endogenous retrovirus long terminal repeats as ready-to-use mobile promoters: the case of primate beta3GAL-T5. Gene 364: 2–12.
8. Sverdlov ED (2000) Retroviruses and primate evolution. Bioessays 22: 161–171.

9. Katzourakis A, Rambaut A, Pybus OG (2005) The evolutionary dynamics of endogenous retroviruses. Trends Microbiol 13: 463–468.

10. Dewannieux M, Harper F, Richaud A, Letzelter C, Ribet D, et al. (2006) Identification of an infectious progenitor for the multiple-copy HERV-K human endogenous retroelements. Genome Res 16: 1548–1556.

11. Belshaw R, Pereira V, Katzourakis A, Talbot G, Paces J, et al. (2004) Long-term reinfection of the human genome by endogenous retroviruses. Proc Natl Acad Sci U S A 101: 4894–4899.

12. Towler EM, Gulnik SV, Bhat TN, Xie D, Gustschina E, et al. (1998) Functional characterization of the protease of human endogenous retrovirus, K10: can it complement HIV-1 protease? Biochemistry 37: 17137–17144.

13. Simpson GR, Patience C, Lower R, Tonjes RR, Moore HD, et al. (1996) Endogenous D-type (HERV-K) related sequences are packaged into retroviral particles in the placenta and possess open reading frames for reverse transcriptase. Virology 222: 451–456.

14. Seifarth W, Baust C, Murr A, Skladny H, Krieg-Schneider F, et al. (1998) Proviral structure, chromosomal location, and expression of HERV-K-T47D, a novel human endogenous retrovirus derived from T47D particles. J Virol 72: 8384–8391.

15. Jha AR, Nixon DF, Rosenberg MG, Martin JN, Deeks SG, et al. (2011) Human endogenous retrovirus K106 (HERV-K106) was infectious after the emergence of anatomically modern humans. PLoS One 6: e20234.

16. Macfarlane C, Simmonds P (2004) Allelic variation of HERV-K(HML-2) endogenous retroviral elements in human populations. J Mol Evol 59: 642–656.

17. Belshaw R, Dawson AL, Woolven-Allen J, Redding J, Burt A, et al. (2005) Genomewide screening reveals high levels of insertional polymorphism in the human endogenous retrovirus family HERV-K(HML2): implications for present-day activity. J Virol 79: 12507–12514.

18. Subramanian RP, Wildschutte JH, Russo C, Coffin JM (2011) Identification, characterization, and comparative genomic distribution of the HERV-K (HML-2) group of human endogenous retroviruses. Retrovirology 8: 90.

19. Mayer J, Stuhr T, Reus K, Maldener E, Kitova M, et al. (2005) Haplotype analysis of the human endogenous retrovirus locus HERV-K(HML-2.HOM) and its evolutionary implications. J Mol Evol 61: 706–715.

20. Hughes JF, Coffin JM (2002) A novel endogenous retrovirus-related element in the human genome resembles a DNA transposon: evidence for an evolutionary link? Genomics 80: 453–455.

21. Hughes JF, Coffin JM (2005) Human endogenous retroviral elements as indicators of ectopic recombination events in the primate genome. Genetics 171: 1183–1194.

22. Brady T, Lee YN, Ronen K, Malani N, Berry CC, et al. (2009) Integration target site selection by a resurrected human endogenous retrovirus. Genes Dev 23: 633–642.

23. Lee J, Cordaux R, Han K, Wang J, Hedges DJ, et al. (2007) Different evolutionary fates of recently integrated human and chimpanzee LINE-1 retrotransposons. Gene 390: 18–27.

24. Carter AB, Salem AH, Hedges DJ, Keegan CN, Kimball B, et al. (2004) Genome-wide analysis of the human Alu Yb-lineage. Hum Genomics 1: 167–178.

25. Otieno AC, Carter AB, Hedges DJ, Walker JA, Ray DA, et al. (2004) Analysis of the Human Alu Ya-lineage. J Mol Biol 342: 109–118.

26. Kim TH, Jeon YJ, Yi JM, Kim DS, Huh JW, et al. (2004) The distribution and expression of HERV families in the human genome. Mol Cells 18: 87–93.

27. Antony JM, van Marle G, Opii W, Butterfield DA, Mallet F, et al. (2004) Human endogenous retrovirus glycoprotein-mediated induction of redox reactants causes oligodendrocyte death and demyelination. Nat Neurosci 7: 1088–1095.

28. Ruprecht K, Ferreira H, Flockerzi A, Wahl S, Sauter M, et al. (2008) Human endogenous retrovirus family HERV-K(HML-2) RNA transcripts are selectively packaged into retroviral particles produced by the human germ cell tumor line Tera-1 and originate mainly from a provirus on chromosome 22q11.21. J Virol 82: 10008–10016.

29. Boller K, Schonfeld K, Lischer S, Fischer N, Hoffmann A, et al. (2008) Human endogenous retrovirus HERV-K113 is capable of producing intact viral particles. J Gen Virol 89: 567–572.

30. Turner G, Barbulescu M, Su M, Jensen-Seaman MI, Kidd KK, et al. (2001) Insertional polymorphisms of full-length endogenous retroviruses in humans. Curr Biol 11: 1531–1535.

31. Depil S, Roche C, Dussart P, Prin L (2002) Expression of a human endogenous retrovirus, HERV-K, in the blood cells of leukemia patients. Leukemia 16: 254–259.

32. Wang-Johanning F, Frost AR, Johanning GL, Khazaeli MB, LoBuglio AF, et al. (2001) Expression of human endogenous retrovirus k envelope transcripts in human breast cancer. Clin Cancer Res 7: 1553–1560.

33. Clerici M, Fusi ML, Caputo D, Guerini FR, Trabattoni D, et al. (1999) Immune responses to antigens of human endogenous retroviruses in patients with acute or stable multiple sclerosis. J Neuroimmunol 99: 173–182.

34. Hishikawa T, Ogasawara H, Kaneko H, Shirasawa T, Matsuura Y, et al. (1997) Detection of antibodies to a recombinant gag protein derived from human endogenous retrovirus clone 4–1 in autoimmune diseases. Viral Immunol 10: 137–147.

35. Srikanta D, Sen SK, Huang CT, Conlin EM, Rhodes RM, et al. (2009) An alternative pathway for Alu retrotransposition suggests a role in DNA double-strand break repair. Genomics 93: 205–212.

36. Chu G (1997) Double strand break repair. J Biol Chem 272: 24097–24100.

37. Hall TA (1999) BioEdit: a user-friendly biological sequence alignment editor and analysis program for Windows 95/98/NT. Nucleic Acids Symposium Series: 95–98.

38. Tamura K, Peterson D, Peterson N, Stecher G, Nei M, et al. MEGA5: molecular evolutionary genetics analysis using maximum likelihood, evolutionary distance, and maximum parsimony methods. Mol Biol Evol 28: 2731–2739.

39. Sperber GO, Airola T, Jern P, Blomberg J (2007) Automated recognition of retroviral sequences in genomic data–RetroTector. Nucleic Acids Res 35: 4964–4976.

40. Hughes JF, Coffin JM (2001) Evidence for genomic rearrangements mediated by human endogenous retroviruses during primate evolution. Nat Genet 29: 487–489.

41. Sugimoto J, Matsuura N, Kinjo Y, Takasu N, Oda T, et al. (2001) Transcriptionally active HERV-K genes: identification, isolation, and chromosomal mapping. Genomics 72: 137–144.

42. Ono M, Kawakami M, Ushikubo H (1987) Stimulation of expression of the human endogenous retrovirus genome by female steroid hormones in human breast cancer cell line T47D. J Virol 61: 2059–2062.

43. Costas J (2001) Evolutionary dynamics of the human endogenous retrovirus family HERV-K inferred from full-length proviral genomes. J Mol Evol 53: 237–243.

44. Medstrand P, Mager DL (1998) Human-specific integrations of the HERV-K endogenous retrovirus family. J Virol 72: 9782–9787.

45. Tonjes RR, Czauderna F, Kurth R (1999) Genome-wide screening, cloning, chromosomal assignment, and expression of full-length human endogenous retrovirus type K. J Virol 73: 9187–9195.

46. Kurdyukov SG, Lebedev YB, Artamonova, II, Gorodentseva TN, Batrak AV, et al. (2001) Full-sized HERV-K (HML-2) human endogenous retroviral LTR sequences on human chromosome 21: map locations and evolutionary history. Gene 273: 51–61.

14

Human Coding Synonymous Single Nucleotide Polymorphisms at Ramp Regions of mRNA Translation

Quan Li[1], Hui-Qi Qu[2]*

1 Endocrine Genetics Lab, The McGill University Health Center (Montreal Children's Hospital), Montréal, Québec, Canada, 2 Division of Epidemiology, Human Genetics and Environmental Sciences, The University of Texas School of Public Health, Houston, Texas, United States of America

Abstract

According to the ramp model of mRNA translation, the first 50 codons favor rare codons and have slower speed of translation. This study aims to detect translational selection on coding synonymous single nucleotide polymorphisms (sSNP) to support the ramp theory. We investigated fourfold degenerate site (FFDS) sSNPs with A↔G or C↔T substitutions in human genome for distribution bias of synonymous codons (SC), grouped by CpG or non-CpG sites. Distribution bias of sSNPs between the 3rd ~50th codons and the 51st ~ remainder codons at non-CpG sites were observed. In the 3rd ~50th codons, G→A sSNPs at non-CpG sites are favored than A→G sSNPs [$P = 2.89 \times 10^{-3}$], and C→T at non-CpG sites are favored than T→C sSNPs [$P = 8.50 \times 10^{-3}$]. The favored direction of SC usage change is from more frequent SCs to less frequent SCs. The distribution bias is more obvious in synonymous substitutions CG(G→A), AC(C→T), and CT(C→T). The distribution bias of sSNPs in human genome, i.e. frequent SCs to less frequent SCs is favored in the 3rd ~50th codons, indicates translational selection on sSNPs in the ramp regions of mRNA templates.

Editor: Zhi Wei, New Jersey Institute of Technology, United States of America

Funding: L.Q. holds a fellowship from Eli Lilly Canada, and H.Q.Q. is supported by intramural funding from the University of Texas School of Public Health. The funders had no role in study design, data collection and analysis, decision to publish, or preparation of the manuscript.

Competing Interests: L.Q. holds a fellowship from Eli Lilly Canada.

* E-mail: huiqi.qu@uth.tmc.edu

Introduction

Synonymous DNA variations may affect mRNA function through the change of mRNA secondary structure, mRNA stability, synonymous codon (SC) usage, or co-translational protein folding [1–4]. With empirical evidence, synonymous single nucleotide polymorphisms (sSNP) in the *COMT* gene (encoding Catechol-O-Methyltransferase) may modulate pain sensitivity through the effect on mRNA secondary structure and efficiency of protein expression [5–7]. Examples of associations of sSNPs and human complex traits like the *COMT* sSNPs in pain sensitivity are rare. Most probably, although not functionally neutral, the functional effects of sSNPs are largely minor, while the minor effects are not readily identifiable by traditional genetic association study. SC usage bias is a widespread phenomenon across biological species [8]. A sSNP changing codon usage may be expected to fine-tune translational efficiency based on the availability of rare tRNAs [9,10]. According to the ramp model of mRNA translation, except the second codon, the first 50 codons of mRNAs tend to favor rarer codons and have slower speed of translation [10–12]. This "ramp" mechanism is important in determining translation efficiency, preventing ribosome congestion, and allowing proper co-translational folding of proteins [3]. Based on the ramp theory, human sSNPs at ramp regions may confront selection pressure because of their functional effect on codon usage. To identify the translational effect of an individual SNP is difficult. Instead, we tried to identify the overall selection effect on sSNPs in human genome in this study. We investigated

the incidences of sSNPs in the 3rd~50th codons vs. those in the remainder codons after the 51st codon.

Methods

Fourfold degenerate site (FFDS, i.e. the four nucleotides A/C/G/T at this site encode the same amino acid) sSNPs with A↔G or C↔T substitutions in human genome were extracted from the NCBI dbSNP database build 134 (http://www.ncbi.nlm.nih.gov/projects/SNP/). Altogether, 39,276 sSNPs in 12,568 genes were collected. All SNP alleles were corresponding to the nucleotides in coding sequences. Among these FFDS sSNPs, 20,122 were A↔G sSNPs, and 19,154 were C↔T sSNPs. Of the 20,122 A↔G FFDS sSNPs, 43 at second codons of coding regions were removed from further analysis; of 19,154 C↔T sSNPs, 25 at second codons were removed from further analysis. The FFDS sSNPs were annotated as $N_1 \rightarrow N_2$, while N_1 represents the ancestral allele and N_2 represents the variant allele. Ancestral alleles of sSNPs were inferred by human-chimpanzee genomic alignment according to the SeattleSeq Annotation 134 (http://snp.gs.washington.edu/SeattleSeqAnnotation134/index.jsp). All sSNPs were differentiated by CpG sites *versus* non-CpG sites, while a CpG site has the pattern of YpG or CpR (Y represents C↔T substitution, and R represents A↔G substitutions).

Results

Our results showed that the fraction of FFDS sSNPs is significantly lower in the ramp (the 3rd ~50th codons) than the

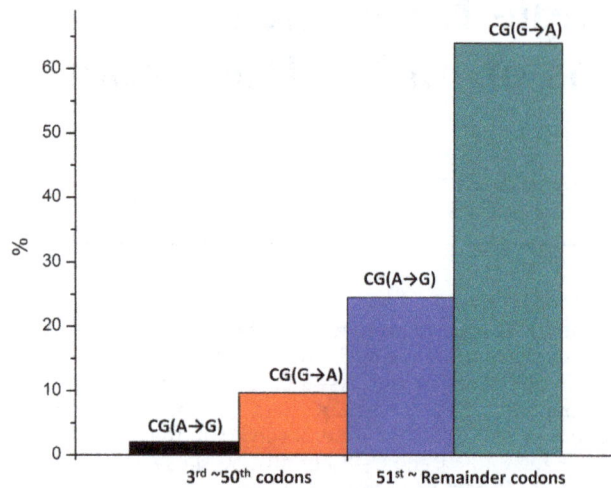

Figure 1. The distribution bias of CG(G→A) and CG(A→G) at the ramp regions. The ratio of CG(G→A)/CG(A→G) at the ramp regions is larger than that at the reminder coding regions ($P = 0.040$). CG(A↔G) synonymous substitutions are all at non-CpG sites.

(95% CI) $= 1.272(1.063, 1.523)$]. In both cases of G→A and C→T, the favored direction of SC usage is the change from more frequent SCs to less frequent SCs. The reference data of human codon usage (Table S1) was calculated by the EMBL human coding sequences (CDS) data release 115 (ftp://ftp.ebi.ac.uk/pub/databases/embl/cds/). By further investigation, our study disclosed that the G→A bias was mainly seen in synonymous substitution CG(G→A) at non-CpG sites [OR (95% CI) $= 1.861(1.020, 3.395)$] (Table 2, Figure 1); the C→T bias was mainly seen in AC(C→T) [OR (95% CI) $= 2.275 (1.255, 4.124)$] and CT(C→T) [OR (95% CI) $= 1.780 (1.053, 3.010)$] at non-CpG sites (Table 3, Figure 2). In all these three types of biased synonymous substitutions [i.e. CG(G→A), AC(C→T), and CT(C→T)], the favored change at the ramp region is from more frequent SCs to less frequent SCs.

To further characterize the distribution bias of FFDS sSNPs, we examined distributions of FFDS sSNPs stepwise by comparing the 3rd~n^{th} (n = 20, 21, …,60) codons vs. the remainder codons (Table S2). The overall C→T bias at non-CpG sites was most significant in the first 46 codons. The codon-specific AC(C→T) bias at non-CpG sites was most significant in the first 50 codons, and the codon-specific CT (C→T) bias at non-CpG sites was most significant in the first 45 codons. The overall G→A bias at non-CpG sites was most significant in the first 55 codons, and the codon-specific CG(G→A) bias at non-CpG sites was most significant in the first 39 codons. Therefore, the ramp region may not have a clear border in term of codon number. As a side note, the GG(G→A) bias at non-CpG sites also showed nominal significance in the first 57 codons ($P = 0.021$), and the CT(G→A) bias at non-CpG sites was nominal significant in the first 46 codons ($P = 0.026$). The change of codon usage of CT(G→A) has also the direction from more frequent SC to less frequent SC. The change of codon usage of GG(G→A) is unobvious. One exception is the statistical significance of GC(G→A) bias ($P = 1.85 \times 10^{-3}$) in the first 25 codons. These GC(G→A)s have the codon usage change from less frequent GCG to more frequent GCA. The GC(G→A) bias disappeared when more codons (≥45 codons) in the ramp region are considered.

rest regions (after the 50th codon) [0.23% *vs.* 0.32%, odds ratio OR (95% confidence interval CI) $= 0.708 (0.684, 0.734)$, $P = 1.60 \times 10^{-81}$), corrected by the FFDS codon usages calculated by the European Molecular Biology Laboratory (EMBL) Human CDSs (Coding sequences) Release 115 (ftp://ftp.ebi.ac.uk/pub/databases/embl/cds/). We identified significant distribution bias of sSNPs between the 3rd ~50th codons and the 51st ~ remainder codons at non-CpG sites (Table 1). This distribution bias at non-CpG sites is consistent with our previous study on the asymmetry pattern of complementary sSNPs at FFDS, which was seen in non-CpG sSNPs only, but not sSNPs at CpG sites. This context-specific distribution bias is related to lower mutation rates and longer periods of evolutionary selection at non-CpG sites [13]. In the 3rd ~50th codons, G→A sSNPs are favored than A→G sSNPs at non-CpG sites [OR (95% CI) $= 1.353 (1.108, 1.652)$], and C→T sSNPs are favored than T→C sSNPs at non-CpG sites [OR

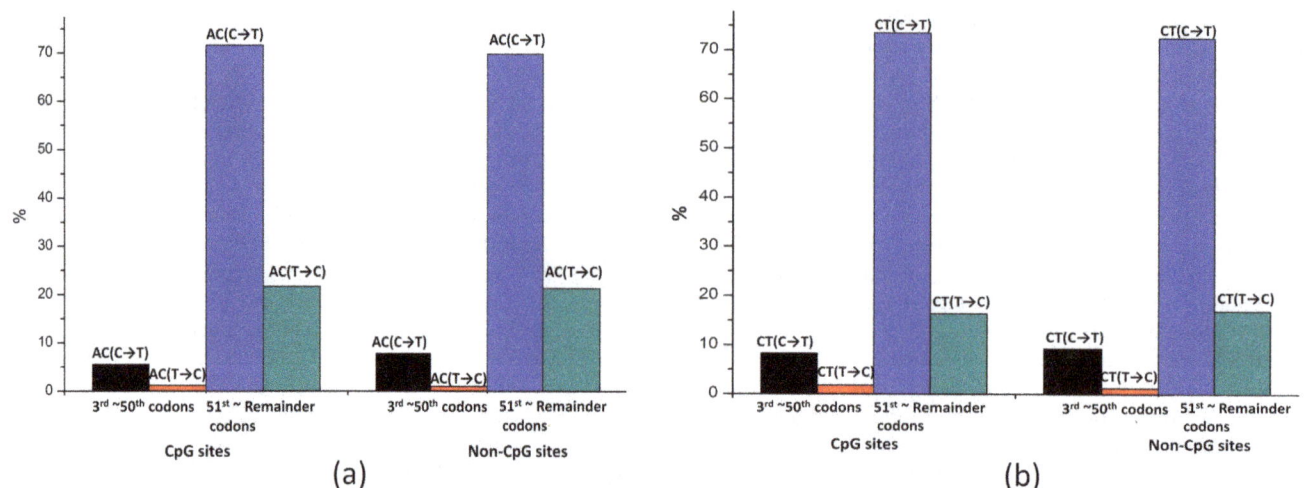

Figure 2. The distribution bias of (C→T) and (T→C) at non-CpG sites of the ramp regions. (a) The ratio of AC(C→T)/AC(T→C) at non-CpG sites of the ramp regions is larger than that at the reminder coding regions($P = 0.006$). (b) The ratio of CT(C→T)/CT(T→C) at non-CpG sites of the ramp regions is larger than that at the reminder coding regions ($P = 0.029$).

Table 1. A↔G and C↔T fourfold degenerate site sSNPs.

Substitution type	A→G n(%)	G→A n(%)	C→T n(%)	T→C n(%)	Total (count)
NonCpG site					
3rd ~50th codons	137(9.3%)	453(30.7%)	727(49.3%)	158(10.7%)	1475
51st ~ Remainder codons	1632(11.6%)	3988(28.5%)	6575(46.9%)	1818(13.0%)	14013
3rd ~50th codons vs. remainders [a]	p = 2.89×10⁻³ **		p = 8.50×10⁻³ **		
CpG site					
3rd ~50th codons	283(15.2%)	760(40.9%)	642(34.6%)	173(9.3%)	1858
Remainder codons	3614(16.5%)	9212(42.1%)	7046(32.2%)	1990(9.1%)	21862
3rd ~50th codons vs. remainders [a]	p = 0.471		p = 0.599		

[a]χ^2 test of the difference of substitution direction between the first 50 codons and the remainder codons; * $P<0.05$; ** $P<0.01$.
Significant distribution bias of sSNPs between the 3rd ~50th codons and the 51st ~ remainder codons was identified at non-CpG sites.

Discussion

Our previous study showed genome-wide discrepancy of human sSNPs between two complementary DNA strands, and suggested widespread selective pressure due to functional effects of sSNPs related to gene transcription [13]. The asymmetry pattern of complementary sSNPs in human genome may be related to transcription-coupled mutation and repair [13]. In this study, we identified another type of distribution bias of sSNPs in human genome related to mRNA translation. Biased directions of SC substitutions between the 3rd ~50th codons and the 51st ~ remainder codons at non-CpG sites were observed. In the 3rd ~50th codons, G→A sSNPs at non-CpG sites are favored than A→G sSNPs, and C→T at non-CpG sites are favored than T→C sSNPs. In both cases, the change from more frequent SCs to less frequent SCs is favored in the 3rd ~50th codons over the remainder codons. This finding is supportive to the ramp model of SC uage in mRNA translation [10,11]. The change from more frequent SCs to less frequent SCs may enhance the function of ramp regions to prevent subsequent ribosome congestion and improve the efficiency of protein synthesis. On the other hand, if a synonymous substitution has the change of a less frequent SC to a more frequent SC, it may impair ramp function and cause

ribosomal traffic jams during protein synthesis. The potential deleterious effect of these sSNPs may be subjected to larger evolutionary selection pressure, and tend to be removed by purifying selection.

By investigating 13,798 common sSNPs genotyped by the HapMap3 project, Waldman et al. demonstrated evolutionary selection for translation efficiency on sSNPs [14]. By investigating all human sSNPs, our study identified the obvious bias in the ramp region for synonymous substitutions CG(G→A), AC(C→T), and CT(C→T), indicating codon-specific effect on gene translation efficiency. As a limitation of this study, the specific SC changes that we identified didn't reach the significance level after correction of multiple testing by Bonferroni correction, which warrants for further study. On the other hand, empirically, codon-specific translation efficiency has been observed in model organisms, e.g. the strongly inhibitory effect of the CGA codon in yeast [15]. The intriguing exception of the GC(G→A) bias may suggest that the hypermutable GCG through methylation-induced deamination of 5-methyl cytosine on the antisense strand [16] meets less negative selection in the first half of the ramp region, but stronger negative selection in the second half of the ramp region which compensates the GC(G→A) bias in the first half of the ramp region. The lack of negative selection on GC(G→A) in the first 25

Table 2. A↔G fourfold degenerate site sSNPs.

First two codon positions	CpG site (yes or no)	3rd ~50th codons		51st ~ Remainder codons		3rd ~50th codons vs. remainders (P value) [a]
		A→G n(%)	G→A n(%)	A→G n(%)	G→A n(%)	
AC	Yes	68(1.8%)	164(4.4%)	1021(27.1%)	2508(66.7%)	0.902
CC	Yes	84(2.1%)	226(5.7%)	1039(26.2%)	2621(66%)	0.628
CG	No	14(2%)	68(9.7%)	172(24.5%)	449(63.9%)	0.040*
CT	No	50(2.2%)	171(7.4%)	601(26%)	1491(64.5%)	0.055
GC	Yes	72(2%)	246(6.8%)	907(25.2%)	2369(65.9%)	0.054
GG	No	40(2.6%)	96(6.3%)	499(32.8%)	887(58.3%)	0.125
GT	No	33(2%)	118(7.1%)	360(21.5%)	1161(69.4%)	0.616
TC	Yes	59(2.3%)	124(4.9%)	647(25.4%)	1714(67.4%)	0.159

[a]χ^2 test of the difference of substitution direction between the first 50 codons and the remainder codons; *Uncorrected $P<0.05$. By Bonferroni correction for multiple comparisons, the threshold for statistical significance is $P<0.00625$.
The G→A bias was mainly explained by the CG(G→A) substitution at non-CpG sites.

Table 3. C↔T fourfold degenerate site sSNPs.

First two codon positions	CpG site (yes or no)	3rd ~50th codons		51st ~ Remainder codons		3rd ~50th codons vs. remainders (P value) [a]
		C→T n(%)	T→C n(%)	C→T n(%)	T→C n(%)	
AC	Yes	79(5.5%)	17(1.2%)	1027(71.6%)	311(21.7%)	0.212
AC	No	97(7.8%)	13(1%)	866(69.8%)	264(21.3%)	0.006**
CC	Yes	106(6%)	31(1.8%)	1249(70.8%)	377(21.4%)	0.882
CC	No	81(6.2%)	26(2%)	906(69.1%)	298(22.7%)	0.917
CG	Yes	42(7.1%)	8(1.4%)	437(73.9%)	104(17.6%)	0.578
CG	No	48(9.6%)	6(1.2%)	377(75.7%)	67(13.5%)	0.435
CT	Yes	69(8.3%)	15(1.8%)	613(73.5%)	137(16.4%)	0.927
CT	No	129(9.3%)	17(1.2%)	1006(72.5%)	236(17%)	0.029*
GC	Yes	127(6%)	35(1.7%)	1514(71.9%)	430(20.4%)	0.879
GC	No	126(7.8%)	27(1.7%)	1165(72.2%)	296(18.3%)	0.442
GG	Yes	116(8.8%)	25(1.9%)	958(72.7%)	219(16.6%)	0.800
GG	No	103(8.8%)	20(1.7%)	834(71.2%)	215(18.3%)	0.267
GT	Yes	33(5.7%)	13(2.2%)	408(70.2%)	127(21.9%)	0.491
GT	No	40(5%)	14(1.8%)	593(74.7%)	147(18.5%)	0.285
TC	Yes	70(5.7%)	29(2.4%)	840(68.6%)	285(23.3%)	0.387
TC	No	103(8.2%)	35(2.8%)	828(65.7%)	295(23.4%)	0.819

[a] χ^2 test of the difference of substitution direction between the first 50 codons and the remainder codons; *Uncorrected $P<0.05$; **Uncorrected $P<0.01$. By Bonferroni correction for multiple comparisons, the threshold for statistical significance is $P<0.003125$.
The C→T bias was mainly explained by the AC(C→T) and CT(C→T) substitutions at non-CpG sites.

codons may suggest a functional heterogeneity of the ramp region, which warrants further study. In addition, Tuller et al. recently highlighted that stronger mRNA folding may also be involved in the ramp function [17]. Different effect of these SCs on mRNA secondary structure is an interesting issue deserving further inquiry.

Author Contributions

Conceived and designed the experiments: HQQ. Performed the experiments: QL HQQ. Analyzed the data: QL HQQ. Contributed reagents/materials/analysis tools: QL HQQ. Wrote the paper: QL HQQ.

References

1. Duan J, Antezana M (2003) Mammalian mutation pressure, synonymous codon choice, and mRNA degradation. J Mol Evol 57: 694–701.
2. Chamary J, Hurst L (2005) Evidence for selection on synonymous mutations affecting stability of mRNA secondary structure in mammals. Genome Biol 6: R75.
3. Sauna ZE, Kimchi-Sarfaty C (2011) Understanding the contribution of synonymous mutations to human disease. Nat Rev Genet 12: 683–691.
4. Tsai CJ, Sauna ZE, Kimchi-Sarfaty C, Ambudkar SV, Gottesman MM, et al. (2008) Synonymous mutations and ribosome stalling can lead to altered folding pathways and distinct minima. J Mol Biol 383: 281–291.
5. Mannisto PT, Kaakkola S (1999) Catechol-O-methyltransferase (COMT): Biochemistry, Molecular Biology, Pharmacology, and Clinical Efficacy of the New Selective COMT Inhibitors. Pharmacol Rev 51: 593–628.
6. Diatchenko L, Slade GD, Nackley AG, Bhalang K, Sigurdsson A, et al. (2005) Genetic basis for individual variations in pain perception and the development of a chronic pain condition. Hum Mol Genet 14: 135–143.
7. Nackley AG, Shabalina SA, Tchivileva IE, Satterfield K, Korchynskyi O, et al. (2006) Human Catechol-O-Methyltransferase Haplotypes Modulate Protein Expression by Altering mRNA Secondary Structure. Science 314: 1930–1933.
8. Behura SK, Severson DW (2012) Codon usage bias: causative factors, quantification methods and genome-wide patterns: with emphasis on insect genomes. Biological Reviews: no-no.

9. Cannarozzi G, Schraudolph NN, Faty M, von Rohr P, Friberg MT, et al. (2010) A Role for Codon Order in Translation Dynamics. Cell 141: 355–367.
10. Fredrick K, Ibba M (2010) How the Sequence of a Gene Can Tune Its Translation. Cell 141: 227–229.
11. Ingolia NT, Ghaemmaghami S, Newman JRS, Weissman JS (2009) Genome-Wide Analysis in Vivo of Translation with Nucleotide Resolution Using Ribosome Profiling. Science 324: 218–223.
12. Tuller T, Carmi A, Vestsigian K, Navon S, Dorfan Y, et al. (2010) An Evolutionarily Conserved Mechanism for Controlling the Efficiency of Protein Translation. Cell 141: 344–354.
13. Qu HQ, Lawrence SG, Guo F, Majewski J, Polychronakos C (2006) Strand bias in complementary single-nucleotide polymorphisms of transcribed human sequences: evidence for functional effects of synonymous polymorphisms. BMC Genomics 7: 213.
14. Waldman YY, Tuller T, Keinan A, Ruppin E (2011) Selection for Translation Efficiency on Synonymous Polymorphisms in Recent Human Evolution. Genome Biology and Evolution 3: 749–761.
15. Letzring DP, Dean KM, Grayhack EJ (2010) Control of translation efficiency in yeast by codon–anticodon interactions. RNA 16: 2516–2528.
16. Strachan T RA (1999) Human molecular genetics. Oxford: BIOS Scientific.
17. Tuller T, Veksler-Lublinsky I, Gazit N, Kupiec M, Ruppin E, et al. (2011) Composite effects of gene determinants on the translation speed and density of ribosomes. Genome Biol 12: R110.

15

CGAP-Align: A High Performance DNA Short Read Alignment Tool

Yaoliang Chen[1], Ji Hong[1], Wanyun Cui[1], Jacques Zaneveld[2], Wei Wang[1], Richard Gibbs[2], Yanghua Xiao[1]*, Rui Chen[2]*

1 School of Computer Science, Fudan University, Shanghai, China, **2** Human Genome Sequencing Center, Department of Molecular and Human Genetics, Baylor College of Medicine, Houston, Texas, United States of America

Abstract

Background: Next generation sequencing platforms have greatly reduced sequencing costs, leading to the production of unprecedented amounts of sequence data. BWA is one of the most popular alignment tools due to its relatively high accuracy. However, mapping reads using BWA is still the most time consuming step in sequence analysis. Increasing mapping efficiency would allow the community to better cope with ever expanding volumes of sequence data.

Results: We designed a new program, CGAP-align, that achieves a performance improvement over BWA without sacrificing recall or precision. This is accomplished through the use of Suffix Tarray, a novel data structure combining elements of Suffix Array and Suffix Tree. We also utilize a tighter lower bound estimation for the number of mismatches in a read, allowing for more effective pruning during inexact mapping. Evaluation of both simulated and real data suggests that CGAP-align consistently outperforms the current version of BWA and can achieve over twice its speed under certain conditions, all while obtaining nearly identical results.

Conclusion: CGAP-align is a new time efficient read alignment tool that extends and improves BWA. The increase in alignment speed will be of critical assistance to all sequence-based research and medicine. CGAP-align is freely available to the academic community at http://sourceforge.net/p/cgap-align under the GNU General Public License (GPL).

Editor: Haixu Tang, Indiana University, United States of America

Funding: Exome and whole genome sequencing was performed at the BCM-FGI core facility supported by NIH shared instrument grant 1S10RR026550 to R.C. This work was largely supported by Microsoft Research Asia with sponsorship number FY10-RES-OPP-010 and partially supported by the National Natural Science Foundation of China under grant No 61003001; Specialized Research Fund for the Doctoral Program of Higher Education No. 20100071120032; Key Program of National Natural Science Foundation of China under grant No. 61033010; NIH training grant T32 EY007102; National Science and Technology Major Project of the Ministry of Science and Technology of China under grant No. 2010ZX01042-003-004. This work is also partially supported by grants from the Retinal Research Foundation and National Eye Institute (R01EY018571) to R.C. and the National Human Genome Research Institute to R.G. (U54HG003273). The funders had no role in study design, data collection and analysis, decision to publish, or preparation of the manuscript.

Competing Interests: This work was largely supported by Microsoft Research Asia with sponsorship number FY10-RES-OPP-010. There are no patents, products in development or marketed products to declare.

* E-mail: shawyh@fudan.edu.cn (YX); ruichen@bcm.edu (RC)

Introduction

In recent years, advances in sequencing have led to the production of unprecedented amount of sequence data. Alignment, which maps reads to the reference sequence, is one of the most computationally demanding tasks performed in typical sequence data processing. Accurate sequence alignment is critical for SNP calling [1], structural variation detection [2] and further downstream analysis.

In order to efficiently and accurately map large numbers of short reads many new alignment programs have been developed. The algorithms underlying most of these tools can be classified into two major categories [3]. The first category uses hash tables to hash either read sequences, as in MAQ [1], SeqMap [4] and CloudBurst [5], or the genome reference, as in SOAPv1 [6], PASS [7], MOM [8] and ProbeMatch [9]. Although this technique can be easily parallelized, the major drawback of using hash tables is that either they must scan the whole genome, even when few reads are aligned, or they require a large amount of memory to build an index for the

reference. The second category is based on string matching using a representation of prefix/suffix trie [2], including suffix tree, suffix array [10], enhanced suffix array [11] and FM-index [12]. The first three representations are used by TRELLIS [13], MUMmer [14] and Vmatch (http://www.vmatch.de). Unfortunately, these programs have poor performance on large-scale references including the human genome due to memory constraints. The FM-index is a type of substring index based on the Burrows-Wheeler transform, with some similarities to suffix array. Owing to its small memory requirement, it is utilized by numerous state of the art programs including SOAPv2 [15], Bowtie [16], Bowtie2 [17] and BWA [18].

BWA has become one of the most widely used alignment tools owing to its efficiency and accuracy. BWA possesses higher recall and precision than SOAPv2 or Bowtie. However, both Bowtie and SOAP2 are significantly faster than BWA. Therefore, it is highly desirable to improve the speed of BWA while maintaining its alignment quality. In this paper, we describe a new efficient alignment tool, CGAP-align, following the framework of BWA. As part of short read mapping in BWA, possible mismatches and gaps are enumerated during the traversal of the FM-index of the reference sequence. The time efficiency of BWA is mainly impacted by two factors. The first is the efficiency of locating a substring in the reference. The second is the ability of the program to bypass segments of the reference where the read would contain a large number of mismatches, avoiding the need to enumerate all possible sets of mismatches and gaps, through a process called pruning. Improvements to these strategies have been implemented in CGAP-align to optimize the efficiency of both the reference querying and the pruning steps.

In this report, we first introduce a novel data structure, Suffix Tarray, which speeds up the process of locating a read on the reference. Second, we present an effective pruning strategy that more accurately predicts the number of mismatches in a read prior to alignment. Pruning is significantly improved by using a set of training reads in advance to identify and study frequent patterns of nucleotides. The performance of CGAP-align was evaluated on both simulated data and several sets of real paired-end sequence data. Our results indicate that by implementing both of these improvements alignment speed is significantly increased without sacrificing recall or precision.

Methods

1.1 Suffix Tarray (STA): Improving Reference Queries

To improve the alignment speed, we utilized a data structure, Suffix Tarray (STA), to index reference sequences in CGAP-align. STA uses a new index structure that is a hybrid of trie (inspired by suffix tree) and suffix array data structures. Before we present the concept of STA, we first briefly review the two most widely-used data structures for sequence indexing: suffix array (SA) and suffix tree (ST).

1.1.1 Background: Suffix Tree (ST) & Suffix Array (SA). Suffix tree is a data structure that encodes all of the suffixes of a given string in the form of tree, allowing quick location of substrings in $O(|W|+N)$ time, where $|W|$ is the length of the substring and N is the number of occurrences of that substring. More specifically, ST for a string X is a tree whose edges are labeled with strings, such that each suffix of X corresponds to exactly one path from the tree's root to a leaf. In the case of sequence data, a suffix tree for string X of n characters is queried and constructed in linear time [19]. Owing to its query efficiency, ST was once the predominant data structure for read alignment [2]. However, ST usually requires $O(n)$ memory with a large constants leading to a large memory requirement. State-of-art ST methods like TDD [20] and TRELLIS [13] cannot currently scale up to the entire human genome without a disk-based strategy that results in a suffix tree consuming tens of gigabytes of space.

Suffix array (SA) is an array of integers each of which gives the start positions of suffixes of a string X in alphabetical order. The alphabetic ordering of SA enables each substring of X to be queried through a binary search on SA. Given a string W is a substring of X, we define the interval $[k, l]$ as the SA interval of W. In particular, if W is empty, the corresponding SA interval is $[1, n-1]$, where n is the length of X. The set of positions of all occurrences of W in X is thus $\{SA[i] : k \le i \le l\}$, where $SA[i]$ maps the SA position i to a reference coordinate. A SA, if optimally implemented, consumes $O(n)$ space with a small constant and can be constructed in linear time [21](Ko et al., 2003). Further improving the space efficiency, FM-index [12](Ferragina et al., 2000), a compressed representation of SA, was developed. In FM-index, SA intervals are retrieved by the summing corresponding C and Occ values based on Burrows-Wheeler transform (BWT) [22] (Burrows et al. 1994). Given a substring W and a character c,

Ferragina and Manzini [12](2000) shows that

$$k(cW) = C(c) + Occ(c, k(W)-1) + 1$$

$$l(cW) = C(c) + Occ(c, l(W))$$

and that $k(cW) \le l(cW)$ iff cW is a substring of X, where $[k(W), l(W)]$ and $[k(cW), l(cW)]$ are the SA intervals of the substrings W and cW respectively. During each query, a backward search is performed on FM-index so that the substring W is scanned from the end to the beginning. FM-index is constructed in $O(n)$ time from the reference X (as it can be constructed in linear time from a SA) and used to query a substringfig of length m in $O(m)$ time. Due to its space efficiency, FM-index has recently been adopted by many widely-used mapping tools, including SOAPv2 [15], Bowtie [16] and BWA [18]. However, such space efficiency comes at the cost of reduced query performance. Empirically, ST is much faster than FM-index for queries on substrings [23].

1.1.2 Overview of the Suffix Tarray (STA) Structure. In an FM-index, for a character c, Occ is a function of the SA position y, which counts the number of the occurrences of c within the suffix interval $[1, y]$ of the SA. To accelerate the calculation of Occ values, the SA is divided into buckets with a fixed size. For the SA positions y that define the left bucket boundaries, the corresponding Occ values are pre-computed and stored in a table. This allows all other Occ values to be efficiently calculated by counting the occurrences of the character c from the start of their corresponding buckets instead of from the beginning of the SA.

However, retrieving a pre-computed Occ value is still much faster than retrieving an Occ value in the middle of a bucket. This observation implies that pre-computing frequent Occ values has the potential to speed up the mapping process. The basic idea of Suffix Tarray (STA) is to first find those most frequently visited Occ values (and their C values) and then organize them in a ST so that they can be efficiently accessed during alignment. With this novel approach, we can take the advantage of both the time efficiency of the ST and the space efficiency of FM-index. At a high-level, STA can be viewed as a truncated suffix tree (TST) encoding corresponding SA intervals of a FM-index at each leaf (Figure 1). As the height of the tree is bounded to a reasonably small value, we adopt trie as a light implementation of the TST.

There are two major steps to construct a STA for a reference string X. In the first step, a FM-index is built for the entire reference sequence to support the SA queries from the suffix tree [18]. In the second step, we construct a truncated ST based on the FM-index. All possible suffix strings of the reversed reference sequence $\bar{x}$ are enumerated. Instead of building the whole ST for $\bar{x}$, we truncate the ST according to the frequency values of the nodes. The frequency value of a node in ST for $\bar{x}$ is defined as the number of occurrences of the substring which it represents in $\bar{x}$. In our example, the frequency value of the node "CT" is 2 since there are 2 prefixes valued "CT" among all of the suffixes of $\bar{x}$. The nodes in tries will be discarded if their frequency values are smaller than a threshold ε which is calculated by St. We select a maximal ε such that the resulting trie has a size less than a user specified trie size St. This is accomplished by a binary-search style enumeration of the possible values of ε. Initially, the possible range of ε is set to $[1, |X|]$. The program first tries to construct a trie using a value P in the middle of possible range, such that in the first iteration P is approximately $|X|/2$. If, during construction, the trie size exceeds the size limit St, then the maximum possible value for ε is reduced to P-1. If, on the other hand, the trie is constructed without exceeding St then the minimum possible value of ε is increased to P. This process is iterated until the possible range of ε is reduced to a single value, which we assign to ε. At last, a truncated trie with frequency threshold ε and size less than St is built. Figure 1 is an example of a STA.

In the current version of CGAP-align, each trie node takes 36 bytes. One concern when building a STA is how to choose an appropriate truncated trie size. If the size is too small, then the index will not see a significant increase in query speed since the FM-index will dominate the query process. Otherwise, memory consumption will increase. Fortunately, since the frequency values of novel nodes decreases rapidly with increasing trie size, a STA with a moderate-sized trie can perform nearly as well as a STA with a large trie. We denote ST_{100} as a full ST for truncated at depth 100 and the frequency ratio of a trie as the values of all nodes in that trie divided by the sum of frequency value of all nodes in ST_{100}. According to our experiment, when indexing the human reference genome, a 300 MB sized truncated ST has a frequency ratio of 13.32% while 1 GB has 14.68%.

1.1.3 Matching Substrings to Suffix Tarray.

Algorithm 1 :

STA index construction : (X,b) //reference string X and the length of candidatestring b

Calculate BWT string B for reference string X;

Calculate array $C(\cdot)$ and $Occ(\cdot,\cdot)$ from B;

Build ST for all the reversed substrings occurring in X with frequency $> = \varepsilon$;

Procedures :

    InexactSearch(W,z)

        return InexRecur(W,$|W|-1$,z,1,$|X|-1$,ST.root);

    InexRecur(W, i, z,k, l, $node$)

*   **if** $z < D[i]$ **then**

*   **return** $\emptyset$

        if $i < 0$ **then**

            return $\{[k,l]\}$

        //deletion, edit cost : 1

        $I \leftarrow$ InexRecur(W,$i-1$,$z-1$, k, l, $node$)

        for each $c \in \{A,C,G,T\}$ **do**

            if $node$ **not null then**

                $k \leftarrow node.k$; $l \leftarrow node.l$;

                $node \leftarrow node.child[c]$;

            else;

                $k \leftarrow C(c) + Occ(c,k-l) + 1$;

                $l \leftarrow C(c) + Occ(c,l)$;

            if $k \leq l$ **then**

                //insertion, edit cost : 1

                $I \leftarrow I \cup$ InexRecur(W,i,$z-1$, k, l, $node$)

                if $c = W[i]$ **then**

                    //no edit operation, edit cost : 0

                    $I \leftarrow I \cup$ InexRecur(W,$i-1$,z, k, l, $node$)

                else

                  //replace, edit cost : 1

                    $I \leftarrow I \cup$ InexRecur(W,$i-1$,$z-1$, k, l, $node$)

    return I

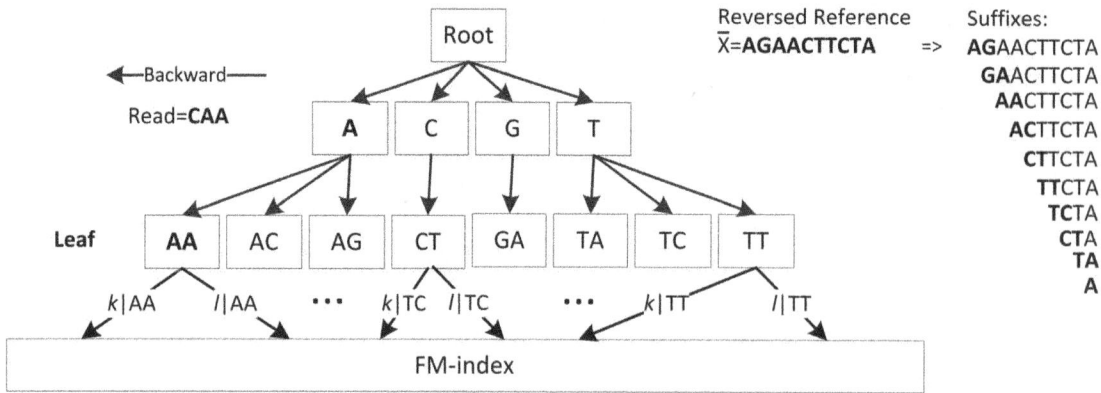

Figure 1. An example Suffix Tarray for the reference sequence "ATCTTCAAGA". A trie (truncated suffix tree) for "AGAACTTCTA" is built on the top of an FM-index. Each node of the trie stores k and l, where $[k, l]$ is the SA interval of the substring that corresponds to the leaf node on the FM-index.

The procedure to query a substring from STA is given in Algorithm 1. Procedure InexactSearch(W, z) takes substring W and maximal edit distance z as input, and returns the SA intervals of substrings that match W with no more than z differences. The algorithm framework is similar to that used in BWA, but with a different index structure. The l and k values are re-calculated based on STA. For a given node of ST, we denote node.*child* as its child-list, while node.k and node.l define its SA interval values. The lines with an asterisk mark the D-array pruning strategy used in BWA, which will be discussed in section 1.2.

The actual implementation of CGAP-align differs slightly from the pseudo code given in Algorithm 1. CGAP-align inherits the logic of BWA's inexact mapping, where "gap extensions" do not add to the number of edits that have been performed (i.e. IndexRecur will be recursively called with maximum edit distance value z). This helps to detect relatively long gaps. However, it also causes a potential issue resulting in false negative alignments in both BWA and CGAP-align. This problem is addressed in section 2.2.

1.2 Data-Conscious D-Array Calculation (DCDC): Improving Pruning

To improve the efficiency of mapping, a pruning step is implemented in BWA. A lower bound is calculated for the number of differences between $W[0,i]$ (i.e. the i-length prefix of the read W) and any substrings of the reference sequence and stored as the ith element of array D, all prior to mapping. Reads with D[i] greater than the maximum tolerated number of mismatches defined by the user are excluded from further alignment. If calculating D-array is faster than the time it would take to map the pruned part of the search trees, pruning improves mapping efficiency. The degree of improvement depends on both the speed at which the D-array can be calculated and the accuracy of the estimation of D-array. A more accurate estimation of D results in a smaller search space and more efficient mapping. Figure 2 gives an example of how D-array would work during a backward search on either FM-index or STA. D-array prunes superfluous enumerations before mapping, providing a significant boost in performance.

1.2.1 D-array: Background. BWA proposes a method to estimate the D-array by splitting W into several small strings. Let e(W) be the minimal number of the edit operations required to make W exactly align onto the reference X. BWA divides W into segments $w_1w_2...w_t$, where e(w_p) = 1 for $1 \leq p < t$ and e(w_p) $\leq$ 1 for $p = t$. Then D[i] is approximated as p-1 for $1 \leq i < |W|$, where w_p

contains the $(i+1)$th element of W, or t for $i = |W|$. For example, given a reference X = "AACGTATCGACG" and a read W = "AACTGA", BWA segments the W as "(AACT)(GA)" and thus produces the D-array "000111". The time to calculate the D-array for read W is in O($|W|$) when the FM-index of the reverse of X is used. In CGAP-align, the calculation of D-array is further accelerated by using STA.

1.2.2 A Tighter Lower Bound for D-array. Identifying a w_p with e(w_p)>1 when splitting W generates a D-array with a tighter lower bound. In the example described above, if we consider segments with an e(w_p) equal to 2, then we derive the segmentation "(AACTG)(A)" with the corresponding D-array "000122". By allowing segments with more mismatches we derive a tighter bound for D.

However, the above tighter bound comes at the cost of longer computations. The time cost of verifying e(W) = k is exponential to the value of k. Therefore, it is too expensive to calculate each e(w_i) value on the fly unless e(w_i) is strictly restricted to be 1 (in this case calculation of e(w) is O($|w|$)). To address this issue, we add a pre-processing step, in which we identify the frequent substrings from the training read set with a e(w) = 2. This approach is feasible as genomic sequences from the same species are similar among individuals. Thus, by providing a training read set from each species, it is possible to generate a list of species specific frequently occurred substrings (w) with e(w) = 2.

In a pre-processing phase, an appropriate number of patterns with the greatest frequency values are identified by a depth-first visit on an FM-index built upon the training reads and then organized into an Aho-Corasick (AC) automaton [24]. The number of patterns is determined by the AC automaton size specified by the user. Given a read W, an AC automaton helps to find all substrings of W (denoted as w) that match to any of the indexed patterns (denote as f) in linear time. A description of AC automaton is given in section 1.2.3. For the sake of the efficiency, we only index patterns with e(f) = 2 into AC automatons.

1.2.3 Aho-Corasick Automaton. In general, millions of frequent patterns (FP) with e(f) = 2 are found in the pre-processing phase. We use a AC automaton G_T to organize these FPs so that when given a read W, we are able to quickly query all of the FPs that are contained in W in a single scan. This allows for efficient segmentation of W according to the FPs it contains. We adopt the AC algorithm [24] in which the FPs are organized in an AC automaton. Instead of searching for occurrences of a single string f within a main text string W, AC automaton supports searching for occurrences of a set of strings F within W. An AC automaton makes use of the information embodied by the string set F itself to determine where to begin then next matching attempt if a mismatch happens, thus bypassing re-examination of previously matched characters in W. There is a particular type of state called a "leaf state", each of which corresponds to a string f. The transition to a leaf state L_f indicates an occurrence of f in W. When a match is detected, we reset the automaton to its initial state and continue the scanning on W. These steps are iteratively performed until we meet the end of W. A simple example is shown in Figure 3.

1.2.4 Calculating the Better D-array.

Algorithm 2 CalculateD(W, X, G_T)

Input : A read W, the reference X and the AC automaton G_T

Output : The $D-$array for W

$j \leftarrow 0; z \leftarrow 0; k \leftarrow 0$

for $i=0$ **to** $|W|-1$ **do**

if $W[j, i]$ is not a substring of X **then** $//e(W(j, i))=1$

$z \leftarrow z+1$

$k \leftarrow j$

$j \leftarrow i+1$

elseif any suffix of $W[k, i]$ matches a patternstring f in G_T **then** $//e(W[k,i])=2$

$z \leftarrow z+1$

$k \leftarrow i+1$

$j \leftarrow i+1$

$D(i) \leftarrow z$

Algorithm 2 shows the detailed procedure to estimate a tighter D-array in CGAP-align. We use a greedy strategy to iteratively find the first matched pattern string f of W from the AC automaton G_T. If no such f is found, we segment W by e($W[j, i]$) = 1 instead (the first "if" condition). As described above, when we find a match in G_T at position i, we reset the automaton, scanning matches within $W[i+1, |W|-1]$ in the following comparison.

Results

As described above, two major changes have been added to the current version of BWA. First, instead of FM-index we utilize STA, a novel data structure that speeds up string matching by constructing a trie on top of an FM-index. Second, we implement a new pruning method DCDC. As mapping time increases exponentially with the number of mismatches, it is highly desirable to prune reads with large numbers of mismatches. To achieve this goal, a data training process has been implemented to identify a set of frequent substrings with two mismatches from the genome. This pre-processed data allows for accurate calculation of the minimal

number of mismatches between a prefix and the genome, resulting in further reduction of the search space.

Both functionalities have been implemented in C. To facilitate the usage of CGAP-align, it offers a command line interface that is almost identical to BWA and outputs SAM files (Sequence Alignment/Map format). CGAP-align is distributed under the GNU General Public License (GPL) with detailed documentation and source code freely available through Fudan University, BCM-HGSC and the Sourceforge web site. The pre-built indices for some public references (eg. Hg19) are also provided.

2.1 CGAP-align Evaluation

To evaluate the performance of CGAP-align, we have benchmarked it against BWA (version 0.5.9), SOAPv2 (version 2.20) and Bowtie2 (version 2.0.6), three of the most commonly used alignment programs. Other tools like Bowtie are not included because gapped alignments are currently not supported. All tools run on 4 threads. Both BWA and CGAP-align were evaluated on their ability to map 100 base pair long reads using either default settings or relatively loose settings that allow up to 5 edits and a gap extension of 3 base pairs (-n 5 -e 3 -l 25). The same data set is used for SOAPv2, allowing up to 5 mismatches and a gap size of 4 (-v 5 -g 4). This setting is looser than SOAPv2's default setting, which enables it to find more alignments. For bowtie2, default parameters are used (–sensitive) since no gap settings can be adjusted. In addition, two modes of CGAP-align, the first with STA alone (denoted CGAP-align) and the second with both DCDC and STA (denoted CGAP-align*), were evaluated. STA was built based on the human reference genome hg19 with 600 MB size. All of the experiments utilized the same AC

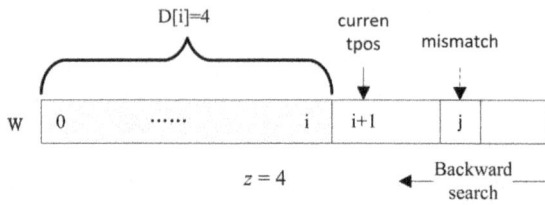

Figure 2. An example of D-Array based pruning. The maximum allowed number of differences, z, is 4. When backward search reaches position $i+1$, the algorithm detects that at least 4 differences must exist (as indicated in $D[i]$), while only three are allowed according to z due to the previous mismatch at position j.

automaton with a size of 1000 MB trained from independent WGS data for DCDC.

2.2 Evaluation on simulated data

The performance of BWA, CGAP-align, Bowtie2, and SOAPv2 on a simulated data set was examined. Wgsim, a program included in SAMtools [25], was used to generate 20 million simulated human genome shotgun 100 bp reads with a 2% error rate. As the exact position of each simulated read in the genome is known, both the sensitivity and specificity of the alignments could be precisely calculated.

As shown in Table 1, a significant speedup was achieved by CGAP-align while maintaining high mapping quality. We report both the absolute running time (Hours) and the percentage relative to BWA for all programs considered. When STA was used alone, CGAP-align was about 10% to 20% faster than BWA. Since the implementation of STA only accelerates the string matching step without affecting alignment, identical results were obtained between CGAP-align and BWA. Both programs produced mapping results with high accuracy and sensitivity. Sensitivity was measured by the true positive rate (TPR), which is defined as the overall percentage of correctly mapped reads relative to the number of input reads. In parallel, mapping specificity was measured by the positive predictive value (PPV), which indicates the percentage of correctly mapped reads relatively to the number of the reads reported as mapping correctly by the algorithm. Under default settings, 97.9% of the reads were accurately mapped with a PPV of 98.7%. The effect of STA was also evaluated under loose settings. A similar speed up was observed under this condition, and identical mapping results obtained by both CGAP-align and BWA as expected. For SOAPv2 and Bowtie, lower TPRs were observed than for CGAP-align although both run faster than CGAP-align.

Surprisingly, despite the use of training data, little speedup observed after adding DCDC. As shown in Table 1, an almost identical speed was observed for CGAP-align* under default settings. We found that in this case only 5% of the reads' D-arrays were improved by DCDC. This is because, in our simulated data set, most reads were aligned to the reference with only 1 or 2 edit operations, leaving little optimization space for DCDC. On the other hand, the DCDC itself consumed more time than the original D-array strategy, which led to a poorer overall performance. However, when we evaluated the programs on real data, where more mismatches and gaps exist, DCDC resulted in a significant speedup (Table 2).

Slightly different mapping results were obtained for CGAP-align* compared with BWA under loose settings, due to the inconsistency of the cost metrics used during read alignment and D-array calculation. During the alignment of each read, BWA and

CGAP-align consider the edit costs gap extensions as 0, as described in section 1.1.3. However, during the D-array calculation, these gap extension costs have to be counted as 1 because otherwise any read could be regarded as a single gap, and no value greater than 1 would be obtained in the D-array. As a result, pruning based on D-array misses some true positive alignments, a phenomenon that occurs in both BWA and CGAP-align. CGAP-align*, which computes a better D-array, omits a little more than BWA. In Discussion, we show a study case to illustrate such omissions. Fortunately, according to our results, in most cases CGAP-align* produces the same alignment as BWA, and only infrequently omits marginal results. Such omissions are even more trivial as only the best candidates are presented in the final alignment. In this experiment, there were 8,544 reads with which BWA obtained an extra 33,105 hits in comparison to CGAP-align*. Of these reads, 8,306 of them (97.21%) contained multiple hits, and only 607 of them were correct.

When mapping to the human genome, SOAPv2 uses 5.4 GB of memory while BWA only uses about 3 GB. In our experiment, CGAP-align without DCDC consumed 3.6 GB of memory, and used 4.6 GB when DCDC was integrated. However, the memory consumption can be controlled by adjusting the index sizes for both STA and DCDC.

The relationship between the sequence error rate and the performance of STA and DCDC is shown in Figure 4. STA only optimizes the counting problem of the FM-index (i.e. determining the number of matches of a substring occurring in a reference) without modifying the locating problem (determining the positions in the reference where the matches occur). A maximal speed up of 26.0% was observed when no errors were introduced as almost all reads mapped precisely to their correct positions, reducing locating costs. As the error rates goes up, multiple hits may occur for a single read, which increases locating costs. The lowest speed up, 10.2%, was observed at an error rate of 0.04. With an error rate between 0.04 and 0.1, the performance of STA improved as the more and more reads became un-mappable, which again led to a decrease in locating costs. On the other hand, DCDC had poor performance with an error rate below 0.04, after which it rapidly improved as most of the missing alignments were avoided through pruning.

2.3 Evaluation on Real Data

We tested CGAP-align using whole exome sequence (WES) data from two individuals (29.2 million read pairs for WES1, 31.6

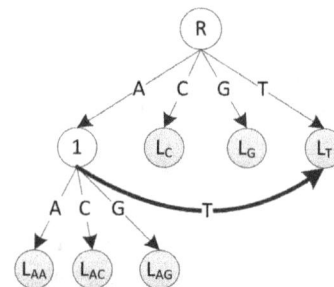

Figure 3. An example AC automaton. The frequent pattern string set F is {AA, AC, AG, C, G, T}. The grey nodes represent the leaf states. The bold "T" edge is connected as among all the prefixes of the 6 patterns, "T" is the longest suffix of "AT". Given a short read W="CAT", the state transition sequence when querying W from the automaton is $<R, L_C, R, 1, L_T>$, which indicates the occurrence of frequent patterns 'C' and 'T'.

Table 1. Evaluation on simulated data.

	Default Settings				Loose Settings			
Program	Hours	%	TPR (%)	PPV (%)	Hours	%	TPR (%)	PPV (%)
BWA	**2.08**	-	99.84	98.51	**3.70**	-	99.92	98.54
CGAP-align	1.77	85.10	99.84	98.51	3.29	88.92	99.92	98.54
CGAP-align*	1.80	86.54	99.84	98.51	2.89	78.11	99.92	98.54
Bowtie2	1.57	75.48	95.05	95.57	-	-	-	-
SOAPv2	0.95	45.67	84.95	98.21	-	-	-	-

The 10 million read pairs were mapped to the human genome. We recorded the run time (BWA in hours, CGAP-align and SOAP in percentage relative to BWA) on a 2.4 GHz Dual-Core AMD Opteron Processor 2216 HE with 4 threads running simultaneously (Hours & percentage relative to BWA), true positive rate (TPR) and positive predictive value (PPV). CGAP-align gave identical results to BWA with a shorter run time.

million read pairs for WES2) and whole genome sequence (WGS) data from a separate pair of individuals (107.8 million read pairs for WGS1 and 86.3 million read pairs for WGS2). All read pairs were produced by Illumina HiSeq with a 100 bp read length. Both the consumed time (Hours) and the frequency with which reads were confidently mapped (Conf) are reported in Table 2. For both CGAP-align and BWA, we use mapping quality threshold 10 to

Table 2. Evaluation on real data.

	Default Settings			Loose Settings		
Program	Hours	%	Conf (%)	Hours	%	Conf (%)
BWA-WES1	**3.350**	-	97.31	**7.130**	-	97.39
CGAP-align-WES1	2.887	86.18	97.31	6.375	89.41	97.39
CGAP-align*-WES1	2.698	80.54	97.31	3.813	53.48	97.33
Bowtie2-WES1	3.117	93.04	98.43	-	-	-
SOAPv2-WES1	1.076	32.12	93.85	-	-	-
BWA-WES2	**4.555**	-	96.84	**11.158**	-	96.92
CGAP-align-WES2	3.857	84.68	96.84	9.699	86.92	96.92
CGAP-align*-WES2	3.829	84.06	96.84	5.598	50.17	96.85
Bowtie2-WES2	3.700	81.23	98.02	-	-	-
SOAPv2-WES2	1.314	28.85	92.94	-	-	-
BWA-WGS1	**22.818**	-	91.84	**71.349**	-	92.26
CGAP-align-WGS1	20.817	91.23	91.84	66.262	92.87	92.26
CGAP-align*-WGS1	19.717	86.41	91.84	30.159	42.27	92.19
Bowtie2-WGS1	8.433	36.96	83.31	-	-	-
SOAPv2-WGS1	10.311	45.19	82.03	-	-	-
BWA-WGS2	**8.504**	-	93.69	**20.244**	-	93.82
CGAP-align-WGS2	7.629	89.71	93.69	19.218	94.93	93.82
CGAP-align*-WGS2	7.047	82.87	93.69	9.379	46.33	93.76
Bowtie2-WGS1	5.283	61.12	84.45	-	-	-
SOAPv2-WGS2	4.647	54.64	88.34	-	-	-

29.2 million read pairs (WES1), 31.6 million read pairs (WES2), 107.8 million read pairs (WGS1) and 86.3 million read pairs (WGS2) were mapped to the human genome. The run time (BWA in hours, CGAP-align and SOAP in percentage relative to BWA) on a 2.4 GHz Dual-Core AMD Opteron Processor 2216 HE with 4 threads running simultaneously (Hours & percentage relative to BWA), percent of confidently mapped reads including paired mapping (Conf) are shown.

determine confident mappings. CGAP-align, utilizing STA alone, obtained a speed up of approximately 10% to 20% compared with BWA. Surprisingly, CGAP-align*, while achieving a small speed up with default settings, shortened the running time by more than 50% with loose settings, due to the high error rates of the reads in real data. While SOAPv2 was faster than all other candidates including CGAP-align*, it found less confidently mapped reads than BWA, especially for the WGS cases. On the WGS1 data, where a significant number of mismatches between the reads and reference were observed, CGAP-align and BWA confidently mapped almost 10% more reads than SOAPv2 did. In comparison to SOAPv2, CGAP-align* further shortens the performance gap while maintaining a relatively high mapping rate. Overall, it possesses impressive mapping rates in WES while the performances when mapping WGS data still need to be improved.

Discussion

Due to the enormous number of reads generated by the next generation sequencing technologies, the efficiency of read alignment has become a critical problem. In this article, we present CGAP-align, a variant of BWA, which doubles the speed of the alignment process. While SOAPv2 is also very fast, it is unable to

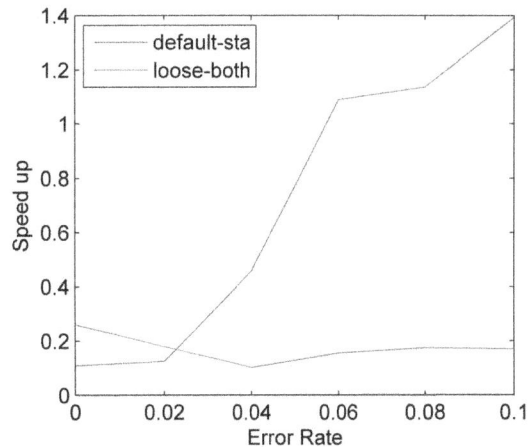

Figure 4. The error rates of the simulated reads vs. the speed up of CGAP-align over BWA, defined as $((T_{BWA}-T_{CGAP})/T_{CGAP})$. Six sets of simulated human genome shotgun 100 bp reads with error rates from 0 to 0.1 were generated. The speed up when using STA under default settings and when using both STA and DCDC under loose settings is reported.

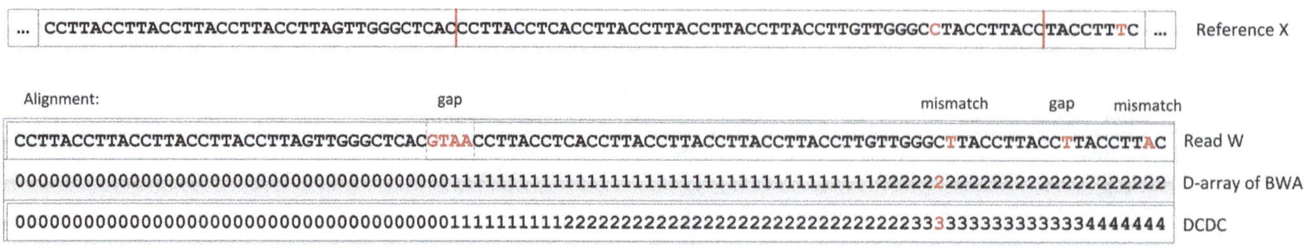

Figure 5. A case study for the inconsistency of the alignment results with loose setting (5 edit costs at most) where BWA successfully finds an alignment for the read *W* while CGAP-align* fails. The red color on the first line indicates the edit operations in the alignment found by BWA. In this particular alignment, we have 2 gaps and 2 mismatches. According to the cost metrics used during the alignment stage where the cost of gap extension is considered as 0, we need 4 edit costs to map *W* to *X*, satisfying the loose setting. The D-arrays calculated by BWA and DCDC are listed below. Both of them are strictly correct under D-array's cost metrics where the cost of a gap extension is 1.

map as many reads as CGAP-align. CGAP-align retains its high recall and precision even when the error rate of the reads is high. CGAP-align also outputs SAM files, allowing easy use of various downstream analysis tools.

CGAP-align requires a tunable amount of memory beyond that required by BWA, which equals to the size of the additional indices for STA and DCDC. To make DCDC work, a set of training reads are needed for CGAP-align to build the DCDC index. In practice, any FASTA file containing reads that are supposed to be mapped to the reference is qualified to be the training dataset for that reference. However, we recommend not using FASTA files larger than 2 GB to reduce memory cost and time of DCDC index construction.

It is very easy to migrate from using BWA to CGAP-align. For each reference sequence, only a single command line is needed to build the additional indices used by CGAP-align. After that, CGAP-align has an identical interface to that of BWA. In addition, CGAP-align is backwards compatible with BWA, so BWA can do read alignment using the FM-index produced by CGAP-align.

The main concern with using CGAP-align* is that it occasionally misses alignments called by BWA. Figure 5 further investigates a real example of the cause of these differences. Consider CGAP-align* mapping during the alignment depicted in Figure 5 under loose settings, where at most 5 edit costs are allowed. The backward search starts from right to left. When the search goes to the second mismatch, 3 edit costs have been counted. The D-array element at the next position (red number on the forth row) also has a value of 3, indicating that at least 6 edit costs are needed to map *W* to *X* according to DCDC's D-array. As a result, this alignment is pruned by CGAP-align*. However, at the same position in BWA's case, the D-array value is 2 (telling us that at least 5 edit costs are needed), which does not trigger the pruning, leading to the discrepancy between the alignments. From the mechanism behind this example, we can see that such discrepancies are quite random and partially depend on the gap positions in the alignment. As previously noted, only a very small fraction of reads encounter this problem and the pruned alignments are typically quite marginal.

Author Contributions

Conceived and designed the experiments: YC YX RC WW. Performed the experiments: YC JH WC. Analyzed the data: YC YX RC. Contributed reagents/materials/analysis tools: YC JH WC WW YX RG. Wrote the paper: YC JZ YX RC. Provided data: RG.

References

1. Li H, Ruan J, Durbin R (2008) Mapping short DNA sequencing reads and calling variants using mapping quality scores. Genome Res 18: 1851–1858.
2. Li H, Homer N (2010) A survey of sequence alignment algorithms for next-generation sequencing. Brief Bioinform 11: 473–483.
3. Ritz A, Bashir A, Raphael BJ (2010) Structural variation analysis with strobe reads. Bioinformatics 26:1291–1298.
4. Jiang H, Wang WH (2008) SeqMap: mapping massive amount of oligonucleotides to the genome. Bioinformatics 24: 2395–2396.
5. Schatz MC (2009) Cloudburst: highly sensitive read mapping with MapReduce. Bioinformatics 25: 1363–1369.
6. Li R, Li Y, Kristiansen K, Wang J (2008) SOAP: short oligonucleotide alignment program. Bioinformatics 24: 713–714.
7. Campagna D, Albiero A, Bilardi A, Caniato E, Forcato C, Manavski S, Vitulo N, Valle G (2009) PASS: a program to align short sequences. Bioinformatics 25: 967–968.
8. Eaves HL, Gao Y (2009) MOM: maximum oligonucleotide mapping. Bioinformatics 25: 969–970.
9. Kim YJ, Teletia N, Ruotti V, Maher CA, Chinnaiyan AM, et al. (2009) ProbeMatch: a tool for aligning oligonucleotide sequences. Bioinformatics 25: 1424–1425.
10. Manber U, Myers G (1990) Suffix arrays: a new method for on-line string searches. In Proceedings of the first annual ACM-SIAM SODA 90: 327.
11. Abouelhoda MI, Kurtz S, Ohlebusch E (2004) Replacing suf- fix treed with enhanced suffix arrays. Journal of Discrete Algorithms 2: 53–86.
12. Ferragina P, Manzini G (2000) Opportunistic Data Structures with Applications. Proceedings of the 41st Annual Symposium on FOCS. pp 390–398.
13. Phoophakdee B, Zaki MJ (2007) Genome-scale disk-based suffix tree indexing. SIGMOD. ACM. pp. 833–844.
14. Kurtz S, Phillippy A, Delcher AL, Smoot M, Shumway M, et al. (2004) Versatile and open software for comparing large genomes. Genome Biol 5: R12.
15. Li R, Yu C, Li Y, Lam TW, Yiu SM, et al. (2009) SOAP2: an improved ultrafast tool for short read alignment. Bioinformatics 25: 1966–1967.
16. Langmead B, Trapnell C, Pop M, Salzberg SL (2009) Ultrafast and memory-efficient alignment of short DNA sequences to the human genome. Genome Biol 10: R25.
17. Langmead B, Salzberg SL (2012) Fast gapped-read alignment with Bowtie 2. Nat Methods 9: 357–359.
18. Li H, Durbin R (2009) Fast and accurate short read alignment with Burrows–Wheeler. Bioinformatics 25:1754–60.
19. McCreight EM (1976) A Space-Economical Suffix Tree Construction Algorithm. Journal of the ACM 23: 262–272.
20. Tata S, Hankins R, Patel J (2003) Practical Suffix Tree Construction. Proceedings of the 30th International Conference on VLDB. pp. 36–47.
21. Ko P, Aluru S (2003) Space efficient linear time construction of suffix arrays. In Combinatorial Pattern Matching. LNCS 2676, Springer, pp 203–210.
22. Burrows M, Wheeler D (1994) A block sorting lossless data compression algorithm. 124, Palo Alto, CA, Digital Equipment Corporation.
23. Hon WK, Lam TW, Sung WK, Tse W, Wong CK, et al. (2004) Practical Aspects of Compressed Suffix Arrays and FM-index. Proceedings Of The Sixth Workshop On ALENEX And The First Workshop On ANALCO. pp.31–38.
24. Aho AV, Corasick MJ (1975) Efficient string matching: An aid to bibliographic search. Communications of the ACM 18: 333–340.
25. Li H, Handsaker B, Wysoker A, Fennell T, Ruan J, et al. (2009) The Sequence alignment/map (SAM) format and SAMtools. Bioinformatics 25: 2078–2079.

16

Outlier-Based Identification of Copy Number Variations Using Targeted Resequencing in a Small Cohort of Patients with Tetralogy of Fallot

Vikas Bansal[1,2☾], **Cornelia Dorn**[1,3☾], **Marcel Grunert**[1], **Sabine Klaassen**[4,5,6], **Roland Hetzer**[7], **Felix Berger**[6,8], **Silke R. Sperling**[1,3]*

1 Department of Cardiovascular Genetics, Experimental and Clinical Research Center, Charité - Universitätsmedizin Berlin and Max Delbrück Center (MDC) for Molecular Medicine, Berlin, Germany, 2 Department of Mathematics and Computer Science, Free University of Berlin, Berlin, Germany, 3 Department of Biology, Chemistry, and Pharmacy, Free University of Berlin, Berlin, Germany, 4 For the National Register for Congenital Heart Defects, Berlin, Germany, 5 Experimental and Clinical Research Center, Charité - Universitätsmedizin Berlin and Max Delbrück Center (MDC) for Molecular Medicine, Berlin, Germany, 6 Department of Pediatric Cardiology, Charité - Universitätsmedizin Berlin, Berlin, Germany, 7 Department of Cardiac Surgery, German Heart Institute Berlin, Berlin, Germany, 8 Department of Pediatric Cardiology, German Heart Institute Berlin, Berlin, Germany

Abstract

Copy number variations (CNVs) are one of the main sources of variability in the human genome. Many CNVs are associated with various diseases including cardiovascular disease. In addition to hybridization-based methods, next-generation sequencing (NGS) technologies are increasingly used for CNV discovery. However, respective computational methods applicable to NGS data are still limited. We developed a novel CNV calling method based on outlier detection applicable to small cohorts, which is of particular interest for the discovery of individual CNVs within families, *de novo* CNVs in trios and/or small cohorts of specific phenotypes like rare diseases. Approximately 7,000 rare diseases are currently known, which collectively affect ~6% of the population. For our method, we applied the Dixon's Q test to detect outliers and used a Hidden Markov Model for their assessment. The method can be used for data obtained by exome and targeted resequencing. We evaluated our outlier- based method in comparison to the CNV calling tool CoNIFER using eight HapMap exome samples and subsequently applied both methods to targeted resequencing data of patients with Tetralogy of Fallot (TOF), the most common cyanotic congenital heart disease. In both the HapMap samples and the TOF cases, our method is superior to CoNIFER, such that it identifies more true positive CNVs. Called CNVs in TOF cases were validated by qPCR and HapMap CNVs were confirmed with available array-CGH data. In the TOF patients, we found four copy number gains affecting three genes, of which two are important regulators of heart development (*NOTCH1*, *ISL1*) and one is located in a region associated with cardiac malformations (*PRODH* at 22q11). In summary, we present a novel CNV calling method based on outlier detection, which will be of particular interest for the analysis of *de novo* or individual CNVs in trios or cohorts up to 30 individuals, respectively.

Editor: Chunyu Liu, University of Illinois at Chicago, United States of America

Funding: This work was supported by the European Community's Seventh Framework Programme contracts ("CardioGeNet") 2009-223463 and ("CardioNet") People-2011-ITN-289600 (all to SRS), a Marie Curie PhD fellowship to VB, a PhD scholarship to CD by the Studienstiftung des Deutschen Volkes, and the German Research Foundation (Heisenberg professorship and grant 574157 to SRS). This work was also supported by the Competence Network for Congenital Heart Defects funded by the Federal Ministry of Education and Research (BMBF), support code FKZ 01GI0601. The funders had no role in study design, data collection and analysis.

Competing Interests: The authors have declared that no competing interests exist.

* E-mail: silke.sperling@charite.de

☾ These authors contributed equally to this work.

Introduction

Many genomic studies have revealed a high variability of the human genome, ranging from single nucleotide variations and short insertions or deletions to larger structural variations and aneuploidies. Structural variations include copy number variations (CNVs), which cause gains (duplications) or losses (deletions) of genomic sequence. These copy number changes are usually defined to be longer than ~500 bases, including large variations with more than 50 kilobases [1,2]. Recent studies have identified CNVs associated with a number of complex diseases such as Crohn's disease, intellectual disability and congenital heart disease [3–6].

Congenital heart disease (CHD) are the most common birth defect in human with an incidence of around 1% in all live births [7,8]. They comprise a heterogeneous group of cardiac malformations that arise during heart development. The most common cyanotic form of CHD is Tetralogy of Fallot (TOF), which accounts for up to 10% of all heart malformations [9]. TOF is characterized by a ventricular septal defect with an overriding aorta, a right ventricular outflow tract obstruction and a right ventricular hypertrophy [10]. It is a well-recognized subfeature of syndromic disorders such as DiGeorge syndrome (22q11 deletion),

Down syndrome, Holt-Oram syndrome and Williams-Beuren syndrome [11]. Deletions at the 22q11 locus account for up to 16% of TOF cases [12] and copy number changes at other loci were identified in several syndromic TOF patients [13–15]. However, the majority of TOFs are isolated, non-syndromic cases caused by a multifactorial inheritance with genetic-environmental interactions, which is also the situation for the majority of CHDs [16]. Using SNP arrays, three recent studies also identified CNVs in large cohorts of non-syndromic TOF patients [17–19]. Observing the overlap between these studies with hundreds of cases revealed only one locus (1q21.1) affected in 11 patients (Figure 1), which underlines the heterogeneous genetic background of non-syndromic TOF.

As an alternative to the conventional SNP arrays, next-generation sequencing (NGS) technologies have been widely used to detect single or short sequence variations. The obtained sequence data can also be used to find larger CNVs. Depending on the sequencing technologies, there are different computational approaches for detecting copy numbers from NGS data. For exome sequencing or targeted resequencing, the read-depth or depth of coverage approach is widely used. It assumes that the mapped reads are randomly distributed across the reference genome or targeted regions. Based on this assumption, the read-depth approach analyses differences from the expected read distribution to detect duplications (higher read depth) and deletions (lower read depth) [20]. Applying this approach, several tools have been developed to identify CNVs from exome sequencing data, such as FishingCNV, CONTRA, ExomeCNV, ExomeDepth, XHMM, CoNVEX and CoNIFER [21–27].

Here, we aimed to identify copy number alterations in a small cohort of non-syndromic TOF patients based on targeted resequencing data. Assuming a heterogeneous genetic background with individual disease-relevant CNVs, we developed a novel CNV calling method based on outlier detection using Dixon's Q test and assessment of outliers using a Hidden Markov Model (HMM). For evaluation, we applied our method to a small cohort of HapMap samples and compared it to results obtained with ExomeDepth and CoNIFER. Subsequently, our method and CoNIFER were used to detect CNVs in the TOF patients. Two copy number gains were identified by both methods and are duplications in the *PRODH* gene located at the 22q11 locus. In addition, our outlier-based method found a gain in *NOTCH1* as well as in *ISL1*. All four CNVs could be validated by quantitative real-time PCR.

Materials and Methods

Ethics Statement

Studies on TOF patients were performed according to institutional guidelines of the German Heart Institute in Berlin, with approval of the ethics committee of the Charité Medical Faculty and informed written consent of patients and/or parents, kin, caretakers, or guardians on the behalf of the minors/children participants involved in our study.

TOF Samples and DNA Targeted Resequencing

Targeted resequencing was performed for eight TOF patients, which are unrelated sporadic cases with a well-defined coherent

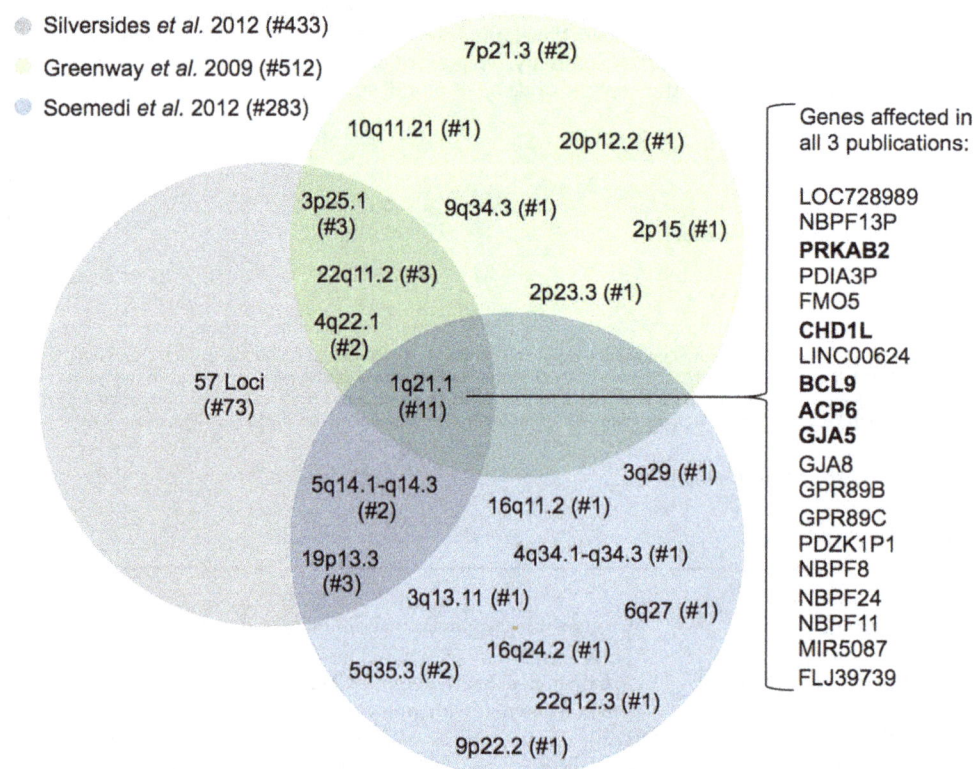

Figure 1. Overlap of three recent CNV studies in TOF patients. All three studies are based on SNP arrays. Loci with detected CNVs are depicted according to their respective cytoband. For 1q21.1, which was identified in all three studies, the RefSeq genes that are affected in at least one patient in each of the publications are listed in the order of their genomic position. Genes that are expressed in mouse heart development (E8.5–E12.0, Mouse Atlas of Gene Expression at http://www.mouseatlas.org/mouseatlas_index_html) are marked in bold. # denotes the number of individuals.

Table 1. Number and quality of 36 bp paired-end reads obtained from targeted resequencing in TOF patients using Illumina's Genome Analyzer IIx platform.

| Sample | Number of reads | Number of read pairs | Captured regions | | | |
			Phred quality score	Median coverage	Mean coverage	Target bases with ≥10x coverage
TOF-01	31,942,782	15,971,391	33.3	40	47	93.85%
TOF-02	26,970,680	13,485,340	32.7	66	76	97.70%
TOF-18	25,476,308	12,738,154	35.4	71	80	98.35%
TOF-23	20,885,192	10,442,596	35.0	60	69	97.41%
TOF-24	25,483,166	12,741,583	34.7	51	58	96.72%
TOF-25	30,551,674	15,275,837	34.6	84	92	98.91%
TOF-26	27,878,750	13,939,375	34.7	75	84	98.34%
TOF-27	24,118,022	12,059,011	34.6	78	90	98.00%

phenotype and no further anomalies. Blood samples (TOF-23, TOF-24, TOF-25, TOF-26, TOF-27) and cardiac tissue from the right ventricle (TOF-01, TOF-02, TOF-18) were collected in collaboration with the German Heart Institute in Berlin and the National Registry of Congenital Heart Disease in Berlin and used for the extraction of genomic DNA. 3–5 μg of genomic DNA were used for Roche NimbleGen sequence capturing using 365 K arrays. For array design, 867 genes and 167 microRNAs (12,910 exonic targets representing 4,616,651 target bases) were selected based on knowledge gained in various projects [28–30]. DNA enriched after NimbleGen sequence capturing was sequenced using the Illumina Genome Analyzer (GA) IIx (36 bp paired-end reads). Sequencing was performed by Atlas Biolabs (Berlin) according to manufacturers' protocols.

On average, sequencing resulted in 13,331,661 read pairs per sample (Table 1). Average read depths of 75× and base quality

scores of 34 (Phred scores) were reached in the captured regions over all samples (Table 1 and Figure 2).

HapMap Samples

We used exome sequencing data from eight HapMap individuals (NA18507, NA18555, NA18956, NA19240, NA12878, NA15510, NA18517, NA19129). The exomes were captured using Roche NimbleGen EZ Exome SeqCap Version 1 and sequencing was performed using an Illumina HiSeq 2000 platform with 50 bp paired-end reads. The exome sequence data are available from the Short Read Archive at the NCBI (SRA039053). The reads were further trimmed to 36 bp.

Outlier-based CNV Calling Method

Our CNV calling method was developed for exome or targeted resequencing data of small sets of samples (at least 3 and at most 30) assuming that the bias in the captured regions is similar in all samples enriched and sequenced with the same technology. Based on a heterogeneous genetic background in the cohort, it was further assumed that a unique disease-related copy number change is only present in very few samples.

First, read mapping and calculation of copy number values were performed for each sample separately. The sequenced reads were mapped to the targeted regions of the reference genome using BWA v.0.5.9 in paired-end mode ('sampe') with default parameters [31]. Up- and downstream, the targeted regions (usually exons) were extended by 35 bp (read length minus one base pair) to correctly capture the coverage at the start and end of a region. After mapping, the extended regions with their mapped reads were joined chromosome-wise and the tool mRCaNaVaR v0.34 [32] was used to split the joined regions into non-overlapping windows of 100 bp in length. The copy number value C for each window $W \in \{1,...,n\}$ of a sample $S \in \{1,...,n\}$ was then calculated by mRCaNaVaR using the following formula:

$$C_W^S = \frac{\text{Number of reads mapped to W}}{\text{Average number of reads mapped over all windows}} \times 2,$$

with additional GC correction [32] (Figure 3A). Reads spanning the border of two windows were assigned to the left window. In general, our method calculates a copy number value using

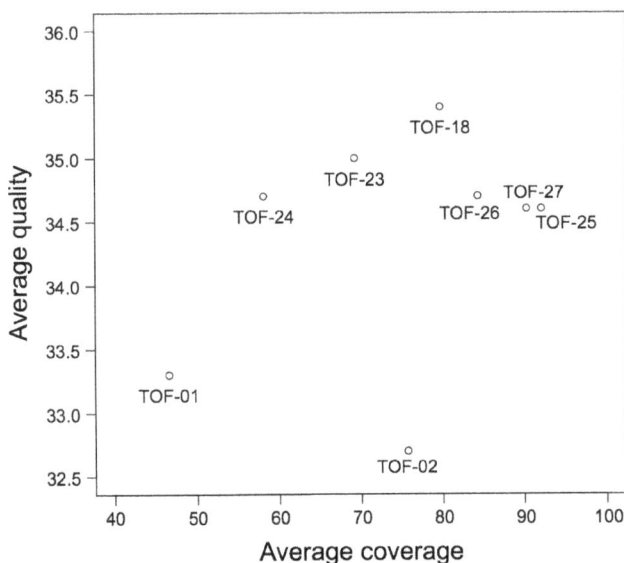

Figure 2. Base qualities versus coverage values. Scatterplot indicates the average base qualities (Phred scores) and depths of coverage for samples targeted resequenced by Illumina's Genome Analyzer IIx platform (36 bp paired-end reads).

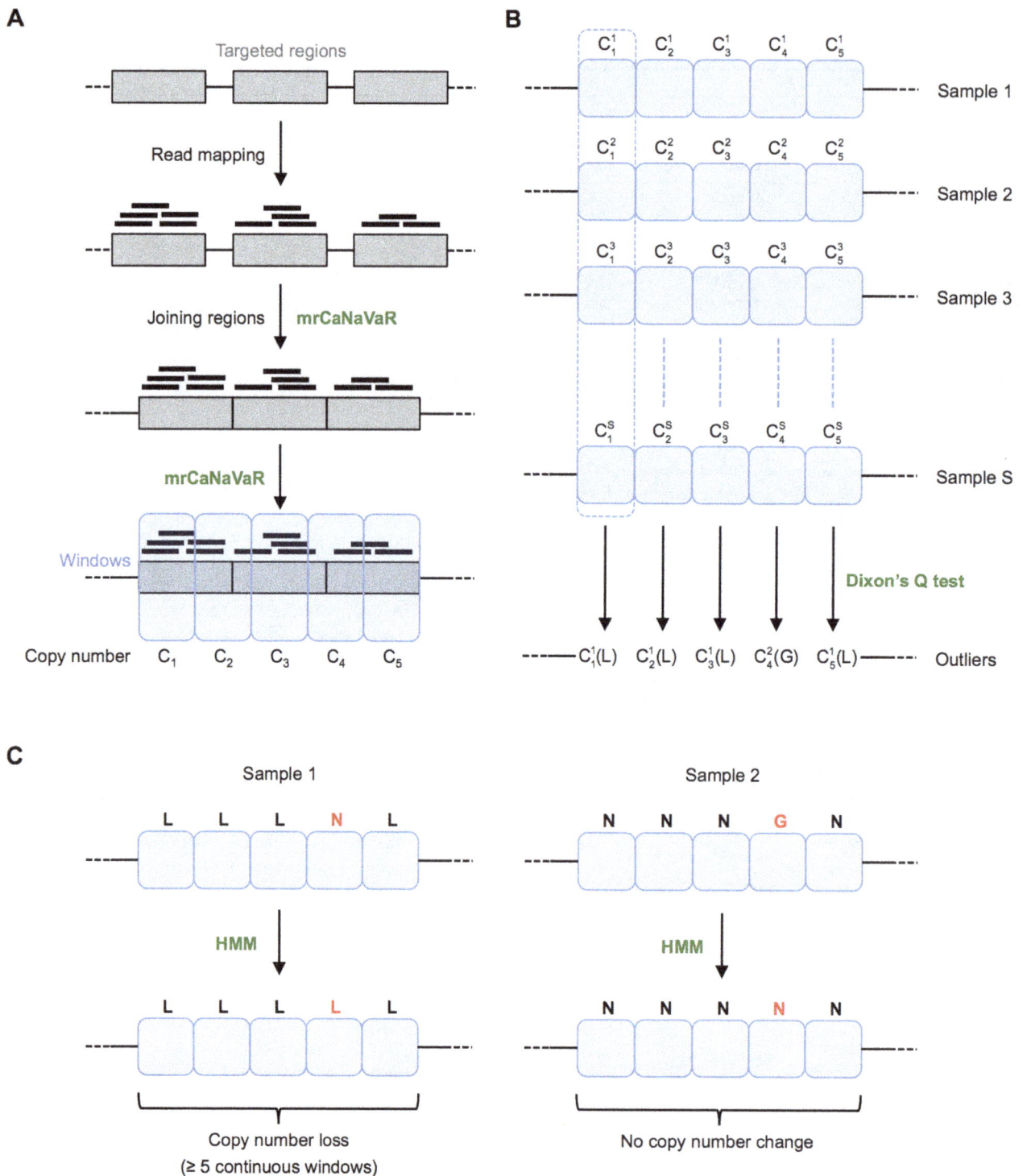

Figure 3. Outlier-based CNV calling method. (A) Read mapping and calculation of copy number value per window. Reads are mapped to extended targeted regions, which are then joined chromosome-wise. mrCaNaVaR is used to split the joined regions into windows. For each window, its copy number value is calculated by mrCaNaVaR, where C_W^S represents the value for window W in sample S. (B) Dixon's Q test is applied for each window over all samples to identify outliers. Here, sample 1 represents an outlier (loss, L) for the first, second, third and fifth window, while sample 2 represents an outlier (gain, G) for the fourth window. (C) Assessment of outliers using a Hidden Markov Model (HMM). In the given example, the fourth window of sample 1 is considered as normal (N). After applying the HMM, it will also be considered as a loss. Similarly, the fourth window of sample 2 is considered as normal after applying the HMM. A region is called as a copy number alteration, if at least five continuous windows show the same kind of change, i.e. either gain or loss.

mrCaNaVaR, which can accurately predict CNVs with at least 4x coverage [32].

Second, Dixon's Q test was applied for each window at the same position over all samples to identify gains or losses considered

Table 2. Exome sequencing-based CNV calls in HapMap samples.

Method	Number of CNVs	Validation dataset	Number of overlapping CNVs	Positive predictive value	Sensitivity
Outlier-based calling method with type10	40	3,330 arrayCGH calls	37	93%	1.1%
Outlier-based calling method with type20 including type10	65		55	85%	1.7%
CoNIFER	32		26	81%	0.8%
ExomeDepth	1,555		253	16%	7.6%

as outliers (Figure 3B). This test was introduced in 1950 for the analysis of extreme values and for the rejection of outlying values [33]. We used the formulas for r_{10} and r_{20} [34], also known as type10 and type20 in the R package 'outliers' v0.14 (http://www.R-project.org). Type10 (recommended for 3–7 samples) can only detect a single outlying window at the same genomic position over all samples, while type20 (recommended for 8–30 samples) can identify exactly two outlying windows, meaning the Q test will not detect outliers if more than 2 outliers are present. For each window, we first applied type20, however, if no two significant outliers (samples) were found, type10 was used to detect at most one outlier. Note that our method can also be applied using type10 and type20 independently. Outliers were regarded as significant with a p-value of less than or equal to 0.01. In general, the higher the p-value cutoff, the higher the number of detected outliers but also the number of false positives, i.e. the p-value is a tuning parameter for sensitivity of our method.

In the third and final step, the samples were again considered separately. For each sample, a Hidden Markov Model [35] was applied to get the most likely state of each window (i.e. gain, loss or normal). The initial transition and emission probabilities of the HMM are given in Table S1 and the values were recomputed using the Baum-Welch algorithm [36] implemented in the R package 'HMM' v1.0. The most likely sequence of the hidden states was then found by the Viterbi algorithm [37] also implemented in the R package 'HMM'. Finally, a region was called as copy number gain or loss if at least five continuous windows were considered as a gain or loss, respectively (Figure 3C). This results in a minimum size of 500 bp for detectable CNVs.

We have included a script, written in R 2.15.1 (http://www.R-project.org), for our CNV calling method based on outlier detection in exome and/or targeted resequencing data (Script S1).

CNV Validation

Genomic DNA was extracted from whole blood or cardiac biopsies using standard procedures. Quantitative real-time PCR was carried out using GoTaq qPCR Master Mix (Promega) on an ABI PRISM 7900HT Sequence Detection System (Applied Biosystems) according to the manufacturer's instructions and with normalization to the *RPPH1* gene. Primer sequences are available on request. As a reference, genomic DNA from the HapMap individual NA10851 was obtained from the Coriell Cell Repositories (New Jersey, USA).

Results and Discussion

We applied our outlier-based CNV calling method to eight HapMap control samples and intersected our exome-based calls from five of the samples with previously generated calls from high-resolution microarray-based comparative genomic hybridization (array-CGH) [2]. In addition to our method, we used the two publicly available tools ExomeDepth and CoNIFER [23,27]. Other tools such as CONTRA, FishingCNV, CoNVEX and ExomeCNV could not be applied to this dataset since they need either matched or non-matched controls.

CoNIFER (copy number inference from exome reads) is a method that combines the read-depth approach with singular value decomposition (SVD) normalization to identify rare and common copy number alterations from exome sequencing data [27]. Applying our method with type10 Dixon's Q test (assuming at most one outlier), we found 40 CNVs over the five HapMap controls (Table S2), out of which 37 regions were also identified in the array-CGH data, showing a high positive predictive value of 93%. With type20 (assuming at most two outliers), we found 65 copy number changes (Table S3), out of which 55 regions are present in the array-CGH data, resulting in a positive predictive value of 85%. Using CoNIFER, 32 CNVs were identified in the

Table 3. Targeted resequencing-based CNV calls in TOF patients.

Method	Type of variation	Position (hg19)	Length in bp	Gene	Sample
Outlier-based calling method with type20 including type10	Gain	chr5:50,689,340–50,689,940	601	*ISL1*	TOF-23
	Gain	chr9:139,402,477–139,404,228	1,752	*NOTCH1*	TOF-01
	Gain	chr22:18,900,412–18,901,127	716	*PRODH*	TOF-02
	Gain	chr22:18,910,691–18,918,575	7,885	*PRODH*	TOF-02
CoNIFER	Gain	chr22:18,900,414–18,905,939	5,526	*PRODH*	TOF-02
	Gain	chr22:18,910,575–18,923,866	13,292	*PRODH*	TOF-02

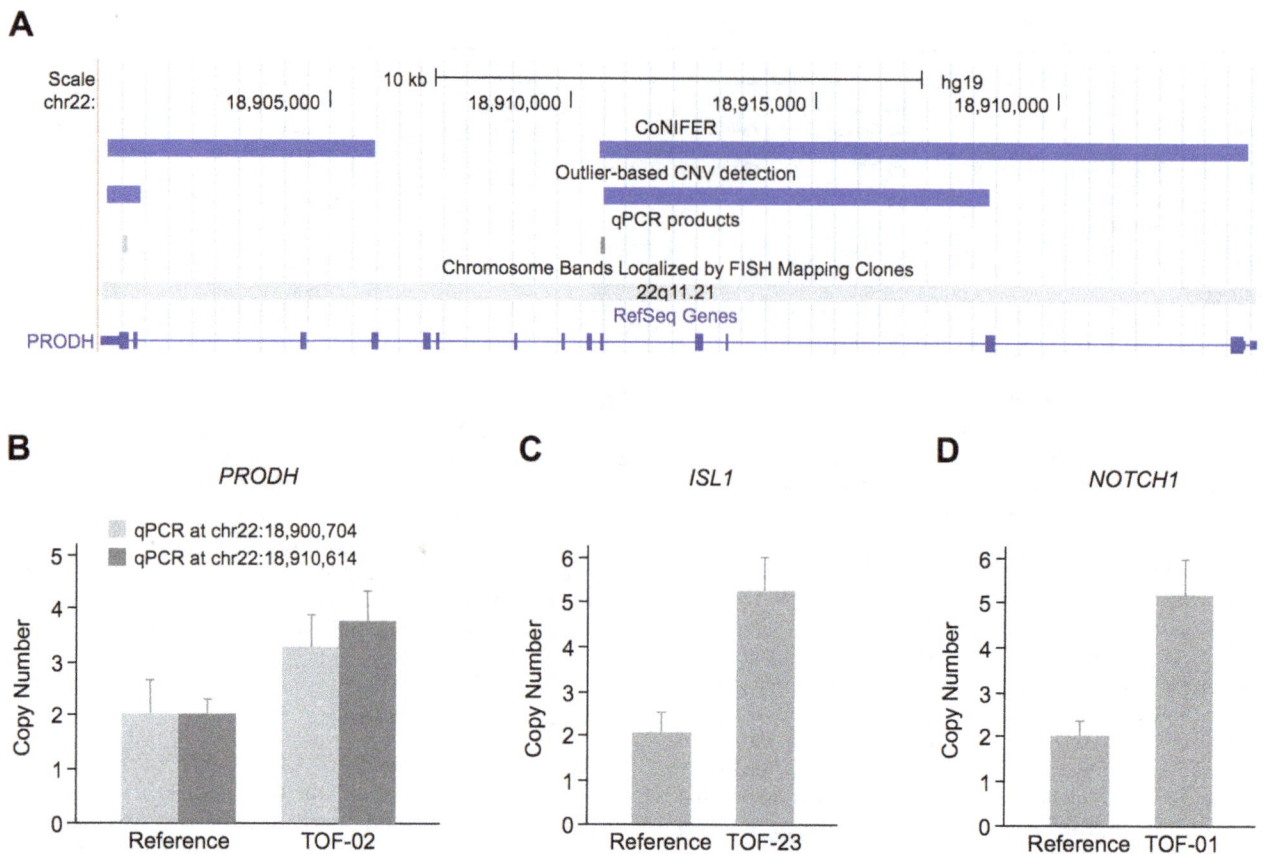

Figure 4. CNVs in TOF patients. (A) CNVs detected in *PRODH* by CoNIFER and our outlier-based CNV calling method. The duplications are depicted in the UCSC Genome Browser as blue bars. The positions of the two quantitative real-time PCR products selected for validation are shown as light and dark grey bars, respectively. (B) Quantitative real-time PCR validation of *PRODH* copy number gains. Measurement was performed at two different positions (light and dark grey bars, respectively) and normalized to the *RPPH1* gene. The HapMap individual NA10851 was used as a reference. The plot shows a representative of two independent measurements, which were each performed in triplicates. (C–D) Validation of copy number gains in *ISL1* and *NOTCH1*, respectively, that were only identified by our outlier-based CNV calling method.

five HapMap exome controls and only 26 of these regions are also present in the array-CGH data [27], which corresponds to a positive predictive value of 81% (Table 2). Comparing our results to those obtained from CoNIFER, we found that with type10 16 out of 40 regions (40%) are overlapping with regions called by CoNIFER by at least one base pair. Vice versa, 11 out of 32 regions (34%) overlap with our calls. With type20, 24 out of our 65 called regions (37%) overlap with those from CoNIFER and oppositely, 47% of the regions (15 out of 32) overlap with our calls. In general, CNV regions identified by CoNIFER are longer than those found by our method, meaning that regions called by CoNIFER can correspond to more than one of our CNVs, which explains the different overlap proportions.

Overall, our method was able to detect more copy number changes and has a higher proportion of true positives compared to CoNIFER. However, there is still a large number of CNVs observed in the array-CGH data, which were identified by neither of the two exome-based methods (Table 2). This can for example be explained by their location in segmental duplications and polymorphic but not duplicated regions [27].

ExomeDepth uses a beta-binomial model for the read count data to identify CNVs from exome sequencing data [24]. We applied ExomeDepth with default parameters to the eight HapMap samples and intersected the found CNVs from five of the samples with previously generated calls from array-CGH. In

summary, ExomeDepth found 1,555 CNVs in the five samples (median number of 286 CNVs per sample). Out of these, only 253 CNVs overlapped with 3,330 array-CGH calls, which suggest a positive predictive value of 16% and sensitivity of 7.6% (Table 2).

Interestingly, all the five rare CNVs in the five HapMap samples (see Krumm *et al.* 2012, Table S2 [27]) were found by our method, CoNIFER and ExomeDepth. Moreover, ExomeDepth identified more CNVs as compared to CoNIFER and to our method (Table 2), however; the positive predictive value is very low. Therefore, we decided not to use ExomeDepth for detecting CNVs in the TOF patients.

To identify copy number alterations in TOF patients, we applied our outlier-based method as well as CoNIFER to targeted resequencing data of our eight cases. Using our method, we found four copy number gains in three genes, namely *ISL1*, *NOTCH1* and *PRODH*. CoNIFER only identified two gains in *PRODH*, which overlap with the two regions found by our method (Table 3 and Figure 4A). We further validated all four regions identified by our method using quantitative real-time PCR (Figure 4B–D). ISL1 is a homeobox transcription factor that marks cardiovascular progenitors [38] and is known to be associated with human congenital heart disease [39]. NOTCH1 is a transmembrane receptor involved in the NOTCH signaling pathway, which plays a crucial role in heart development [40]. Mutations in *NOTCH1* are associated with a spectrum of congenital aortic valve anomalies

[41,42] and a copy number loss was identified in a patient with TOF [17] (locus 9q34.3, Figure 1). The mitochondrial protein PRODH catalyzes the first step in proline degradation and is located in the 22q11.2 locus. Deletions in this region are associated with the DiGeorge syndrome and 80% of cases harbor cardiovascular anomalies [43]. A copy number gain and two losses in the 22q11.2 locus overlapping *PRODH* were also identified in sporadic TOF patients [17,18] (Figure 1).

In summary, we developed an outlier-based CNV calling method for a small cohort size of up to 30 individuals. The exploration of the human phenotype and its genetic and molecular background is the challenge of the next century and it is already clear that more precise phenotyping will lead to smaller cohort sizes. Here, novel approaches will be of exceptional relevance. Moreover, analyzing small patient cohorts is of special interest for rare diseases with only few available patient samples. Approximately 7,000 rare diseases are currently known and together affect about 6% of the population [44]. Our method is based on the assumption that individual CNVs (outliers) are disease-relevant and can be applied to exome as well as targeted resequencing data. Both sequencing techniques achieve a high read coverage over the targeted regions. Nevertheless, there are non-uniform patterns in the read depth resulting mainly from repetitive regions. Thus, the detection of copy number alterations is limited in these genomic regions, which is shown by the high number of false negatives compared to array-CGH [27].

We evaluated our method using publicly available data of eight HapMap samples and subsequently applied it to a small number of TOF patients. Compared to CoNIFER we identified more CNVs in both the HapMap samples as well as in our TOF cohort. In general, our method assumes a uniform read distribution over all exons of all individuals enriched and sequenced with the same technology to compare read counts between all samples to detect outliers. In contrast, CoNIFER considers the read depth across all individuals after SVD normalization. This difference is also reflected by the overlap of their calls in the eight HapMap samples. Although the general overlap is relatively low, we were able to identify all rare CNVs detected by CoNIFER. In addition to searching for rare CNVs, we also found a subset of common CNVs called by CoNIFER. This might be explained by variations present in only one or two of the eight individuals, but defined as common based on their frequency in a larger population.

In our TOF cohort comprising eight cases, we found four copy number gains in three patients, while CoNIFER only detected two

of the gains in one patient. All four gains could be validated and in addition, the three genes affected by the CNVs are important regulators of heart development (*NOTCH1*, *ISL1*) or are located in a region associated with cardiac malformations (*PRODH*). Two of the variations also overlap with copy number alterations in TOF patients previously identified by array-CGH [17,18]. Taken together, this illustrates the advantage of using an outlier-based detecting method in a small cohort with a heterogeneous genetic background. Thus, our method is of special interest for small cohorts of specific phenotypes like rare diseases. Moreover, it can be used for the discovery of individual CNVs within families and *de novo* CNVs in trios.

Acknowledgments

We are deeply grateful to the TOF patients and families for their cooperation. We thank the German Heart Institute Berlin (Berlin, Germany) and the National Registry of Congenital Heart Disease (Berlin, Germany) for sample contribution. We further thank Ilona Dunkel for sample preparation. We also thank Biostar (www.biostars.org) and Cross Validated Stack Exchange (www.stats.stackexchange.com) for providing supporting discussion platforms.

Author Contributions

Conceived and designed the experiments: SRS. Performed the experiments: CD. Analyzed the data: VB MG. Contributed reagents/materials/analysis tools: FB RH SK SRS. Wrote the paper: CD MG VB.

References

1. Feuk L, Carson AR, Scherer SW (2006) Structural variation in the human genome. Nat Rev Genet 7: 85–97. doi:10.1038/nrg1767.

2. Conrad DF, Pinto D, Redon R, Feuk L, Gokcumen O, et al. (2010) Origins and functional impact of copy number variation in the human genome. 464: 704–712. Available: http://eutils.ncbi.nlm.nih.gov/entrez/eutils/elink.fcgi?dbfrom=pubmed&id=19812545&retmode=ref&cmd=prlinks.

3. Fellermann K, Stange DE, Schaeffeler E, Schmalzl H, Wehkamp J, et al. (2006) A chromosome 8 gene-cluster polymorphism with low human beta-defensin 2 gene copy number predisposes to Crohn disease of the colon. 79: 439–448. Available: http://eutils.ncbi.nlm.nih.gov/entrez/eutils/elink.fcgi?dbfrom=pubmed&id=16909382&retmode=ref&cmd=prlinks.

4. de Vries BBA, Pfundt R, Leisink M, Koolen DA, Vissers LELM, et al. (2005) Diagnostic genome profiling in mental retardation. 77: 606–616. Available: http://eutils.ncbi.nlm.nih.gov/entrez/eutils/elink.fcgi?dbfrom=pubmed&id=16175506&retmode=ref&cmd=prlinks.

5. Thienpont B, Mertens L, de Ravel T, Eyskens B, Boshoff D, et al. (2007) Submicroscopic chromosomal imbalances detected by array-CGH are a frequent cause of congenital heart defects in selected patients. Eur Heart J 28: 2778–2784. doi:10.1093/eurheartj/ehl560.

6. Erdogan F, Larsen LA, Zhang L, Tümer Z, Tommerup N, et al. (2008) High frequency of submicroscopic genomic aberrations detected by tiling path array comparative genome hybridisation in patients with isolated congenital heart disease. J Med Genet 45: 704–709. doi:10.1136/jmg.2008.058776.

7. Hoffman JIE, Kaplan S (2002) The incidence of congenital heart disease. J Am Coll Cardiol 39: 1890–1900.

8. Reller MD, Strickland MJ, Riehle-Colarusso T, Mahle WT, Correa A (2008) Prevalence of congenital heart defects in metropolitan Atlanta, 1998–2005. J Pediatr 153: 807–813. doi:10.1016/j.jpeds.2008.05.059.

9. Ferencz C, Rubin JD, McCarter RJ, Brenner JI, Neill CA, et al. (1985) Congenital heart disease: prevalence at livebirth. The Baltimore-Washington Infant Study. American journal of epidemiology 121: 31–36. Available: http://eutils.ncbi.nlm.nih.gov/entrez/eutils/elink.fcgi?dbfrom=pubmed&id=3964990&retmode=ref&cmd=prlinks.

10. Apitz C, Webb G (2009) ScienceDirect.com - The Lancet - Tetralogy of Fallot. Available: http://www.sciencedirect.com/science/article/pii/s0140-6736(09)60657-7.

11. Fahed AC, Gelb BD, Seidman JG, Seidman CE (2013) Genetics of congenital heart disease: the glass half empty. Circ Res 112: 707–720. doi:10.1161/CIRCRESAHA.112.300853.

12. Goldmuntz E, Clark BJ, Mitchell LE, Jawad AF, Cuneo BF, et al. (1998) Frequency of 22q11 deletions in patients with conotruncal defects. J Am Coll Cardiol 32: 492–498.

13. Cuturilo G, Menten B, Krstic A, Drakulic D, Jovanovic I, et al. (2011) 4q34.1-q35.2 deletion in a boy with phenotype resembling 22q11.2 deletion syndrome. 170: 1465–1470. Available: http://eutils.ncbi.nlm.nih.gov/entrez/eutils/elink.fcgi?dbfrom=pubmed&id=21833498&retmode=ref&cmd=prlinks.

14. Luo H, Xie L, Wang S-Z, Chen J-L, Huang C, et al. (2012) Duplication of 8q12 encompassing CHD7 is associated with a distinct phenotype but without duane anomaly. 55: 646–649. Available: http://eutils.ncbi.nlm.nih.gov/entrez/eutils/elink.fcgi?dbfrom = pubmed&id = 22902603&retmode = ref&cmd = prlinks.

15. Luo C, Yang Y-F, Yin B-L, Chen J-L, Huang C, et al. (2012) Microduplication of 3p25.2 encompassing RAF1 associated with congenital heart disease suggestive of Noonan syndrome. 158A: 1918–1923. Available: http://eutils. n c b i . n l m . n i h . g o v / e n t r e z / e u t i l s / e l i n k . fcgi?dbfrom = pubmed&id = 22786616&retmode = ref&cmd = prlinks.

16. Nora JJ (1968) Multifactorial inheritance hypothesis for the etiology of congenital heart diseases. The genetic-environmental interaction. Circulation 38: 604–617.

17. Greenway SC, Pereira AC, Lin JC, DePalma SR, Israel SJ, et al. (2009) De novo copy number variants identify new genes and loci in isolated sporadic tetralogy of Fallot. Nat Genet 41: 931–935. doi:10.1038/ng.415.

18. Silversides CK, Lionel AC, Costain G, Merico D, Migita O, et al. (2012) Rare copy number variations in adults with tetralogy of Fallot implicate novel risk gene pathways. PLoS Genet 8: e1002843. doi:10.1371/journal.pgen.1002843.

19. Soemedi R, Wilson IJ, Bentham J, Darlay R, Töpf A, et al. (2012) Contribution of Global Rare Copy-Number Variants to the Risk of Sporadic Congenital Heart Disease. The American Journal of Human Genetics 91: 489–501. doi:10.1016/j.ajhg.2012.08.003.

20. Alkan C, Coe BP, Eichler EE (2011) Genome structural variation discovery and genotyping. Nat Rev Genet 12: 363–376. doi:10.1038/nrg2958.

21. Shi Y, Majewski J (2013) FishingCNV: a graphical software package for detecting rare copy number variations in exome-sequencing data. Bioinformatics 29: 1461–1462. doi:10.1093/bioinformatics/btt151.

22. Li J, Lupat R, Amarasinghe KC, Thompson ER, Doyle MA, et al. (2012) CONTRA: copy number analysis for targeted resequencing. Bioinformatics 28: 1307–1313. doi:10.1093/bioinformatics/bts146.

23. Sathirapongsasuti JF, Lee H, Horst BAJ, Brunner G, Cochran AJ, et al. (2011) Exome sequencing-based copy-number variation and loss of heterozygosity detection: ExomeCNV. 27: 2648–2654. Available: http://eutils.ncbi.nlm.nih. g o v / e n t r e z / e u t i l s / e l i n k . fcgi?dbfrom = pubmed&id = 21828086&retmode = ref&cmd = prlinks.

24. Plagnol V, Curtis J, Epstein M, Mok KY, Stebbings E, et al. (2012) A robust model for read count data in exome sequencing experiments and implications for copy number variant calling. Bioinformatics 28: 2747–2754. doi:10.1093/bioinformatics/bts526.

25. Fromer M, Moran JL, Chambert K, Banks E, Bergen SE, et al. (2012) Discovery and statistical genotyping of copy-number variation from whole-exome sequencing depth. Am J Hum Genet 91: 597–607. doi:10.1016/j.ajhg.2012.08.005.

26. Amarasinghe KC, Li J, Halgamuge SK (2013) CoNVEX: copy number variation estimation in exome sequencing data using HMM. BMC Bioinformatics 14 Suppl 2: S2. doi:10.1186/1471-2105-14-S2-S2.

27. Krumm N, Sudmant PH, Ko A, O'Roak BJ, Malig M, et al. (2012) Copy number variation detection and genotyping from exome sequence data. 22: 1525–1532. Available: http://eutils.ncbi.nlm.nih.gov/entrez/eutils/elink.fcgi?dbfrom = pubmed&id = 22585873&retmode = ref&cmd = prlinks.

28. Kaynak B, Heydebreck von A, Mebus S, Seelow D, Hennig S, et al. (2003) Genome-wide array analysis of normal and malformed human hearts. Circulation 107: 2467–2474. Available: http://eutils.ncbi.nlm.nih.gov/entrez/ e u t i l s / e l i n k . fcgi?dbfrom = pubmed&id = 12742993&retmode = ref&cmd = prlinks.

29. Toenjes M, Schueler M, Hammer S, Pape UJ, Fischer JJ, et al. (2008) Prediction of cardiac transcription networks based on molecular data and complex clinical phenotypes. Mol Biosyst 4: 589–598. Available: http://eutils.ncbi.nlm.nih.gov/ e n t r e z / e u t i l s / e l i n k . fcgi?dbfrom = pubmed&id = 18493657&retmode = ref&cmd = prlinks.

30. Schlesinger J, Schueler M, Grunert M, Fischer JJ, Zhang Q, et al. (2011) The cardiac transcription network modulated by Gata4, Mef2a, Nkx2.5, Srf, histone modifications, and microRNAs. PLoS Genet 7: e1001313. doi:10.1371/journal.pgen.1001313.

31. Li H, Durbin R (2009) Fast and accurate short read alignment with Burrows-Wheeler transform. 25: 1754–1760. Available: http://eutils.ncbi.nlm.nih.gov/ e n t r e z / e u t i l s / e l i n k . fcgi?dbfrom = pubmed&id = 19451168&retmode = ref&cmd = prlinks.

32. Alkan C, Kidd JM, Marques-Bonet T, Aksay G, Antonacci F, et al. (2009) Personalized copy number and segmental duplication maps using next-generation sequencing. Nat Genet 41: 1061–1067. doi:10.1038/ng.437.

33. Dixon WJ (1950) Analysis of extreme values. Available: http://www.jstor.org/stable/10.2307/2236602.

34. Rorabacher DB (1991) Statistical treatment for rejection of deviant values: critical values of Dixon's 'Q' parameter and related subrange ratios at the 95% confidence level - Analytical Chemistry (ACS Publications). Available: http://pubs.acs.org/doi/abs/10.1021/ac00002a010.

35. Rabiner LR (1989) A tutorial on hidden Markov models and selected applications in speech recognition. 77: 257–286. Available: http://ieeexplore. ieee.org/lpdocs/epic03/wrapper.htm?arnumber = 18626.

36. Baum LE, Petrie T, Soules G, Weiss N (1970) A maximization technique occurring in the statistical analysis of probabilistic functions of Markov chains. Available: http://www.jstor.org/stable/10.2307/2239727.

37. Viterbi A (1967) Error bounds for convolutional codes and an asymptotically optimum decoding algorithm. Available: http://ieeexplore.ieee.org/xpls/abs_all.jsp?arnumber = 1054010.

38. Bu L, Jiang X, Martin-Puig S, Caron L, Zhu S, et al. (2009) Human ISL1 heart progenitors generate diverse multipotent cardiovascular cell lineages. 460: 113–117. Available: http://pubget.com/site/paper/19571884?institution = .

39. Stevens KN, Hakonarson H, Kim CE, Doevendans PA, Koeleman BPC, et al. (2009) Common Variation in ISL1 Confers Genetic Susceptibility for Human Congenital Heart Disease. 5: e10855–e10855. Available: http://pubget.com/site/paper/20520780?institution = .

40. Nemir M, Pedrazzini T (2008) Functional role of Notch signaling in the developing and postnatal heart. 45: 10–10. Available: http://pubget.com/site/paper/18410944?institution = .

41. Garg V, Muth AN, Ransom JF, Schluterman MK, Barnes R, et al. (2005) Mutations in NOTCH1 cause aortic valve disease. Nature 437: 270–274. doi:10.1038/nature03940.

42. Mohamed SA, Aherrahrou Z, Liptau H, Erasmi AW, Hagemann C, et al. (2006) Novel missense mutations (p.T596M and p.P1797H) in NOTCH1 in patients with bicuspid aortic valve. 345: 1460–1465. Available: http://eutils.ncbi.nlm. n i h . g o v / e n t r e z / e u t i l s / e l i n k . fcgi?dbfrom = pubmed&id = 16729972&retmode = ref&cmd = prlinks.

43. Momma K (2010) Cardiovascular anomalies associated with chromosome 22q11.2 deletion syndrome. Am J Cardiol 105: 1617–1624. doi:10.1016/j.amjcard.2010.01.333.

44. Humphreys G (2012) Coming together to combat rare diseases. Bull World Health Organ 90: 406–407. doi:10.2471/BLT.12.020612.

17

From Days to Hours: Reporting Clinically Actionable Variants from Whole Genome Sequencing

Sumit Middha[9], Saurabh Baheti[9], Steven N. Hart, Jean-Pierre A. Kocher*

Division of Biomedical Statistics and Informatics, Department of Health Sciences Research, Mayo Clinic, Rochester, Minnesota, United States of America

Abstract

As the cost of whole genome sequencing (WGS) decreases, clinical laboratories will be looking at broadly adopting this technology to screen for variants of clinical significance. To fully leverage this technology in a clinical setting, results need to be reported quickly, as the turnaround rate could potentially impact patient care. The latest sequencers can sequence a whole human genome in about 24 hours. However, depending on the computing infrastructure available, the processing of data can take several days, with the majority of computing time devoted to aligning reads to genomics regions that are to date not clinically interpretable. In an attempt to accelerate the reporting of clinically actionable variants, we have investigated the utility of a multi-step alignment algorithm focused on aligning reads and calling variants in genomic regions of clinical relevance prior to processing the remaining reads on the whole genome. This iterative workflow significantly accelerates the reporting of clinically actionable variants with no loss of accuracy when compared to genotypes obtained with the OMNI SNP platform or to variants detected with a standard workflow that combines Novoalign and GATK.

Editor: Charles Y. Chiu, University of California, San Francisco, United States of America

Funding: The funding was provided by the Center for Individualized Medecine at Mayo Clinic. The funders had no role in study design, data collection and analysis, decision to publish, or preparation of the manuscript.

Competing Interests: The authors have declared that no competing interests exist.

* E-mail: kocher.jeanpierre@mayo.edu

9 These authors contributed equally to this work.

Introduction

Whole Genome Sequencing has the potential to transform diagnostic testing in the very near future. As the cost of sequencing continues to decrease, the broader adoption of this protocol by clinical laboratories is expected. Sequencing platforms are being redesigned to accelerate the sequencing of whole genomes. For instance, the Illumina Hi-Seq 2500 platform can perform this task in 24 hours, shifting the rate-limiting step to data processing. The computationally-expensive step of aligning millions of short reads to the whole genome could be prohibitive for routine use of WGS in a clinical setting where the speed of analysis can impact patient outcome. Clinical applicability can be improved by prioritizing WGS variant reporting based on relevance for clinical decision-making. Currently, most of the clinically relevant genomics information is related to protein-coding exome regions [1] where the impact of coding variants can be interpreted in the context of proteins and their function [2,3]. This current focus opens opportunities to develop new bioinformatics algorithms that prioritize and swiftly report clinically relevant findings.

Recently, an ultra-fast preprocessing workflow was published: ISAAC [4]. This workflow completes the whole genome alignment and variant calling in 7–8 hours. Although, ISAAC is the fastest solution currently to our knowledge, its deployment requires specific hardware and is, at least for now, limited to Illumina sequencing data.

In this manuscript we explore another approach that does not require specific hardware or software solution and is independent of the next generation sequencing platform used. Instead of expediting the whole alignment and calling process, our proposed approach prioritizes read alignment and variant calling in genomic regions of clinical relevance (referred to as the Target Reference Genome) before reporting variants in genomic regions of lower clinical significance. The proposed workflow operates in three steps. First, clinically relevant reads are selected by aligning all the sequencing data to the Target Reference Genome. Then, this reduced set of aligned reads is aligned to the whole reference genome to correct for alignment artifacts. These artifacts arise from reads forcibly aligned to the Target Reference Genome that align more accurately to non-targeted regions. After the second alignment step, reads that remain aligned on the Target Reference Genome are re-aligned and recalibrated followed by variant calling. Variants are immediately reported to clinical experts for interpretation and decision support. The final step, which can be deferred or executed at a slower pace, handles the remaining reads that are aligned on the whole reference genome.

The gain of reporting speed obtained with this iterative workflow is due to the significantly smaller size of the Target Reference Genome compared to the whole reference genome. If the targeted region corresponds to the whole exome, read alignment in the first step would be limited to less than 2% of the reference genome. Similarly, assuming even coverage, only 2% of the reads will be aligned on the whole reference genome in the second step.

Although conceptually very simple and straightforward to implement, the question of results accuracy remains to be addressed. In this manuscript, we compare results obtained by the target workflow with a generic whole genome sequencing workflow. In the process, we have also compared the impact on our iterative workflow of two aligners, BWA [5,6] and Novoalign (http://www.novocraft.com/), on results accuracy.

Materials and Methods

Datasets

To test out approach, we selected a CEPH family trio from the 1000 Genomes project [7] consisting of NA12878 (Child), NA12891 (Father), and NA12892 (Mother). Each sample was sequenced using the Illumina Next Generation Sequencing Platform (HiSeq 2000) with the pair-end protocol that produced on average 397 bp long sequence fragments from which 100 bp were sequenced at both ends. Sequencing of these samples resulted in more than 2.4 billion 100 bp long reads with an average coverage of 80x across the entire genome. The Binary Alignment Map (BAM) files obtained for these samples were converted to FASTQ reads format for further analysis. The same individuals have been genotyped with a combination of Illumina and Affymetrix SNP chips for HapMap Phase III [8]. This genotype data was used to validate variants calls from sequencing data.

Data Availability

– Sequencing data: ftp://ftp-trace.ncbi.nih.gov/1000genomes/ftp/technical/working/20120117_ceu_trio_b37_decoy/.

– Genotyping data: ftp://ftp.ncbi.nlm.nih.gov/hapmap/genotypes/2010-08_phaseII+III/forward/.

Target Reference Genome

We arbitrarily selected a set of 2638 clinically relevant genes from the Clinical Genomic Database [9]. It should be noted a smaller set similar to clinical gene panels could have been selected as well. The 2638 genes include 42048 unique exons in the UCSC RefFlat annotation. The average length of these exons is 280 bp with a standard deviation of 635 bp. The boundary of each exon was extended by 550 bp to account for the sequencing protocol that produced 100 bp long paired-end read from about 400 bp long sequence fragments. From the fragment length distribution, we estimated that an average of 0.39% read pairs would not be fully aligned with the 550 bp cutoff. The extended sequence of each exon was extracted from the Human Reference Genome (Build 37) and concatenated into a single Target Reference Genome fasta file.

Standard sequence alignment and variant calling workflow

As the standard whole genome alignment workflow, we used Novoalign for initial alignment of sequence reads followed by GATK for re-alignment, re-calibration and variant calling (Figure 1).

Iterative workflow

The different steps of the iterative workflow are displayed in Figure 1. The first step filters out the reads that do not map on the Target Reference Genome while the second step refines the alignment of the mapped reads by aligning them on to the Human Reference Genome. As previously explained, this step eliminates reads that have been forcibly mapped on the Target Reference Genome but would have aligned more accurately to another location of the Human Reference Genome. Since the first alignment step produced a BAM file with mapped reads

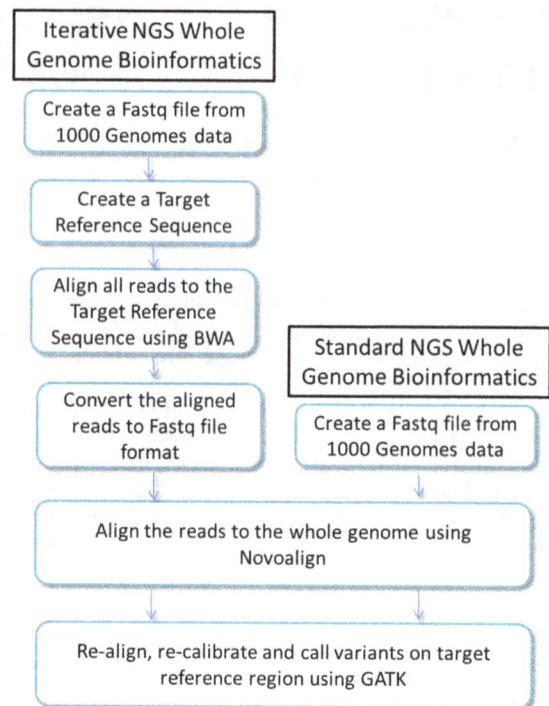

Figure 1. Basic components of the iterative workflow as compared to a standard NGS whole genome analysis.

information, the BAM file was converted in FASTQ format to perform the second alignment step.

We tested the iterative workflow with two aligners BWA and Novoalign. BWA is known as being faster that Novoalign however, from our internal benchmark Novoalign produces slightly better read alignments. Since the workflow includes two alignment steps, we ran the workflow with different combinations of the two aligners. For any investigated combination of aligners, GATK was used to call variants.

Results

The CEPH family FASTQ files were processed with both the standard and iterative workflows. The two sets of results were compared with the genotypes obtained from the OMNI SNP platform reported by 1000 Genomes project. No additional processing was done on these reported data that were used as the gold standard in this study. 18634 OMNI genotypes included in the Target Reference Genome were used for accuracy estimates.

Results accuracy estimated from genotype calls

Using the 18634 genotypes of the OMNI SNP platform as 'truth', we assessed the accuracy of genotypes called by the standard workflow and the iterative workflow (Table 1). The iterative workflow results were produced with different combinations of aligners. Apart from one SNP on chromosome Y, all genotypes had adequate coverage. Results in Table 1 highlight that the BWA-Novoalign workflow has slightly higher performance accuracy than the Novoalign-Novoalign workflow. Although not necessarily significant, this result suggests that the accuracy difference that we have observed between BWA and Novoalign in the first alignment steep has little impact on the quality of the final results. However, since BWA is significantly

Table 1. Concordance of SNP data with variants from standard and iterative workflows for sample NA12878.

Workflow	Aligner used in step 1	Aligner used in step 2	Number of Concordant SNVs	Number of Discordant SNVs	% Concordance	Execution time (hrs)
Standard	Novoalign	-	18344	130	99.29	73.86
Iterative	BWA	BWA	17459	947	94.88	3.09
Iterative	**BWA**	**Novoalign**	**18435**	**129**	**99.30**	**4.98**
Iterative	Novoalign	Novoalign	18435	129	99.30	14.09
Iterative	Novoalign	BWA	18324	172	99.07	10.03

faster than Novoalign, the BWA-Novoalign workflow completes the task more than 5 times faster than the Novoalign-Novoalign workflow. Based on these findings, our remaining analysis is limited to the results obtained with the BWA-Novoalign workflow.

Genotyping calls missed by the standard and iterative workflows

About 99.3% of the genotypes called accurately by both workflows. When comparing the overlap between the 0.7% miscalled genotypes (i.e. 130 with the standard workflow and 129 with the iterative workflow), all but one of the genotypes were identical. This result reinforces the very similar performance of the two workflows and suggests that no significant bias was introduced by the iterative approach.

Performance accuracy of iterative and standard workflows on SNVs and indels

We demonstrated that both the standard and iterative workflows had similar accuracy when compared to the OMNI SNP genotype calls. We then investigated the overlap between all the variants reported by the standard and the iterative workflow. These variants include single nucleotide variants (SNVs) and insertions/deletions (indels). The large majority of the variants were called by both workflows (Table 2). We further analyzed discordant variants not called by the two workflows. Using the basic quality metrics of quality-by-depth (QD), strand bias and low read-depth coverage of less than 10 we observed that the majority of discordant variants had poor quality. For reference, less than 3% (236 out of 8754) of the concordant SNVs had QD<5 or strand bias. We reviewed the 35 (out of 118) exclusive SNV/indel variant calls with QD>5. Out of these 35 variants, 19 have a clear strand bias. Of the remaining 16, 9 have a low coverage depth of less than 10 reads and 6 fall in a region with multiple (>=5) homologous regions in the whole genome. This leaves just one exclusive variant of good quality that was called by our iterative workflow but not called by the standard workflow. Thus, we concluded that variants exclusively called by only one of the approaches are of low quality.

Importance of the second alignment step

We explore the contribution of the second alignment step to the accuracy of variant calling. When using our iterative workflow, 27.5% of the reads aligned from 1st step to the CGD genes are aligned to a different location in the 2nd step. When calling variants directly after the first alignment step, only 83.25% concordance is obtained with the SNP chip data compared to 99.3% concordance when the reads are processed by the second alignment step. We also observed that more than 15,000 exclusive variants are reported after the 1st alignment step, this number

Table 2. Evaluation of SNVs and Indels called by the iterative and standard workflow.

Workflow	Variant	Type	NA12878	NA12891	NA12892
Iterative	SNVs	Shared	8754	8506	8809
Standard	SNVs	Shared	8754	8506	8809
Iterative	SNVs	Exclusive	38	34	39
Standard	SNVs	Exclusive	62	57	70
Iterative	INDELs	Shared	975	902	905
Standard	INDELs	Shared	975	902	905
Iterative	INDELs	Exclusive	5	5	9
Standard	INDELs	Exclusive	13	11	14

dropping to 100 after the second alignment step. The second alignment step in the iterative workflow is therefore critical for accurate variant calling.

Reporting speed of clinically relevant variants

As shown in Table 1, the preferred iterative workflow takes less than 5 CPU hours to complete the alignment on the target reference genome and calling of the variants. The alignment of the remaining reads and variant calling took ~71 CPU hours. A total of ~76 CPU hours was therefore needed to complete the full preprocessing of the whole genome experiment. In comparison, it also took ~76 CPU hours for the standard workflow to complete. We believe that this CPU overhead is acceptable in a clinical setup where the fast reporting of clinical variants could have a critical impact on patient's fate.

As a test, we extended Target Reference Genome to include all gene exons. The variants calls were reported in ~15 CPU hours, still an acceptable time compared to the 76 CPU hours needed for alignment of the whole genome using standard workflow.

Discussion

We have developed and tested an iterative whole genome sequencing workflow designed to rapidly report variants in target genomic locations. The approach first focuses on aligning all the sequence reads on the target genomic locations and then realigning this subset of mapped reads to the reference genome. We benchmarked the accuracy of the iterative workflow against genotype data used a gold standard and also compared reported SNVs to those reported by our standard whole genome sequencing workflow. Our results indicate that the standard and iterative workflows performed similarly well, with 99.3% accurate genotypes called The overlap between any variants (SNVs and Indels) called by the standard and iterative workflow is also very high (98.8%), with most of the non-concordant calls being of low confidence (low QD score).

From this analysis, we can conclude that the iterative approach does not introduce significant noise or bias that would have a negative impact on the downstream calling of variants. With regards to time, using the Target Reference Genome, which included 2638 genes, allowed for the reporting of variants called in these regions in less than 5 hours. When extending the alignment to the whole exome, results were obtained in ~76 hours.

This iterative workflow can be particularly useful clinically when only a limited set of actionable variants need to be rapidly reported to clinicians. As compared to other published approaches, our iterative workflow does not require any additional investment in software or hardware. It is independent of the sequenced organism and the sequencing platform used as long as a reference genome is used to align the reads. Moreover, the iterative workflow can be implemented with any aligner or target reference region to swiftly report variants in those regions from whole genome sequencing data.

Finally, the third step of the alignment, which consists of aligning the remaining reads, is the most time consuming. Interestingly, in our example, these reads are now naturally organized in independent islands covering the intergenic and intronic regions of the genome, facilitating the parallel processing of read realignment in these regions. Parallelization could be a means to significantly accelerate this final step. This option, however, was not investigated in this study.

Acknowledgments

The authors thank the Center for Individualized Medicine at Mayo Clinic for funding this work.

Author Contributions

Conceived and designed the experiments: JPK SH. Performed the experiments: SM SB. Analyzed the data: JPK SM. Wrote the paper: JPK SM.

References

1. Choi M, Scholl UI, Ji W, Liu T, Tikhonova IR, et al. (2009) Genetic diagnosis by whole exome capture and massively parallel DNA sequencing. Proc Natl Acad Sci U S A 106: 19096–19101.
2. Berg JS, Khoury MJ, Evans JP (2011) Deploying whole genome sequencing in clinical practice and public health: Meeting the challenge one bin at a time. Genet Med 13: 499–504.
3. Kohane IS, Masys DR, Altman RB (2006) The incidentalome: a threat to genomic medicine. JAMA: the journal of the American Medical Association 296: 212–215.
4. Raczy C, Petrovski R, Saunders CT, Chorny I, Kruglyak S, et al. (2013) Isaac: ultra-fast whole-genome secondary analysis on Illumina sequencing platforms. Bioinformatics 29: 2041–2043.
5. Li H, Durbin R (2010) Fast and accurate long-read alignment with Burrows-Wheeler transform. Bioinformatics 26: 589–595.
6. Li H, Durbin R (2009) Fast and accurate short read alignment with Burrows-Wheeler transform. Bioinformatics 25: 1754–1760.
7. Durbin RM, Abecasis GR, Altshuler DL, Auton A, Brooks LD, et al. (2010) A map of human genome variation from population-scale sequencing. Nature 467: 1061–1073.
8. Altshuler DM, Gibbs RA, Peltonen L, Dermitzakis E, Schaffner SF, et al. (2010) Integrating common and rare genetic variation in diverse human populations. Nature 467: 52–58.
9. Solomon BD, Nguyen A–D, Bear KA, Wolfsberg TG (2013) Clinical Genomic Database. Proceedings of the National Academy of Sciences. 110: 9851–5.

18

CUSHAW3: Sensitive and Accurate Base-Space and Color-Space Short-Read Alignment with Hybrid Seeding

Yongchao Liu[1]*, Bernt Popp[2], Bertil Schmidt[1]*

1 Institut für Informatik, Johannes Gutenberg Universität Mainz, Mainz, Germany, **2** Institute of Human Genetics, University of Erlangen-Nuremberg, Erlangen, Germany

Abstract

The majority of next-generation sequencing short-reads can be properly aligned by leading aligners at high speed. However, the alignment quality can still be further improved, since usually not all reads can be correctly aligned to large genomes, such as the human genome, even for simulated data. Moreover, even slight improvements in this area are important but challenging, and usually require significantly more computational endeavor. In this paper, we present CUSHAW3, an open-source parallelized, sensitive and accurate short-read aligner for both base-space and color-space sequences. In this aligner, we have investigated a hybrid seeding approach to improve alignment quality, which incorporates three different seed types, i.e. maximal exact match seeds, exact-match *k*-mer seeds and variable-length seeds, into the alignment pipeline. Furthermore, three techniques: weighted seed-pairing heuristic, paired-end alignment pair ranking and read mate rescuing have been conceived to facilitate accurate paired-end alignment. For base-space alignment, we have compared CUSHAW3 to Novoalign, CUSHAW2, BWA-MEM, Bowtie2 and GEM, by aligning both simulated and real reads to the human genome. The results show that CUSHAW3 consistently outperforms CUSHAW2, BWA-MEM, Bowtie2 and GEM in terms of single-end and paired-end alignment. Furthermore, our aligner has demonstrated better paired-end alignment performance than Novoalign for short-reads with high error rates. For color-space alignment, CUSHAW3 is consistently one of the best aligners compared to SHRiMP2 and BFAST. The source code of CUSHAW3 and all simulated data are available at http://cushaw3.sourceforge.net.

Editor: Oliver Hofmann, Harvard School of Public Health, United States of America

Funding: The authors have no support or funding to report.

Competing Interests: The authors have declared that no competing interests exist and thank the Novocraft Technologies Company for granting a trial license of Novoalign.

* E-mail: liuy@uni-mainz.de (YL); bertil.schmidt@uni-mainz.de (BS)

Introduction

The emergence and rapid progress of next-generation sequencing (NGS) technologies has driven a substantial amount of research efforts into the development of short-read alignment algorithms. To date, a variety of short-read aligners have been developed, which can be further classified into two generations in terms of functionality. The first-generation aligners are usually designed and optimized for very short reads (typically ≤ 100 bps). These aligners usually postulate that the short-reads have very small deviations from the genome, and thus typically only allow mismatches. Even though some aligners provide support for gaps, the maximum allowable number of gaps is also quite limited (typically one gap) for the sake of speed. Example first-generation aligners include RMAP [1], MAQ [2], BFAST [3], Bowtie [4], BWA [5], CUSHAW [6] and SOAP3 [7].

With the progress of NGS, the maximum or average read lengths are steadily increasing beyond 100 for Illumina sequencing, which is most widely used. However, these longer short-reads usually come at the expense of higher sequencing error rates. On the other hand, these reads are prone to have more true insertions or deletions (indels) to the genome. These new features make the first-generation aligners become inefficient to align such longer reads in terms of alignment quality, speed or even both, and thus motivate the development of second-generation aligners that allow

for fully gapped alignments with more mismatches and indels supported.

Several second-generation aligners have been developed recently, including BWA-SW [8], GASSST [9], Bowtie2 [10], CUSHAW2 [11], GEM [12], SeqAlto [13], SOAP3-dp [14] and BWA-MEM [15]. All these aligners are designed based on the seed-and-extend paradigm. In this paradigm, a read is aligned by first identifying seeds, i.e. short ungapped/gapped alignments, on the genome and then extending the alignment to the rest of the read using dynamic programming. Constraints and filtrations are often exerted on alignment extensions to further reduce search space. Different seeding polices may be employed by different aligners. BWA-SW employs variable-length gapped seeds, and Bowtie2 extracts fixed-length ungapped seeds (inexact matches). Both GASSST and SeqAlto employ fixed-length exact-match *k*-mer (a *k*-mer is a substring of *k* bases) seeds, while CUSHAW2 and BWA-MEM respectively identifies variable-length maximal exact match (MEM) seeds and super MEM seeds. SOAP3-dp is an aligner based on graphics processing unit (GPU) computing and adopts a similar seeding approach to Bowtie2, while GEM adopts a filtration-based approximate string matching approach to extract relevant candidate matches by suitable pigeonhole-like rules. In addition, Novoalign (http://www.novocraft.com) is a proprietary short-read aligner for fully gapped alignments. However, its method has not been published. Although these aligners can efficiently align the majority of short-reads at high speed, they still have difficulties in correctly aligning all reads, even for simulated ones, to large genomes such as the human genome [11] [16]. Hence, it is of great significance to design new short-read aligners to further improve alignment quality.

In this article, we present CUSHAW3, an open-source sensitive and accurate short-read aligner for both base-space and color-space sequences. In our aligner, we have investigated a hybrid seeding approach to improve alignment quality, which incorporates three different seed types: MEM seeds, exact-match k-mer seeds and variable-length seeds derived from local alignments, into the alignment pipeline. Furthermore, three techniques: weighted seed pairing heuristic, paired-end (PE) alignment pair ranking and read mate rescuing, have been proposed to facilitate accurate PE alignment. It needs to be stressed that the concept of hybrid seeding has already been implied in some other implementations for short-read alignment. One example is Stampy [17], an aligner for Illumina sequencing, which first aligns reads with BWA (based on inexact-match seeds) and then processes unmapped reads with another seed-and-extend-based approach using exact-match k-mers. Another example is TMAP (https://github.com/iontorrent/TMAP), an aligner for Ion Torrent sequencing, which incorporates the alignment approaches from SSAHA (fixed-length k-mer seeds) [18], BWA, BWA-SW and BWA-MEM.

The performance of CUSHAW3 has been assessed by aligning both simulated and real short-reads to the human genome in terms of single-end (SE) and PE alignment. For base-space alignment, our aligner is further compared to Novoalign, CUSHAW2, BWA-MEM, Bowtie2 and GEM. The experimental results reveal that CUSHAW3 is consistently superior to CUSHAW2, BWA-MEM, Bowtie2 and GEM for both SE and PE alignments. Furthermore, our aligner achieves better PE alignment quality than Novoalign for short-reads with higher error rates. As for the speed, CUSHAW3 is inferior to CUSHAW2, BWA-MEM, Bowtie2 and BWA-MEM, but nearly always faster than Novoalign. As for color-space alignment, our aligner is consistently one of the best aligners in terms of alignment quality at superior speed, compared to SHRiMP2 [19] and BFAST.

Results

Evaluation on Base-space Reads

We have evaluated the performance of CUSHAW3 (v3.0.2) by aligning both simulated and real short-reads to the human genome (hg19). This performance is further compared to that of CUSHAW2 (v2.1.10), Novoalign (v3.00.04), BWA-MEM (v0.7.3a), Bowtie2 (v2.1.0) and GEM (v 1.376). All tests are conducted in a workstation with a dual hex-core Intel Xeon X5650 2.67 GHz CPUs and 96 GB RAM, running Linux (Ubuntu 12.04 LTS).

To measure alignment quality, we have used the sensitivity metric, which is calculated by dividing the number of aligned reads by the total number of reads, for both simulated and real reads. For simulated reads, as the true mapping positions are known beforehand, we have further used the recall metric, which is defined as dividing the number of correctly aligned reads by the total number of reads. For simulated reads, an alignment is deemed to be correct if the mapping position has a distance of ≤10 to the true position on the genome. Considering that GEM reports all detected alignments and BWA-MEM might produce multiple primary alignments for a read, we define that a read is deemed to be correctly aligned if any of its reported alignments is correct. To provide fair comparisons, we have configured CUSHAW3, CUSHAW2 and Bowtie2 to report a maximum of

10 alignments for each read and Novoalign to report all repetitive alignments. Detailed alignment parameters of all evaluated aligners can be obtained from Tables S1, S2, S3, S4 and S5 in File S1. In addition, all best values in the following tables have been highlighted in bold.

On simulated data. We have simulated three Illumina-like PE datasets from the human genome (hg19) using the wgsim simulator in SAMtools v0.1.18 [20]. All datasets have the same read lengths of 100, but with different mean base error rates: 2%, 4% and 6%. Each dataset comprises one million read pairs with insert-sizes drawn from a normal distribution N(500, 50).

Firstly, we have compared the alignment quality of all evaluated aligners by considering all reported alignments (see Table 1), by setting the minimum mapping quality score (MAPQ) to 0. For the SE alignment, Novoalign yields the best sensitivity and recall for each dataset. CUSHAW3 holds equally best sensitivity for the dataset with 2% error rate, and is consistently the second best for all other datasets. With the increase of error rates, each aligner has experienced some performance drops in terms of both measures. Novoalign has the smallest sensitivity (recall) decrease by 0.02% (2.95%), whereas Bowtie2 shows the most significant sensitivity (recall) decrease by 18.10% (21.66%). CUSHAW3 gives the second smallest performance drop with a sensitivity (recall) decrease by 0.74% (3.76%). With PE information, each aligner gets the alignment quality improved over the SE alignment in terms of both measures. In terms of sensitivity, CUSHAW3, Novoalign and BWA-MEM are consistently the top three aligners for all datasets, and Bowtie2 is the worst. In terms of recall, CUSHAW3 is superior to other aligners for the dataset with 6% error rate, while Novoalign performs best for all remaining datasets. CUSHAW3 outperforms CUSHAW2, BWA-MEM, Bowtie2 and GEM for each dataset. Similar to the SE alignment, the error rates also have significant impact on the sensitivity and recall for each aligner. As the error rate grows, Novoalign gives the least significant performance drop and CUSHAW3 the second least in terms of sensitivity. However, in terms of recall, CUSHAW3 has the smallest performance decrease.

Secondly, we have further assessed all evaluated aligners by only considering the first alignment occurrence per read in the SAM file, with the minimum MAPQ set to 0. This alignment sampling does not affect the sensitivity of each aligner, but might change the recall. Hence, all aligners are only compared in terms of recall in this evaluation (see Table 1). In terms of SE alignment, Novoalign achieves the best recall for each dataset and CUSHAW3 performs second best. In terms of PE alignment, CUSHAW3 is superior to all other aligners for the dataset with 6% error rate, while BWA-MEM performs best for the remaining datasets. Bowtie2 is the worst for each case. Some readers may argue that for a read with multiple alignments, we can choose the alignment with the largest MAPQ instead of the first alignment. Actually, we can explain that for each evaluated aligner, our evaluation by choosing the first alignment occurrence per read is consistent with that by selecting the alignment with the largest MAPQ from amongst the multiple alignments. Firstly, GEM does not compute MAPQs, but stratifies all identified alignments of a read in ascending order of string distance (Hamming distance or edit distance) [12]. This suggests that GEM implicitly considers the first alignment occurrence as the best candidate in terms of the specified distance metric. Secondly, when enabling multiple alignments per read (by option "-k"), Bowtie2 assigns a pseudo MAPQ to the identified alignments of a read and then reports them in descending order of alignment score (see Bowtie2 manual and command-line help). Thirdly, CUSHAW3, CUSHAW2 and BWA-MEM rank the alignments and build a sorted list with the alignments ordered

Table 1. Alignment quality on simulated reads (in %).

Aligner	2%			4%			6%		
	Sensitivity	Recall*	Recall**	Sensitivity	Recall*	Recall**	Sensitivity	Recall*	Recall**
SE									
CUSHAW3	**100.00**	99.04	95.96	99.92	97.85	94.81	99.26	95.28	92.32
CUSHAW2	99.95	99.00	95.96	99.33	97.61	94.64	95.45	92.84	90.04
Novoalign	**100.00**	**99.59**	**96.20**	**99.97**	**98.81**	**95.42**	**99.98**	**96.65**	**93.33**
BWA-MEM	99.99	95.95	95.95	99.59	94.33	94.33	97.38	89.86	89.86
Bowtie2	99.30	95.69	92.98	93.64	87.59	85.20	81.20	74.03	72.03
GEM	99.76	99.02	95.46	97.08	92.28	89.09	90.46	77.64	75.11
PE									
CUSHAW3	**100.00**	99.54	97.35	**100.00**	99.14	96.99	99.96	**98.06**	**96.28**
CUSHAW2	99.73	99.43	97.27	99.36	98.71	96.61	96.47	95.07	93.16
Novoalign	**100.00**	**99.87**	97.57	**100.00**	**99.23**	96.93	**100.00**	97.13	94.88
BWA-MEM	**100.00**	97.59	**97.59**	**100.00**	97.11	**97.11**	99.88	95.55	95.55
Bowtie2	99.45	98.53	96.41	93.54	91.52	89.54	80.29	77.37	75.68
GEM	**100.00**	99.20	96.85	99.79	98.06	95.77	97.99	93.24	91.15

*means the recall is calculated from all reported alignments per read and ** means the recall is calculated form the first alignment occurrence per read.

from best to worst. Both CUSHAW3 and CUSHAW2 produce the same MAPQ for the alignments (possibly with slight differences depending on the degree of soft clipping). BWA-MEM computes one MAPQ for one alignment, but ensures that the MAPQ of each alignment must not exceed that of the best alignment, i.e. the first alignment in the sorted list (refer to the source code). Finally, for a read with multiple alignments, Novoalign first ranks the multiple alignments and then determines the significance of the alignments based on the alignment score difference between the best alignment and the rest of the alignments (see the Novoalign manual). Since the source code of Novoalign is closed, we are not able to reveal more details about the mapping quality score computation and alignment reporting. However, after having examined the alignments on the simulated data, we found that the first alignment occurrences hold the largest MAPQs for each dataset.

Thirdly, we have generated the receiver operating characteristic (ROC) curves by plotting the true positive rate (TPR) against the false positive rate (FPR) in terms of MAPQ, where for each dataset all alignments are sorted in descending order of MAPQ. For any MAPQ q, we compute TPR by dividing the number of correctly aligned reads, whose MAPQs are not less than q, by the total number of reads, and FPR by dividing the number of incorrectly aligned reads, whose MAPQs are not less than q, by the number of aligned reads whose MAPQs are also not less than q. In this evaluation, we have merely taken into account the alignments whose MAPQs are greater than 0. As GEM does not compute MAPQs, it has been excluded. For Bowtie2, we have disabled the option "-k" to enable meaningful MAPQ and have used the default setting to report at most one alignment per read. CUSHAW2 and CUSHAW3 have both been configured to report at most one alignment per read for the SE and PE alignments. For Novoalign, we have used the "-r Random" option to report at most one alignment for a single read. Figure 1 shows the ROC curves for all evaluated aligners on the simulated data. We can see that Novoalign produces the most significant MAPQs for each case.

On real data. Finally, we have assessed all aligners using three real PE datasets produced from the Illumina sequencing. All datasets are publicly available and named after their accession numbers in the NCBI sequence read archive (see Table 2). The performance of each aligner is evaluated from two aspects: one is to calculate the sensitivity from all reported alignments; and the other is to calculate the sensitivity after removing the alignments with low aligned base proportion per read. This is because we have observed that some alignments, produced by Novoalign, BWA-MEM, Bowtie2 and GEM, have low aligned base proportion per read (typically <50%) due to soft clipping. Intuitively, such short alignments to the genome are supposed to have higher probabilities to be false positives compared to those aligned to the reference in full lengths (or with high aligned base proportion per read). However, this is not surly the case, especially when there are large indels at the end of the read. In such cases, the correct alignments of the read may be shortened with soft-clipping. However, it is still of great significance to re-evaluate the sensitivity of each aligner by removing the alignments with <50% aligned base proportion per read, which may more truly reflect the alignment quality of an aligner on real data.

Table 3 shows the alignment quality of all evaluated aligners with or without alignment removal, where the minimum MAPQ threshold is set to 0. For each value x/y in the table, x is the sensitivity calculated from all reported alignments and y is the sensitivity after removing the alignments with <50% aligned base proportion per read. Without alignment removal, in terms of SE alignment, CUSHAW3 aligned the most reads for each dataset and GEM is the worst. In terms of PE alignment, BWA-MEM gives the best sensitivity and CUSHAW3 is the second best for all datasets. However, after alignment removal, the sensitivities of both BWA-MEM and Novoalign significantly drop for all datasets in terms of both SE and PE alignment. Bowtie2 keeps its SE sensitivity, but has a slight decrease in PE sensitivity. GEM has also experienced significant PE sensitivity drops for all datasets. Both CUSHAW2 and CUSHAW3 keep their sensitivities unchanged for each case. With alignment removal, CUSHAW3 is consistently superior to all other aligners for each dataset in

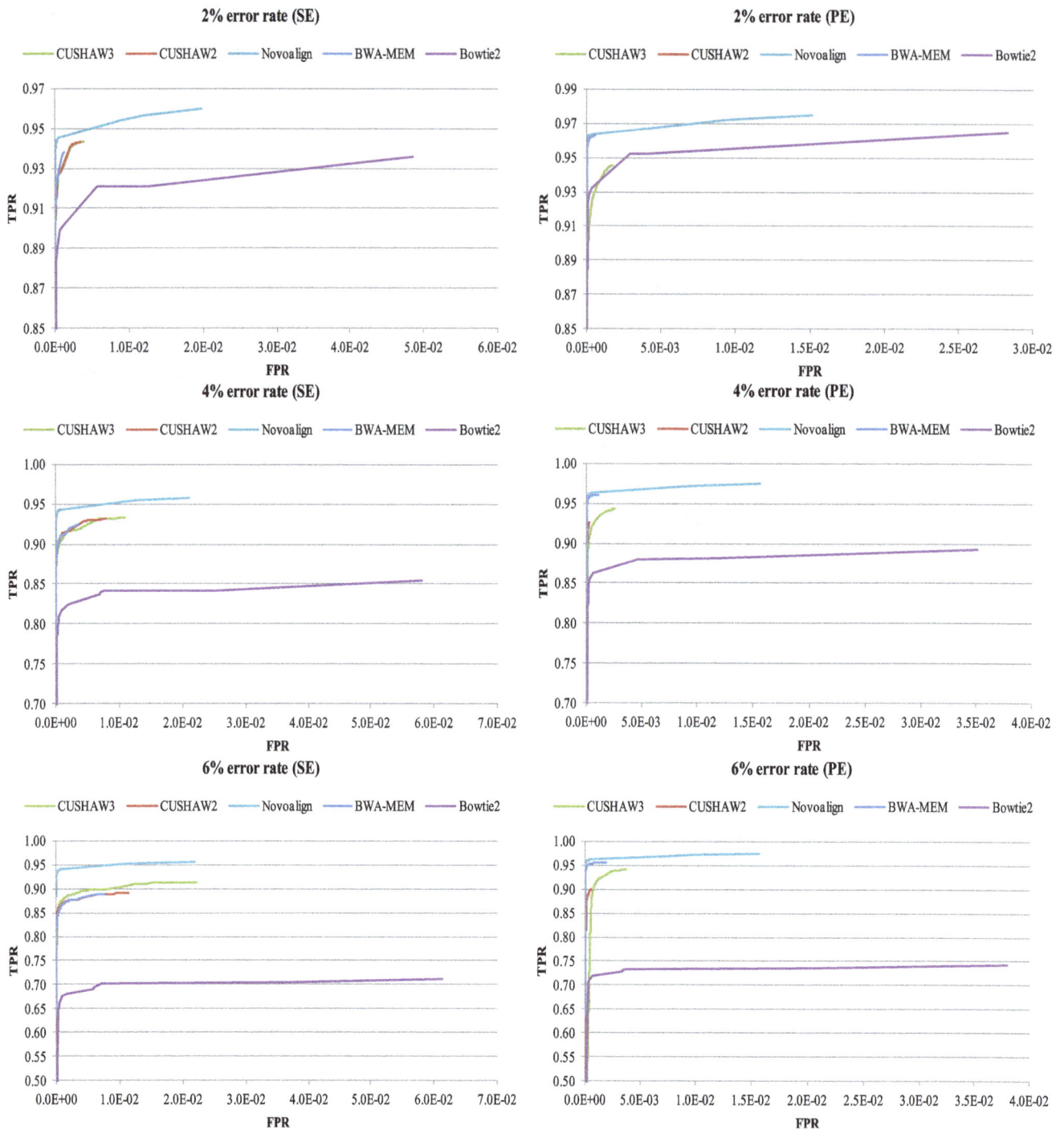

Figure 1. ROC curves of all evaluated aligners on the simulated data with the minimum MAQP>0.

Table 2. Real dataset information.

Name	Type	Length	No. of Reads	Mean Insert
SRR034939	PE	100	36,201,642	525*
SRR211279	PE	100	50,937,050	302*
ERR024139	PE	100	53,653,010	313*

*estimated using CUSHAW3.

terms of both SE and PE alignment. In addition, we have shown how the sensitivity (without alignment removal) varies as MAPQ changes (see Figure S1 in File S1). In this evaluation, all alignments are first sorted in descending order of MAPQs and then the sensitivity corresponding to any MAPQ q ($0 \leq q \leq 255$) is computed by taking into account all alignments whose MAPQs are not less than q.

Speed and memory comparison. Besides alignment quality, the speed of each aligner has been evaluated using the aforementioned simulated and real data. We have run each aligner

Table 3. Alignment quality on real reads (in %).

Aligner	SRR034939	SRR211279	ERR024139
SE			
CUSHAW3	**98.48/98.48**	**99.25/99.25**	**99.12/99.12**
CUSHAW2	93.86/93.86	96.76/96.76	96.74/96.74
Novoalign	96.80/91.27	98.44/98.28	98.49/97.50
BWA-MEM	98.30/97.14	99.17/98.58	99.07/98.50
Bowtie2	95.56/95.56	97.13/97.13	97.20/97.20
GEM	93.69/93.69	95.10/95.10	94.82/94.81
PE			
CUSHAW3	98.92/**98.92**	99.46/**99.46**	99.33/**99.33**
CUSHAW2	94.38/94.38	96.94/96.94	96.92/96.92
Novoalign	98.00/94.23	99.25/98.85	99.13/97.87
BWA-MEM	**99.06**/97.14	**99.49**/98.58	**99.36**/98.50
Bowtie2	96.23/95.56	97.31/97.13	97.39/97.20
GEM	95.52/93.69	96.16/95.10	96.15/94.81

For each value x/y, x is the sensitivity calculated from all reported alignments and y is the sensitivity after removing the alignments with <50% aligned base proportion per read.

Table 4. Runtimes (in minutes) on simulated and real base-space reads.

Simulated	2%		4%		6%	
	SE	PE	SE	PE	SE	PE
CUSHAW3	3.4	6.2	3.7	8.1	3.9	10.7
CUSHAW2	2.5	2.5	2.8	2.9	2.9	3.1
Novoalign	6.7	6.6	38.1	7.0	131.7	12.6
BWA-MEM	**1.4**	**2.3**	**1.9**	**1.9**	2.0	**2.1**
Bowtie2	2.1	3.6	2.0	2.7	**1.7**	2.2
GEM	5.7	2.4	5.9	**1.9**	5.4	2.0

Real	SRR034939		SRR211279		ERR024139	
	SE	PE	SE	PE	SE	PE
CUSHAW3	62.0	292.4	78.6	317.9	85.1	264.1
CUSHAW2	38.0	38.5	47.2	49.0	51.4	50.5
Novoalign	862.1	497.6	2,024.0	1,243.8	754.2	460.3
BWA-MEM	**25.2**	**25.9**	**24.6**	**26.1**	**27.7**	**30.9**
Bowtie2	50.4	55.9	79.1	69.5	78.0	72.7
GEM	53.0	34.4	72.2	44.7	68.3	51.0

with 12 threads on the aforementioned workstation. For fair comparisons, GEM has counted in the SAM format conversion time (sometimes takes >50% of the overall runtime), as every other aligner reports alignments in SAM format. In addition, all runtimes are measured in wall clock time.

Table 4 shows the runtime (in minutes) of all evaluated aligners on both simulated and real data. For the simulated data, Novoalign is the slowest for nearly all cases, with an exception that CUSHAW3 performs worst in terms of PE alignment for the dataset with 4% error rate. For the SE alignment, BWA-MEM runs fastest on the datasets with 2% and 4% error rates, while Bowtie2 performs best for the dataset with 6% error rate. For the PE alignment, BWA-MEM is superior to all other aligners for each dataset, with an exception that GEM has a tie with BWA-MEM for the dataset with 4% error rate. In addition, the runtimes of both Novoalign and CUSHAW3 are more sensitive to the error rates compared to other aligners. For the real data, BWA-MEM is consistently the fastest for each case and Novoalign is the worst.

As for memory consumption, the peak resident memory of each aligner has been calculated by performing PE alignment on the dataset with 2% error rate using a single CPU thread (see Figure 2). Bowtie2 takes the least memory of 3.2 GB and Novoalign consumes the most memory of 7.9 GB. CUSHAW3 and CUSHAW2 have a memory footprint of 3.3 GB and 3.5 GB, respectively. For BWA-MEM and GEM, the peak resident memory is 5.2 GB and 4.1 GB, respectively.

Evaluation on Color-space Reads

In addition to base-space alignment, we have evaluated the performance of CUSHAW3 for color-space alignment, and have further compared our aligner to SHRiMP2 (v2.2.3) and BFAST (v0.7.0a). In this evaluation, we have simulated two mate-paired datasets (read lengths are 50 and 75) from the human genome using the ART (v1.0.1) simulator [21]. Each dataset has 10% coverage of the human genome (resulting in 6,274,322 reads in the 50-bp dataset and 4,182,886 reads in the 75-bp dataset) and has an insert-size 200±20.

Both CUSHAW3 and SHRiMP2 are configured to report up to 10 alignments per read and BFAST to report all alignments with the best score for each read. Each aligner conducts mate-paired alignments and runs with 12 threads on the aforementioned workstation. Table 5 shows the alignment quality and the runtimes of the three aligners. In terms of sensitivity, CUSHAW3 outperforms both SHRiMP2 and BFAST for the 50-bp dataset, while BFAST is the best for the 75-bp dataset. When considering all reported alignments, SHRiMP2 produces the best recall and CUSHAW3 performs second best for every dataset. When only considering the first alignment occurrence per read, CUSHAW3 is superior to both SHRiMP2 and BFAST for each dataset. In terms of speed, CUSHAW3 is the fastest for each case. On average, CUSHAW3 achieves a speedup of 9.5 (and 11.9) over SHRiMP2 (and BFAST). In particular, for the 75-bp dataset, our aligner runs 13.5× and 19.9× faster than SHRiMP2 and BFAST, respectively. In addition, for each aligner, the recall gets improved as the read length increases.

Evaluation on GCAT Benchmarks

Finally, we have evaluated the performance of our aligner using the public benchmarks at GCAT (http://www.bioplanet.com/gcat), which is a free collaborative platform for comparing multiple genome analysis tools across a standard set of metrics. In this evaluation, we have compared CUSHAW3 to CUSHAW2, BWA-MEM and Novoalign in terms of alignment quality and variant calling. The evaluation results for each aligner can also be obtained from our CUSHAW3 homepage (http://cushaw3.sourceforge.net).

In terms of alignment quality, two Illumina-like SE datasets as well as two Illumina-like PE datasets have been used. For the two datasets of each alignment type, one has small indels in reads (the small-indel dataset) and the other contains large indels (the large-indel dataset). All of the four datasets are simulated from the human genome and have read length 100, where there are 11,945,249 reads in each SE dataset and 11,945,250 reads in each

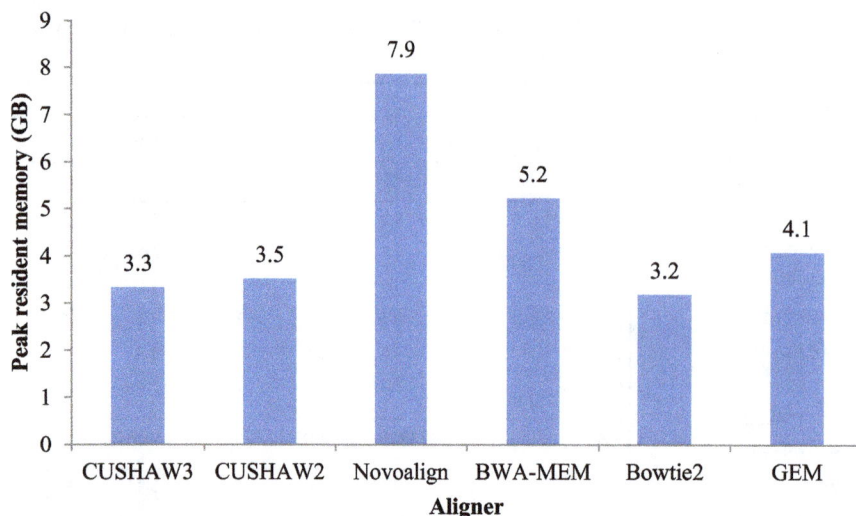

Figure 2. Peak resident memory of all evaluated aligners.

PE dataset. To be consistent with the GCAT standard evaluations, both CUSHAW2 and CUSHAW3 are configured to report at most one alignment per read for both the SE and PE alignments. Table 6 shows the alignment results of all evaluated aligners. In terms of SE alignment, CUSHAW3 yields the best sensitivity for both datasets. The best recall is achieved by CUSHAW3, CUSHAW2 and BWA-MEM on the small-indel dataset and by BWA-MEM on the large-indel dataset. CUSHAW3 performs better than Novoalign for each case. CUSHAW2 outperforms Novoalign on the small-indel dataset in terms of both sensitivity and recall, while yielding smaller recall on the large-indel dataset. In terms of PE alignment, BWA-MEM performs best for each case and CUSHAW3 is the second best. On the small-indel dataset, CUSHAW2 outperforms Novoalign in terms of both sensitivity and recall. On the large-indel dataset, CUSHAW2 yields better sensitivity than Novoalign, while Novoalign gives better recall. In addition, the alignment accuracy of different aligners has been further compared by plotting the percentage of incorrectly aligned reads against the percentage of correctly aligned reads with respect to MAPQs. In this plotting, all alignments of each aligner are first sorted in descending order of MAPQ and then the correct and incorrect percentages are calculated. Figures S2, S3, S4 and S5 in File S1 show the alignment accuracy comparison for both SE and

PE alignments. In terms of SE and PE alignment, Novoalign is superior to all other aligners and BWA-MEM is the second best with respect to the plotting. CUSHAW2 and CUSHAW3 have demonstrated nearly identical curves, and still need further improvement on the calculation of MAPQs compared to Novoalign and BWA-MEM.

In terms of variant calling, a real exome sequencing dataset has been used in this test. This dataset is comprised of Illumina 100-bp PE reads and has $30\times$ coverage of the human exome. In this test, we have used SAMtools as the variant caller. Table 7 shows the variant calling results, where the novel single nucleotide polymorphisms (SNPs) in the dbSNP database are not taken into account. BWA-MEM yields the maximum sensitivity and Novoalign performs second best. In terms of specificity, Novoalign achieves the best performance, while CUSHAW2 and CUSHAW3 tie for the second place. As for Ti/Tv ratio, CUSHAW2 produces the maximum value of 2.323 and Novoalign gives the second best value of 2.289. CUSHAW3 and BWA-MEM are joint third. BWA-MEM identifies the most correct SNPs, while Novoalign

Table 5. Alignment quality and runtimes on color-space reads.

Dataset	Measure	CUSHAW3	SHRiMP2	BFAST
50-bp	Sensitivity	**92.13**	91.55	88.94
	Recall*	86.28	**88.58**	81.01
	Recall**	**84.72**	84.22	81.01
	Time(min)	**41**	227	160
75-bp	Sensitivity	92.27	92.33	**93.44**
	Recall*	91.16	**91.24**	86.14
	Recall**	**89.31**	88.15	86.14
	Time(min)	**20**	263	389

Same as Table 1.

Table 6. Alignment results on GCAT benchmarks.

Dataset	Measure	CUSHAW3	CUSHAW2	Novoalign	BWA-MEM
SE					
Small indels	Sensitivity	**100.00**	99.86	97.56	99.99
	Recall	**97.52**	**97.52**	97.47	**97.52**
Large indels	Sensitivity	**100.00**	99.50	97.56	99.99
	Recall	97.37	97.04	97.35	**97.40**
PE					
Small indels	Sensitivity	**100.00**	99.99	98.85	**100.00**
	Recall	99.06	99.05	98.83	**99.22**
Large indels	Sensitivity	**100.00**	99.71	98.84	**100.00**
	Recall	98.91	98.62	98.69	**99.08**

Table 7. Variant calling results on a GCAT benchmark.

Aligner	Sensitivity	Specificity	Ti/Tv	Correct SNP	Correct Indel
CUSHAW3	83.74	99.9930	2.285	115,709	5,974
CUSHAW2	83.51	99.9930	**2.323**	112,727	5,841
Novoalign	84.10	**99.9951**	2.289	121,992	**9,416**
BWA-MEM	**85.30**	99.9926	2.285	**124,459**	9,232

Sensitivity = TP/(TP+FN), specificity = TN/(TN+FP) and Ti/Tv is the ratio of transitions to transversions in SNPs.

yields the most correct indels. Compared to CUSHAW2, CUSHAW3 holds a smaller Ti/Tv ratio, but has an improved sensitivity as well as identifies more correct SNPs and indels. In addition, we have given a Venn diagram (see Figure S6 in File S1) to show the variant concordance between the evaluated aligners.

Discussion

In this article, we have presented CUSHAW3, an open-source tool for sensitive and accurate short-read alignment to large genomes, such as the human genome. This aligner is designed based on the well-known seed-and-extend heuristic and has introduced a hybrid seeding approach to improve alignment quality for both SE and PE alignments. This hybrid seeding approach works by incorporating three different seed types, namely MEM seeds, exact-match k-mer seeds and variable-length seeds derived from local alignments, into our alignment pipelines. Furthermore, we have proposed three critical bioinformatics techniques: weighted seed-paring heuristic, PE alignment pair ranking and read mate rescuing, to facilitate accurate PE alignments. CUSHAW3 accepts short-reads represented in FASTA, FASTQ, SAM/BAM [20] format, which can be uncompressed or zlib-compressed, and provides an easy-to-use and well-structured interface as well as a more detailed documentation about the installation and usage. In addition, our aligner produces PHRED [22] compliant MAPQs for all alignments and reports them in SAM format. This enables seamless integration of our aligner with established downstream analysis tools like SAMtools [20] and GATK [23].

CUSHAW3 provides support for both base-space and color-space alignments. For base-space alignment, we have assessed the performance of CUSHAW3 and other top-performing short-read aligners: Novoalign, CUSHAW2, BWA-MEM, Bowtie2 and GEM using simulated as well as real reads from the human genome. For both simulated and real data, we have employed the sensitivity measure. Additionally, the recall measure has been further used on simulated data, as the ground truth of alignments is known beforehand. On simulated data, CUSHAW3 achieves consistently better alignment quality (by considering all reported alignments) than CUSHAW2, BWA-MEM, Bowtie2 and GEM in terms of both SE and PE alignment. Compared to Novoalign, CUSHAW3 has comparable PE alignment performance for short-reads with low error rates, but performs better for short-reads with high error rates. On real data, CUSHAW3 achieves the highest SE and PE sensitivities for each dataset. As for speed, CUSHAW3 does not have any advantage over CUSHAW2, BWA-MEM, Bowtie2 and GEM, but shows to be nearly always faster than Novoalign. In terms of color-space alignment, we have evaluated and compared the performance of CUSHAW3, SHRiMP2 and BFAST using simulated mate-paired color-space reads. The results

show that CUSHAW3 is consistently one of the best color-space aligners in terms of alignment quality. Moreover, on average CUSHAW3 is one order-of-magnitude faster than both SHRiMP2 and BFAST on the same hardware configurations. From our evaluations, we have observed that a considerable number of alignments, reported by Novoalign, BWA-MEM, Bowtie2 and GEM, have low aligned base proportion per read (<50%), especially for the PE alignments. Furthermore, even though both CUSHAW3 and Novoalign are shown to have higher alignment accuracy, some simulated reads still missed their correct alignments. Moreover, this situation becomes even worse as the error rate grows larger. Hence, more research efforts are still required in order to better align short-reads with high error rates. Finally, as shown in our evaluations, the hybrid seeding approach does improve accuracy, but at the expense of speed. To significantly reduce the runtime, one promising solution is the use of GPU computing, as some pioneer work (e.g. [6] [7] [14]) has shown that short-read alignment can significantly benefit from the GPU computing with respect to speed. This acceleration based on special hardware can be considered as part of our future work.

Methods

Hybrid Seeding

Our hybrid seeding approach incorporates MEM seeds, exact-match k-mer seeds, and variable-length seeds at different phases of the alignment pipeline. For a single read, the alignment pipeline generally works as follows (see Figure 3).

First, we produce the MEM seeds for both strands of the read based on Burrows-wheeler transform [24] and FM-index [25]. Secondly, from each seed we determine on the genome a potential mapping region for the read, and then perform the Smith-Waterman algorithm [26] to gain the optimal local alignment score between the read and the mapping region. All seeds are subsequently ranked in terms of optimal local alignment score, where greater scores mean higher ranks.

Thirdly, dynamic programing is employed to identify the optimal local alignment of the read to the genome from the highest-ranked seeds. If satisfying the local-alignment constraints,

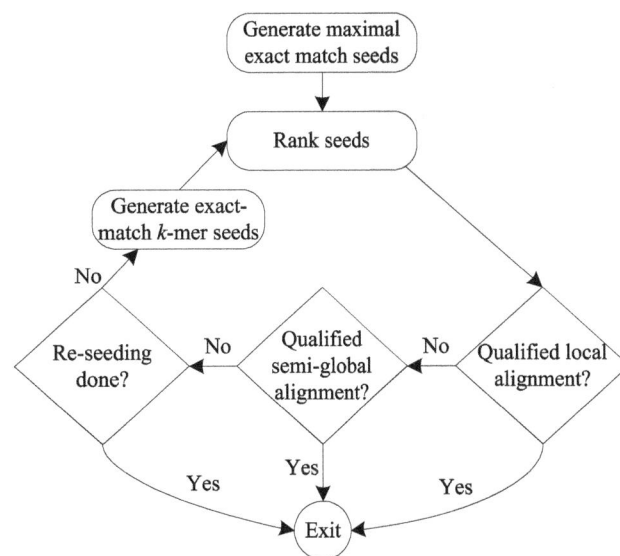

Figure 3. Program workflow of the single-end alignment using hybrid seeding.

including minimal percentage identity (default = 90%) and aligned base proportion per read (default = 80%), the optimal local alignment will be considered as qualified. Otherwise, we will attempt to rescue the read using a semi-global alignment approach. As an optimal local alignment usually indicates the most similar region on the genome, our semi-global alignment approach takes the optimal local alignment as a variable-length seed, re-computes a new mapping region on the genome and then performs semi-global alignment between the read and the new mapping region to obtain an optimal semi-global alignment. If the optimal semi-global alignment satisfies the global-alignment constraints, including minimal percentage identity (default = 65%) and aligned base proportion per read (default = 80%), this alignment will be deemed to be qualified. This double-alignment approach enables us to capture the alignments with more continuous mismatches and longer gaps. This is because we might fail to get good enough optimal local alignments in such cases, as the positive score for a match is usually smaller than the penalty charged for mismatches and indels.

Finally, when we still fail to get any qualified alignment, this likely means that the true alignment is implied by none of the evaluated MEM seeds. In this case, we attempt to rescue the alignment by re-seeding the read using exact-match k-mer seeds. To improve speed, we search all non-overlapping k-mers of the read against the genome to identify seed matches. Subsequently, we employ the k-mer seeds to repeat the aforementioned alignment process to rescue the read. If we still fail to gain a qualified alignment, we will stop the alignment process and then report this read as unaligned.

Paired-end Mapping

In comparison with SE alignment, the long-range positional information contained in PE reads usually allow for more accurate short-read alignment, by either disambiguating alignments when one of the two ends aligns to repetitive regions or rescuing one end from its aligned mate. In addition, for aligners based on the seed-and-extend heuristic, the PE information, such as alignment orientations and insert-size of both ends, can aid to significantly reduce the number of noisy seeds prior to the time-consuming alignment extensions. This filtration can be realized through a seed-paring heuristic [11], as a seed determines the alignment orientation of a read and the mapping distance constraint on seed pairs can be inferred from the insert-size of read pairs.

For a read pair S_1 and S_2, our PE alignment pipeline generally works as follows (see Figure 4). First, we generate and rank the MEM seeds, following the same procedure as in SE alignment. Secondly, a weighted seed-paring heuristic is introduced to pair seeds, where only high-quality seeds, whose scores are not less than a minimal score threshold (default = 30), will be taken into account. This heuristic enumerates each high-quality seed pair of S_1 and S_2 to identify all qualified seed pairs that meet the alignment orientation and insert-size requirements. To distinguish all qualified seed pairs in terms of quality, we have calculated a weight for each qualified seed pair and further ranked all of them by a max-heap data structure. This quality-aware feature allows for us to visit all qualified seed pairs in the descending order of quality. Thirdly, if failed to find any qualified seed pair, we will resort to the re-seeding based on exact-match k-mers by sequentially checking both ends to see if either of them has not yet been re-seeded. If so, the k-mer seeds will be produced for that end and all new seeds will be ranked in the same way as for MEM seeds. Subsequently, we merge all high-quality k-mer seeds with the high-quality MEM seeds, and then re-rank all seeds. The seed merge is used because some significant alignments, which are not

covered by MEM seeds, may be reflected by k-mer seeds, and vice versa. After getting the new list of seeds, we repeat the weighted seed-paring heuristic to gain qualified seed pairs. The seed-paring and re-seeding process will be repetitively continued until either both ends have been re-seeded or any qualified seed pair has been identified. Fourthly, we compute the real alignments of both ends from the qualified seed pairs. An alignment pair will be considered as qualified if their mapping position distance satisfies the insert-size constraint. Similar to the weighted seed-pairing approach, we have also ranked all qualified alignment pairs by means of a max-heap data structure. In this manner, we would expect better alignment pairs to come out earlier in the output. Finally, we attempt to rescue read mates from the best alignments of each end, when failed to pair reads in previous steps.

Weighted Seed-pairing Heuristic and Alignment Pair Ranking

To guide the production of real PE alignments in a quality-aware manner, we introduce a weighted seed-paring heuristic computing a weight w for each qualified seed pair as follows.

$$w = \frac{2w_1 w_2}{w_1 + w_2}$$
$$w_i = \frac{q_i}{|S_i| \times m}, i = 1,2 \tag{1}$$

where q_i is the optimal local alignment score between read S_i ($1 \leq i \leq 2$) and the mapping region derived from the seed, and m is the positive score for an alignment match. To rank all qualified seed pairs, we employ w as the key of each entry in the max-heap.

In addition to seed pairs, all qualified alignment pairs have been further ranked in terms of weight and edit distance. For an alignment pair, we calculate its weight following Equation (1) with

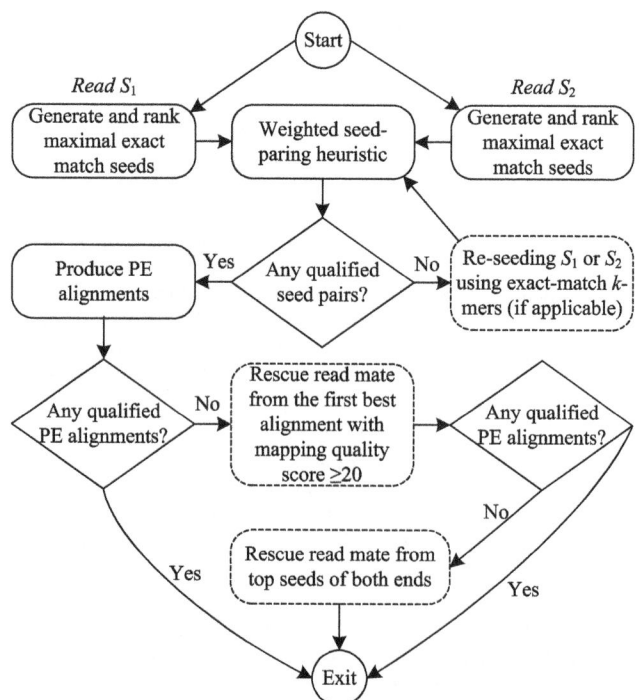

Figure 4. Program workflow of the paired-end alignment with hybrid seeding.

the difference that q_i is not definitely the optimal local alignment score, but might be the optimal semi-global alignment score. This is because an alignment is possibly produced from a semi-global alignment as mentioned above. Furthermore, when two qualified alignment pairs hold the same weights, we further rank them by comparing the sums of the edit distances of each alignment pair. In this case, smaller edit distance sums mean higher ranks.

Read Mate Rescuing

For unpaired reads, we have employed a read mate rescuing procedure, which attempts to rescue one read from the top hits of its aligned mate by using the paired-end long-range distance information. In general, our rescuing procedure works as follows.

First, the best alignments of the two reads are computed (if available). The read, whose best alignment has a MAPQ exceeding a minimum threshold (default = 20), will be used to rescue its mate. If an optimal alignment satisfying the aforementioned constraints has been gained for the mate, the two reads are considered as paired. Otherwise, we will continue the rescuing process using the alignments with smaller MAPQs. Secondly, if the two reads have not yet been properly paired, we will attempt to pair them from more top hits of both reads. The rescuing process will not stop until the two reads have been properly paired or having reached the maximum number (default = 100) of top seeds for each read. Finally, for unpaired reads, we will report their best alignments (if available) in a SE alignment mode.

This read mate rescuing is usually time-consuming mainly due to two factors. One is the dynamic-programing-based alignment with quadratic time complexity. The other is the maximal insert-size of a read pair, which basically determines the mapping region size of the mate on the genome. In sum, the more reads are paired by seed-pairing heuristic; the less time is taken by the read mate rescuing procedure.

Color-space Alignment

Most existing color-space aligners encode a nucleotide-based genome as a color sequence and then identify potential short-read alignment hits in color space. However, different approaches may be used to produce the final base-space alignments. For a color-space read, one approach is to identify a final color-space alignment and then convert the color sequence to nucleotides under the guidance of the alignment using dynamic programming [5]. An alternative is to directly perform color-aware dynamic-programming-based alignment by simultaneously aligning all four possible translations [3] [19].

In our aligner, we also convert a nucleotide-based genome to a color sequence and perform short-read alignment in color-space basically following the same workflow as the base-space alignment (mentioned above). For a color-space read, after obtaining a qualified color-space alignment, we must convert the color sequence into a nucleotide sequence. This conversion is accomplished by adopting the dynamic programming approach proposed by Li and Durbin [5]. Subsequently, the translated nucleotide sequence will be re-aligned to the nucleotide-based genome using either local or semi-global alignment depending on how its parent alignment has been produced.

Acknowledgments

We thank the editor and the three reviewers for their helpful and constructive comments which helped to improve the manuscript.

Author Contributions

Conceived and designed the experiments: YL BS BP. Performed the experiments: YL BP. Analyzed the data: YL BS BP. Contributed reagents/materials/analysis tools: YL BP. Wrote the paper: YL.

References

1. Smith AD, Xuan Z, Zhang MQ (2008) Using quality scores and longer reads improves accuracy of Solexa read mapping. BMC Bioinformatics, 9: 128.
2. Li H, Ruan J, Durbin R (2008) Mapping short DNA sequencing reads and calling variants using mapping quality scores. Genome Res., 18: 1851–1858.
3. Homer N, Merriman B, Nelson SF (2009) BFAST: an alignment tool for large scale genome resequencing. Plos One, 4: e7767.
4. Langmead B, Trapnell C, Pop M, Salzberg SL (2009) Ultrafast and memory-efficient alignment of short DNA sequences to the human genome. Genome Biol., 10: R25.
5. Li H, Durbin R (2009) Fast and accurate short read alignment with Burrows–Wheeler transform. Bioinformatics, 25: 1755–1760.
6. Liu Y, Schmidt B, Maskell DL (2012) CUSHAW: a CUDA compatible short read aligner to large genomes based on the Burrows-Wheeler transform. Bioinformatics, 28: 1830–1837.
7. Liu CM, Wong T, Wu E, Luo R, Yiu SM, et al. (2012) SOAP3: ultra-fast GPU-based parallel alignment tool for short reads. Bioinformatics, 28: 878–879.
8. Li H, Durbin R (2010) Fast and accurate long-read alignment with Burrows–Wheeler transform. Bioinformatics, 26: 589–595.
9. Rizk G, Lavenier D (2010) GASSST: global alignment short sequence search tool. Bioinformatics, 26: 2534–2540.
10. Langmead B, Salzberg S (2012) Fast gapped-read alignment with Bowtie 2. Nature Methods, 9: 357–359.
11. Liu Y, Schmidt B (2012) Long read alignment based on maximal exact match seeds. Bioinformatics, 28, i318–i324.
12. Marco-Sola S, Sammeth M, Guigó R, Ribeca P (2012) The GEM mapper: fast, accurate and versatile alignment by filtration. Nature Methods, 9: 1885–1888.
13. Mu JC, Jiang H, Kiani A, Mohiyuddin M, Bani Asadi N, et al. (2012) Fast and accurate read alignment for resequencing. Bioinformatics, 28: 2366–2373.

14. Luo R, Wong T, Zhu J, Liu CM, Zhu X, et al. (2013) SOAP3-dp: fast, accurate and sensitive GPU-based short-read aligner. PLOS One, 8: e65632.
15. Li H (2013) Aligning sequence reads, clone sequences and assembly contigs with BWA-MEM. arXiv: 1303.3997 [q-bio.GN].
16. Li H, Homer N (2010) A survey of sequence alignment algorithms for next-generation sequencing. Brief Bioinform., 11: 473–483.
17. Lunter G, Goodson M (2011) Stampy: a statistical algorithm for sensitive and fast mapping of Illumina sequence reads. Genome Res., 21: 936–939.
18. Ning Z, Cox AJ, Mullikin JC (2001) SSAHA: a fast search method for large DNA databases. Genome Res., 11: 1725–1729.
19. David M, Dzamba M, Lister D, Ilie L, Brudno M (2011) SHRiMP2: sensitive yet practical short read mapping. Bioinformatics, 27: 1011–1012.
20. Li H, Handsaker B, Wysoker A, Fennell T, Ruan J, et al. (2009) The sequence alignment/map format and SAMtools. Bioinformatics, 25: 2078–2079.
21. Huang W, Li L, Myers JR, Marth GT (2012) ART: a next-generation sequencing read simulator. Bioinformatics, 28: 593–594.
22. Ewing B, Green P (1998) Base-calling of automated sequencer traces using phred. II. Error probabilities. Genome Res 8: 186–194.
23. McKenna A, Hanna M, Banks E, Sivachenko A, Cibulskis K, et al. (2010) The Genome Analysis Toolkit: a MapReduce framework for analyzing next generation DNA sequencing data. Genome Res 20: 1297–1303.
24. Burrows M, Wheeler DJ (1994) A block sorting lossless data compression algorithm. Technical Report 124, Digital Equipment Corporation, Palo Alto, CA.
25. Ferragina P, Manzini G (2005) Indexing compressed text. Journal of the ACM, 52: 4.
26. Smith TF, Waterman MS (1981) Identification of common molecular subsequences. J. Mol. Biol., 147: 195–197.

19

PolyaPeak: Detecting Transcription Factor Binding Sites from ChIP-seq Using Peak Shape Information

Hao Wu[1]*, Hongkai Ji[2]*

1 Department of Biostatistics and Bioinformatics, Emory University, Atlanta, Georgia, United States of America, **2** Department of Biostatistics, Johns Hopkins University, Baltimore, Maryland, United States of America

Abstract

ChIP-seq is a powerful technology for detecting genomic regions where a protein of interest interacts with DNA. ChIP-seq data for mapping transcription factor binding sites (TFBSs) have a characteristic pattern: around each binding site, sequence reads aligned to the forward and reverse strands of the reference genome form two separate peaks shifted away from each other, and the true binding site is located in between these two peaks. While it has been shown previously that the accuracy and resolution of binding site detection can be improved by modeling the pattern, efficient methods are unavailable to fully utilize that information in TFBS detection procedure. We present PolyaPeak, a new method to improve TFBS detection by incorporating the peak shape information. PolyaPeak describes peak shapes using a flexible Pólya model. The shapes are automatically learnt from the data using Minorization-Maximization (MM) algorithm, then integrated with the read count information via a hierarchical model to distinguish true binding sites from background noises. Extensive real data analyses show that PolyaPeak is capable of robustly improving TFBS detection compared with existing methods. An R package is freely available.

Editor: Yi Xing, University of California, Los Angeles, United States of America

Funding: Research by HJ is supported by the National Institute of Health grant: R01HG006282. The funders had no role in study design, data collection and analysis, decision to publish, or preparation of the manuscript.

Competing Interests: The authors have declared that no competing interests exist.

* E-mail: hao.wu@emory.edu (HW); hji@jhsph.edu (HJ)

Introduction

One major goal of functional genomics is to comprehensively characterize the regulatory circuitry behind coordinated spatial and temporal gene activities. In order to achieve this goal, a critical step is to monitor downstream regulatory programs of various transcription factors (TFs). ChIP-seq [1,2], a technology that couples chromatin immunoprecipitation with massively parallel sequencing, is capable of mapping genome-wide transcription factor binding sites (TFBSs), and is increasingly used by scientists and nation-wide projects such as ENCODE [3] and modENCODE [4] to annotate functional sequence elements in human genome and genomes of model organisms. ChIP-seq data grow rapidly. The first ChIP-seq studies were published in 2007. Since then, several thousands studies have been performed and data are available in public databases [5]. This highlights the importance of continuous development of robust and powerful ChIP-seq data analysis tools.

The raw data produced by a ChIP-seq experiment are tens of millions of short (usually less than 100 base pairs) DNA sequences called "reads". To identify TFBSs, the reads are first aligned to a reference genome and the uniquely aligned reads are retained. Next, the genome is scanned to identify "peaks", or regions enriched in aligned sequence reads, which are the predicted TF binding sites. Since 2007, a number of peak calling algorithms and software tools have been developed. Examples include BayesPeak [6], CisGenome [7], FindPeaks [8], GPS [9], Hpeak [10], MACS [11], MOSAiCS [12], PeakSeq [13], PICS [14], QuEST [15], SISSRs [16], T-PIC [17], etc. Several benchmark studies have

also been conducted to compare different peak calling tools [18,19].

Early analyses revealed that ChIP-seq data for mapping TFBSs have a characteristic pattern: surrounding each true binding site, sequence reads aligned to the forward and reverse strands of the reference genome are clustered into two distinct peaks that are shifted away from each other, and the binding site is located in between them (Figure 1). This phenomenon is caused by the sequencing protocol which involves cutting chromatin into fragments and reading the sequences from both ends of the TF bound DNA fragments. The cutting points seldom fall within the binding sites since the DNA is protected by the TF. As a result, in most cases the binding sites sit within the DNA fragments, and their flanking sequences are read out by the sequencer. Since the machine reads the DNA sequences in a directional way, reads from one end of the DNA fragments are always aligned to the forward strand of the reference genome, and reads from the other end are always aligned to the reverse strand. This creates the bimodal peak pattern shown in Figure 1.

This pattern has been shown to be useful for improving TFBS detection. For example, SISSRs [16] uses the sign change of the forward and reverse strand read count difference along the genome to identify the true binding sites. CisGenome [7] uses the summits of the two coupled peaks to determine the boundaries of binding sites. QuEST [15] and MACS [11] estimate the offset between the forward and reverse strand peaks, and shift these two peaks together based on the offset. They then merge signals from both peaks to increase accuracy for identifying true binding

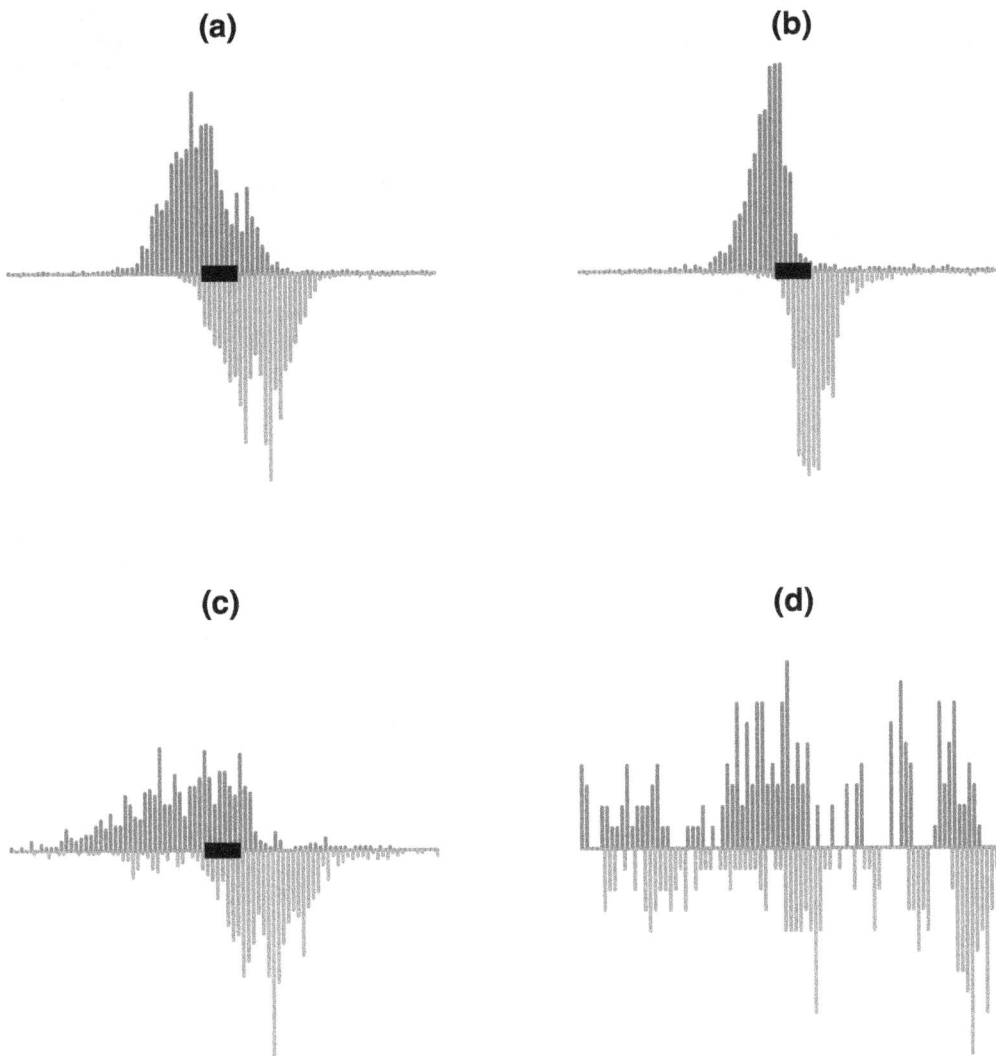

Figure 1. Illustration of peak shapes from three good binding sites (a, b, c) and one false positive (d). X-axis is genomic location centered at the summit of each peak. Y-axis represents ChIP read counts in 10 bp genomic windows. For each peak, data from the IP sample in a 800 bp window surrounding the MACS peak summit is shown. Bars above the middle represent counts from forward strand, and bars below represent the counts from reverse strand. The black rectangles illustrate the TF binding sites. (a) and (b) are two different binding sites from the same ChIP-seq dataset, and (c) is a binding site from a different dataset. They illustrate that peak shapes at true TFBSs are similar but can vary across binding sites and datasets.

location. While many state-of-the-art peak callers use the bimodal pattern in their design, most methods only use the information contained in the offset between the two coupled peaks. Few method fully utilize the information contained in the peak shapes. As pointed out in a recent publication [20], many false positive TFBSs can be "filtered out by visual inspection on the peak sizes and appearance". To demonstrate, Figure 1 shows three examples of good binding sites along with an example of false positive. All examples have large read counts in both strands with certain offsets. Methods using the offset information alone would call all examples as binding sites. However, the example in Figure 1(d) is very likely to be sequencing artifacts. It does not contain DNA motifs for the TF, and clearly has very different shapes from the true binding sites. Thus if the peak shape information is used in the inference, this false positive can be eliminated.

There are two existing methods fully incorporate the peak shape information in a rigorous statistical framework. PICS [14] uses two t-distributions with shifted centers to jointly model the positions of forward and reverse reads. A limitation of this approach is that it assumes that the peak shapes can be described by scaled t-distributions, which may not reflect the true peak shapes as they could be asymmetric. GPS [9] implements a more flexible approach to use an empirical distribution to characterize the peak shapes. The estimation of the shapes and peak calling are iterated until convergence. Both PICS and GPS are computationally intensive, especially when the total read counts is large because they model the position of all aligned reads. T-PIC [17] is another method that partially uses the peak shape information by summarizing it into a one-number statistic. However it only reflects certain aspects of the peak shape and cannot capture the full detail. Moreover, T-PIC does not treat the forward and reverse strand reads separately and ignores the offset between them.

In this article we propose a new method called PolyaPeak to utilize the peak shape information for detecting TFBSs. PolyaPeak models the read counts from equal sized bins around the binding sites and describes the peak shapes using a multivariate Pólya distribution. It then uses a hierarchical model to integrate the peak shape and the read count information to identify binding sites. Compared with PICS and GPS, PolyaPeak models the bin counts so it's more computationally efficient and can be easily embedded into MACS or CisGenome as a downstream peak ranking algorithm. Therefore, PolyaPeak provides a more flexible and efficient model for utilizing the peak shape information. Our extensive real data analyses show that its performance is robustly among the bests compare to several state-of-the-art peak callers.

Materials and Methods

We use a two-step procedure to detect binding sites. In the first step, a simple and fast peak calling algorithm based on smoothing is applied to roughly identify the locations and summits of candidate peaks. In this article we use MACS as the first step peak caller although one can easily replace MACS by other methods. In the second step, the candidate peaks are scored and ranked by a more sophisticated model which considers both the read count and the peak shape information. Since the first step is based on existing algorithms, this article focuses on the second step.

A hierarchical model for peak scoring

Assume there are P candidate peaks obtained from the first step peak calling. For each peak, we take an L ($L = 800$ by default) base pair (bp) window centered at its summit. The window is divided into equal sized non-overlapping bins of S bp long ($S = 50$ by default). The number of reads within each bin is obtained. Reads aligned to the forward and reverse strands are counted separately. Hereinafter, "peak" refers to the L bp window. For peak p, denote the read counts on the forward and reverse strands in bin $0i$ and sample j by Y_{fij}^p and Y_{rij}^p respectively. Here $j = 1$ or 0 refers to IP or control sample, and $i = 1, 2, \ldots, L/S$. Reads from replicate samples are pooled together. Define $\mathbf{Y}^p = \{ Y_{fij}^p, Y_{rij}^p : i = 1, 2, \ldots, L/S; j = 0, 1 \}$ to be read counts from all bins and all samples for peak p. Let T_p be the total read count in peak p: $T_p = \sum_i \sum_j (Y_{fij}^p + Y_{rij}^p)$.

Let Z_p indicate whether peak p is a true binding site ($= 1$) or not ($= 0$). We assume that *a priori*, the probability for a candidate peak being a true binding site is q, or $P(Z_p = 1) = q$. Then the observed bin-level read counts $\mathbf{Y}^p$ for each peak are assumed to be generated hierarchically. First, a total read count T_p is drawn from certain distribution. Second, the T_p reads are randomly allocated to different bins based on a probability distribution that specifies the peak shape.

The distributions for generating T_p conditioned on Z_p are:

$$T_p | Z_p = 1 \sim Unif[a, b], \quad T_p | Z_p = 0 \sim (1 - \epsilon) NB(\alpha, \beta) + \epsilon Unif[a, b] \quad (1)$$

In other words, if $Z_p = 1$, the candidate peak p is a true binding site and T_p is assumed to follow a uniform distribution. If $Z_p = 0$, the candidate peak p represents background noise and T_p is assumed to follow a mixture of a negative binomial distribution and a uniform distribution. Negative binomial distribution is a popular choice for modeling the background read counts [7,12]. Compared with Poisson, it allows over-dispersion and provides better fit to the data. Our choice of using a mixture of negative binomial and uniform for background is motivated by real data observation. It implies that for most background regions, the total counts follow a negative binomial distribution. However, some non-binding regions may have unusually large read counts due to artifacts. These outliers are modeled by the uniform mixing

component. Technically, these model assumptions guarantee that the likelihood ratio $P(T_p | Z_p = 1) / P(T_p | Z_p = 0)$ increases monotonically with T_p, but is bounded at $1/\epsilon$ in interval $[a, b]$. Thus the inference will not be overly influenced by the outliers, and the results will be more robust. In PolyaPeak, we set ε to 0.001 and treat it as fixed and known.

Given T_p, the bin counts $\mathbf{Y}^p$ are assumed to follow multinomial distributions with random bin level probabilities $\mathbf{\Theta}^p$: $\mathbf{Y}^p | T_p, \mathbf{\Theta}^p \sim MN(T_p, \mathbf{\Theta}^p)$. Here $\mathbf{\Theta}^p = \{ \theta_{fij}^p, \theta_{rij}^p : i = 1, 2, \ldots, L/S; j = 0, 1 \}$, and $\sum_i \sum_j (\theta_{fij}^p + \theta_{rij}^p) = 1$. $\mathbf{\Theta}^p$ characterizes the peak shape at peak p. We further assume that the peak-specific multinomial probabilities $\mathbf{\Theta}^p$ follow Dirichlet distributions with different parameters at background and binding regions:

$$\mathbf{\Theta}^p | Z_p = k \sim Dir(\mathbf{a}_k), k = 0, 1.$$

Here $\boldsymbol{\alpha}_k = \{ \alpha_{fij}^k, \alpha_{rij}^k : i = 1, 2, \ldots, L/S; j = 0, 1; k = 0, 1 \}$. Conceptually, at a true binding site the proportions of multinomial distribution, e.g., $\mathbf{\Theta}^p$, describe the peak shape. Using a Dirichlet prior for $\mathbf{\Theta}^p$ is based on observation that peaks in real data vary in widths, heights, and shapes, possibly due to various biological and technical factors. The Dirichlet prior provides a flexible model to allow the heterogeneity and variation in the peak shapes.

Integrating out the peak specific multinomial probabilities $\mathbf{\Theta}^p$, one can obtain the marginal bin counts distribution conditional on the total counts:

$$\mathbf{Y}^p | T_p, Z_p = 0 \sim MP(T_p, \boldsymbol{\alpha}_0), \quad \mathbf{Y}^p | T_p, Z_p = 1 \sim MP(T_p, \boldsymbol{\alpha}_1) \quad (2)$$

Here MP represents multivariate Pólya distribution, also known as Dirichlet-multinomial compound distribution which generalizes the Beta-binomial distribution.

In the model above, the bin counts $\mathbf{Y}^p$ from all candidate peaks are the observed data. Unknown model parameters include q, a, b, α, β, $\boldsymbol{\alpha}_0$ and $\boldsymbol{\alpha}_1$. These parameters can be either specified or estimated from the data. Given the parameters, one can compute the posterior probability for a candidate peak being a true binding site. Such posterior probability can be decomposed into three components:

$$P(Z_p | \mathbf{Y}^p) \propto P(Z_p) \times P(T_p | Z_p) \times P(\mathbf{Y}^p | T_p, Z_p) \quad (3)$$

The first component is the prior probability for peak p to be a true binding site. The second component is the information from the total read count. The third component is the allocation of read counts in different bins conditional on the total count. This component characterizes the peak shape around the binding site and was ignored by many existing peak callers. PolyaPeak first determines the unknown parameters, then compute $P(Z_p = 1 | \mathbf{Y}^p)$ for each candidate peak and rank the candidate peaks accordingly.

Choice of parameters

Choosing a, b, α and β. Given a list of candidate peaks, the distributional parameters in Equation 1 are determined as follows. For each candidate peak, we obtain the total read count T_p. The minimum and maximum of T_p from all candidate peaks are taken as a and b for the uniform distribution.

To estimate the negative binomial parameters α and β, we first obtain genomic regions not covered by candidate peaks. These regions are cut into L bp non-overlapping windows and read counts for each window are obtained. A large proportion of the genome (e.g., the repetitive regions) is unmappable. As a result, many background windows have zero count. In practice, if there are five consecutive regions with zero count, we exclude these regions from the analysis. The real data analyses show that the counts from the remaining regions can be fitted well by a negative binomial. Using these regions, we estimate α and β by a moment estimator. A negative binomial random variable $X \sim NB(\alpha, \beta)$ has $E[X] = \alpha\beta$ and $Var(X) = \alpha\beta + \alpha\beta^2$. Let m and v be the sample mean and variance for X, we get: $\hat{\alpha} = m^2/(v-m)$ and $\hat{\beta} = (v-m)/m$.

Choosing q. The first step peak calling algorithm provides a FDR estimate for the candidate peaks. We simply estimate q by $(1-FDR) * L_b/L_g$. Here L_b is the total length of the candidate peaks, and L_g is the length of the genome after excluding the unmappable regions. Since the FDR estimate provided by the initial peak calling may be biased, the estimate of q may be inaccurate. However, q only affects the value of the posterior probability $P(Z_p = 1|\mathbf{Y}^p)$. It will not change the final peak ranking provided by PolyaPeak. For this reason, we can tolerate the bias in estimating q, since it does not compromise our main goal of improving the peak ranking.

Estimating α_0 and α_1. The procedures for estimating the parameters of multivariate Pólya distribution has been proposed previously. Here we implement the MM algorithm (reviewed in File S1) introduced by [21] to estimate α_0 and α_1.

For α_0, we first randomly sample N ($= 1000$ by default) genomic intervals of L bps from the non-peak regions. Each interval is divided into equal sized bins of S bps, and the bin read counts are obtained. Using these counts as input, α_0 can be estimated via an MM algorithm. For α_1, we first obtain N top ranked candidate peaks from the initial peak calling. These high-quality peaks are most likely to be true binding sites, and will be used as the training data to learn peak shapes. For each of the N top peaks, we obtain the bin read counts $\mathbf{Y}^p$. Using these bin counts as data, α_1 can be estimated using MM algorithm.

Ranking the peaks

With all parameters determined, PolyaPeak will compute $P(Z_p = 1|\mathbf{Y}^p)$, the posterior probability that each candidate peak is a true binding site, and use these posterior probabilities to rank the peaks. Some peaks may have the same posterior probabilities due to constraints of numerical precision. For these peaks, we use the log likelihood ratio $\log\{P(\mathbf{Y}^p|Z_p = 1)/P(\mathbf{Y}^p|Z_p = 0)\}$ to break the ties.

Implementation

The proposed method is implemented as an R package titled "PolyaPeak" for re-ranking a list of peaks reported from another peak calling software. The computational intensive part (MM algorithm) was implemented in C for efficiency. The software package can be freely downloaded from *http://web1.sph.emory.edu/users/hwu30/polyaPeak.html*.

Results

Data

We tested PolyaPeak on a large number of publicly available datasets and obtained similar results. Here we present representative results using six datasets generated by three different labs. The first two datasets were generated by [22] for mapping binding sites of mouse TFs OCT4 (POU5F1) and MYCN in mouse embryonic stem cells (mESC). The reads aligned to mouse genome build mm8 were downloaded from the Gene Expression Omnibus (GEO) (accession number GSE11431). The number of uniquely

aligned reads for each sample ranged from 3 to 9 million. Each dataset contained one IP and one control sample. Datasets 3 and 4 were generated by Snyder lab at Yale/Stanford University for mapping the binding sites of MYC and MAX TFs in human K562 cell line. The data were generated as part of the ENCODE project. Sequence reads aligned to human genome build hg18 were downloaded from the ENCODE data coordination center. Each dataset contained two IP and one Input control samples. There were 10 to 20 million uniquely aligned reads in each sample. Datasets 5 and 6 were generated by the HudsonAlpha Institute, also for ENCODE project, for mapping the binding sites of GABP and NRSF TFs in human HepG2 cell line (GEO accession numbers GSM803343 for GABP and GSM803344 for NRSF). We downloaded the raw sequence reads aligned to human genome build hg19. Each dataset contained two IP and one Input control samples. Each sample had 50 to 100 million uniquely aligned sequence reads.

The diversity of the test data in terms of lab origin, cell type, species, reference genome and sequencing depth demonstrates the general applicability of PolyaPeak.

Exploratory analysis

First, we explored various characteristics of the real data. Figure 2(a)–(c) shows histograms of background window read counts T_p in the mESC OCT4, K562 MYC, and HepG2 GABP data. In each plot, the density curve of a negative binomial distribution with parameters estimated from the proposed method of moment estimator is also shown. It shows that the background total counts can be approximately modeled by a negative binomial distribution, and the estimation procedure for choosing negative binomial parameters works well.

Figure 2(d) shows the estimated peak shapes from the K562 MYC data. Plotted in the figure are the mean shapes for binding sites, e.g., $\alpha^1_{fij}/|\alpha_1|$ and $\alpha^1_{rij}/|\alpha_1|$ for $j = 0,1$. The estimated parameters are able to capture the location shift between the forward and reverse strand peaks, and the enrichment of IP compared with control samples. Figure 2 are only representative examples to illustrate key data characteristics. Analyses of the other datasets produced similar results.

Comparison with other methods

Next, we compared PolyaPeak with several existing peak calling methods: MACS, CisGenome, PICS, GPS and T-PIC. While a number of peak calling methods have been developed, two comprehensive and independent benchmark studies showed that MACS robustly performs among the best in terms of peak calling sensitivity and overall receiver operating characteristics [18,19]. For this reason, we used MACS as the baseline to benchmark our method. CisGenome is another popular software tool for peak calling with relatively high specificity. Currently, MACS and CisGenome are the two most cited ChIP-seq peak calling tools according to both the ISI Web of Science and Google Scholar using "ChIP-seq" as the query keyword. PICS, GPS and T-PIC are recently developed peak calling algorithms which attempt to use the peak shape information. This makes them different from other peak callers that do not fully utilize the shape information. Since the main point of this paper is to use peak shape to improve TFBS detection, we included PICS, GPS and T-PIC into our comparisons to test the effectiveness of our model.

For each dataset and each peak calling method except T-PIC, the analysis produced a ranked peak list. We compared different methods based on the enrichment of DNA motifs in the reported peaks. To avoid biases caused by peak lengths, we truncated or

(a) mES OCT4

(b) K562 MYC

(c) HepG2 GABP

(d) Estimated peak shape

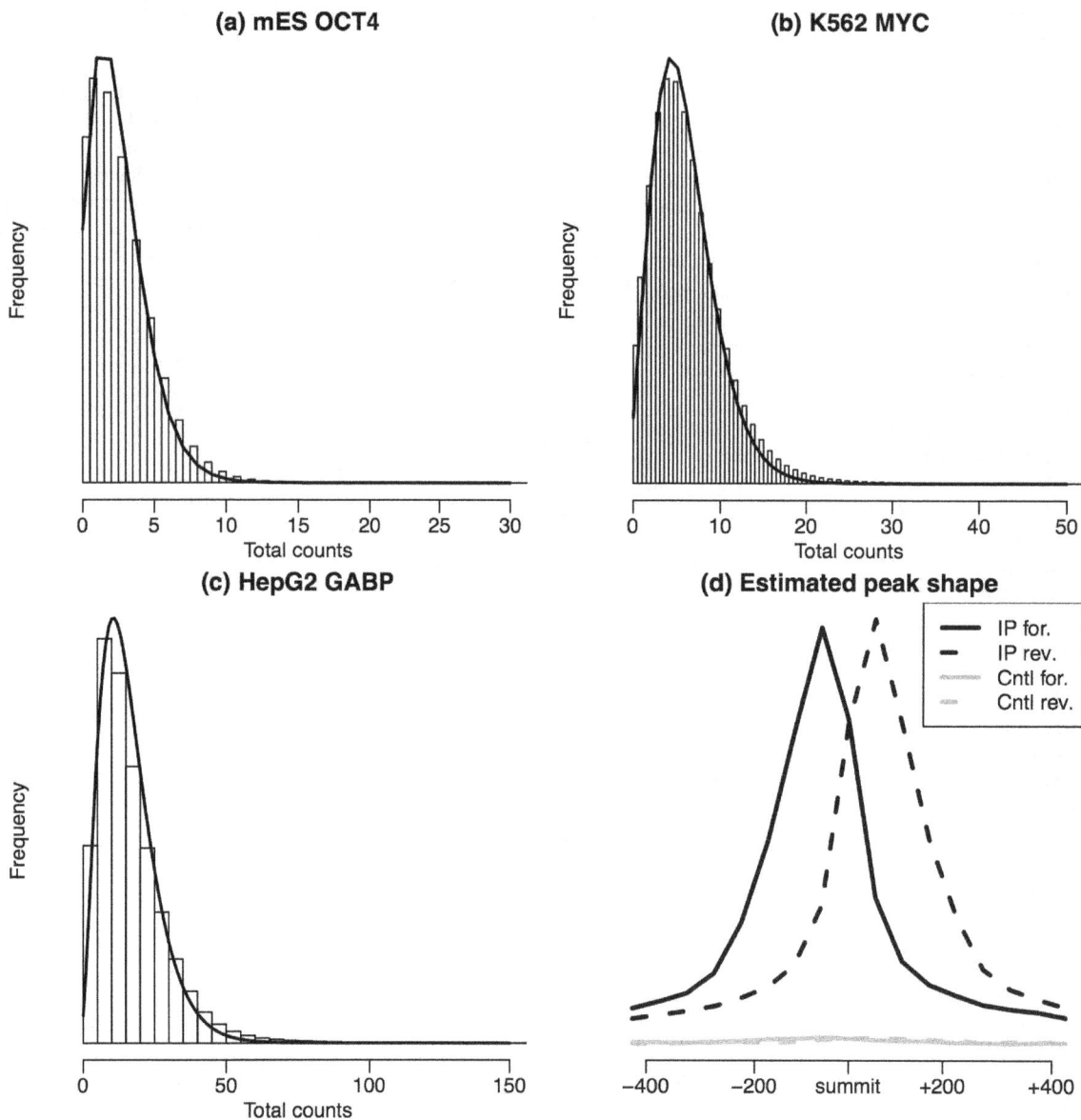

Figure 2. (a)–(c): histogram of total read counts in background regions from different datasets: (a) TF OCT4 in mouse ES cells; (b) TF MYC in K562 cell line; and (c) TF GABP in HepG2 cell line. The black solid curve is the theoretical density for the negative binomial distribution with parameters estimated from the method of moment procedure. It can be seen that the total counts in background regions can be approximated well by a negative binomial distribution, and the method of moments for estimating negative binomial parameters works well. (d) Estimated peak shape from polyaPeak for K562 MYC data.

extended all peaks to make them having the same length (200 bps) before the motif analysis. For each ranked peak list, we evaluated the percentage of top N peaks containing at least one DNA motif site for the TF. We then plotted this percentage as a function of N. The DNA motif sites were obtained using the motif mapping function matchPWM in the Biostrings packages in BioConductor. The position weight matrices (PWMs) were obtained from TRANSFAC [23].

Figure 3 compares the motif contents in peaks ranked by different methods. The results show that top peaks reported by PolyaPeak consistently have higher or comparable motif enrichment level compared with peaks reported from MACS, CisGenome or PICS, and the improvement could be substantial. For example, in the HepG2 GABP data, 56% of of the top 1000 peaks

reported by PolyaPeak contained at least one GABP motif, whereas the percentages was 40%, 38% and 40% for MACS, CisGenome and PICS respectively. Thus PolyaPeak improved MACS by 40%. Compared to GPS, PolyaPeak outperformed in mESC OCT4 and MYCN, as well as K562 MAX data. GPS outperformed PolyaPeak in K562 MYC and HepG2 GABP data. The two performed similarly in HepG2 NRSF data. These results showed that overall, the performance of PolyaPeak is slightly better than GPS in terms of motif enrichment. Among all existing methods, we found that overall GPS and PICS provides better results than MACS and CisGenome, which is not surprising since it has incorporated peak shape information. In all the data we have tested, PolyaPeak robustly performed among the bests.

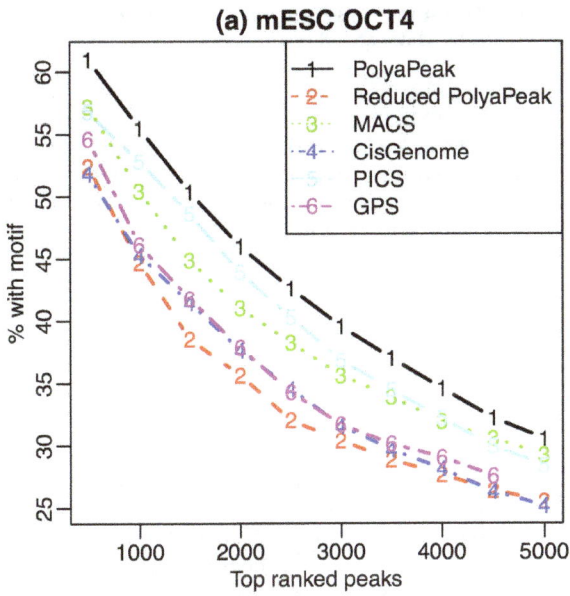

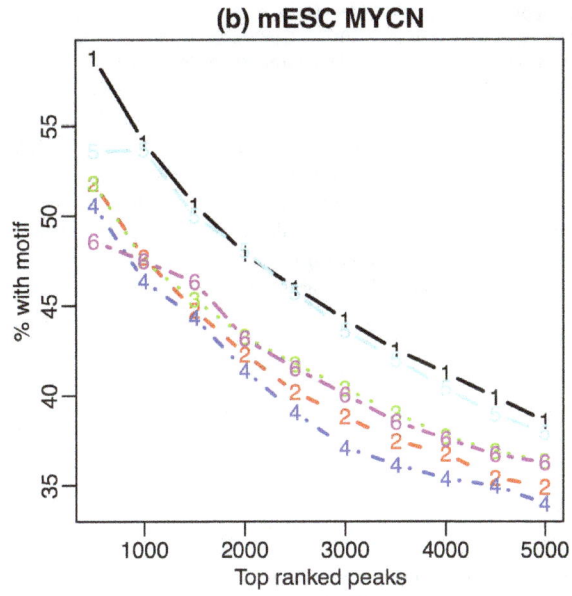

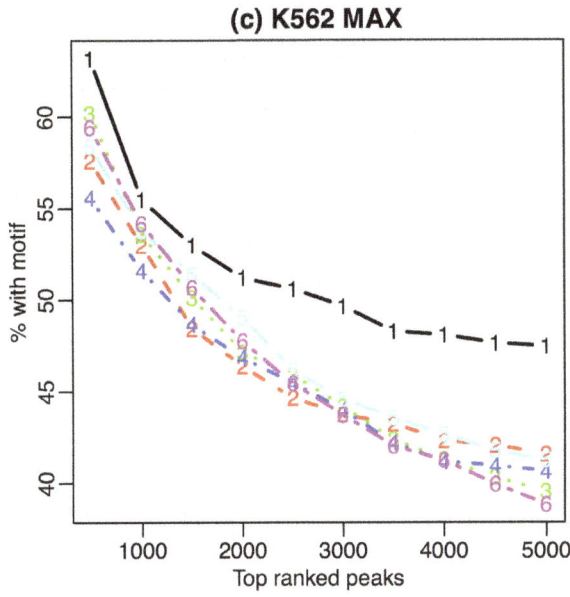

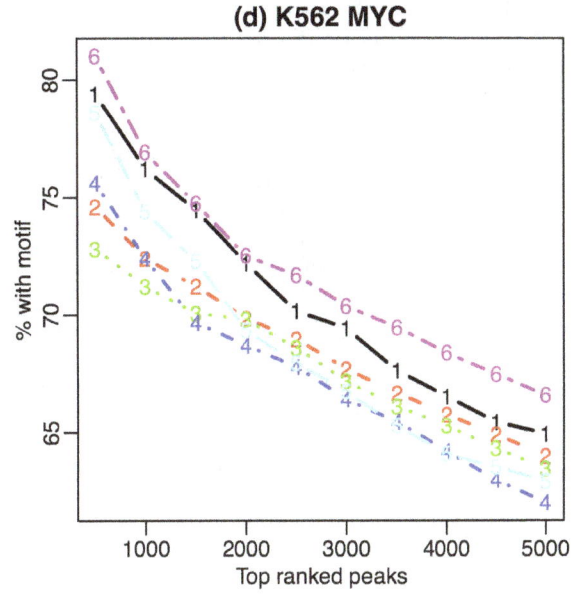

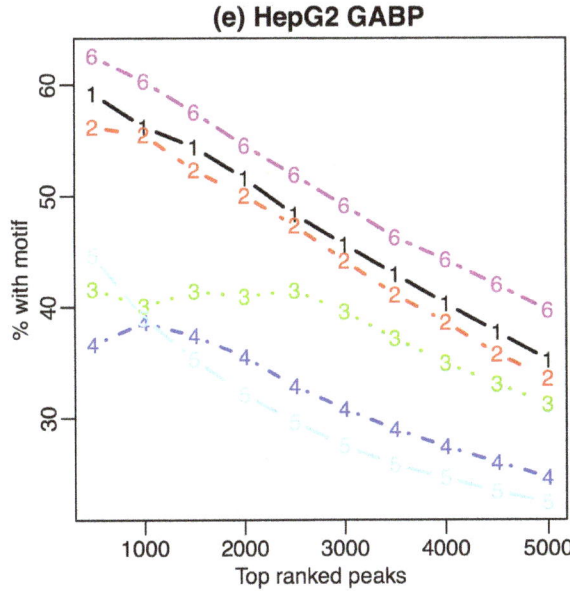

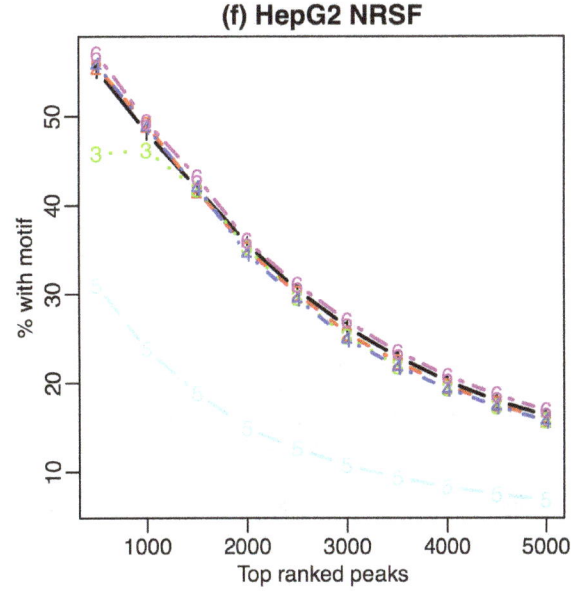

Figure 3. Comparison of motif enrichment in top ranked peaks reported from different methods (PolyaPeak, reduced PolyaPeak without peak shape information, MACS, CisGenome, PICS and GPS) for six public datasets. X-axis represents peak ranks. Y-axis represents the percentage of reported peaks containing at least one DNA motif site.

The comparison to T-PIC was more difficult since T-PIC runs very slow and usually reports larger number of peaks without providing peak rankings. Therefore we were only able to roughly compare the performance based on the overall motif contents of all reported peaks. Details of this comparison are provided in File S1, and the results show that PolyaPeak again performed better than T-PIC.

The net gain by modeling peak shapes

The observed differences between PolyaPeak and the other methods can be caused by many different factors, such as differences in statistical models, parameter estimation methods, or implementation details. MACS was used as the first step peak caller for PolyaPeak. Even for the relatively well-controlled comparison between PolyaPeak and MACS, the observed differences could be due to differences in (1) modeling the total read counts, i.e., $P(T_p|Z_p)$, or (2) using the peak shape information, i.e., $P(\mathbf{Y}^p|T_p, Z_p)$. In order to determine whether the peak shape information really played an important role in improving the peak ranking performance, we developed a reduced PolyaPeak model by removing the peak shape information, and compared the reduced model with the original PolyaPeak that uses the peak shapes. To develop the reduced PolyaPeak model, we did not break peaks into L/S bins. Instead, for each peak we only counted the number of IP and control reads in the L bp window centered at the peak summit. As a result, we obtained two numbers per peak: $\mathbf{Y}^p = \{Y_0^p, Y_1^p\}$ where Y_1^p is the read count from the IP sample, and Y_0^p is the read count from the control sample. $T_p = Y_0^p + Y_1^p$ is the total read counts. We then modeled $\mathbf{Y}^p$ similarly as in Formulas 1–3. However, rather than having $4 * L/S$ numbers per peak as the observed data, now we only have two numbers per peak. The reduced model still allows one to use the information from the IP and control read count differences, but the peak shape information is lost. We fit the reduced model and applied it to rank peaks in the same way as the original PolyaPeak. Figure 3 shows that the original PolyaPeak consistently outperformed the reduced model. This carefully controlled comparison clearly demonstrates that incorporating the peak shape information is able to substantially improve the accuracy for detecting true binding sites.

The change of peak rankings

We further looked at the peaks with large rank changes between the results from MACS and PolyaPeak. Figure 4 shows several examples of MACS peaks from the K562 MYC data that are down- or up-ranked by PolyaPeak. Figure 4 (a) and (b) show two peaks with high ranks from MACS and low ranks from PolyaPeak. This type of peaks often contain large read counts but do not have a clear shape. None of the peaks shown here contained MYC motif sites. Therefore the high read counts in these regions very likely represent artifacts. Figure 4 (c) and (d) are two examples of peaks up-ranked by PolyaPeak (ranked low from MACS but high from PolyaPeak). Both of them have nice shapes, and contain MYC motif.

Furthermore, it is found that the peaks ranked high in MACS could have huge rank reduction in PolyaPeak. On the other hand, the top ranked peaks from PolyaPeak usually are also ranked high in MACS and seldom have dramatic rank reduction. These results

suggest that PolyaPeak mainly worked as a false positive filter to down-rank low-quality MACS peaks. In general, MACS peaks down-ranked by PolyaPeak have lower motif content. As a result, the overall motif enrichment level is higher from PolyaPeak results, as shown in Figure 3.

Discussion

Previous benchmark studies have shown that MACS robustly performs among the best compared to other peak callers. Our results show that by fully utilizing the peak shape information, PolyaPeak is able to robustly outperform MACS. Compared with the other two packages that uses peak shape information (PICS and GPS), PolyaPeak also outperforms based on our tests. PolyaPeak is computationally efficient because it models the read counts from equal sized bins, whereas both PICS and GPS model the positions of all aligned reads so their computational burden grows with total reads. While we deliver PolyaPeak as an R package, the computation intensive parts were written in C. On a computer with 2.6 GHz CPU and 32 GB RAM running Linux, the total computation time (using MACS to call peaks then PolyaPeak to rank peaks) for an experiment with 25 million total aligned reads is around 15 minutes. As a comparison, GPS takes over an hour and PICS takes more than two hours on a single CPU.

The performance difference between PolyaPeak and other peak callers can be caused by different factors. Even for the relatively well-controlled comparison between PolyaPeak and MACS, we do not have enough information to tell whether peak shape really helps. For this reason, and also because MACS has already been shown by others to have favorable performance, we did not include more peak callers into our comparisons as these comparisons will not produce generalizable principle for future algorithm design. Instead, we performed a well-controlled comparison between PolyaPeak and the reduced PolyaPeak. The only difference is that the reduced PolyaPeak does not use the peak shape information. This comparison is more informative than the comparison between PolyaPeak and the other peak callers since it clearly shows that peak shape information brings improvement. This produces a generalizable principle which in the future may be used together with other established principles (e.g., adjusting for GC content [12]) to continually improve the peak calling algorithms.

PolyaPeak is specifically designed to re-rank the peaks using the shape information. It is important to point out that PolyaPeak does not model the clustering of multiple peaks within a small region, like in GPS and PICS. As a result, PolyaPeak could down rank regions with several closely spaced peaks. Nevertheless we found that overall the improvements overweigh the sacrifices as shown in the real data results.

When modeling a shape, it is often helpful to encourage some smoothness. In our model, this can be achieved by introducing some regularization procedures such as to penalize the second-order derivative of the parameter α_1 with respect to the bin locations. Such a model could potentially further improve the results, especially when the peaks are wider. This is a research topic worth exploring in the near future.

PolyaPeak is able to improve TFBS detection because there is strong shape information at the true binding sites in TF ChIP-seq

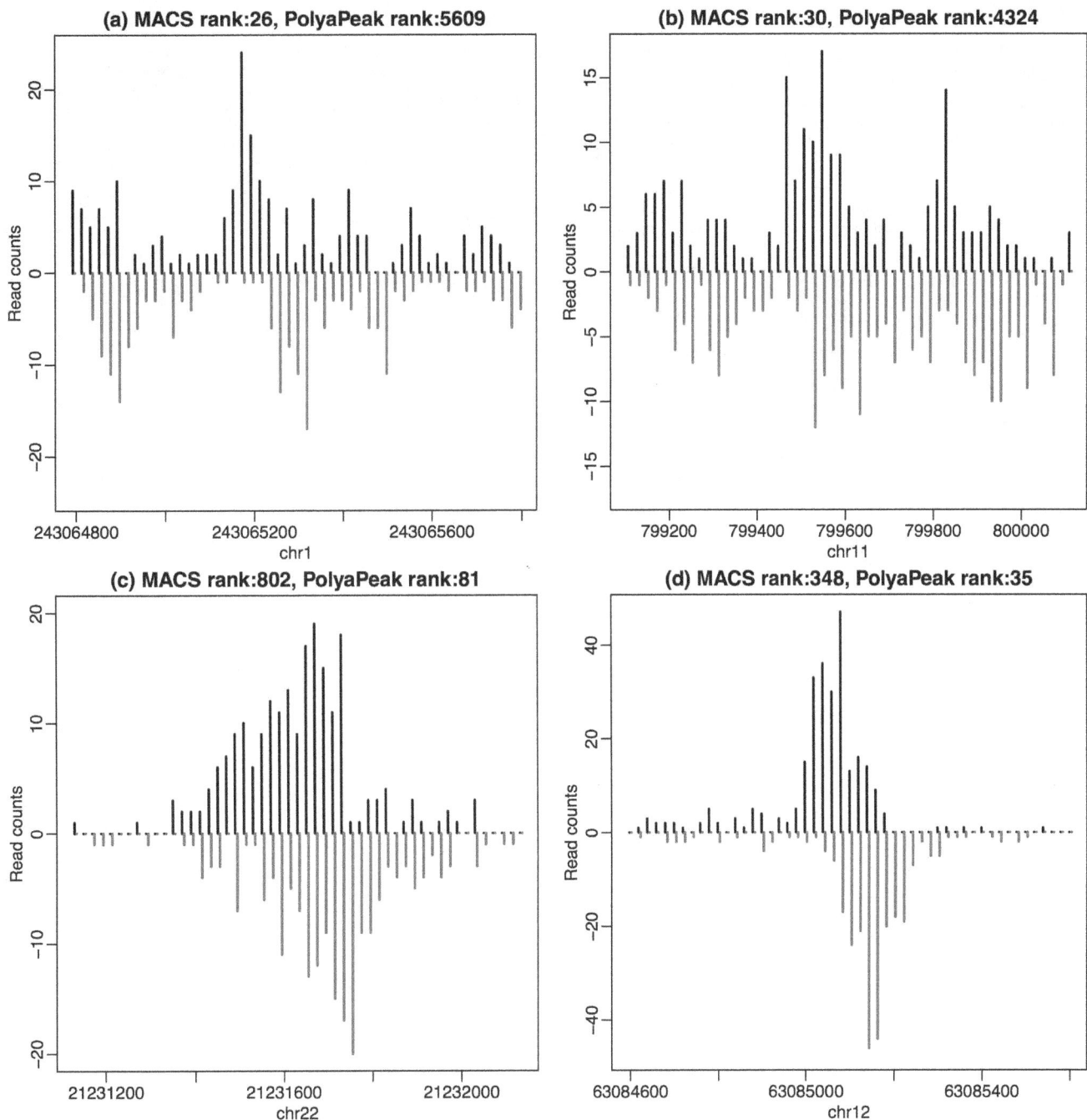

Figure 4. Examples of peaks showing large differences in ranks from MACS and PolyaPeak. (a) and (b): Peaks ranked high in MACS but low in PolyaPeak. (c) and (d): Peaks ranked low in MACS but high in PolyaPeak.

experiments. Examples shown in Figure 4 shows that peaks containing large number of reads but didn't show the desired pattern are often down-ranked. On the other hand, peaks containing relatively less reads but showing strong pattern will be up-ranked. ChIP-seq can also be used to study histone modifications (HMs). However, HM ChIP-seq signals usually are highly variable in terms of their widths and shapes. Therefore, a future research topic is to explore whether peak shape is helpful for analyzing HM ChIP-seq data as well.

Author Contributions

Conceived and designed the experiments: HW. Analyzed the data: HW. Wrote the paper: HW HJ.

References

1. Johnson D, Mortazavi A, Myers R, Wold B (2007) Genome-wide mapping of in vivo protein-dna interactions. Science 316: 1497.

2. Robertson G, Hirst M, Bainbridge M, Bilenky M, Zhao Y, et al. (2007) Genome-wide profiles of stat1 dna association using chromatin immunoprecipitation and massively parallel sequencing. Nature methods 4: 651–657.

3. Thomas D, Rosenbloom K, Clawson H, Hinrichs A, Trumbower H, et al. (2006) The encode project at uc santa cruz. Nucleic acids research 35: D663.

4. Celniker SE, Dillon LA, Gerstein MB, Gunsalus KC, Henikoff S, et al. (2009) Unlocking the secrets of the genome. Nature 459: 927–930.

5. Chen L, Wu G, Ji H (2011) hmchip: a database and web server for exploring publicly available human and mouse chip-seq and chip-chip data. Bioinformatics 27: 1447–1448.

6. Spyrou C, Stark R, Lynch A, Tavaré S (2009) Bayespeak: Bayesian analysis of chip-seq data. BMC bioinformatics 10: 299.

7. Ji H, Jiang H, Ma W, Johnson D, Myers R, et al. (2008) An integrated software system for analyzing ChIP-chip and ChIP-seq data. Nature Biotechnology 26: 1293–1300.

8. Fejes AP, Robertson G, Bilenky M, Varhol R, Bainbridge M, et al. (2008) Findpeaks 3.1: a tool for identifying areas of enrichment from massively parallel short-read sequencing technology. Bioin-formatics 24: 1729–1730.

9. Guo Y, Papachristoudis G, Altshuler R, Gerber G, Jaakkola T, et al. (2010) Discovering homotypic binding events at high spatial resolution. Bioinformatics 26: 3028–3034.

10. Qin Z, Yu J, Shen J, Maher C, Hu M, et al. (2010) Hpeak: an hmm-based algorithm for de_ning read-enriched regions in chip-seq data. BMC bioinformatics 11: 369.

11. Zhang Y, Liu T, Meyer C, Eeckhoute J, Johnson D, et al. (2008) Model-based analysis of chip-seq (macs). Genome Biol 9: R137.

12. Kuan P, Chung D, Pan G, Thomson J, Stewart R, et al. (2011) A statistical framework for the analysis of chip-seq data. Journal of the American Statistical Association doi:10.1198/jasa.2011.ap09706.

13. Rozowsky J, Euskirchen G, Auerbach R, Zhang Z, Gibson T, et al. (2009) Peakseq enables systematic scoring of chip-seq experiments relative to controls. Nature biotechnology 27: 66–75.

14. Zhang X, Robertson G, Krzywinski M, Ning K, Droit A, et al. (2010) Pics: Probabilistic inference for chip-seq. Biometrics.

15. Valouev A, Johnson D, Sundquist A, Medina C, Anton E, et al. (2008) Genome-wide analysis of transcription factor binding sites based on chip-seq data. Nature methods 5: 829.

16. Jothi R, Cuddapah S, Barski A, Cui K, Zhao K (2008) Genome-wide identification of in vivo protein–dna binding sites from chip-seq data. Nucleic acids research 36: 5221–5231.

17. Hower V, Evans S, Pachter L (2011) Shape-based peak identification for chip-seq. BMC bioinfor-matics 12: 15.

18. Laajala T, Raghav S, Tuomela S, Lahesmaa R, Aittokallio T, et al. (2009) A practical comparison of methods for detecting transcription factor binding sites in chip-seq experiments. BMC genomics 10: 618.

19. Wilbanks E, Facciotti M (2010) Evaluation of algorithm performance in chip-seq peak detection. PLoS One 5: e11471.

20. Rye M, Sætrom P, Drabløs F (2011) A manually curated chip-seq benchmark demonstrates room for improvement in current peak-finder programs. Nucleic acids research 39: e25.

21. Zhou H, Lange K (2010) Mm algorithms for some discrete multivariate distributions. Journal of Computational and Graphical Statistics 19: 645–665.

22. Chen X, Xu H, Yuan P, Fang F, Huss M, et al. (2008) Integration of external signaling pathways with the core transcriptional network in embryonic stem cells. Cell 133: 1106–1117.

23. Matys V, Fricke E, Geffers R, Goessling E, Haubrock M, et al. (2003) Transfac R: transcriptional regulation, from patterns to profiles. Nucleic acids research 31: 374.

20

A Conserved BDNF, Glutamate- and GABA-Enriched Gene Module Related to Human Depression Identified by Coexpression Meta-Analysis and DNA Variant Genome-Wide Association Studies

Lun-Ching Chang[1], Stephane Jamain[2,3,4], Chien-Wei Lin[1], Dan Rujescu[5], George C. Tseng[1,6]*, Etienne Sibille[7]*

1 Department of Biostatistics, University of Pittsburgh, Pittsburgh, Pennsylvania, United States of America, 2 Inserm U955, Psychiatrie Génétique, Créteil, France, 3 Université Paris Est, Créteil, France, 4 Fondation FondaMental, Créteil, France, 5 Department of Psychiatry, University of Halle, Halle, Germany, 6 Department of Human Genetics, University of Pittsburgh, Pittsburgh, Pennsylvania, United States of America, 7 Department of Psychiatry, Center For Neuroscience, University of Pittsburgh, Pittsburgh, Pennsylvania, United States of America

Abstract

Large scale gene expression (transcriptome) analysis and genome-wide association studies (GWAS) for single nucleotide polymorphisms have generated a considerable amount of gene- and disease-related information, but heterogeneity and various sources of noise have limited the discovery of disease mechanisms. As systematic dataset integration is becoming essential, we developed methods and performed meta-clustering of gene coexpression links in 11 transcriptome studies from postmortem brains of human subjects with major depressive disorder (MDD) and non-psychiatric control subjects. We next sought enrichment in the top 50 meta-analyzed coexpression modules for genes otherwise identified by GWAS for various sets of disorders. One coexpression module of 88 genes was consistently and significantly associated with GWAS for MDD, other neuropsychiatric disorders and brain functions, and for medical illnesses with elevated clinical risk of depression, but not for other diseases. In support of the superior discriminative power of this novel approach, we observed no significant enrichment for GWAS-related genes in coexpression modules extracted from single studies or in meta-modules using gene expression data from non-psychiatric control subjects. Genes in the identified module encode proteins implicated in neuronal signaling and structure, including glutamate metabotropic receptors (GRM1, GRM7), GABA receptors (GABRA2, GABRA4), and neurotrophic and development-related proteins [BDNF, reelin (RELN), Ephrin receptors (EPHA3, EPHA5)]. These results are consistent with the current understanding of molecular mechanisms of MDD and provide a set of putative interacting molecular partners, potentially reflecting components of a functional module across cells and biological pathways that are synchronously recruited in MDD, other brain disorders and MDD-related illnesses. Collectively, this study demonstrates the importance of integrating transcriptome data, gene coexpression modules and GWAS results for providing novel and complementary approaches to investigate the molecular pathology of MDD and other complex brain disorders.

Editor: Monica Uddin, Wayne State University, United States of America

Funding: This work was supported by National Institute of Mental Health (NIMH) MH084060 (ES) and MH085111 (ES). The funding agency had no role in the study design, data collection and analysis, decision to publish and preparation of the manuscript. The content is solely the responsibility of the authors and does not necessarily represent the official views of the NIMH or the National Institutes of Health.

Competing Interests: The authors have declared that no competing interests exist.

* E-mail: ctseng@pitt.edu (GCT); sibilleel@upmc.edu (ES)

Introduction

Major depressive disorder (MDD) is a common psychiatric disease with an estimated prevalence of 5.3% for a 12- month period and 13.2% for a lifetime disorder [1], a high rate of recurrence [2], a higher prevalence in women [3], and a heritability of 37% (95% CI = 31%–42%) [4]. Transcriptome (the set of all expressed genes in a tissue sample) and genome-wide association studies (GWAS) have separately provided clues to mechanisms of MDD, although not to the anticipated extent. Transcriptome studies mostly focus on changes in gene expression in disease states (altered expression), but also provide unique opportunities for assessing the less-investigated changes in the coordinated function of multiple genes (altered coexpression) [5]. GWAS seek to identify genetic markers for diseases, and have generated some findings in MDD [6,7,8,9,10], but overall results from GWAS meta-analyses have been disappointing [11,12], potentially due to the complexity of the disease and heterogeneity of patient cohorts. GWAS and transcriptome studies are highly complementary in that they provide unbiased and large scale investigation of DNA structural [single nucleotide polymorphisms (SNP) and other variants] and functional (RNA expression) changes across conditions, although these two approaches are only beginning to be integrated [13,14,15,16,17].

Gene arrays allow for the unbiased quantification of expression (mRNA transcript levels) for 10,000 to 20,000 genes simultaneously. Since gene transcript levels represent the integrated output of many regulatory pathways, the study of all expressed genes provides an indirect snapshot of cellular function under diverse conditions. For instance, using postmortem brain samples, this approach has implicated dysregulated BDNF, GABA, glutamate and oligodendrocyte functions in MDD [18,19,20,21,22]. However, current studies are still few, were performed in heterogeneous cohorts, and utilized early and rudimentary versions of gene arrays. Moreover, gene array studies are subject to similar limitations as early GWA studies, in that large number of genes are tested in few subjects (n = 10–100). Typical analyses identify 1–10% of genes affected in the illness (differentially expressed genes), are characterized by high rates of false discovery, and may be confounded by numerous clinical (drug exposure, subtypes, duration, etc.), demographic (age, sex, race), technical parameters (RNA integrity, brain pH, postmortem interval for brain collection), or other potential co-segregating factors of unknown origin (See [13] for discussion). Conditions of postmortem brain collection also preclude the reliable identification of acute state-dependent gene changes, but are appropriate for investigating stable long-term disease-related homeostatic adaptations.

Gene coexpression studies offer complementary perspectives on gene changes in the context of transcriptome studies. Here, two genes are defined as coexpressed in a dataset if their patterns of expression are correlated across samples. Coexpression reflects possible shared function between genes, and may arise through multiple biological pathways including cellular coexpression and common regulatory pathways (e.g., hormone signaling, transcription factors) [23,24]. Hence, coexpression links have been used to build gene networks, and to identify communities, or modules, of genes with shared functions [25,26]. Notably, by incorporating multiple interactions among a large number of genes, the study of gene coexpression networks provides an approach to tackle the complexity of biological changes occurring in complex polygenic disorders [24]. See [5] for a general review.

Concepts and methods for integrating functional (transcriptome) and structural (DNA polymorphism GWA) studies of the molecular bases of complex neuropsychiatric disorders such as MDD need to be developed to harness the potential of systematic large-scale molecular and genetic investigations of the brain. Here, our central hypothesis states that stable brain co-regulation modules identified through meta-analysis of multiple transcriptome studies may overlap with sets of genes and associated SNPs related to MDD. Based on the continuum of pathological changes between MDD and other brain disorders [27] and co-morbidity with selected medical illnesses including cardiovascular diseases and metabolic syndrome [28,29], we also predicted that MDD coexpression modules may be enriched in genes identified by GWAS for other psychiatric and brain disorders and potentially for medical illnesses related to depression, together identifying functionally-coherent gene sets implicated in MDD-related disease processes.

Materials and Methods

Figure 1 illustrates the meta-clustering and validation methods of the approach. In step I, we identified 50 robust co-regulation modules in human brains by combining 11 transcriptome datasets collected from several brain regions in different cohorts of subjects with MDD and non-affected comparison subjects. Steps II and III were performed to identify MDD-related gene modules, and

exclude other gene modules linked to biological functions not related to MDD. In step II, we collected different sets of genes located nearby SNPs identified by GWAS for MDD, neuropsychiatric disorders, related traits, and for systemic diseases often associated with psychiatric disorders, and performed gene set analysis to identify MDD-related gene modules. In step III, we performed functional annotations of gene module members by using 2,334 gene sets collected from MSigDB (http://www. broadinstitute.org/gsea/msigdb/). We also organized genes identified by SNPs in published GWAS into three categories (cancer studies, human body indices and unrelated diseases) and treated them as a non-MDD-related negative control gene sets in step IV.

Transcriptome Data Sets

Eleven MDD microarray datasets generated in our lab were used here. Cohorts and brain areas investigated are listed in **Table 1** and details were provided in [30,31]. Among these studies, six used Affymetrix Human Genome U133 Plus 2.0 platforms (Affymetrix Inc., Santa Clara, CA), two used Affymetrix Human Genome U133A platforms, and the remaining three used Human HT-12 arrays from Illumina (Illumina Inc, San Diego, CA). **Figure S1** provides a diagram and results of the transcriptome dataset preprocessing procedures. Data has been deposited to the NCBI Geo database with accession numbers: GSE54562, GSE54563, GSE54564, GSE54565, GSE54566, GSE54567, GSE54568, GSE54570, GSE54571, GSE54572 and GSE54575.

For gene matching across studies, when multiple probes or probe sets match to one gene symbol, we choose the probe set with the largest variation (largest interquartile range; IQR) to represent the gene [32]. See below and Figure S3 for probe overlap assessment. For preprocessing, data were log-transformed (base 2). Non-expressed (small mean intensity) and non-informative (small standard deviation) genes were filtered out. To perform such filtering for 11 studies simultaneously, we calculated the ranks of row means and row standard deviations of each gene in each single study. The ranks were summed up across 11 studies and used as criteria to filter out non-expressed and non-informative genes. Note that ideally we should map the probes across platforms to large overlapped locations so we make sure they measure the same signal. There are, however, several reasons that doing so may not be possible or optimal. First, Affymetrix probesets are designed with combination of multiple short probes and Illumina arrays use a single and longer probe. As a result, Affymetrix probes have large "target regions" (044–728 KB, 95% coverage of the 88 genes of module #35 we investigated in Figure S3) which are covered by multiple short probes, while Illumina's probe is only around 50 bp. Secondly, many other factors affect signal detection efficiency, including exact probe sequence, integration of multiple probeset in Affymetrix arrays, hybridization efficiency, GC content, cross hybridization, etc. As a whole these differences can affect the consistency of the results and potentially decrease the final signal. For the purpose of running the meta-analysis (as opposed to single study analysis) it has been recommended to use the probe set with the largest IQR to represent a gene symbol [32]. We want to point out that if the IQR probe matching procedure had introduced large errors, the meta-analyzed modules would not have been detected by chance.

Meta-clustering of Transcriptomic Data to Construct Co-expression Gene Modules

The 11 transcriptome studies were combined to construct co-expression gene modules using a meta-clustering technique described below. We denoted by X_{gsk} the gene expression intensity

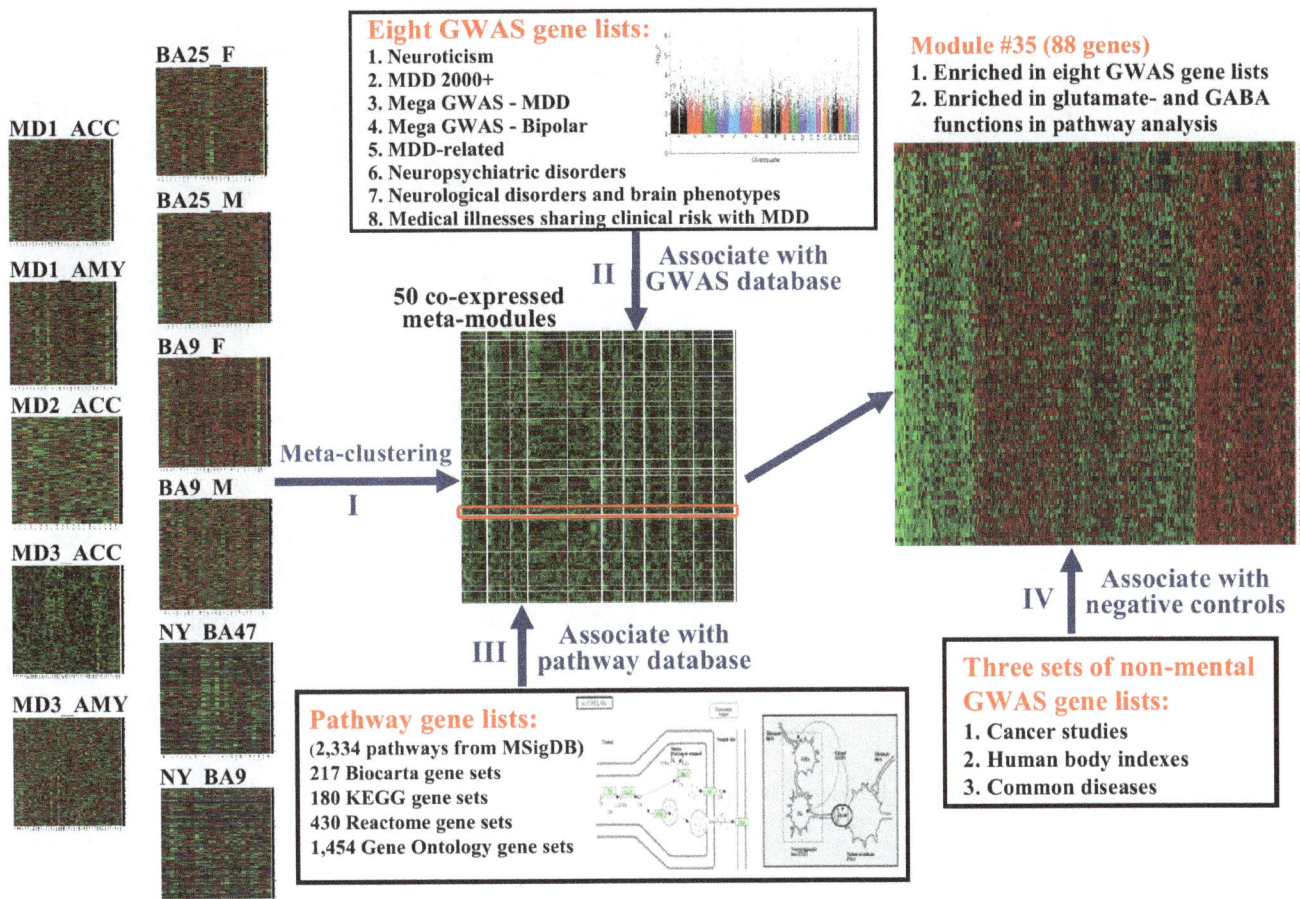

Figure 1. Overall analytical strategy. In step I, 50 co-regulation modules were generated using meta-clustering of gene clusters identified by the "penalized K-medoids" method across 11 transcriptome MDD and matched controls studies. In step II, modules enriched from most of the selected GWAS studies related to MDD, neuropsychiatric disorder and traits, including systemic disease linked to psychiatric disorders were identified. In step III, the biological functions represented by genes included in each module were defined by pathway analysis from 2,334 gene sets of MSigDB (www.broadinstitute.org/gsea/msigdb). In step IV, SNPs from the Catalog of GWAS were organized into three categories: cancer GWAS, human body indices GWAS and GWAS for common diseases and medial illnesses unrelated to MDD or other brain function. Three additional categories were defined as non-MDD-related negative control gene sets. (Note: In order to increase the performance of the heatmap in module #35, we first performed the hierarchical clustering with "complete" agglomeration method to aggregated samples with similar expression among all 88 genes, and the genes were sorted by the correlation from high to low of selected genes in the top.).

Table 1. Description of cohorts in 11 MDD microarray platforms.

Cohort	Region	Code	Platform	# of probe sets	# of genes	# of subjects
1	ACC	MD1_ACC	Affymetrix Human Genome U133 Plus 2.0	40,610	19,466	32
2	AMY	MD1_AMY	Affymetrix Human Genome U133 Plus 2.0	40,610	19,621	28
3	ACC	MD2_ACC	Illumina HumanHT –12 (v3)	48,803	25,159	20
4	ACC	MD3_ACC	Illumina HumanHT –12 (v3)	48,803	25,159	50
5	AMY	MD3_AMY	Illumina HumanHT –12 (v3)	48,803	25,159	42
6	ACC	BA25_F	Affymetrix Human Genome U133 Plus 2.0	53,596	19,572	26
7	ACC	BA25_M	Affymetrix Human Genome U133 Plus 2.0	53,596	19,572	26
8	DLPFC	BA9_F	Affymetrix Human Genome U133 Plus 2.0	53,596	19,572	32
9	DLPFC	BA9_M	Affymetrix Human Genome U133 Plus 2.0	53,596	19,572	28
10	OFC	NY_BA47	Affymetrix Human Genome U133A	20,338	12,703	24
11	DLPFC	NY_BA9	Affymetrix Human Genome U133A	20,338	12,703	26

ACC, anterior cingulate cortex; AMY, amygdala; DLPFC, dorsolateral prefrontal cortex, OFC, orbital ventral prefrontal cortex.

of gene g, sample s and study k, and $X_{gk}=(X_{g1k},\ldots,X_{gSk})$ the vector of gene expression intensities of gene g and study k. We defined the dissimilarity measure between gene i and gene j for a given study k as $d_{i,j}^{(k)}=1-|\mathrm{cor}(X_{ik},X_{jk})|$, where $\mathrm{cor}(X_{ik},X_{jk})$ is the Pearson correlation of the two gene vectors. To combine the dissimilarity information of the $K=11$ studies, we took the mean of meta-dissimilarity measure between gene i and gene j as $d(g_i,g_j)=Mean(d_{ij}^{(1)},d_{ij}^{(2)},\ldots,d_{ij}^{(K)})$. Given the meta-dissimilarity measure, the "Penalized K-medoids" clustering algorithm was then applied to construct co-expression gene modules [33]. The target function to be minimized by Penalized K-medoids is shown below

$$L(C)=\sum_{i=1}^{G}\sum_{g_i\in C_h}d(g_i,\bar{g}_h)+\lambda\cdot|S|$$

where the clustering result $C=(C_1,\cdots,C_H,S)$ contains H non-overlapping gene clusters (i.e. H gene modules $C_1,\cdots,C_H$) and a set of scattered genes S that cannot be clustered into any of the tight gene modules, $\bar{g}_h$ denotes the medoid gene of cluster h such that its average dissimilarity to all other genes in the cluster is minimal, $|S|$ is the size of the scattered gene set S and λ is a tuning parameter controlling tightness of detected gene modules and the number of scattered genes discarded to S. The first term of the target function $L(C)$ calculates the total sum of within-cluster dispersion and is essentially the K-medoids algorithm (an extended form of K-means using arbitrary non-Euclidean dissimilarity measure). The second penalty term allows scattered genes not to be clustered into any gene module. For example, if the distances of a gene g_i to all cluster medoids are greater than λ, minimizing $L(C)$ will assign the gene into the scattered gene set S, instead of into any gene cluster. Intuitively, smaller λ generates tighter clusters and allow more genes into scattered gene set S. The rationale for the choice of this approach was based on finding in the literature, where comparative studies show that many genes are not tightly co-expressed with any gene clusters and methods that allow scattered gene assignment generates tighter gene modules that are biologically more informative [34].

Parameter Selection and Evaluation of Meta-clustering

We tested different parameter settings of $H=50$ or 100 modules, and λ such that $\beta=0\%$, 25% or 50% of genes are left to scattered gene set S. In all performance of the $2\times3=6$ combinations for the meta-clustering method, a biological validation was performed using biological pathway information. We searched ten keywords ("GABA", "Insulin", "Diabetes", "Immune", "Thyroid", "Estrogen", "Depression", "Alzheimer", "Parkinson" and "Huntington") in MSigDB and finally obtained 98 MDD-related pathways. In each clustering result, Fisher's exact test was applied to each module to correlate with each of the 98 MDD-related pathways and eight GWAS gene lists and the p-values were generated. Wilcoxon signed rank test was used to compare any pair of clustering results (from different parameter setting) so that the best parameter setting could be determined.

Evaluation of Robustness and Stability of Meta-clustering Method

To evaluate the robustness of the meta-clustering results, we used the Adjusted Rand Index (ARI) as a measurement of consistency between two clustering results [35]. Specifically, ARI calculates the proportion of concordant gene pairs across two clustering results (i.e. two genes are clustered together in both

clustering results or not clustered together in both) among all possible gene pairs and the index is standardized between 0 and 1, where 0 reflects expected similarity measure of two random clustering and 1 reflects similarity measure between two identical clustering. We randomly selected a subset of studies (n = 8, 9 or 10) from 11 MDD studies and calculated the ARI to assess the similarity of the obtained modules compared to those obtained using the 11 MDD studies. The procedure was repeated 100 times and the average ARI was calculated. For the stability of meta-clustering method, the mean and standard deviation of ARIs were obtained by bootstrapping method [36] (sampled with replacement to obtain the same number of samples for each single study), where the 11 MDD studies were bootstrapped 100 times.

Genome-wide Association Studies (GWAS)-related Gene Categories

Eight neuropsychiatry-related candidate gene lists and three gene lists from presumably unrelated disorders or traits were identified from relevant GWAS. Individual genes were identified by the presence of GWAS significant SNPs within a given nucleotide distance from the coding region of that gene (UCSC hg18 with build 36.3 was used for all GWAS).

- The first gene list was obtained from a published GWAS for neuroticism [37]. Neuroticism is a personality trait that reflects a tendency toward negative mood states, and that is linked to several internalizing psychiatric conditions. That GWAS involved 1,227 healthy individuals with self-report of no diagnosis or treatment for schizophrenia, schizoaffective disorder or bipolar disorder and personality measures of neuroticism. Genotyped data were generated from Affymetrix GeneChip Human Mapping 500 K using BRLMM algorithm. 449 SNPs were selected by p-value less than 0.001, and 155 genes were identified to have contained one or more selected SNPs in the 10 kilobases (kb) up- and down-stream extension of the coding regions.

- The second gene list was obtained from the MDD 2000+ project that included a meta-analysis of MDD studies with 2,431 MDD cases and 3,673 controls [38]. Similarly, 532 SNPs with p-value less than 0.001 were mapped to gene coding regions (including 10 kb upstream and downstream regions) and 159 genes were identified.

- The third gene list was obtained from a mega-analysis of GWAS for MDD [11]. The associated 202 SNPs' p-values were less than 10^{-5} and 52 genes were identified using the University of California Santa Cruz Human Genome Browser, hg18 assembly (UCSC hg18) with build 36.3. Gene symbols from the build version 36.3 in the National Center for Biotechnology Information (NCBI) database were used.

- The fourth candidate gene list was obtained from a mega-GWAS of bipolar disease which contained 7,481 patients and 9,250 controls [39]. 6,887 SNPs were identified when p-value less than 0.001. By mapping the SNPs to gene coding region using SNPnexus software (http://snp-nexus.org/), 602 genes were obtained.

- For the fifth to eighth gene lists, we interrogated the Catalog of Published Genome-Wide Association Studies [40] (http://www.genome.gov/gwastudies/). The database (as of 01/31/13; time of the latest data analysis update) contained 10,183 entries of disease- or trait-associated SNPs with p-values smaller than 10^{-5} in 1,491 GWAS studies. We manually regrouped the disorders and traits into 4 categories: (1) all MDD-related studies, (2) all neuropsychiatric disorder studies,

(3) all neurological disorder and brain phenotypes studies, (4) all medical illnesses sharing increased risk with MDD. Note that the genes in the list #3 were included in the list #2, and genes in the list #2 were included in the list #1, which is the larger category (see more detail list in **Table S1**). Lists #4 is independent and non-overlapping with others. The associated four gene lists were then compiled, and genes were uniquely included when the mapped SNP was within the gene region including a 100 kb upstream and downstream.

– As negative controls, we identified in the catalog of published GWAS three gene sets presumably not related to psychiatric diseases: (a) 65 publications (270 genes) of cancer GWAS studies; (b) 42 publications (459 genes) of human body indices GWAS studies (HBI: genetic phenotypes for human, for example: height, weight, eye color, etc.); and (c) 33 publications (187 genes) of GWAS studies for common disease traits not related to brain function or major mental illnesses (**Table S2**).

Meta-analysis to Aggregate Evidence of Association of each Module with the GWAS Gene Lists

We performed Fisher's exact test to examine the significance of the association of genes within each coexpression module with individual GWAS-derived gene lists, using the 10,000 genes evaluated in transcriptome meta-analysis (**Figure S1**) as background. To assess statistical significance of association of each identified module from meta-clustering method, we applied the Stouffer's method to combine the p-values obtained from Fisher's exact test of the association between gene modules and eight GWAS gene sets. The Stouffer's statistics $T_{Stouffer} = \frac{\sum_{i=1}^{k} \phi^{-1}(P_i)}{\sqrt{k}}$ where ϕ is the cumulative distribution function of a standard normal distribution [41]. The p-values were assessed for each of the 50 modules from non-parametric permutation analysis by randomly selecting the same number of genes from the whole genome without replacement (using genome background 10,000 genes) for each of the 50 modules and the analysis is repeated for 500 times.

Pathway Analysis and Enrichment Analysis of GWAS Gene Lists

For biological association, 2,334 annotated pathways (gene sets) were obtained from MSigDB (www.broadinstitute.org/gsea/msigdb/), which consists of 880 canonical pathways (217 Biocarta gene sets, 180 KEGG gene sets, 430 Reactome gene sets and 53 other gene sets) and 1,454 pathways from Gene Ontology (GO). For each of the gene module, gene set (pathway) analysis was performed for the 2,334 pathways and 11 GWAS gene lists (including 3 negative controls). Fisher's exact test was performed to assess the biological association between gene modules and given gene sets. To account for multiple comparisons, Benjamini and Hochberg procedure was used to control the false discovery rate (FDR) [42].

Results

Data Preprocessing and Parameter Determination

16,443 genes were retained after gene matching across the 11 studies. Cohorts 10 and 11 were from older platforms with fewer probesets representing only 12,703 genes (**Figure S1**). In order to minimize the loss of information from gene matching, we allowed 20% missing values during matching, i.e., we kept genes with at least 9 existing measurements out of 11 studies. 13,500 genes were retained after filtering out lower sum rankings of median row

means, and 10,000 genes after further filtering out lower sum rankings of median row standard deviations. We then tested different parameter settings for the number of modules ($H = 50$ or 100), and genes (tuned the λ values for controlling tightness of detected gene modules and the number of scattered genes set) for $\beta = 0\%$, 25% or 50% of genes left out of the gene set S. In all tests of the Penalized K-medoids meta-clustering method ($2 \times 3 = 6$ combinations), we performed a validation by biological pathway information content. For all clustering results, Fisher's exact test was applied to each module to correlate with each of the 98 MDD pathways and eight GWAS gene lists described in the methods, and p-values were generated. The Wilcoxon signed rank test was used to compare any pair of clustering results (from different parameter settings) so that the best parameter setting could be determined. The result shows that there was no significant difference (by Wilcoxon signed rank test) between $H = 50$ and $H = 100$ clusters except $\beta = 0\%$ (i.e., keep all genes), and the minimum p-value of gene set analysis in $H = 50$ was always lower than that in $H = 100$ in $\beta = 25\%$ and $\beta = 50\%$. It is reasonable to set the noise level in clustering method because noise will increase if we combined more studies. We chose $H = 50$ because the mean of the $-\log10(p)$ in 50 modules (3.2793) was higher than 100 modules (3.0224) in $\beta = 25\%$, and the mean of the $-\log10(p)$ in 50 modules (3.1896) was higher than 100 modules (3.0588) in $\beta = 50\%$. 50 modules also provide adequate number and sizes of gene modules for the purpose of further analyses. Given $H = 50$, we compared the performance with different choices of β. $\beta = 25\%$ performed better than $\beta = 0\%$ (p = 0.0004 using Wilcoxon signed rank test), and there was no significant difference between $\beta = 25\%$ and $\beta = 50\%$ (p = 0.0856). Finally, we selected $H = 50$ and tuning parameter λ such that $\beta = 25\%$ genes are left to scattered gene set S throughout this paper (**Table S3**).

Construction of 50 Meta-modules from 11 MDD Studies

Using the parameters determined above, we performed a meta-analysis of module gene membership to identify the top 50 meta-analyzed coexpression modules across 11 MDD transcriptome studies. A total of 10,000 genes were clustered using the Penalized K-Medoid method. 7,797 genes were clustered into $K = 50$ modules and 2,203 genes ($\beta = \sim 25\%$) were determined as scattered genes with no coherent expression pattern. We performed subsampling and bootstrap methods to assess the stability of the resulting clusters. Subsets (n = 8, 9 or 10) of the 11 studies were randomly selected and the meta-clustering procedure was similarly applied. The resulting meta-modules were compared with the meta-modules obtained using the 11 MDD studies using adjusted Rand index (ARI = 0.47, 0.52 and 0.63 for n = 8, 9, 10). We also generated bootstrapped samples in each study and repeated the meta-clustering procedures. Comparison of meta-modules generated from bootstrapped samples with original samples generated an average ARI = 0.45 (standard deviation 0.025) in 100 repeated bootstrapping simulations. In the literature, an ARI of ~ 0.5 is interpreted as reproducible clustering result [34], hence demonstrating good stability under data perturbation (subsampling and bootstrapping) for the 50 meta-modules obtained by combining 11 studies.

Association of Meta-modules with Eleven GWAS-determined Gene Lists

We examined association of the 50 meta-modules with the eight GWAS gene lists using Fisher's exact test. The results are shown in **Table S4**. Module #35 is found to have significant associations (p<0.05) with the six psychiatric disorder related GWAS gene sets (p = 0.03 for the neuroticism GWAS gene set; p = 0.03 for MDD

2000+ project; p = 0.0001 for Mega-GWAS MDD; p = 0.03 for Mega-GWAS of bipolar disorder; p = 0.008 for the catalog of GWAS studies of neuropsychiatric disorder; p = 0.03 for the catalog of GWAS studies of neurological disorders and brain phenotypes) and two studies with borderline p-values (p = 0.05 for the catalog of MDD-related GWAS studies; p = 0.05 for the catalog of GWAS studies of Medical illnesses sharing clinical risk with MDD). We combined the p-values of the eight psychiatric disorder related GWAS gene sets by Stouffer meta-analysis method. The p-value of module #35 is $4e^{-05}$ after the permutation test (25,000 resamples). In contrast, there was no association with cancer (p = 1.00), human body indices (p = 0.18) and other control diseases (p = 0.46) GWAS gene sets. **Figure 2 (a)** shows the heatmaps of log-transformed p-values from enrichment analysis for the 50 modules obtained from MDD cases and controls combined analysis. It shows that module #35 (highlighted in green) from the combined cases and controls analysis is enriched in genes contained in six MDD-related GWAS gene sets, but not enriched in the three negative control GWAS gene sets. None of the other 49 modules showed such consistent pattern.

During the review process, a new GWAS meta-analysis for schizophrenia was published by the Psychiatric Genomics Consortium (PGC) using 1,000 genome Project imputation [43]. Accordingly, we independently examined the reported 52,509 SNPs spanning 2,507 genes under the p-value threshold of 10^{-3} (Table S9). Module #35 was significantly enriched in genes associated with this new study (p = 0.0013).

Pathway Analysis of Meta-module #35

Table S6 lists detailed information of the 88 genes in module #35 and their overlap with the eight GWAS gene lists. Many GWAS-hit genes were related to synaptic function, signal transduction, and neuronal development and morphogenesis (**Table 2**). Of specific interest, and consistent with current hypotheses for the molecular pathology of MDD, was the inclusion of brain-derived neurotrophic factor (BDNF) and other factors implicated in development and maintenance of cell circuits (Ephrin receptors EPHA3 and EPHA 5; Netrin G1 (NTNG1); SLITRK3 and SLITRK5), of GABA-related genes (GABBR2, GABRA4 and CALB1), glutamate receptors (GRM1 and GRM7) and other signaling neuropeptides previously implicated in mechanisms of psychiatric disorders [reelin (RLN) and gastrin-releasing peptide (GRP)] (**Table 2**). Results from a pathway enrichment analysis confirmed the role of genes in module #35 in overall signaling mechanisms (**Table 3**). Together, these results suggest that module #35 may include multiple components of functionally-relevant local cell circuits.

Control Studies

To demonstrate the improvement of meta-clustering versus single study clustering, we compared the histograms of p-values obtained under those different conditions. In **Figure 3**, the histogram of the minus log-transformed p-values of the Stouffer statistic was first plotted for the 50 meta-modules obtained from the case and control combined analysis. Module #35 with 88 genes is shown to have an aggregated minus log-transformed p-value at 4.4 (i.e. p = 4e-05). We then applied the penalized K-medoid method with the same parameter setting (K = 50 clusters and 25% of scattered genes) for each single study. The 11 single

(a) MDD cases and matched controls

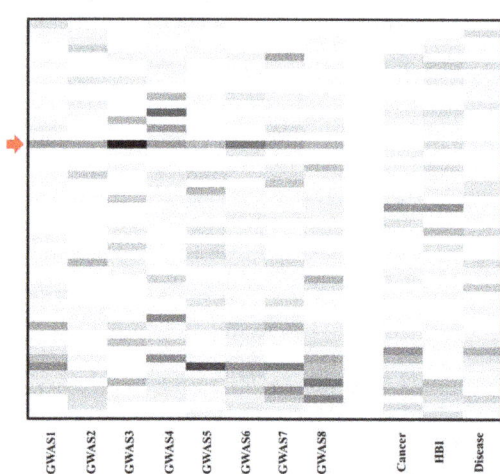

(b) Controls only

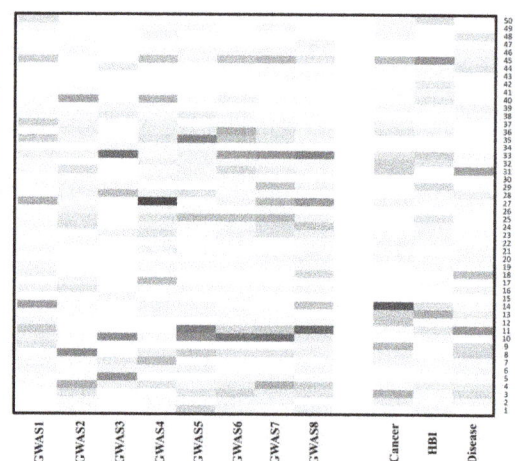

Color key

$-\log_{10}(p)$

GWAS Studies
GWAS1: Neuroticism
GWAS2: GWAS MDD 2000+
GWAS3: Mega GWAS MDD
GWAS4: Mega GWAS Bipolar

Catalog of GWAS
GWAS5: MDD-related
GWAS6: Neuropsychiatric disorder
GWAS7: Neurological disorders and brain phenotypes
GWAS8: Medical illnesses sharing clinical risk with MDD

Catalog of GWAS
Cancer: Cancer GWAS
HBI: Human body indices GWAS
Disease: Common disease traits
GWAS not related to brain function

Figure 2. Consistent association of genes in module #35 with MDD-related gene categories. (a) Heatmap of $\log_{10}$-transformed p-values from Fisher's exact test for 50 modules obtained from MDD cases and matched controls and 8 MDD related GWAS and 3 negative controls. **(b)** Heatmap of $\log_{10}$-transformed p-values from Fisher's exact test for 50 modules obtained from controls and 8 MDD related GWAS and 3 negative controls. The green rectangle identifies module #35.

Table 2. Functional groups of 88 gene in module #35.

Functional groups	Gene Symbols
Transmembrane cellular localization	CLSTN2, SYT4, LRRC8B, GPR6, TMEM158
	ST8SIA3, GABBR2, NRN1, ST6GALNAC5
	GLT8D2, MPPE1, GNPTAB, PVRL3, SLC35B4
	SLC35F3, KCNG3, SLC30A9, PTGER4, CYP46A1
	GABRA4, UST, LOC646627, NTNG1, TMEM200A
	TMEM70, RFTN1, GRM1, TMEM132D, KCNV1
	EPHA3, CDH12, EPHA5, BEAN, SLITRK3
	FREM3, GRM7, CD82, SLITRK5, VLDLR
Neuronal development and morphogenesis	BDNF, SLITRK3, RPGRIP1L, MAEL, NTNG1,
	RELN, LAMB1, SLITRK5, MYCBP2d
GABA and glutamate	GRM1, GRM7, GABBR2, GABRA4
Cell adhesion	PPFIA2, CDH12, FREM3, CLSTN2, PVRL3, RELN
	LAMB1
Transcription regulation	EGR3, DACH1, HDAC9, ATOH7, SLC30A9
	ATF7IP2, ZNF436, MYCBP2

Annotations are based on Gene Ontology. See Table 3 for a separate analysis of pathway enrichment.

study histograms of Stouffer p-values showed overall much weaker statistical significance than for module #35. Particularly, none of the 550 modules from 11 single study cluster analysis was enriched (p-value threshold 0.05) in more than three GWAS results (**Figure 3**). Only four out of the 550 modules had more than 14 genes (~15% of the 88 genes; indicated by blue arrows in **Figure 3**) that overlapped with module #35. Hence, the meta-clustering approach efficiently combined weak signals in single studies to identify a stable and biologically more meaningful gene module. In other words, module #35 would not have been discovered without combining 11 studies.

We next tested the meta-clustering approach using transcriptomic data from control subjects only (i.e., removing all MDD

subjects) from the same 11 studies. Out of the 50 modules generated, no module was enriched in more than two GWAS studies (p-value threshold 0.05) among the eight GWAS results (see **Table S5** and heatmap in **Figure 2 (b)**), indicating that the inclusion of the MDD cases was necessary for the detection of significant module/GWAS overlap (i.e. module #35). We note that this comparison is not a "proof" of the significance of module #35 since the "control-only" analysis contains only half the sample size of the "cases+controls". To investigate the impact of the sample size, we randomly sampled half cases and half controls to perform "cases+controls" versus "control-only" comparisons. We meta-analyzed the p-values of the eight enrichment analyses (using Stouffer's method) in each module and retain the most

Table 3. Top 15 enriched pathways in module #35.

Pathways	P-values
METABOTROPIC_GLUTAMATE_GABA_B_LIKE_RECEPTOR_ACTIVITY	0.0003
REACTOME_CLASS_C3_METABOTROPIC_GLUTAMATE_PHEROMONE_RECEPTORS	0.0005
G_PROTEIN_SIGNALING_COUPLED_TO_CAMP_NUCLEOTIDE_SECOND_MESSENGER	0.002
CAMP_MEDIATED_SIGNALING	0.002
GLUTAMATE_RECEPTOR_ACTIVITY	0.003
G_PROTEIN_COUPLED_RECEPTOR_PROTEIN_SIGNALING_PATHWAY	0.003
G_PROTEIN_SIGNALING_COUPLED_TO_CYCLIC_NUCLEOTIDE_SECOND_MESSENGER	0.008
CYCLIC_NUCLEOTIDE_MEDIATED_SIGNALING	0.01
NEUROPEPTIDE_HORMONE_ACTIVITY	0.015
REACTOME_GPCR_LIGAND_BINDING	0.02
KEGG_NEUROACTIVE_LIGAND_RECEPTOR_INTERACTION	0.03
G_PROTEIN_COUPLED_RECEPTOR_ACTIVITY	0.03
SECOND_MESSENGER_MEDIATED_SIGNALING	0.04
HORMONE_ACTIVITY	0.04
REACTOME_EICOSANOID_LIGAND_BINDING_RECEPTORS	0.04

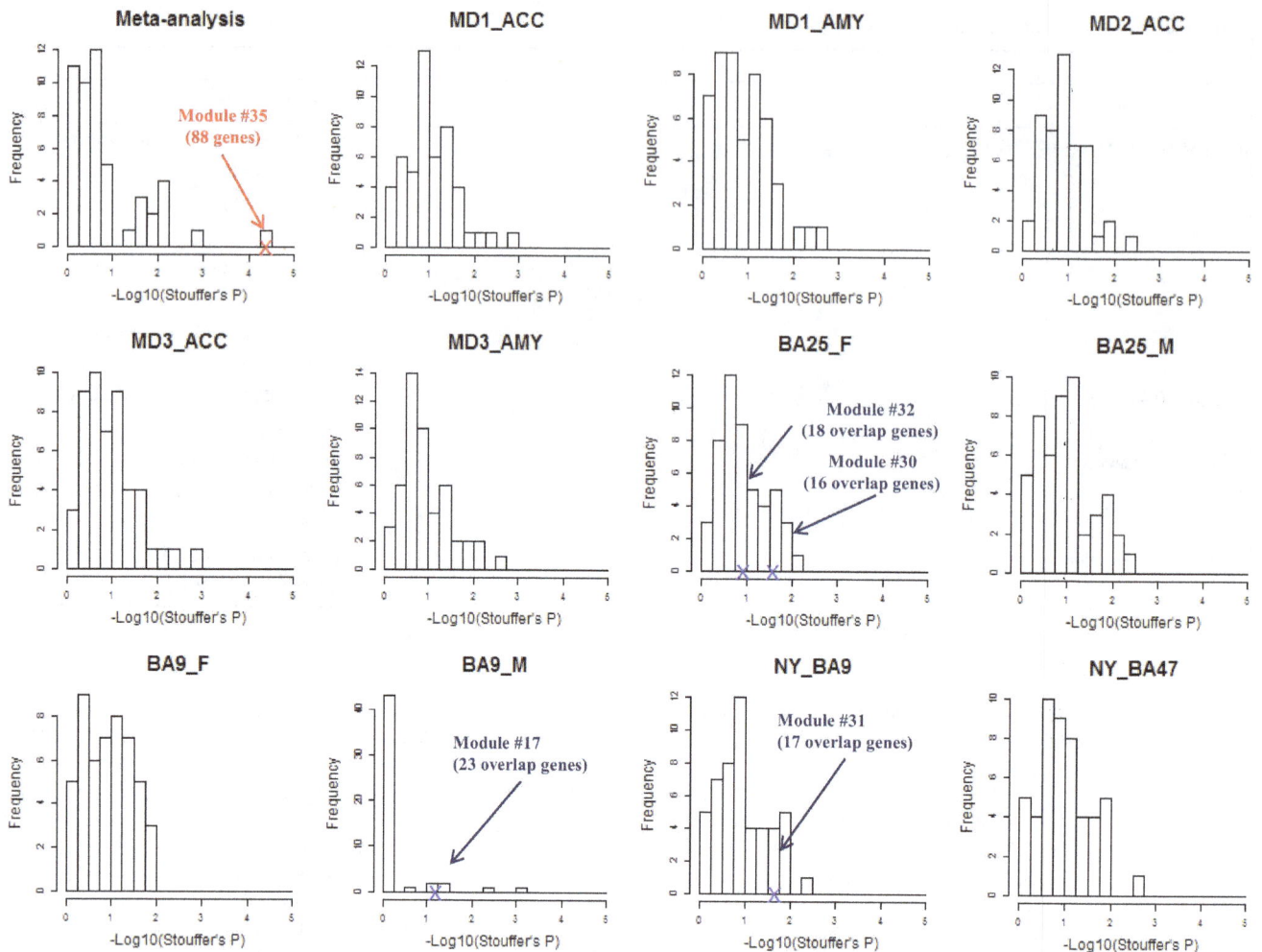

Figure 3. Histograms of the $-\log_{10}(p)$ of the Stouffer statistic from 50 modules of meta-analysis of 11 MDD studies and each single study. Module #35 with 88 genes (red arrow and double-cross) have largest $-\log_{10}$ transformed p-value of Stouffer's statistic 4.4. The other four blue arrows and double crosses indicated that these four modules in all single studies have more than 14 (15% of the 88 genes in module #35) overlapped with module #35. See detailed description in text.

significant one (i.e. smallest p value) among the 50 modules. The procedure was repeated for 20 times. The result shows that meta-coexpression analysis using case and control samples combined has better detection power to identify modules associated with neuropsychiatric diseases. In **Figure S2**, the red cross shows result of "control-only" and the histogram shows the 20 subsampled "cases+controls". The result from the full "cases+controls" is also shown for reference (blue cross). We also tested the meta-clustering approach using transcriptomic data from MDD subjects only (i.e., removing all control subjects) from the same 11 studies. Among the 50 modules generated, one module (module #15 with 169 genes) was enriched in six out of the 8 GWAS categories (p<0.05) but notably not in the gene set corresponding to the Mega GWAS MDD (p = 0.29) and to MDD-related studies (p = 0.43) in the catalog of GWAS (**Table S7**). This module only has 3 genes overlapped with the 88 genes (ST8SIA3, GRM7 and MYCBP2) of module #35 extracted from the case and control combined analysis. Pathway analysis of this module indicated an over-representation of signal transduction pathways (**Table S8**). Overall, the statistical significance of results using MDD data only was lower and potentially inconclusive (i.e. at background noise level).

Together these results indicate that combining MDD and control subjects in meta-clustering approaches increased the significance and robustness of the results, as demonstrated by the identification of the tight module of 88 genes with high relevance to current biological knowledge about MDD.

Discussion

Using methods we developed to identify meta-analyzed coexpression modules across transcriptome datasets, we report the identification of a module consisting of 88 genes that is significantly enriched in genetic variants located nearby genes otherwise associated with major depression and related phenotypes. The finding of a significant intersection of two unbiased large-scale approaches (transcriptome and GWAS) provide robust evidence for the putative recruitment and contribution to molecular and cellular mechanisms of MDD of a biological module that is formed by the identified gene set. This module includes numerous genes encoding proteins implicated in neuronal signaling and structure, including glutamate metabotropic receptors (GRM1, GRM7), GABA-related proteins (GABRA2, GABRA4, CALB1), and neurotrophic and development-related molecules [e.g., BDNF, reelin (RELN), Ephrin receptors (EPHA3, EPHA5)]. These findings are consistent with current hypotheses of molecular mechanisms of MDD, notably with the GABA, glutamate and neurotrophic hypotheses of depression [27,44,45,46,47]. This biological "internal validation", combined with control studies showing that these results could not be achieved using single studies (due to weak signal) demonstrates that integrating transcriptome data, gene coexpression modules and GWAS results can provide a novel and powerful framework to improve understanding of MDD and other complex neuropsychiatric disorders. This approach also provided here a set of putative interacting molecular partners, potentially reflecting a core biological module that is recruited and implicated in biological mechanisms of MDD.

The meta-clustering approach in this paper has the following novelty and advantages. (1) *Meta-analysis*: Our result indicated that a meta-analysis of gene clustering to combine multiple transcriptome studies can identify more accurate and robust gene modules, since the same clustering method applied to single studies did not lead to the identification of any significant and/or neuropsychiatry-related module. (2) *Cluster analysis allowing "scattered genes"*: Gene coexpression modules were identified by penalized K-medoid. This clustering technique searches for tight gene modules and allows some genes to be scattered. This means that they are not included in the final set of modules/clusters, unlike other traditional clustering methods, such as hierarchical clustering, K-means or self-organizing maps that force all genes into clusters. In genomic applications, it was shown that allowing scattered genes can improve clustering performance with better biological knowledge discovery [34]. (3) *Integration and validation with external databases*: Integration with rich GWAS and pathway knowledge databases for biological and disease interpretation identified a robust module with 88 genes that is consistent with current knowledge about depression, hence providing some level of "internal control" for the methods. (4) *Case and control combined coexpression analysis*: We showed that the combination of case and control coexpression analysis was necessary to reveal the coexpression perturbation originating from the disease. This is an important observation as coexpression studies rely on subtle differences in expression patterns compared to differential expression between two groups. Hence disease-related coexpression modules could have been predicted to be unique to the disease groups and "diluted" when combined with control data. However, we show that the opposite is true, resulting in increased power in the combined dataset. For technical validation, we have performed the following: First, we fine-tuned the parameters to be used in the final meta-clustering analysis (i.e. number of modules, percentage of allowed scattered genes in penalized K-medoid method) and tested those parameters in three studies using "surrogate" information, i.e. gene families and biological pathways broadly associated with psychiatric disorders (See Methods). Second, subsampling and bootstrap simulation were applied to investigate the stability of the identified gene modules. Third, three non-psychiatric related GWAS gene sets (cancer, human body indexes and disease traits unrelated to mental functions) served as negative controls.

Coexpression links between genes are inferred from microarray expression studies but do not refer to any specific mechanism underlying these correlations. In fact, any mechanism that synchronously regulates transcription of multiple genes may potentially generate coexpression relationships, including biophysical sources (e.g., transcription factors, spatial configuration of chromosomes, mRNA degradation, miRNA or other upstream regulation, histone acetylation and methylation patterns), technical effects (e.g., batch processing, RNA quality), cell biological sources (e.g., cellular admixture of the sampled tissue, brain region), and importantly synchronized activities across cells under homeostatic equilibria corresponding to "control" states, trait conditions, or chronic disease states for instance. Here, results in module #35 identify a set of genes whose products are distributed across cell types, cellular compartments and biological processes (**Tables 2–3**) that together contribute to various and potentially complementary biological processes, and whose collective function may be related to pathological processes implicated in depression.

The biological content of the identified gene module is notable in that it brings together multiple genes that have been otherwise associated with depression and other neuropsychiatric disorders through multiple studies both in humans and animal models, in addition to the genetic links (i.e., GWAS) that were used here to identify them. Such commonly associated genes include those coding for BDNF, and GABA- and glutamate receptors, for instance [18,19,20,21,22]. Prior findings often refer to differential expression, e.g. reduced BDNF [22], or reduction in calbindin (CALB1) positive GABA neurons [48]. Here, reports of conserved co-regulated patterns between these genes suggests that changes in the fine-tuning and synchronization of the function of these gene products across cells and pathways may contribute to pathophysiological mechanisms related to brain dysfunction in MDD. The fact that these results implicate genes that are likely to be expressed across cell types or to regulate ensembles of cells (i.e. neurotrophic and neuro-maintenance factors) is consistent with mechanisms expected for polygenic complex disorders. Moreover, the identification of module #35 through overlap with GWAS findings for traits (i.e. neuroticism) and other neuropsychiatric disorders (**Figure 2**) also suggests that those genes may participate in basic cellular functions that are implicated in a continuum of biological states (i.e., from normal to disease brain functions), consistent with a dimensional understanding of biological mechanisms of brain disorders. The fact that borderline significance in gene overlap was also observed for categories of disorders sharing clinical risk with MDD (i.e. cardiovascular diseases, inflammation and metabolic syndrome) suggest that the same gene sets may also contribute to dysfunctions in peripheral organs through pleiotropic functions of common genes, hence providing putative biological links for the clinical and symptom co-morbidity. Follow-up studies of coexpression patterns obtained in datasets across these disorders may be necessary to further investigate these interesting hints.

So while these studies provide insight into the biology of complex disorders, one may reasonably ask how they may contribute to the generation of novel hypotheses and predictions. Two directions are worth mentioning. First, for the purpose of therapeutic development and target identification, the application of graph theory and other network analysis may help identify critical genes within the identified module or upstream factors, as potential mediators of the function of this module in disease state. Preliminary analyses of the network properties of module #35 did not provide clear insight into hub genes or other parameters of interest (data not shown); however these studies may be confounded by circular analyses within the same datasets. Thus, testing these hypotheses in other large-scale disease related datasets are needed to, firstly, refine gene membership into the identified module, in view of the reasonable and significant conservation of module structure across datasets, although not to absolute levels; and, secondly, to identify key network nodes with conserved cross-studies functions, as potential targets to modulate the functional outcome of the identified gene module. Finally, an additional and important outcome of these studies is that they provide a focused set of genes, which can be used for follow-up genetic association studies, hence potentially mitigating the problem of reduced statistical power of large scale genome-wide studies.

There are several limitations to this study. First, there is a bias when selecting gene sets from the catalog of published GWAS results since the targeted markers (SNPs) are updated every six months, and many more SNPs were reported in the past five years when GWAS have achieved greater sample size (including studies with more than 10,000 participants) and detection of markers with very small effect size. However, large sample sizes will also introduce a bias towards false positive markers. A related limitation is that the choice of markers (or gene) was based on fixed and arbitrary thresholds (i.e., p-value and genomic distance). Moreover, we used only a small fraction of the datasets and pre-defined pathways related to psychiatric disease to decide on the number of clusters and sets of scattered genes during the method development phase, so the result of the clustering approaches may still show some instability and may vary based on different numbers of clusters and applied thresholds. Indeed, although we performed extensive validation analyses to select the clustering parameters and increase stability of modules, the 88 genes in module #35 will inevitably vary slightly under additional data perturbation (e.g., when adding additional MDD or related studies). An additional limitation is that generating gene coexpression modules using cluster analyses is known to be sensitive to small data perturbation. To mitigate these effects, we combined multiple studies and concentrated on tight modules by leaving out scattered genes. While this approach increased the power of the meta-clustering method, it also meant combining datasets from different brain regions, hence potentially diluting the effects of local coregulation patterns that may be important for disease mechanisms. The integration of multiple datasets comes at the expense of variable technical platforms, including inclusion of different probesets across array types. We investigated this potential issue and showed considerable overlap in genomic region targeted by the various probes for a same gene (**Figure S3**), hence lowering the potential impact of this array differences. So these results should be considered proof-of-concept, rather than experimentally and biologically optimized. Finally, it is important to note that changes in gene coexpression are difficult to confirm by independent measures. Indeed coexpression links rely on large sample size and we previously showed that the sample-to-sample variability in array-based measures of expression is typically lower than the variability obtained using alternate measures such as quantitative PCR [24], so the ultimate test of the added value of these meta-coexpression studies will need to come from additional independent studies. Nonetheless, this study allowed the identification of a focused set of genes for use in future genetic association studies, and together demonstrates the importance of integrating transcriptome data, gene coexpression modules and GWAS results, paving the way for novel and complementary approaches to investigate the molecular pathology of MDD and other complex brain disorders.

Supporting Information

Figure S1   Diagram of pre-processing procedure of 11 MDD transcriptiome data sets. Number of samples and number of matched genes in each single (MDD) study. In matching step, we allowed 20% missing studies, then 16,443 genes were identically matched among 11 studies. 13,500 genes were kept by filtering out lower sum ranks of median row means; 10,000 genes were kept by filtering out lower sum ranks of median row standard deviations.

Figure S2   Histogram of minimum $\log_{10}$-transformed p values from Stouffer's statistics for 50 modules obtained from randomly selected MDD cases and matched controls into half for 20 times. Red cross represents the minimum $\log_{10}$-transformed p value for controls only study and blue cross represents the minimum $\log_{10}$-transformed p value for cases plus controls study.

Figure S3   Sequence target overlap between Affymetrix and Illumina array probesets. We have systematically mapped the respective probes that were chosen by our approach and used genes in module #35 to specifically look at overlap in targeted regions. As shown in the individual graphs below, there is overlap in regions for 94% of the genes, indicating that for a few exceptions the same transcript region is used. A histogram represents a chromosomal area of a target sequence in either affymetrix or illumina platform. Wider histogram means the target sequence span over DNA sequence more widely. The height of each histogram shows the number of studies use that specific probe. See main text for additional information.

Table S1   Categories of the GWAS.

Table S2   Disease traits of negative controls from catalog of GWAS.

Table S3   Wilcoxon signed rank test of parameter determination.

Table S4   Meta modules and GWAS gene lists (cases and controls).

Table S5   Meta modules and GWAS gene lists (controls only).

Table S6   GWAS-hit genes in module #35.

Table S7   Meta modules and GWAS gene lists (cases only).

Table S8   Pathway analysis.

Table S9   Meta modules and PGC schizophrenia.

Acknowledgments

We thank Beverly French for careful review of the manuscript.

Author Contributions

Conceived and designed the experiments: LCC GCT ES. Performed the experiments: LCC GCT ES CWL. Analyzed the data: LCC GCT ES CWL. Contributed reagents/materials/analysis tools: SJ DR. Wrote the paper: LCC GCT ES.

References

1. Hasin DS, Goodwin RD, Stinson FS, Grant BF (2005) Epidemiology of major depressive disorder: results from the National Epidemiologic Survey on Alcoholism and Related Conditions. Archives of general psychiatry 62: 1097–1106.

2. Mueller TI, Leon AC, Keller MB, Solomon DA, Endicott J, et al. (1999) Recurrence after recovery from major depressive disorder during 15 years of observational follow-up. The American journal of psychiatry 156: 1000–1006.

3. Weissman MM, Bland R, Joyce PR, Newman S, Wells JE, et al. (1993) Sex differences in rates of depression: cross-national perspectives. Journal of affective disorders 29: 77–84.

4. Sullivan PF, Neale MC, Kendler KS (2000) Genetic epidemiology of major depression: review and meta-analysis. The American journal of psychiatry 157: 1552–1562.

5. Gaiteri C, Ding Y, Tseng GC, Sibille E (2013) Beyond Modules & Hubs: Investigating pathogenic molecular mechanisms of brain disorders through gene coexpression networks. Submitted.

6. Rietschel M, Mattheisen M, Frank J, Treutlein J, Degenhardt F, et al. (2010) Genome-wide association-, replication-, and neuroimaging study implicates HOMER1 in the etiology of major depression. Biological psychiatry 68: 578–585.

7. Shi J, Potash JB, Knowles JA, Weissman MM, Coryell W, et al. (2011) Genome-wide association study of recurrent early-onset major depressive disorder. Molecular psychiatry 16: 193–201.

8. Muglia P, Tozzi F, Galwey NW, Francks C, Upmanyu R, et al. (2010) Genome-wide association study of recurrent major depressive disorder in two European case-control cohorts. Molecular psychiatry 15: 589–601.

9. Lewis CM, Ng MY, Butler AW, Cohen-Woods S, Uher R, et al. (2010) Genome-wide association study of major recurrent depression in the U.K. population. The American journal of psychiatry 167: 949–957.

10. Shyn SI, Shi J, Kraft JB, Potash JB, Knowles JA, et al. (2011) Novel loci for major depression identified by genome-wide association study of Sequenced Treatment Alternatives to Relieve Depression and meta-analysis of three studies. Molecular psychiatry 16: 202–215.

11. MDD GWAS consortium (2012) A mega-analysis of genome-wide association studies for major depressive disorder. Mol Psychiatry.

12. Hek K, Demirkan A, Lahti J, Terracciano A, Teumer A, et al. (2013) A Genome-Wide Association Study of Depressive Symptoms. Biol Psychiatry.

13. Kupfer DJ, Angst J, Berk M, Dickerson F, Frangou S, et al. (2011) Advances in bipolar disorder: selected sessions from the 2011 International Conference on Bipolar Disorder. Ann N Y Acad Sci 1242: 1–25.

14. Cristino AS, Williams SM, Hawi Z, An JY, Bellgrove MA, et al. (2013) Neurodevelopmental and neuropsychiatric disorders represent an interconnected molecular system. Mol Psychiatry.

15. Niculescu AB (2013) Convergent functional genomics of psychiatric disorders. Am J Med Genet B Neuropsychiatr Genet.

16. Le-Niculescu H, Kurian SM, Yehyawi N, Dike C, Patel SD, et al. (2009) Identifying blood biomarkers for mood disorders using convergent functional genomics. Mol Psychiatry 14: 156–174.

17. Le-Niculescu H, Balaraman Y, Patel SD, Ayalew M, Gupta J, et al. (2011) Convergent functional genomics of anxiety disorders: translational identification of genes, biomarkers, pathways and mechanisms. Transl Psychiatry 1: e9.

18. Tripp A, Oh H, Guilloux JP, Martinowich K, Lewis DA, et al. (2012) Brain-derived neurotrophic factor signaling and subgenual anterior cingulate cortex dysfunction in major depressive disorder. Am J Psychiatry 169: 1194–1202.

19. Klempan TA, Sequeira A, Canetti L, Lalovic A, Ernst C, et al. (2009) Altered expression of genes involved in ATP biosynthesis and GABAergic neurotransmission in the ventral prefrontal cortex of suicides with and without major depression. MolPsychiatry 14: 175–189.

20. Sequeira A, Mamdani F, Ernst C, Vawter MP, Bunney WE, et al. (2009) Global brain gene expression analysis links glutamatergic and GABAergic alterations to suicide and major depression. PLoS One 4: e6585.

21. Choudary PV, Molnar M, Evans SJ, Tomita H, Li JZ, et al. (2005) Altered cortical glutamatergic and GABAergic signal transmission with glial involvement in depression. ProcNatlAcadSci USA 102: 15653–15658.

22. Guilloux JP, Douillard-Guilloux G, Kota R, Wang X, Gardier AM, et al. (2012) Molecular evidence for BDNF- and GABA-related dysfunctions in the amygdala of female subjects with major depression. Mol Psychiatry 17: 1130–1142.

23. Lee HK, Hsu AK, Sajdak J, Qin J, Pavlidis P (2004) Coexpression analysis of human genes across many microarray data sets. Genome research 14: 1085–1094.

24. Gaiteri C, Guilloux JP, Lewis DA, Sibille E (2010) Altered gene synchrony suggests a combined hormone-mediated dysregulated state in major depression. PLoS one 5: e9970.

25. Dobrin R, Zhu J, Molony C, Argman C, Parrish ML, et al. (2009) Multi-tissue coexpression networks reveal unexpected subnetworks associated with disease. Genome biology 10: R55.

26. Elo LL, Jarvenpaa H, Oresic M, Lahesmaa R, Aittokallio T (2007) Systematic construction of gene coexpression networks with applications to human T helper cell differentiation process. Bioinformatics 23: 2096–2103.

27. Sibille E, French B (2013) Biological substrates underpinning diagnosis of major depression. Int J Neuropsychopharmacol 16: 1893–1909.

28. Pan A, Keum N, Okereke OI, Sun Q, Kivimaki M, et al. (2012) Bidirectional association between depression and metabolic syndrome: a systematic review and meta-analysis of epidemiological studies. Diabetes Care 35: 1171–1180.

29. Musselman DL, Evans DL, Nemeroff CB (1998) The relationship of depression to cardiovascular disease: epidemiology, biology, and treatment. ArchGenPsychiatry 55: 580–592.

30. Wang X, Lin Y, Song C, Sibille E, Tseng GC (2012) Detecting disease-associated genes with confounding variable adjustment and the impact on genomic meta-analysis: with application to major depressive disorder. BMC Bioinformatics 13: 52.

31. Sibille E, Arango V, Galfalvy HC, Pavlidis P, Erraji-BenChekroun L, et al. (2004) Gene expression profiling of depression and suicide in human prefrontal cortex. Neuropsychopharmacology 29: 351–361.

32. Gentleman R, Carey V, Huber W, Irizarry R, Dudoit S (2005) Bioinformatics and computational biology solutions using R and Bioconductor: Springer New York.

33. Tseng GC (2007) Penalized and weighted K-means for clustering with scattered objects and prior information in high-throughput biological data. Bioinformatics 23: 2247–2255.

34. Thalamuthu A, Mukhopadhyay I, Zheng X, Tseng GC (2006) Evaluation and comparison of gene clustering methods in microarray analysis. Bioinformatics 22: 2405–2412.

35. Hubert L, Arabie P (1985) Comparing Partitions. Journal of the Classification 2: 193–218.

36. Efron B (1979) Bootstrap Methods: Another Look at the Jackknife. The annals of statistics 7: 1–26.

37. van den Oord EJ, Kuo PH, Hartmann AM, Webb BT, Moller HJ, et al. (2008) Genomewide association analysis followed by a replication study implicates a novel candidate gene for neuroticism. Archives of general psychiatry 65: 1062–1071.

38. Wray NR, Pergadia ML, Blackwood DH, Penninx BW, Gordon SD, et al. (2012) Genome-wide association study of major depressive disorder: new results, meta-analysis, and lessons learned. Molecular psychiatry 17: 36–48.

39. Psychiatric GWAS Consortium Bipolar Disorder Working Group (2011) Large-scale genome-wide association analysis of bipolar disorder identifies a new susceptibility locus near ODZ4. Nat Genet 43: 977–983.

40. Hindorff LA, Sethupathy P, Junkins HA, Ramos EM, Mehta JP, et al. (2009) Potential etiologic and functional implications of genome-wide association loci for human diseases and traits. Proceedings of the National Academy of Sciences of the United States of America 106: 9362–9367.

41. Stouffer SA (1949) A study of attitudes. Scientific American 180: 11–15.

42. Benjamini Y, Drai D, Elmer G, Kafkafi N, Golani I (2001) Controlling the false discovery rate in behavior genetics research. Behavioural brain research 125: 279–284.

43. Ripke S, O'Dushlaine C, Chambert K, Moran JL, Kahler AK, et al. (2013) Genome-wide association analysis identifies 13 new risk loci for schizophrenia. Nature genetics 45: 1150–1159.

44. Luscher B, Shen Q, Sahir N (2011) The GABAergic deficit hypothesis of major depressive disorder. Mol Psychiatry 16: 383–406.

45. Belmaker RH, Agam G (2008) Major depressive disorder. NEnglJ Med 358: 55–68.

46. Nestler EJ, Barrot M, DiLeone RJ, Eisch AJ, Gold SJ, et al. (2002) Neurobiology of depression. Neuron 34: 13–25.

47. Duman RS, Monteggia LM (2006) A neurotrophic model for stress-related mood disorders. Biol Psychiatry 59: 1116–1127.

48. Rajkowska G, O'Dwyer G, Teleki Z, Stockmeier CA, Miguel-Hidalgo JJ (2007) GABAergic neurons immunoreactive for calcium binding proteins are reduced in the prefrontal cortex in major depression. Neuropsychopharmacology 32: 471–482.

21

SynChro: A Fast and Easy Tool to Reconstruct and Visualize Synteny Blocks along Eukaryotic Chromosomes

Guénola Drillon[1,2], **Alessandra Carbone**[1,2,3], **Gilles Fischer**[1,2]*

1 Sorbonne Universités, UPMC Univ Paris 06, UMR 7238, Biologie Computationnelle et Quantitative, Paris, France, **2** CNRS, UMR7238, Biologie Computationnelle et Quantitative, Paris, France, **3** Institut Universitaire de France, Paris, France

Abstract

Reconstructing synteny blocks is an essential step in comparative genomics studies. Different methods were already developed to answer various needs such as genome (re-)annotation, identification of duplicated regions and whole genome duplication events or estimation of rearrangement rates. We present SynChro, a tool that reconstructs synteny blocks between pairwise comparisons of multiple genomes. SynChro is based on a simple algorithm that computes Reciprocal Best-Hits (RBH) to reconstruct the backbones of the synteny blocks and then automatically completes these blocks with non-RBH syntenic homologs. This approach has two main advantages: (i) synteny block reconstruction is fast (feasible on a desk computer for large eukaryotic genomes such as human) and (ii) synteny block reconstruction is straightforward as all steps are integrated (no need to run Blast or TribeMCL prior to reconstruction) and there is only one parameter to set up, the synteny block stringency Δ. Benchmarks on three pairwise comparisons of genomes, representing three different levels of synteny conservation (Human/Mouse, Human/Zebra Finch and Human/Zebrafish) show that Synchro runs faster and performs at least as well as two other commonly used and more sophisticated tools (MCScanX and i-ADHoRe). In addition, SynChro provides the user with a rich set of graphical outputs including dotplots, chromosome paintings and detailed synteny maps to visualize synteny blocks with all homology relationships and synteny breakpoints with all included genetic features. SynChro is freely available under the BSD license at http://www.lcqb.upmc.fr/CHROnicle/SynChro.html.

Editor: Cecile Fairhead, Institut de Genetique et Microbiologie, France

Funding: This work was supported by the Agence Nationale de la Recherche ('GB-3G', ANR-10-BLAN-1606-01, http://www.agence-nationale-recherche.fr/) and by an ATIP grant from the CNRS (http://www.cnrs.fr/). The funders had no role in study design, data collection and analysis, decision to publish, or preparation of the manuscript.

Competing Interests: The authors have declared that no competing interests exist.

* E-mail: Gilles.Fischer@upmc.fr

Introduction

Synteny block reconstruction consists on the identification of a series of homologous genes whose order is conserved between two (or more) genomes. Analysis of synteny conservation between different genomes allows to identify similarity patterns and differences in genome structure and content. In practice, genomes with different levels of divergence generate different types of questions and require different analysis methods and different visualization tools. For closely related genomes, synteny conservation can be performed at the DNA level, which can be useful to annotate newly sequenced genomes [1] and to identify conserved non-coding sequences [2–4]. For very distantly related genomes, detection of synteny conservation requires the development of statistical models or the construction of synteny profiles obtained from different genomes [5–7]. In this case, synteny can help to the gene annotation process based on conservation of gene clusters [6,8] or can be used to estimate the number of whole genome duplication events [9]. For genomes sharing intermediate phylogenetic proximity, protein-coding genes may have retained enough sequence similarity and physical collinearity along chromosomes to allow synteny block reconstruction which can help infering the history of chromosomal rearrangements and the structure of ancestral genomes [10].

SynChro falls in this last category. It is designed to define conserved synteny blocks based on the relative order of protein-coding genes along chromosomes, in order to help in rearrangement and ancestral reconstruction studies. Its main properties are the followings:

1. it makes multiple pairwise comparisons and traces information shared by each pair of genomes; it is not suited to reconstruct synteny blocks shared by several genomes at a time but instead provides analysis tools to compare different sets of pairwise synteny blocks.

2. it defines syntenic homologous genes by computing protein sequence similarity (with fastp and blastp [11,12]) and by taking into account the gene order information. It does not require to run additional tools such as blast or tribeMCL [13] prior the synteny reconstruction step (as it is the case for MCScanX [14] and i-ADHoRe [15], respectively).

3. it reconstructs synteny blocks based on syntenic homologous genes and not on DNA alignment. This enables (i) to compare both relatively close and distant genomes and (ii) in a second time, to compare the different pairwise sets of synteny blocks using genes as common denominator.

4. it allows synteny blocks to be overlapping, included in one another or duplicated, in order to (i) support comparison

involving genomes having undergone a whole genome duplication event and (ii) keep the trace of small rearrangements that may be responsible for small overlaps or inclusions between synteny blocks.

SynChro is a simple algorithm that is not meant to bring new theoretical advances over existing and more sophisticated tools in the field of synteny block identification. The interests of SynChro lie in the *all in one* package with few parameters, rapid execution time and several useful visualization tools that are more flexible than that of other existing methods.

Results and Discussion

SynChro Algorithm

In order to preserve good sensitivity (*i.e.* not to lose pairs of divergent orthologs due to stringent homology criteria) and specificity (*i.e.* not to infer false homology between genes), SynChro uses two different criteria of homology to reconstruct synteny blocks between two genomes G_1 and G_2. The reconstruction is achieved through three successive simple steps that are detailed in [16] and quickly recalled here (black frame in Fig. 1):

1. Identification of Reciprocal Best Hits (RBH, also called BDBH for Bi-Directional Best Hits) using Opscan (see Material and Methods). Two genes g_1 and g_2, encoding two proteins p_1 and p_2 and occurring respectively in G_1 and in G_2, are called *RBH* if the best match of p_1 in G_2 is p_2 and, reciprocally, the best match of p_2 in G_1 is p_1. In this case, the pair of genes (g_1,g_2), or equivalently (g_2,g_1), is called a RBH and g_1 and g_2 are called RBH-genes.

2. Definition of the synteny blocks. Synteny blocks are primarily defined by their anchors which correspond to series of RBH that are co-localized along chromosomes in the two compared genomes, G_1 and G_2. RBH are defined as anchors if they are in Δ_{RBH} synteny. A RBH (g_1^1,g_2^1) is in Δ_{RBH} synteny with another RBH (g_1^n,g_2^n) if it exists a chain of n RBH $(g_1^1,g_2^1)(g_1^2,g_2^2)...(g_1^n,g_2^n)$, with $n \geq 2$, such that $\forall i \in [1,n-1]$ there are strictly less than Δ_{RBH} RBH-genes lying between g_1^i and g_1^{i+1} in G_1 and strictly less than Δ_{RBH} RBH-genes between g_2^i and g_2^{i+1} in G_2. By allowing the insertion of an unlimited number of non-RBH genes, this Δ_{RBH} threshold allows to focus on balanced rearrangements such as inversions, translocations and chromosome fusion/fission.

3. Completion of the synteny blocks with non-RBH homologs. Two genes, $g_1 \in G_1$ and $g_2 \in G_2$, are *non-RBH homologs* (non-RBH, in short), if at least one of them does not correspond to a RBH-gene and if their amino-acid sequences share at least 30% of similarity (*i.e.* percentage of positive residues) and if the ratio between the length of the match between the two protein sequences (including internal gaps introduced by blastp) and the length of the smallest protein sequence is larger than 0.5. A pair of non-RBH (g_1,g_2) is in Δ_{gene} synteny with an anchor (g_1',g_2'), and therefore complete the corresponding synteny block, if g_1 and g_1' are at strictly less than Δ_{gene} genes apart in G_1, and g_2 and g_2' are at strictly less than Δ_{gene} genes apart in G_2. Note that in order to keep a single parameter to launch the program, called Δ, the algorithm imposes that $\Delta_{RBH} = \Delta_{gene}$ if only one value is provided by the user. Alternatively, the user can decide to provide two different values to Δ_{RBH} and Δ_{gene}. In the rest of the manuscript we will use the general Δ parameter to account for both Δ_{RBH} and Δ_{gene}.

SynChro Input, Output and Parameter

SynChro is a set of awk and python scripts with graphical outputs supplied using gnuplot. It can be applied to two or more genomes to realize all possible pairwise comparisons.

The minimum input information that must be provided to SynChro is a list of protein-coding genes, ordered along the chromosomes (or scaffolds) and their associated amino-acid sequences. Their coordinates along chromosomes, centromere positions, and other genomic features are useful information but not compulsory for synteny block reconstruction. The indication of the coding strand is also a useful but optional information that is used to orient synteny relationships between genes in the synteny map (if they are not specified, genes are assumed to be all on the same strand). Formats of the input files are detailed in the README file (http://www.lcqb.upmc.fr/CHROnicle/SynChro.html). Allowed formats include EMBL, GenBank and Fasta files and the scripts that convert these files into the expected input format are provided within the package.

For each pairwise comparison, four different outputs are provided (see orange frames in Fig. 1):

1. a detailed synteny map allowing to visualize synteny blocks with all individual homology relationships (including their relative orientation in the two compared genomes) and the breakpoint regions including the protein-coding genes they encompass as well as other genetic features such as tRNA, pseudogene, LTR (Long Terminal Repeats), etc. This synteny map is interactive, the names of the different genetic features pop-up on the screen when the mouse points to their symbols. This map is a vectorial image, therefore it is possible to zoom in and out as necessary. This detailed synteny map represents a true improvement compared to other tools where graphical outputs are often poor, being reduced to dotplots [1,17,18] or chromosomal painting [19,20].

2. text files containing homology relationships (RBH and non-RBH) and synteny blocks description

3. a chromosomal painting representation

4. a genome-wide dotplot of syntenic homologs.

Moreover, for several pairwise comparisons, SynChro provides scripts to compute, correlate and plot relevant information such as the proportion of genes/genome that is conserved in synteny, the average percentage of amino-acid similarity between orthologs, the number of synteny blocks, the average length (in nucleotides or in number of genes) of the breakpoint regions (*i.e.* regions between two contiguous synteny blocks), the average number of genes per synteny block or the proportion of consecutive synteny blocks whose homologous blocks map also on the same chromosome in the other species (see the README file for the complete list).

Another script is also provided to reconstruct families of orthologous genes (*i.e.* syntenic homologs, RBH and non-RBH, shared between multiple genomes inferred by transitivity from the pairwise relationships) containing exactly one gene per genome (all families containing duplicated genes are discarded). More formally, given a graph where vertices represent genes from multiple genomes and edges represent the RBH and the non-RBH homology relationship (deduced from all pairwise comparisons), each connected component (independent group of vertices linked together) containing one and only one gene per genome is defined as a family of orthologous genes. Families of orthologous genes could be very useful. For instance, delineating such families is of primary importance to define a set of genes that can be used in phylogenetic reconstruction.

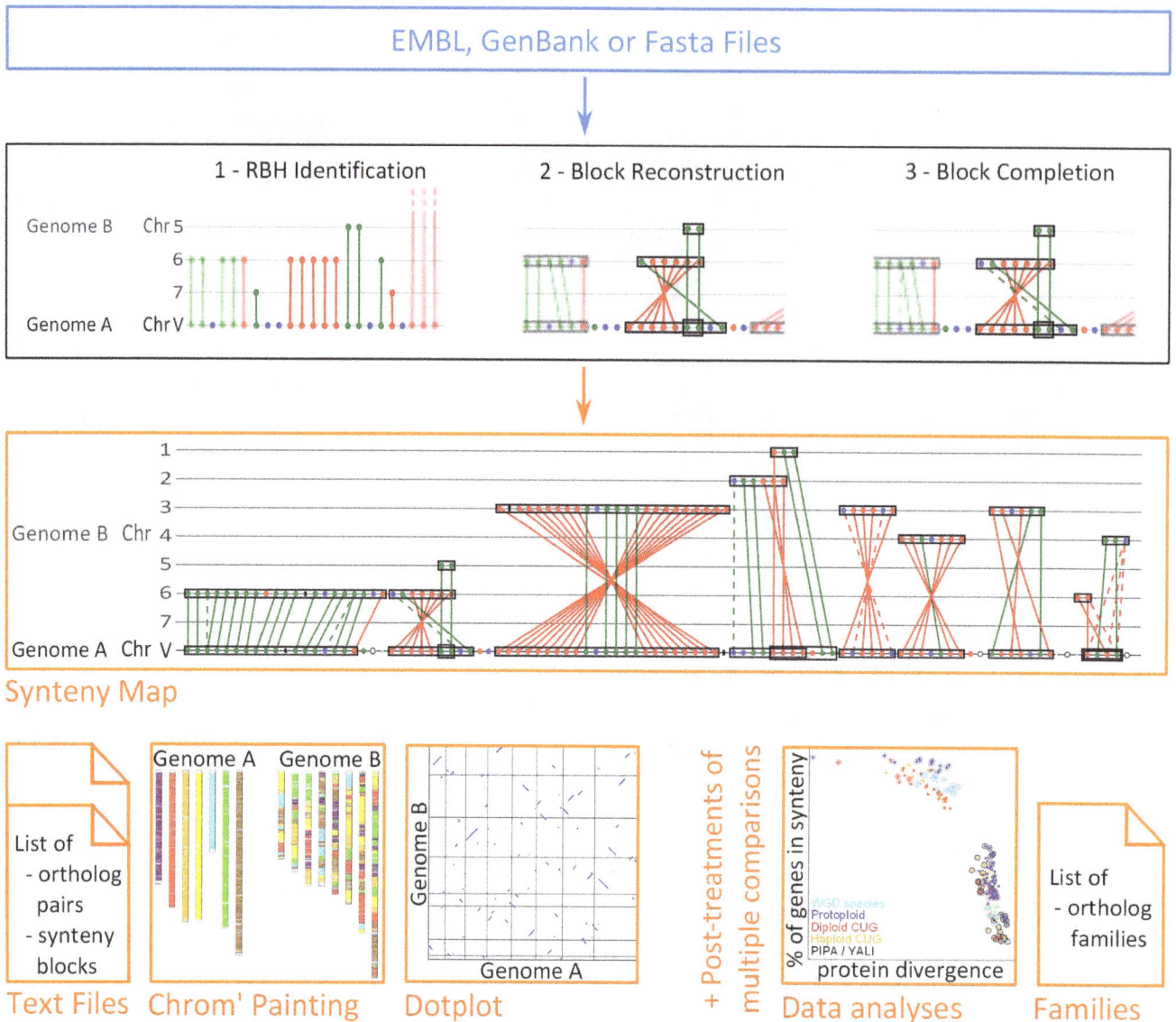

Figure 1. SynChro algorithm, inputs and outputs. The format of input files are indicated in the blue frame. The different steps of the algorithm are illustrated in the black frame (colored dots symbolize genes, green and red plain lines highlight RBH relationships and dotted lines represent non-RBH homologous relationships). In step 1, all RBH gene-pairs are mapped regardless of their chromosomal positions, in step 2 only the syntenic RBH-pairs are mapped and in step 3 the non-RBH syntenic homologs are added to the map. The different types of outputs are shown in the orange frames.

SynChro is very easy to use as there is only one parameter to set up, the synteny block stringency Δ. The Δ parameter is easy to learn and to master: higher values of Δ are more permissive and allow larger micro-rearrangements to be tolerated within synteny blocks while smaller values of Δ are more stringent and split synteny blocks at micro-rearrangement breakpoints. Table 1 illustrates the evolution of the number of reconstructed synteny blocks and the number of syntenic RBH involved in these blocks as a function of the Δ value for three comparisons: *Homo sapiens/Mus musculus, Homo sapiens/Taeniopygia guttata* and *Homo sapiens/Danio rerio*. It shows that for the two first comparisons, the number of syntenic RBH in synteny blocks do not increase drastically, confirming that the main impact of Δ is to split, or merge synteny blocks. However, for more distantly related genomes such as in the third comparison (Human/Zebrafish), the number of syntenic RBH increases with Δ, as do the number of synteny blocks,

meaning that, for larger phylogenetic distances, increasing the Δ value allows, above all, to recover a larger number of synteny blocks.

Benchmarking SynChro on Vertebrate Genomes

To evaluate the performance of our algorithm, we compared the synteny block reconstruction achieved by SynChro to the synteny blocks reconstructed by two other commonly used tools that also reconstruct synteny blocks from annotated genome/ genes: MCScanX [14] and i-ADHoRe [15]. These tools are regularly updated since their first publication [9,21]. The three tools were run on the same dataset composed of three pairwise comparisons of genomes corresponding to three different levels of synteny conservation: Human/Mouse (*Homo sapiens/Mus musculus*), Human/Zebra finch (*Homo sapiens/Taeniopygia guttata*) and Human/Zebrafish (*Homo sapiens/Danio rerio*). SynChro appears to be

Table 1. Evolution of the number of synteny blocks and syntenic homologs as a function of the Δ value.

	Δ	1	2	3	4	5	6	7
Human/	# synteny blocks	1 279	446	377	354	339	331	318
Mouse	# syntenic RBHs	13 786	13 995	14 031	14 035	14 045	14 047	14 054
Human/	# synteny blocks	1 217	727	654	628	604	575	555
Zebra finch	# syntenic RBHs	6 995	7 258	7 311	7 343	7 358	7 372	7 396
Human/	# synteny blocks	1 652	1 812	1 833	1861	1 868	1892	1 900
Zebrafish	# syntenic RBHs	4 206	5 157	5 542	5 791	5 970	6 152	6 317

between 2 and 3 time faster than the two other tools to reconstruct synteny blocks between the three pairwise comparisons (SynChro takes, on a desk computer, on the order of 40 minutes to reconstruct synteny blocks between two vertebrate genomes, Table 2).

In order to quantify the level of consistency between the three tools, we compared the coordinates of the synteny blocks detected by the different tools to quantify the proportion of the human genome that was covered by the same synteny blocks by the different tools (Fig. 2). For each pairwise comparison, this quantification was performed by scanning the human genome to identify the regions where synteny blocks from two different tools are overlapping and by checking if their homologous blocks in the other genome were also overlapping (if so, these synteny blocks are said to be congruent). Only two tools were compared at a time and then the intersection between the three two-way comparisons was realized. This analysis allowed identifying different types of regions in the human genome: regions congruently covered by the three tools, regions covered by the three tools but with some discordances (*i.e.* one or two tools would map different non-overlapping regions in the other genome), regions covered by only one tool, regions not covered by any of the three tools, etc. (in total 15 different types of regions were identified). As an example, Figure 2 shows 8 successive regions representing 6 different types. For each tool, we quantify from these regions the proportions of the human genome where synteny was supported (i) only by this tool (or also by the other tools but not consistently with the considered tool), (ii) consistently by this tool and another one and (iii) consistently by the 3 tools (see the Venn diagram, in Fig. 3). In the case of overlapping synteny blocks (as the two last blocks of MCScanX, or the two last blocks of i-ADHoRe, in Fig. 2), the region is considered to be congruent if at least one of the two overlapping synteny blocks is congruent with a synteny block detected by another tool (see the intersection Syn-Chro∩MCScanX in Fig. 2). In addition, congruence between the different tools was assessed separately for regions covered by successive or partially overlapping synteny blocks (referred as 'Not included' in Fig. 3) and for regions covered by synteny blocks

Table 2. Characteristics of SynChro, MCScanX and i-ADHoRe synteny blocks for three pairwise comparisons.

		SynChro	MCScanX	i-ADHoRe
Human/mouse	time (in minutes)	36 (Opscan)+9 (non-RBH+blocks)	131 (blastp)+1 (blocks)	131 (blastp) +1 (blocks)
	# blocks	339	602	497
	# syntenic homologs	25 000(14 045)	14 624(14 624)	19 349(14 205)
	% syntenic homologs	80.1	69.2	69.0
	% genome within synteny blocks	89.3	89.3	89.3
Human/Zebra finch	time	27+6	65+0	65+0
	# synteny blocks	604	552	767
	# syntenic homologs	10 833(7 358)	8 879(8 879)	10 377(9 489)
	% syntenic homologs	49.2	43.8	46.2
	% genome within synteny blocks	71.3	70.9	71.7
Human/Zebrafish	time	35+10	122+1	122+1
	# synteny blocks	1 868	627	1115
	% syntenic homologs	9 279(5 970)	3 958(3 958)	6 239(5 028)
		39.8	18.1	22.8
	% genome within synteny blocks	49.9	39.3	37.3

The execution time (in minutes) indicates the time used for homolog identification and for synteny block reconstruction (for SynChro, these two steps are not really separable because reconstruction of synteny blocks implies the identification of additional non-RBH homologs by blastp). The number of syntenic homologs represents the total number of homology relationships in the synteny blocks. The numbers between brackets indicate the number of homology relationships when only one relationship per gene per synteny block is allowed (*i.e.* removing the homology relationships corresponding to tandemly duplicated genes within a given synteny block). Note that for MCScanX these 2 values are identical because the program was run with the ' − b 2' option which prevents MCScanX to detect tandemly duplicated genes within a given synteny block.

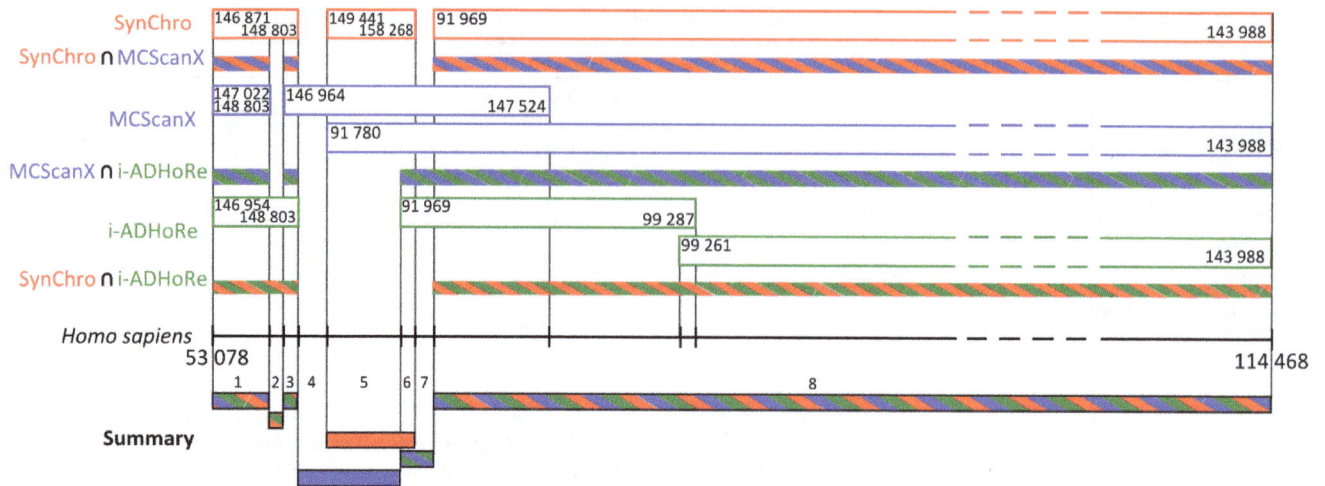

Figure 2. Congruence between the 3 different synteny block reconstructions. An example based on a segment of the *Homo sapiens*' X chromosome (from coordinates 53,078 to 114,468 kb) and the genome of *Mus musculus* is presented. The synteny blocks reconstructed by the three tools, SynChro, MCScanX and iADHoRe are represented by red, blue and green-framed open boxes, respectively. The two coordinates, inside each box, refer to the coordinates in the mouse genome. Synteny blocks from 2 different reconstructions are congruent when overlaping synteny blocks, along the human chromosome X, map overlapping regions in the mouse genome. These congruent synteny blocks are represented by hatched bi-colored boxes and are denoted: SynChro∩MCScanX, MCScanX∩i-ADHoRe and SynChro∩i-ADHoRe. The intersection of these three sets of synteny blocks allows to define regions (such as regions 1, 3 and 8) where the three tools are in agreements (tri-colored hatched boxes) and to deduce regions (such as the other regions) where only one or two tools detect synteny conservation (or are in agreement). The 5 lines at the bottom of the figure summarize these regions. Note that overlapping synteny blocks predicted by MCSanX or i-ADHoRe correspond to regions containing duplicated genes between the blocks. These regions do not necessarily contain many duplicated genes given that a single duplicate is sufficient to produce an overlap.

where one block was included in a larger block (mostly representing duplicated regions and referred as 'Included' in Fig. 3, respectively).

From these analyses, we first estimated the proportion of the human genome that was found to be conserved in synteny by at least one of the three detection tools. This proportion cannot be

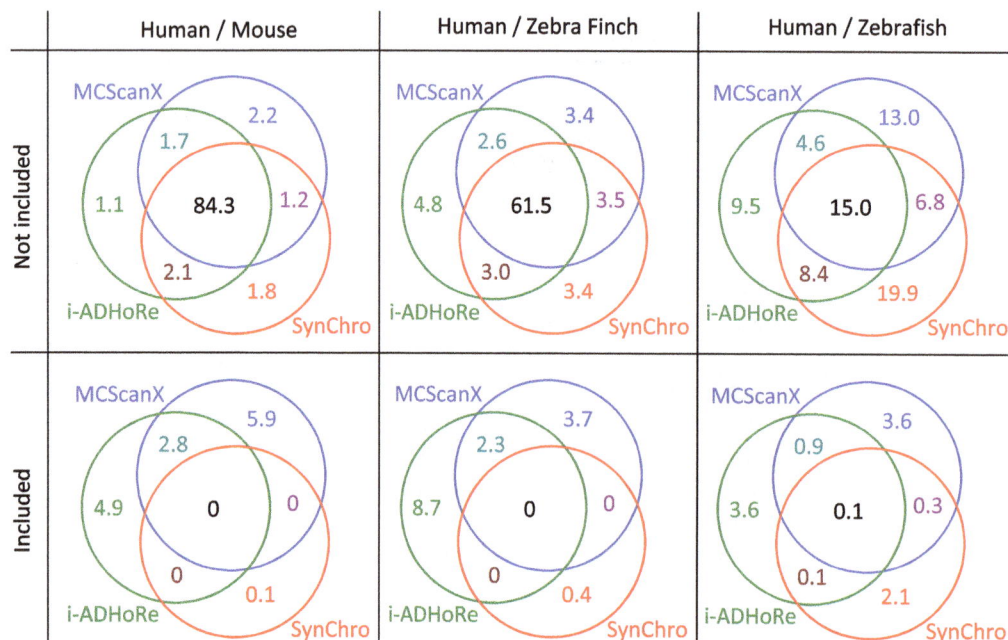

Figure 3. Venn diagrams showing the congruence between the three tools: SynChro, MCScanX and i-ADHoRe. The numbers indicate the percentages of the human genome found in (congruent) synteny (with the mouse, zebra finch, zebrafish genomes) by one, two or three tools. The first row, 'Not included', reports the proportions that are comprised within not-included synteny blocks (consecutive or partially overlapping) in the human genome, whereas the second row, 'Included', shows the proportions of the human genome that is recovered by included synteny blocks which mostly represent duplicated regions.

directly deduced from the Venn diagram by summing up the 7 percentages because regions where two or three tools disagree (such as the regions 5 and 6 in Fig. 2) would be counted two or three times. This proportion is in fact at least equal to the highest proportion of the genome recovered by only one tool (*e.g.* at least 89.4% of the human genome was found in synteny with the mouse genome because $89.4 = max(84.3 + 2.1 + 1.1 + 1.7, 84.3 + 1.7 + 2.2 + 1.2, 84.3 + 1.2 + 1.8 + 2.1))$. This proportion decreases with increasing phylogenetic distances between compared genomes: 89.4% between Human and Mouse, 71.9% between Human and Bird and 50.1% between Human and Fish (Fig. 3, top). In the first two pairwise comparisons involving relatively close genomes (Human/Mouse and Human/Zebra finch), a large proportion of the human genome was congruently recovered by all three tools, 84.3% and 61.5%, respectively. Between 5% and 20% of the genome were recovered either by only one tool or congruently by two tools or even not congruently by two or three tools (Fig. 3). These results, and in particular the proportions specifically found by each of the three methods, show that all three tools can efficiently reconstruct synteny blocks between these genomes and that SynChro performs equally well as the two other tools. For the comparison involving more distant genomes (Human/Zebrafish), the proportions of the genome that is congruently found in synteny by the three methods is much more limited (15%). However, the proportion of the genome that was recovered by only two methods also remains limited (between 4.6 and 8.4%) which shows that the lack of congruent synteny in this comparison does not result from the inability of one tool to correctly reconstruct synteny but rather from a true loss of synteny between these genomes probably due to the accumulation of numerous chromosomal rearrangements [22]. It is interesting to note that a proportion of the human genome co-detected by SynChro and any of the two other programs (8.4 and 6.8%) is higher than the proportion co-detected by MCScanX and iADoRe (4.6%), which suggests that SynChro could be more efficient than the two other tools to detect synteny between divergent genomes (with the parameters used in this work, see Materials and Methods). The relatively high proportion of the genome only covered by SynChro synteny blocks (19.9%, Fig. 3) can be explained by the fact that 508 synteny blocks (over the 1868 identified by SynChro, Table 2) are defined by only two anchors. These small synteny blocks escape detection with MCScanX and i-ADHoRe because of the higher minimal number of anchors that is required to define a block in these programs (5 and 3, respectively). We checked whether small blocks composed of only two genes detected by SynChro corresponded mainly to false positive blocks or if they comprise true synteny information. The probability that two pairs of homologs are found by chance as direct neighbors simultaneously in two different genomes is given by the following formula $\frac{2}{\#genes - 1}$. This probability equals 10^{-4} for the human genome. However, we found that 9 out of the 10 two-gene synteny blocks in the human/mouse comparison were composed of such direct neighbors (90%). For the human/bird comparison we found 26 blocks of direct neighbors out of the 42 two-gene synteny blocks (62%). For the human/fish comparison, we detected 250 blocks of direct neighbors among the 508 blocks of two genes (49%). These results clearly show that an important proportion of the small synteny blocks composed of only two genes that are detected by SynChro, do in fact comprise true synteny signal. This also suggests that the proportion of false positives in these small synteny blocks would increase for comparisons involving more remote species.

The proportion of the human genome that is comprised in included synteny blocks mainly represents the proportion of the genome that is duplicated. From the two first comparisons (Human/Mouse and Human/Zebra finch), it clearly appears that SynChro detects significantly less such regions than the two other tools (0.1 to 0.4% *vs* 3.7 to 8.7%, respectively), which was expected given that SynChro is not designed to predict duplicated regions (due to the RBH step), contrary to MCScanX and i-ADHoRe. It is noteworthy that although MCScanX and i-ADHoRe manage to detect some duplicated regions, the congruence between their predictions is rather limited (2.3 to 2.8%, Fig. 3). In addition, when a genome has undergone a recent whole genome duplication event, as it is the case for the zebrafish genome, SynChro manage to identify a non-negligible fraction of the duplicated regions (2.1% compared to 3.6% for the two other tools).

It is also interesting to note that SynChro detects more syntenic homologs than MCScanX or iADHoRe. For instance, the three tools detect an identical proportion of the genome that is conserved in synteny between Human and Mouse (89.3%, Table 2). However, both the number and the percentage of syntenic homologs in the synteny blocks are much higher for Synchro than for the two other tools (Table 2). Finally, we made the intriguing observation that the number of synteny blocks detected by MCScanX does not increase with increasing phylogenetic distances as it is expected from an increasing number of chromosomal rearrangements and as it is found to be the case with both SynChro and i-ADHoRe (Table 2).

Materials and Methods

For each tool, the same parameters were used for the three comparisons: *Homo sapiens/Mus musculus*, *Homo sapiens/Taeniopygia guttata* and *Homo sapiens/Danio rerio*. The four genomes were downloaded from the *Ensemble* website (http://www.ensembl.org/info/data/ftp/index.html).

SynChro

The RBH identification is achieved with OPSCAN (http://wwwabi.snv.jussieu.fr/public/opscan/), which is based on the FASTA algorithm [12]:

1. For each protein sequence from the query genome, OPSCAN scan the database with a simple version of the fastp algorithm where no gap is allowed and where the alignment is achieved through shifting sequences to maximize the number of matches between the two compared sequences. This step leads to the identification of a set of K most similar genes with $K = 6$ (default value). Other parameters are used with their default values (kuple: 2, fastp diag integ: 0 and fastp lower threshold: 5).

2. For each query gene OPSCAN refines the alignment with its K most similar target genes by performing a dynamic programming alignment (with zero cost end gaps). The parameters used are BestFit (local) and BLOSUM60 scoring matrix.

3. RBH are defined from these refined alignments when the most similar gene to the query gene Gi amongst its K most similar database genes is Gj, and the most similar gene to the database gene Gj is the query gene Gi. The parameters used are Bestfit score threshold for "homologs": 40 (in the 0–100 range) and length ratio threshold (longest sequence divided by the shortest): 1.3.

The reason why we chose to use OPSCAN rather than blast is because this algorithm was optimized for RBH identification. The fastp part permits OPSCAN to quickly scan the database genome (by simply shifting the compared sequences), as a pre-filter for possible RBH, and then, the BestFit algorithm is run only between query genes and their K = 6 most similar homologs. OPSCAN takes only 36 min (on a desk computer) to identify RBH between the human and the mouse genome. By comparison, a single pass of blastp using the human genome as query against the mouse genome takes 131 min and it would be necessary to run blast in the other direction using the mouse genome (or a subset of it) as query to identify RBH.

There is only one parameter to set up in SynChro, the synteny block stringency Δ. This parameter concomitantly sets both the $Δ_{RBH}$ and the $Δ_{gene}$ parameters although each value can also be set separately (see the description of the SynChro algorithm above). The same value Δ = 5 was used for the three pairwise comparisons of genomes described in this work. This Δ value allows a maximum of 4 intervening RBH within a synteny block which is well-suited to identifying synteny blocks between human and zebrafish (even if Table 1 shows that other values would have been fine too: there are no major differences between Δ = 2, 3, 4, 5, 6 or 7).

SynChro uses several other parameters (% of similarity between homologs, length of the alignments, minimal number of anchors per block ($n = 2$, this value has no relationship whatsoever with the Δ value) that have fixed values. These values were shown to be well suited to perform efficient synteny block reconstruction between a large range of organisms sharing various phylogenetic relationships (successfully applied to 18 yeast and 13 vertebrate genomes [21]). Nevertheless, the user can easily change the values of these parameters in the source code (*SynChro.py*) where they are clearly commented at the top of the file.

MCScanX

MCScanX uses as input a file containing pairwise homologous relationships (typically an all-against-all BLAST search). The blastp minimal expectation value (E) was set to 1e−10 (as suggested in the manual). This value impacts the number of reconstructed synteny blocks. In addition, at least 6 parameters need to be set (even if, many of them can be used with their default value):

1. MATCH_SCORE, a final score used to validate a synteny block: we used the default value (50)
2. GAP_PENALTY, we used the default value (−1)
3. MATCH_SIZE, a number of genes required to call a collinear block: as SynChro performs synteny block reconstruction from 2 anchors, we set this parameter to the minimum (*i.e.* 5, the default value)
4. E_VALUE, the synteny block alignment significance: we used the default value (1e−10)
5. MAX_GAPS, the maximum of gaps allowed: default value is 25, which is too much permissive (each regions map tens of regions in the other genome), we used a value of 10 instead.

6. OVERLAP_WINDOW, the maximum distance (in number of genes) to collapse BLAST matches: we used the default value (5).

i-ADHoRe

i-ADHoRe takes, as input a file containing pairwise homologous relationships (typically an all-against-all BLAST search), so we use the same e-value of 1e−10 that for MCScanX (that is why the execution time, in Table 2, corresponding to the execution of blastp, is the same for MCScanX and i-ADHoRe). To run i-ADHoRe, at least 5 additional parameters need to be set:

1. prob_cutoff, indicating the maximum probability for a cluster to be generated by chance: we use the suggested value (0,001)
2. gap_size, indicating the maximum (pseudo-)distance that should exist between points in a cluster: we use the value given as an example (15)
3. cluster_gap, indicating the maximum (pseudo-)distance that should exist between individual base clusters in a cluster: we use the value given as an example (20)
4. q_value, indicating the minimum r^2-value (a measure for the linearity of a series of points) a cluster should have: we use the value given as an example (0.9)
5. anchor_points, the minimum number of anchor points: as SynChro reconstruct synteny blocks from 2 anchors, we set this parameter to the minimum, meaning 3 (the suggested values was comprised between 3 and 6)

Conclusion

We showed in this work that SynChro is a fast, efficient and user-friendly tool to reconstruct synteny blocks between (complex) genomes harboring different levels of synteny conservation. Despite a very simple algorithm, the reconstruction is highly congruent with reconstructions obtained with more sophisticated tools. The main advantages of SynChro are the following: (i) it is fast (it takes, on a desk computer, on the order of 40 minutes to compare two vertebrate genomes); (ii) it is easy to use (a unique parameter Δ, which is really simple to handle, needs to be set) and (iii) it provides a rich set of graphic outputs (notably an interactive synteny map that allows zooming in breakpoint regions).

Author Contributions

Conceived and designed the experiments: GD AC GF. Performed the experiments: GD. Analyzed the data: GD AC GF. Contributed reagents/materials/analysis tools: AC GF. Wrote the paper: GD AC GF.

References

1. Soderlund C, Bomhoff M, Nelson WM (2011) SyMAP v3.4: a turnkey synteny system with appli-cation to plant genomes. Nucleic Acids Research 39: e68.
2. Pan X, Stein L, Brendel V (2005) SynBrowse: a synteny browser for comparative sequence analysis. Bioinformatics 21: 3461–3468.
3. Lyons E, Pedersen B, Kane J, Alam M, Ming R, et al. (2008) Finding and comparing syntenic regions among arabidopsis and the outgroups papaya, poplar, and grape: Coge with rosids. Plant physiology 148: 1772–81.
4. Dong X, Fredman D, Lenhard B (2009) Synorth: exploring the evolution of synteny and long-range regulatory interactions in vertebrate genomes. Genome Biology 10: R86.
5. Hampson S, McLysaght A, Gaut B, Baldi P (2003) LineUp: Statistical Detection of Chromosomal Homology With Application to Plant Comparative Genomics. Genome Research 13: 999–1010.

6. Simillion C, Vandepoele K, Saeys Y, Van de Peer Y (2004) Building genomic profiles for uncovering segmental homology in the twilight zone. Genome Research 14: 1095–1106.
7. Wang X, Shi X, Li Z, Zhu Q, Kong L, et al. (2006) Statistical inference of chromosomal homology based on gene colinearity and applications to arabidopsis and rice. BMC Bioinformatics 7: 447.
8. Ng MP, Vergara I, Frech C, Chen Q, Zeng X, et al. (2009) OrthoClusterDB: an online platform for synteny blocks. BMC Bioinformatics 10: 192.
9. Tang H, Bowers JE, Wang X, Ming R, Alam M, et al. (2008) Synteny and collinearity in plant genomes. Science 320: 486–488.
10. Ma J, Zhang L, Suh BB, Raney BJ, Burhans RC, et al. (2006) Reconstructing contiguous regions of an ancestral genome. Genome Research 16: 1557–1565.

11. Altschul S, Madden T, Schaffer A, Zhang J, Zhang Z, et al. (1997) Gapped BLAST and PSI-BLAST: a new generation of protein database search programs. Nucleic Acids Research 25: 3389–3402.

12. Lipman DJ, Pearson WR (1985) Rapid and sensitive protein similarity searches. Science 227: 1435–1441.

13. Enright AJ, Van Dongen S, Ouzounis CA (2002) An efficient algorithm for large-scale detection of protein families. Nucleic Acids Research 30: 1575–1584.

14. Wang Y, Tang H, DeBarry JD, Tan X, Li J, et al. (2012) MCScanX: a toolkit for detection and evolutionary analysis of gene synteny and collinearity. Nucleic Acids Research 40: e49.

15. Proost S, Fostier J, De Witte D, Dhoedt B, Demeester P, et al. (2012) i-ADHoRe 3.0|fast and sensitive detection of genomic homology in extremely large data sets. Nucleic Acids Research 40: e11.

16. Drillon G, Carbone A, Fischer G (2013) Combinatorics of chromosomal rearrangements based on synteny blocks and synteny packs. Journal of Logic and Computation 23: 815–838.

17. Cannon S, Kozik A, Chan B, Michelmore R, Young N (2003) DiagHunter and GenoPix2D: programs for genomic comparisons, large-scale homology discovery and visualization. Genome Biology 4: R68.

18. Haas BJ, Delcher AL, Wortman JR, Salzberg SL (2004) DAGchainer: a tool for mining segmental genome duplications and synteny. Bioinformatics 20: 3643–3646.

19. Sinha A, Meller J (2007) Cinteny: exible analysis and visualization of synteny and genome rear-rangements in multiple organisms. BMC Bioinformatics 8: 82.

20. Zeng X, Nesbitt MJ, Pei J, Wang K, Vergara IA, et al. (2008) In: OrthoCluster: a new tool for mining synteny blocks and applications in comparative genomics. Proceedings of the 11th international conference on Extending database technology: Advances in database technology. New York, NY, USA: ACM, EDBT '08, pp. 656 – 667. Available: http://doi.acm.org/10.1145/1353343.1353423. doi:10.1145/1353343.1353423

21. Vandepoele K, Saeys Y, Simillion C, Raes J, Van de Peer Y (2002) The automatic detection of homologous regions (adhore) and its application to microcolinearity between arabidopsis and rice. Genome Research 12: 1792–1801.

22. Drillon G, Fischer G (2011) Comparative study on synteny between yeasts and vertebrates. Comptes rendus biologies 334: 629–638.

22

MOSAIK: A Hash-Based Algorithm for Accurate Next-Generation Sequencing Short-Read Mapping

Wan-Ping Lee[1]*, Michael P. Stromberg[1], Alistair Ward[1], Chip Stewart[1,2], Erik P. Garrison[1], Gabor T. Marth[1]

1 Department of Biology, Boston College, Chestnut Hill, Massachusetts, United States of America, **2** Broad Institute of Harvard and Massachusetts Institute of Technology, Cambridge, Massachusetts, United States of America

Abstract

MOSAIK is a stable, sensitive and open-source program for mapping second and third-generation sequencing reads to a reference genome. Uniquely among current mapping tools, MOSAIK can align reads generated by all the major sequencing technologies, including Illumina, Applied Biosystems SOLiD, Roche 454, Ion Torrent and Pacific BioSciences SMRT. Indeed, MOSAIK was the only aligner to provide consistent mappings for all the generated data (sequencing technologies, low-coverage and exome) in the 1000 Genomes Project. To provide highly accurate alignments, MOSAIK employs a hash clustering strategy coupled with the Smith-Waterman algorithm. This method is well-suited to capture mismatches as well as short insertions and deletions. To support the growing interest in larger structural variant (SV) discovery, MOSAIK provides explicit support for handling known-sequence SVs, e.g. mobile element insertions (MEIs) as well as generating outputs tailored to aid in SV discovery. All variant discovery benefits from an accurate description of the read placement confidence. To this end, MOSAIK uses a neural-network based training scheme to provide well-calibrated mapping quality scores, demonstrated by a correlation coefficient between MOSAIK assigned and actual mapping qualities greater than 0.98. In order to ensure that studies of any genome are supported, a training pipeline is provided to ensure optimal mapping quality scores for the genome under investigation. MOSAIK is multi-threaded, open source, and incorporated into our command and pipeline launcher system GKNO (http://gkno.me).

Editor: Chuhsing Kate Hsiao, National Taiwan University, Taiwan

Funding: NIH: 5R01HG004719-04; NIH: 3U01HG006513-02S1. The funders had no role in study design, data collection and analysis, decision to publish, or preparation of the manuscript.

Competing Interests: The authors have declared that no competing interests exist.

* E-mail: wanping.lee@bc.edu

Introduction

The widespread availability of next-generation sequencing platforms has revolutionized the life sciences through the ever more accessible ultra-high throughput DNA sequencing efforts [1]. Next-generation sequencing technologies including Illumina, Complete Genomics, and Applied Biosystems (AB) SOLiD have been driving the current market forward, whereas Pacific Biosciences SMRT [2], Ion Torrent [3], and Nanopores [4] are leading the development of third-generation sequencing instruments. These technologies bring novel opportunities for many applications including genetic variant discovery, epigenomic variant discovery, RNA-Seq and ChIP-Seq, but also provide complex computational challenges. The short reads generated by these technologies are generally aligned to a reference genome as an early step in many of the current analysis workflows and the alignment quality limits the accuracy of any downstream analysis. Large sequencing projects often use sequencing machines from multiple manufacturers for data generation and can also make use of legacy data. It is desirable that any researcher tasked with analyzing the available data need not learn the intricacies of multiple alignment software packages to utilize all of the available data. This is unnecessary, since, MOSAIK can, uniquely, accurately align sequencing data from all current and legacy platforms.

Current sequencing technologies typically generate on the order of hundreds of millions of short reads (of the order of a few hundred nucleotides or shorter) on a single run. In order to analyze all of these reads in a reasonable amount of computational time, the performance of reference-guided alignment programs is paramount. The memory footprint of these algorithms must also be well managed to allow their deployment beyond institutions with extremely expensive computational infrastructure. These goals must be met without compromising the accuracy of resulting alignments. Most existing aligners utilize hashing algorithms or the Burrows-Wheeler transform [5,6] to search exact matches (algorithms may be modified to allow few mismatches) as their first step to achieve high performance and optimize memory usage. Theoretically, hashing method outperforms BWT method for DNA database searching [7]. The hash-based aligners, Eland (AJ Cox, Illumina, San Diego), MAQ [8], mrFAST/mrsFast [9,10], SHRiMP [11,12], and ZOOM [13,14] hash reads and fit these hashes to the reference genome, while MOM [15], MOSAIK, PASS [16], ProbeMatch [17], SOAP [18], SRmapper [19], and STAMPY [20] hash the reference genome and store this for comparison with reads. Major Burrows-Wheeler transform (BWT) based aligners include BWA [21], Bowtie [22,23], segemehl [24] and SOAP2 [25]. In general, BWT-based aligners are sensitive but include a slow query operation (each FM-index query is slower than a hash query [26,27]). In regions with genomic variation (e.g. those regions in which the investigator is usually most interested), maintaining good performance generally leads to lower sensitivity [19,28]. In addition, the Burrows-Wheeler transform method is less flexible than hash based methods. For example, it is more difficult for the Burrows-Wheeler transform to consider ambiguities by using IUPAC [29] ambiguity codes representing, for example, known SNPs. The main drawback of hash-based aligners is that they usually consume more memory than BWT-based aligners; however, as high-memory machines become cheaper, this becomes less of a problem. Currently, MOSAIK can be operated in a low-memory mode that keeps the memory footprint small (~8Gb for the human genome), ensuring that even for lower memory machines, MOSAIK can still be used with confidence.

Here, we introduce a reference-guided aligner, MOSAIK, that is highly sensitive, stable and flexible, whose utility on a range of different sequencing technologies has been demonstrated in the context of the 1000 Genomes Project [30,31]. In addition to MOSAIK's ability to map data from all major sequencing technologies, it has been developed to address many of the issues currently facing genome researchers. These developments are outlined here. The primary goal of any mapping software is to minimize alignment artefacts and increase alignment sensitivity and accuracy. To achieve this, MOSAIK uses a Smith-Waterman algorithm and is able to align reads to a genome using IUPAC ambiguity codes, ensuring that alignments against known *single-nucleotide polymorphisms* (SNPs) are not penalized. Using this method, MOSAIK achieves positive predictive values (PPVs) of 99.5% for all alignments and 100.0% for high confidence alignments (those with a mapping qualities larger than 20) in experiments on simulated data. In addition to providing the genomic coordinates of the read mapping, it also important to provide a measure of the confidence in this coordinate. For this purpose, MOSAIK uses a neural-network based training scheme to provide well-calibrated mapping quality scores. In our experiments, the correlation coefficient between the quality scores assigned by MOSAIK and the actual scores is 0.97. To ensure that studies of any genome are supported, MOSAIK provides a training pipeline to ensure optimal mapping quality scores for the genome under investigation. A major area of active investigation is the study of structural variation (SV). MOSAIK has been designed to aid and simplify the discovery of such variants. In particular, known-insertion sequences, for example, mobile element insertions (MEIs), can be included as part of the reference genome. This helps to minimize alignment artefacts, but MOSAIK also provides a host of valuable information to user on the paired-end reads that map to one of these sequences. When requested, MOSAIK also outputs all possible mapping locations for every read in a separate BAM file. This is essential for determining the mappability of the genome under study. The most recent versions of BWA, BOWTIE and MOSAIK are comparable in their run times, and STAMPY is approximately six times slower. Finally, MOSAIK is implemented in C++ as a modular suite of programs that is dual licensed under the GNU General Public License and MIT License. It is multi-threaded, open source, and incorporated into our command and pipeline launcher system GKNO (http://gkno.me).

Results

Alignments from all sequencing technologies

All of the available sequencing technologies use different techniques for library preparation, paired-end read protocols and DNA sequencing, resulting in a range of read lengths, fragment lengths, base quality assignments, as well as different error profiles.

Additionally, not all technologies report their sequencing reads in the conventional basespace (strings of the A, G, C and T nucleotides) format. Notably, AB SOLiD uses a di-base encoding scheme known as colorspace and single-molecule sequencing technologies use dark bases [32] for bases not registered by the instrument. These facts mean that all of the currently available aligners are tailored for use on data from one, or a small number of the available technologies. MOSAIK is the only aligner that can be used in a consistent manner across most of these technologies.

In addition to the second-generation technologies, Illumina, Roche 454 and AB SOLiD, MOSAIK can also be deployed on third-generation technologies, in particular, Pacific Biosciences and Ion Torrent reads. MOSAIK uses the same algorithmic approach for all sequencing technologies, however, since the characteristics of each technology are different, the resultant alignment rates vary, as shown in Table 1. These alignment rates were generated using Illumina paired-end (PE), single-end (SE) and Roche 454 SE reads generated using the MASON read simulator (http://www.seqan.de/projects/mason/) as well as Illumina and AB SOLiD reads from the Han Chinese in Beijing (CHB) population from the 1000 Genomes Project. For the third-generation technologies, we used *E. coli* reads provided by Ion Torrent (http://www.iontorrent.com/applications-pgm-accuracy/) and *V. cholerae* reads provided by Pacific Biosciences (ftp://ftp.ncbi.nlm.nih.gov/sra/Submissions/SRA026/SRA026766/provisional/SRX032454/SRR075103/).

In general, sequencing reads containing fewer sequencing errors have higher alignment rates, e.g. Illumina reads, and longer or paired-end reads require more time to align. That paired-end reads take additional time is not unexpected. If one of the reads in a pair cannot be mapped unambiguously, additional searches are performed guided by the mapped mate in the pair. The additional processing time results in more accurate alignments as well as a lower fraction of unaligned reads. AB SOLiD reads are aligned in colorspace (converting to basespace prior to alignment loses all of the benefits of colorspace), but additional processing is required due to the required conversion of the alignments into basespace post-alignment. These experiments show that MOSAIK works well for existing sequencing technologies.

Highly accurate alignments on simulated data

To investigate the accuracy of reads aligned using MOSAIK, we simulated a total of 12 million Illumina paired-end reads from chromosome 20 of the Hg19 human genome using the MASON read simulator. Reads of length 76 and 100 basepairs were simulated with a haplotype SNP rate of 0.1%. The reads were aligned against the entire human genome using BWA-0.5.9, BOWTIE-2.0-beta5, STAMPY-1.0.13, and MOSAIK-2.1.78. The default parameter settings were used for all of the aligners. The positive predictive value of each aligner was then calculated as the number of correctly placed reads (the genomic coordinate of the mapped read agreed with the known location of the read from MASON) divided by the total number of mapped reads. Notice that an alignment is considered incorrect as the aligned position is out the 20 bp tolerant window and thus alignments with more than 20 bp unmapped bases may be considered as incorrect. We choose 20 bp as the tolerant window since on the dataset most of alignments contain fewer than 20 bp clipped bases (see supplemental Figure S1).

Figure 1 shows the positive predictive value (PPV, the number of correctly mapped reads divided by the total number of mapped reads) of the aligners as a function of mapping quality cutoffs (complete information is shown in Figure S2). At a mapping quality cutoff of twenty, for example, the PPV is calculated using

Table 1. Summary of the alignment accuracies achieved by MOSAIK for reads generated from different sequencing technologies.

Technologies	Aligned (%)	Speed (reads/second)	Read lengths [min;max]	Reference genome	Dataset
Illumina; PE	99.98	83.95	100; 50	Human hg19	MASON simulated
Illumina; SE	99.75	153.98	100; 76; 50	Human hg19	MASON simulated
Illumina; PE/SE	91.48	147.42	81; 76; 51; 45; 41	Human hg19	CHB population in 1000G
454; SE	99.42	8.018	400.673 [266;529]	Human hg19	MASON simulated
Ion Torrent	77.02	20.85	223.99 [59;398]	E. coli strain 536	Ion Torrent released
SOLiD	55.64	126.81	50	Human hg19	CHB population in 1000G
Pacific Biosciences*	85.79	0.69	698.61 [48;6084]	V. cholerae 4,033,464 bp.	Pacific Biosciences released

*The parameter set "-hs 10 -mmp 0.5 -act 15" was used as opposed to the default values "-hs 15 -mmp 0.15 -act 55".
With the exception of the Pacific Biosciences data, all alignments were generated using MOSAIK's default parameters.

only those reads with mapping quality values greater than or equal to twenty. It can be seen that the PPVs of BWA and MOSAIK are comparable and are significantly better than those achieved by BOWTIE and STAMPY. For mapping quality cutoffs smaller than five, BWA is more accurate (fewer incorrect alignments among total mapped alignments) than MOSAIK, however, MOSAIK is the most accurate as the mapping quality cutoff is increased. For a mapping quality cutoff of twenty (a common cutoff employed by downstream analysis tools that only wish to consider confidently aligned reads), the PPVs of MOSAIK, BWA, BOWTIE and STAMPY are 100.00%, 99.99%, 99.79% and 99.63% respectively. These results are summarized in Table 2.

Figure 2 shows receiver operating characteristic (ROC) curves for the same data. The total number of mapped reads (x axis) is plotted against the number of incorrectly mapped reads (y axis). Each point on the curve represents the number of alignments whose mapping qualities are greater than or equal to the indicated value. MOSAIK has a relatively smooth curve, ensuring that downstream tools that employ mapping quality cutoffs (i.e. ignoring all reads with mapping qualities less than the cutoff) do not incur extremely large changes in the number of reads while

progressively increasing the cutoff. Conversely, the other aligners do not share this property. For example, consider the BWA alignments. By decreasing the mapping quality cutoff from 30 to 29, the number of incorrectly mapped reads increases by 308.56% while for MOSAIK, the increase is a much more modest 6.25%. Downstream analysis tools require a useful mapping quality scale, so that excluding lower quality reads improves the specificity of the analysis results. The dynamic range demonstrated by MOSAIK is therefore a very valuable result for these tools.

Mapping quality calibration

The Phred mapping quality score present in the standard SAM/BAM format represents the probability that the read was mapped incorrectly and is defined as:

$$Q = -10\log_{10}P, \qquad \text{(Equation 1)}$$

where Q is the Phred score and P is the probability that the read was misaligned. For example, a read assigned a Phred mapping quality score of 30 has a 1 in 1000 chance of being misaligned.

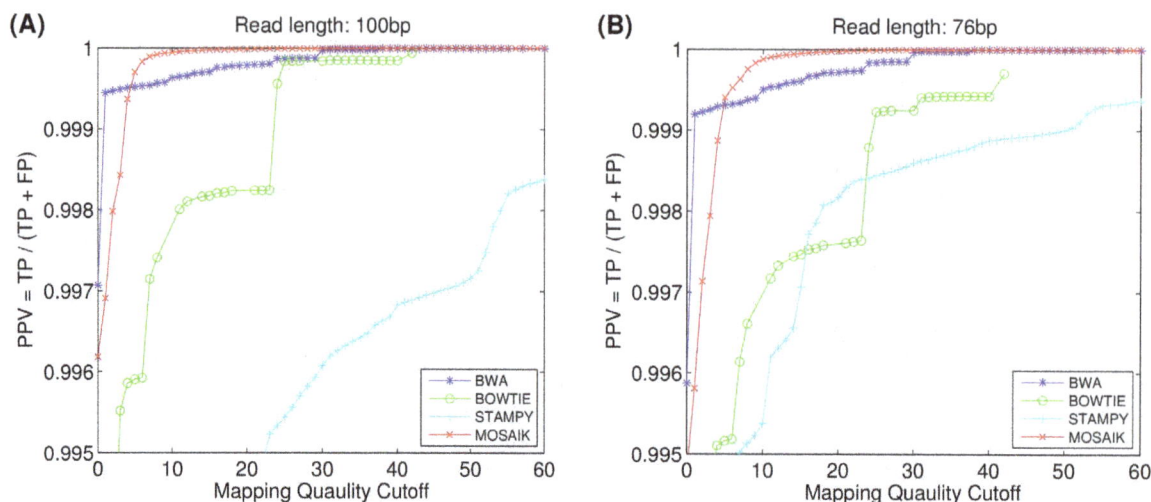

Figure 1. The positive predictive value of aligners (the number of correctly mapped reads divided by the total number of mapped reads) as a function of mapping quality threshold. Datasets in (A) 100 bp and (B) 76 bp read lengths. PPV, TP, and FP stand for positive predictive value, true positive and false positive, respectively.

Table 2. The positive predictive values (the number of correctly mapped reads divided by the total number of mapped reads) in terms of mapping quality cutoffs.

MQ cutoffs	30		20		10		0	
Read lengths	100	76	100	76	100	76	100	76
BWA	1	1	0.9998	0.9997	0.9996	0.9995	**0.9971**	**0.9959**
BOWTIE	0.9998	0.9992	0.9982	0.9976	0.9980	0.9972	0.9823	0.9819
STAMPY	0.9961	0.9986	0.9945	0.9982	0.9897	0.9954	0.9813	0.9909
MOSAIK	1	1	1	1	**0.9999**	**0.9999**	0.9962	0.9947

MOSAIK's mapping qualities are obtained using a neural network that approximates the error function when provided with features such as best and second best Smith-Waterman alignment scores, read entropy, number of potential mapping locations and hashes. For paired-end reads, the fragment length of mapped paired end reads is also used in the neural network to produce more precise mapping quality calculations. MOSAIK embeds the Fast Artificial Neural Network (FANN) library (http://leenissen.dk/fann/wp/), which implements multilayer artificial neural networks in C, supporting both fully connected and sparsely connected networks, to calculate Phred score for each alignment.

The default neural network provided with MOSAIK was generated by training on the human genome. The first step involves simulating reads and then aligning them to the human reference genome to obtain MOSAIK's behaviour such as best and second best Smith-Waterman scores, and numbers of obtained mappings and hashes. Then, the neural network was trained based on MOSAIK's behaviour.

Figures 3(A) and 3(B) compare the actual (calculated using Equation (1)) and the assigned mapping quality scores. Both, BOWTIE and MOSAIK produce very accurate Phred score mapping qualities across the whole quality score spectrum. The Pearson correlation coefficients between the assigned and actual quality scores are shown in Table 3. MOSAIK has an average (across all read lengths investigated) correlation coefficient of

0.9698, compared with 0.9061, 0.9207, and 0.8652 for BWA, BOWTIE, and STAMPY respectively.

Retraining Mapping-Quality Neural Network for *E. coli* Alignment

The genomes of different species differ in many respects including sequence content (base composition as well as relative frequency of repeat or low-complexity sequence) as well as the size of the genome. Most aligners, including MOSAIK, are general programs that can operate on any given reference genome, however, in general, the properties of the genome under investigation are ignored. MOSAIK provides a retrainable mapping-quality pipeline to generate applicable neural networks for different genomes or sequencing technologies. This means that the calibration of the mapping quality scores remains of a very high quality, regardless of reference genomes.

To demonstrate the merit of the retrainable mapping-quality pipeline, we used 6 million simulated paired-end reads from the *E. coli* genome to train a neural network (see supplemental method (A): Retraining Mapping Quality Neural Network). An additional independent set of 6 million simulated *E. coli* paired-end reads were then generated and aligned to the *E. coli* genome using multiple aligners. The assigned and actual mapping quality scores are plotted for all aligners in Figure 3(C). There are two sets of

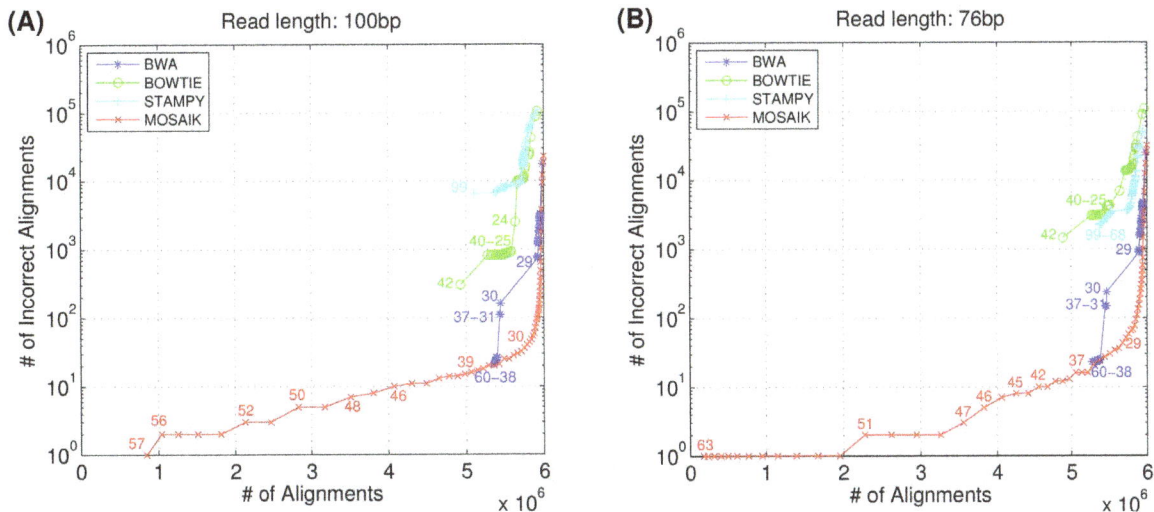

Figure 2. The receiver operating characteristic (ROC) curves; Datasets in (A) 100 bp and (B) 76 bp read lengths. Each point represents the total numbers of alignments whose mapping qualities are greater than the indicated value. MOSAIK has a relatively smooth curve, ensuring that downstream tools that employ mapping quality cutoffs (i.e. ignoring all reads with mapping qualities less than the cutoff) do not incur extremely large changes in the number of reads while progressively increasing the cutoff.

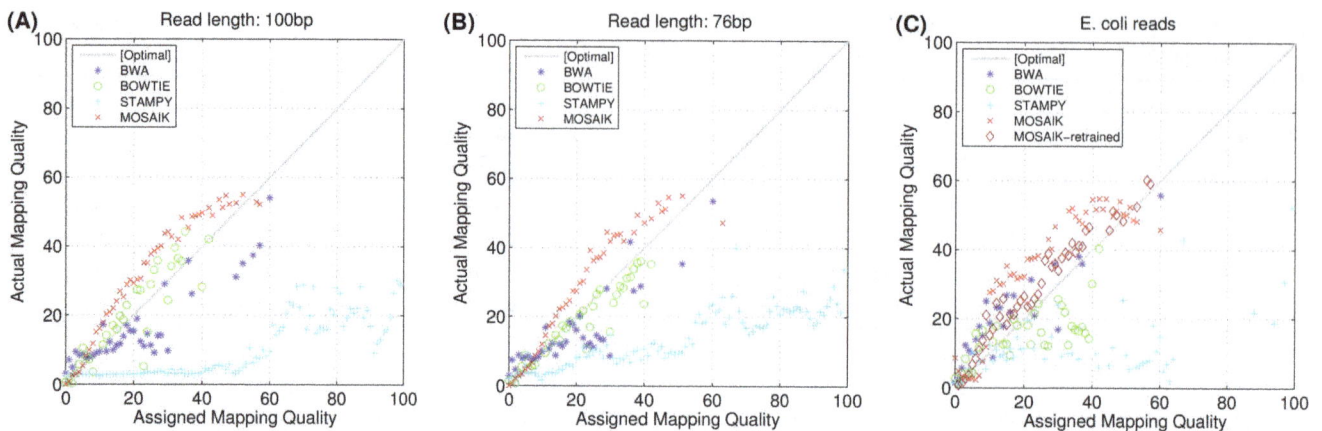

Figure 3. The correlations between the aligners' assigned and actual mapping qualities. Phred score scheme. (A) and (B) simulated datasets in 100 bp and 76 bp read lengths. (C) *E. coli* simulated dataset in which "MOSAIK" is MOSAIK's default mapping-quality network trained by human genome while "MOSAIK-retrained" is the retrained mapping-quality network by using *E. coli* simulation and *E. coli* genome. The detailed numbers of the Pearson's correlation coefficients are given in Table 3.

data for MOSAIK: the first (red crosses) is generated using the default neural network trained on the human genome, and the second (dark red diamond) uses the neural network retrained on the *E. coli* genome. It is clear that the mapping qualities generated by the retrained neural network for MOSAIK are the best calibrated, although the data using the human genome trained neural network is still of a high quality. Also of note, Figures 3(A) and 3(B) show that BOWTIE has quite well calibrated mapping qualities for mapping to the human genome, however, when applied to *E. coli*, the calibration is noticeably worse.

MOSAIK accurately accounts for short INDELs

MOSAIK uses a Smith-Waterman (SW) algorithm as the final polishing step to produce pairwise read alignments, which is the preferred choice for aligning gapped (short INDELs) sequences since it seeks all possible frames of alignment with all possible gaps. To assess the sensitivity of different aligners to short INDELs, we simulated Illumina paired-end reads containing 1-14 bp INDEL events that are generated by a genome simulator, MUTATRIX (https://github.com/ekg/mutatrix). For each INDEL length, we introduced an average of 100 events, with approximately 800 spanning reads (see supplemental Figure S4).

Figures 4(A) and 4(B) plot the sensitivity (number of correctly mapped reads divided by the total number of simulated reads) as a function of the INDEL length. An alignment is considered correct when it is mapped to the correct position as well as contains the simulated variant. Alignments containing the correct variants can facilitate downstream variant detectors detecting variants depend-

ing on alignments and need no any realignment step which is timing consuming. MOSAIK is the most sensitive aligner considered here when considering deletions. When considering insertions, MOSAIK's sensitivity is comparable to, but slightly worse than those of STAMPY and BOWTIE. It is clear from Figures 4(A) and 4(B) that MOSAIK is the only mapper considered here that is highly sensitive to both insertion and deletion polymorphisms. We understand that some aligners tend to report partial alignments that may not contain variants but are mapped to right places. Those alignments still provide values for variant detections. We thus change the criteria of correct alignments used in Figures 4(A) and 4(B). In figures 4(C) and 4(D), an alignment is considered a correct mapping when it is entirely or partially mapped to the correct positions. The four aligners achieve 96% sensitivity based on the criteria.

Effect of mapping errors on SNP studies

Aligners provide information on where reads map in the human genome along with information on the confidence of the mapping, however, they do not themselves weigh evidence for genetic variants in the genome being studied. Dedicated variant callers use the information provided by mapper in statistical models to determine if there is enough evidence to report a difference with respect to the reference genome. To determine the effect of the mapping on single nucleotide polymorphism (SNP) discovery, we simulated 1,486 SNPs on the human genome chromosome 20 using MUTATRIX. We then used MASON to generate 12 million reads (with read lengths of 76 and 100 basepairs) from this mutated chromosome generated by MUTATRIX. The same four aligners were then used to align these reads back to the entire human reference genome and the variant callers FREEBAYES [33] and SAMTOOLS [34] were used to call SNPs. Figure 5 shows the variant callers sensitivity to SNPs as a function of the false discovery rate (FDR) (the complete information is shown in Figure S3). The points on the curves are generated by only considering SNP calls with variant quality scores (provided by the variant caller) greater than a specific cutoff. Moving from lower-left to upper-right, SNP calls with lower quality scores are cumulatively being included. Both FREEBAYES and SAM-TOOLS produce lower sensitivity calls on the BOWTIE alignments and have a lower FDR on BWA and MOSAIK alignments. It is clear from both Figures 5(A) and 5(B) that the

Table 3. Pearson's correlation coefficients of mapping qualities.

Read lengths	100	76	*E. coli*
BWA	0.8987	0.8625	0.8936
BOWTIE	0.9027	0.9449	0.6989
STAMPY	0.8317	0.8818	0.5262
MOSAIK	**0.9609**	**0.9497**	0.8881
MOSAIK-retrained	–	–	**0.9749**

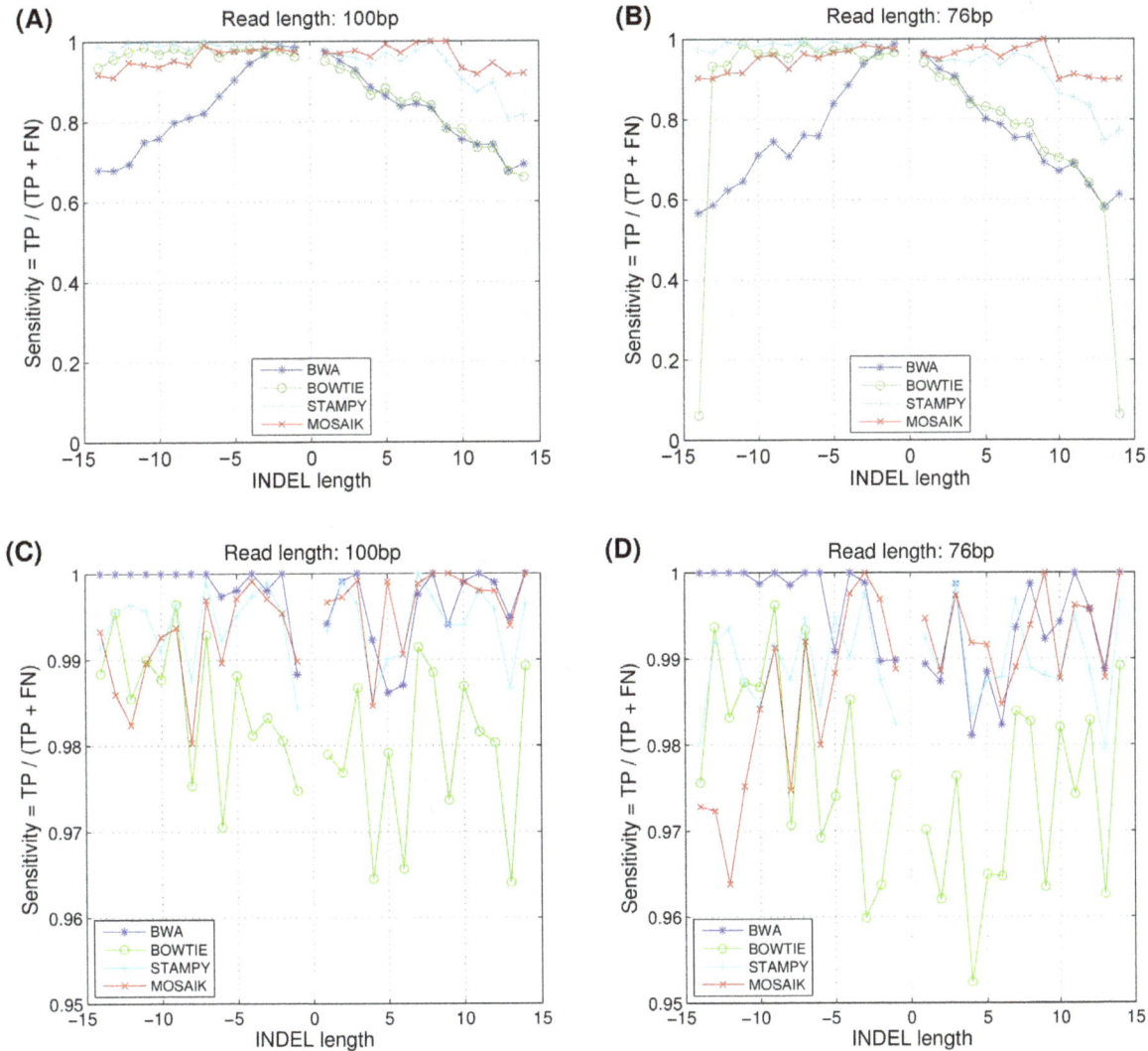

Figure 4. The sensitivities of simulated reads spanning INDELs, which is defined as the number of correct mapped reads divided by the number of simulated reads for each INDEL length. In (A) and (B), the alignments are considered correct as they cross INDELs, while in (C) and (D), the alignments are considered correct as they are entirely or partially mapped to the correct positions. TP and FN are "true positive" and "false negative" respectively.

most sensitive SNP calls are produced when using the MOSAIK alignments, although the BWA alignments are also of a high quality. It is also worth noting that the SNP calls produced by FREEBAYES are more sensitive than those produced by SAMTOOLS regardless of the mapper used.

Support for mobile element insertion

Detecting structural variations using NGS data is a more complex task than detection of short variants and often requires or would benefit from information over and above that ordinarily required for small variant detection. An increasing number of SV detection algorithms are being developed and, in order to increase the effectiveness of these algorithms, MOSAIK has been developed to provide as much relevant and useful information as possible.

There are many genetic sequences that can be considered distinct from the standard set of chromosomes in the genome under investigation. These can include repetitive sequences such as mobile elements [35], viruses (e.g. human endogenous

retroviruses [36]), known novel insertions [37,38] or bacterial contaminants [39] amongst others. MOSAIK provides support for an additional reference genome file containing any genetic sequences provided by the investigator. The advantages of this are two-fold: a) reads originating from contaminants will map to the additional sequences, rather than a lower quality mapping to the best location in the standard reference genome. These sequences essentially act as a sink to catch all the reads that do not originate from the standard reference, reducing the number of mismapped reads that variant detectors have to contend with. b) Reads mapping to repetitive elements (e.g. ALU or LINE elements) are identified as mapping to the additional reference sequence. MOSAIK reports the coordinates of the best mapping in reference genome coordinates, but also includes an additional tag in the BAM file (appearing as ZA in the BAM file), indicating that the read maps to one of the additional reference sequences. Our MEI detector, TANGRAM (https://github.com/jiantao/Tangram) looks for read pairs with one mate uniquely aligned to the genome and the other mate falls within a mobile element

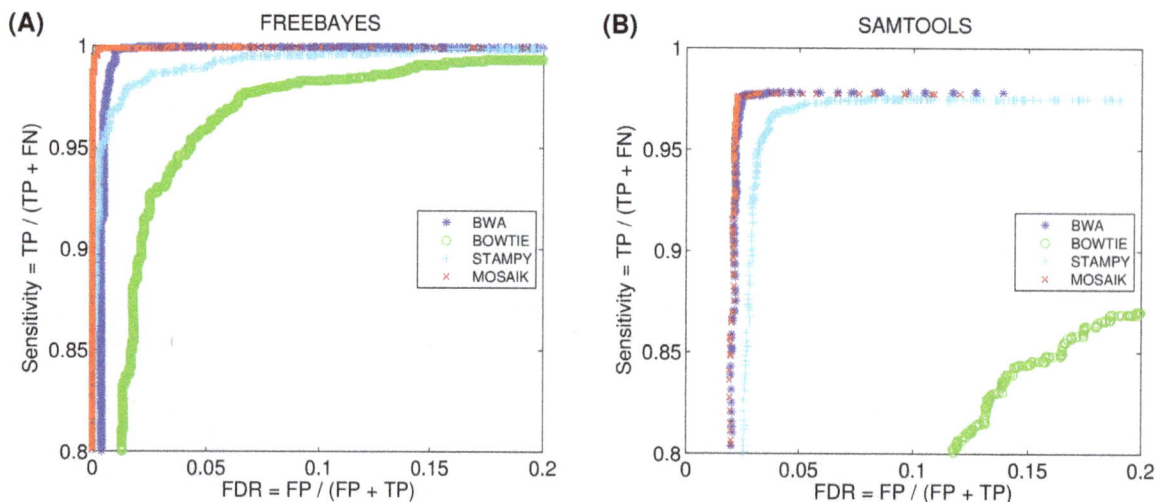

Figure 5. The receiver operating characteristic (ROC) curves of SNPs called by FREEBAYES and SAMTOOLS. The points on the curves are sorted by called qualities and the points closer to the upper-right corner have higher called qualities. The true positive (TP), false positive (FP), and false negative (FN) are calculated by intersecting SNPs called on each aligner's alignments and gold SNPs called on the simulated alignments.

reference sequence. Only relying on this information provided by MOSAIK, the sensitivity of MEI detection can achieve 84%.

Applications

SNP and INDEL Analyses in the 1000 Genomes Project. The 1000 Genomes Project is in the process of using second-generation sequencing instruments to study human genetic variation at the population level. The Phase I [31] release, based on a population of 1,092 sample individuals in 14 populations includes approximately 38 million SNPs, 1.4 million bi-allelic INDELs and 14,000 large deletions. These calls were generated from approximately 966 billion reads and 64 trillion base pairs of human DNA and were sequenced using Illumina, AB SOLiD and Roche 454 for both low-coverage whole-genome and exome targeted sequencing data. A collaborative effort between Boston College and the National Center for Biotechnology Information (NCBI) used MOSAIK to align all of the reads from all of these machines, and served as the official primary alignment set for the exome sequencing data [40] and an alternative alignment set for the low-coverage.

Based on the MOSAIK alignments, SNP, MNP (multi-nucleotide polymorphism) and INDEL calls were generated using the FREEBAYES Bayesian variant calling software. 33,324,407 SNPs were detected in the autosomes of the 1,092 samples, of which, only 23.8% were previously known sites (contained in dbSNP). The transition/transversion (ts/tv) ratio for these sites was 2.12 (2.1 for novel sites and 2.17 for known sites). The Illumina exome data yielded 344,781 SNPs with a ts/tv ratio of 3.18 (3.09 for the novel sites and 3.52 for the known sites) and 22.1% of the exome sites were previously known. The SOLiD exome data yielded 176,637 SNPs with a ts/tv ratio of 3.34 (3.22 for novel sites and 3.58 for known sites). The ts/tv ratios are in accordance with expectations for both the low-coverage and the exome SNPs.

Other SNP Studies. In addition to the 1000 Genomes Project, MOSAIK is widely used for other human clinical genome studies, such as human cancer studies [41–46]. MOSAIK is also used for other species genome studies including model species [47,48], HIV [49–52], parasites [53–55], plants [56–58], and other animals [59,60].

Human Mobile Element Insertion Discovery

In addition to short variants, the 1000 Genomes Project aims to characterize larger structural variations present in the human population. By augmenting the reference genome with known mobile element insertions (MEI), the MOSAIK alignments were able to provide a host of information about their distribution in the human population. As part of the pilot phase of the project, 7,380 MEI polymorphisms were detected using the whole-genome sequencing data [61]. This sample set included 60 samples of European origin (CEU), 59 African (YRI) and 60 Asian samples from Japan and China (CHB/JPT). The FDRs for Alu, L1, and SVA insertions were 2%, 17%, and 27% respectively.

Discussion

MOSAIK is a highly sensitive, stable and flexible reference-guided read mapper which supports most existing sequencing technologies. While MOSAIK is extremely accurate (positive predictive values achieve 99.5% for all alignments and 100.0% for alignments whose mapping qualities are larger than 20 on simulated data), not all reads are aligned with equal confidence. The mapping qualities that MOSAIK provides are generated using a retrainable neural network and are a very good representation of the probability of the alignment being incorrect. In fact, the correlation coefficient between MOSAIK assigned and the actual mapping qualities is 0.97. The retraining pipeline ensures optimized mapping quality score schemes for any genome being studied. For example, when considering aligning against the *E. coli* genome, the correlation coefficient increases from 0.89 to 0.97 when using the human and the *E. coli* neural nets respectively. By using the Smith-Waterman algorithm, MOSAIK is very effective at mapping reads containing short INDELs and the experiments demonstrate that the sensitivity of INDEL mappings is greater than 90%. Additionally, MOSAIK provides explicit support for SV detections.

Most SV detectors make extensive use of information from paired end reads [62–64]. If the two mates in a pair map to greatly separated locations (often the case when the read pair spans or falls within a structural variant), multiple searches through the BAM files are required to assemble all of the information about both mates. This can be a lengthy task, severely impacting the performance of SV detectors. The ZA tag provides a host of information about the reads mate, including the location, mapping quality, number of mappings for the mate, which ensures that these searches are not required, created vast increases in the efficiency of the SV detectors using this information.

The other utility for SV detections is reporting all possible mappings. Many genomes contain regions that are considered unmappable, usually due to the presence of low complexity DNA. Depending on the algorithms employed, NGS reads can still map to these regions; however, it is often prudent to omit these reads from variant detection. Instead of discarding reads mapping to multiple locations or picking the best quality alignment, MOSAIK records all locations to which a read maps (given the constraints imposed by the selected parameters) and records them in a separate BAM file. Since the number reads mapping to multiple locations as well as the number of entries for each multiply aligned read can be extremely large, the resulting BAM file has the potential to be excessively large. By default, MOSAIK omits much of the read specific information (e.g. read name, sequence and error information), allowing for effective compression of the file after positional sorting, resulting in very small BAM files. The information contained in these BAM files allows easy identification of genomic regions where many individual reads are aligning. These regions are those that can be considered unmappable, since reads hitting these regions are also able to align to other genomic regions. Thus they provide a guide to the mappable genome which can greatly aid in variant discovery.

The default parameters used by MOSAIK were optimized using simulated Illumina datasets from the human genome. They were generated to provide a balance between mismatches and gaps in the alignments, leading to balanced calling of SNPs and INDELs by variant callers. For the experimenter only interested in a specific variant type, it is possible to modify the parameters to provide alignments more sensitive for the variant type of interest. For example, if INDEL discovery is paramount, reducing the Smith-Waterman penalty for the creation and extension of gaps in alignments will lead to a greater likelihood that INDELs will be discovered.

MOSAIKs memory footprint depends on the size of the reference hash-table which, in turn, depends on the hash (k-mer) size as well as the length of the reference sequence. For the human genome using the default value of $k = 15$, MOSAIK requires approximately 20Gb of memory. For machines with less available RAM, MOSAIK can be run in a low-memory mode that performs alignments chromosome by chromosome. This reduces the required memory to 7Gb, which makes MOSAIK accessible to most machines.

Improvements in the computational performance can be achieved at the expense of decreased sensitivity, but ongoing development (including replacing the traditional Smith-Waterman algorithm with a *single-instruction-multiple-data* (SIMD) Smith-Waterman algorithm [65,66]) provides significant performance improvements. Initial testing of the SIMD Smith-Waterman algorithm demonstrate a twofold speed up [65]. Further improvement is achieved by reducing the number of applications of the Smith-Waterman algorithm for each read. Reads that originate in highly repetitive sequence can produce tens of thousands of candidate loci (see supplemental Figure S6) in the genome and the Smith-Waterman algorithm is applied to each one of these regions. This is extremely computationally intensive with very little benefit to the alignment sensitivity. As a result, if there are greater than a preset number (the default is 200) of potential mapping loci, MOSAIK only invokes the Smith-Waterman algorithm on the top

200 loci. MOSAIK then reports the most confident alignment from all of the regions in which the Smith-Waterman algorithm was applied. Supplemental Table S1 demonstrates that MOSAIK 2.2.3 (with these modifications) is of the order of five times faster than version 2.1.78 (without the modifications). Importantly, these modifications do not adversely impact the sensitivity of MOSAIK.

Methods

Overview

MOSAIK is a hash-based aligner and it hashes reference sequences as its first step. MOSAIK splits the reference sequences into overlapping contiguous k-mers (hashes) and stores the positions of each hash in a hash table data structure that guarantees $O(1)$ lookups. Then, MOSAIK hashes each read in the same hash size and looks hashes up in the hash table to obtain the genomic positions of the hashes of a read. Next, nearby hash positions are consolidated as a hash region (hashes of a read may be clustered as several hash regions) where a Smith-Waterman algorithm is applied to align the read to the local region of a genome reference sequence as a final "polishing" step. For paired-end reads, each end-mate of a read is mapped separately. For some cases, that may be one end-mate aligned well and the other one failing to be aligned. The well-aligned mate can be used to try and rescue the unaligned mate using knowledge of the approximate fragment length used in the paired-end read generation.

Processing Reference Sequences

MOSAIK can handle a nearly unlimited number of reference sequences, however, the maximum aggregated reference length is four billion bases. Alignments to the human transcriptome using more than 95,000 individual reference sequences are easily handled. The available hash sizes are 4–32.

MOSAIK supports the full set of IUPAC ambiguous nucleotide characters. This allows users to use reference sequences that have been masked by confirmed dbSNP (http://www.ncbi.nlm.nih. gov/projects/SNP/) calls. The ambiguity codes minimize the alignment bias that might be caused when aligning to reference sequences containing SNPs. For considering IUPAC, MOSAIK substitutes ambiguous codes with all of the alternative bases represented by the ambiguity code and stores the resulting hashes in the hash table. In order to avoid increasing the size of the jump database dramatically, the ambiguity codes N and X are not considered when hashing the reference sequences.

Clustering Hashes

MOSAIK supports various read formats (SRF, FASTA, FASTQ, Bustard, and Gerald). In each case, the reads are split into a set of overlapping hashes and the genomic positions of each hash are queried from the stored reference hash table. A modified AVL tree [67] is employed to handle and cluster nearby hash positions to form a hash region. The clustering algorithm considers sequencing errors, SNPs and single base INDELs. For example, consider a 35 bases read split into hashes of 15 bases. The first hash consists of the first 15 bases in the read. The second hash consists of bases 2–16 in the read and so on. The read consists of 22 individual hashes, each of which is associated with positions within the reference genome. If the read can be aligned perfectly to the somewhere in the reference genome (i.e. there are no sequencing errors or variations), each of the 22 hashes will have a reference genome position offset by a single base (i.e. if the first hash in the read is associated with the reference position x, the second hash with the reference position $x+1$ etc.). The AVL tree will consolidate those hits into a single alignment candidate region (see Supplemental Figure S5(A)). The presence of a single sequencing error will ensure that 15 of the hashes (each hash overlapping the error), will not be associated with the correct genomic coordinate. Since the clustering algorithm considers sequencing errors, however, an alignment candidate region is still present in the AVL tree (see Supplemental Figure S5(B)).

Applying Smith-Waterman Algorithm

After identifying alignment candidate regions, MOSAIK employs a Smith-Waterman algorithm to align reads to the alignment candidate regions. The Smith-Waterman algorithm, which was invented over 30 years ago, is still regarded as the most accurate pairwise alignment algorithm and the preferred choice for aligning gapped sequences since it seeks all possible frames of alignment with all possible gaps. Specifically, the alignments are performed using the Smith-Waterman-Gotoh alignment algorithm [68,69].

The time complexity of the Smith-Waterman algorithm is $O(n^2)$, which may render the mapper useless due to poor performance. To address this, a banded Smith-Waterman algorithm [70] has been implemented to improve the performance. According to our experiments, the runtimes for aligning Illumina and Roche 454 data are reduced by approximately $3\times$ and $8\times$ respectively. The further development of using SIMD SW promises significant performance improvements.

Rescuing Paired-End Mates

Each mate in a paired-end read is initially aligned individually. There are various factors that lead to some reads failing to be aligned to the reference. In the case of paired-end reads, the aligned mate can be used to try and rescue the unaligned mate using knowledge of the approximate fragment length used in the paired-end read generation. A local alignment search algorithm has been implemented which performs a Smith-Waterman algorithm in the region proximal to the aligned mate. If the read exhibits the expected strand, orientation, and fragment length, the read is considered rescued. Even if both mates in the pair are successfully aligned, the local alignment search may still be triggered, if the alignments are inconsistent with the expected fragment length.

The number of mates rescued by the local alignment search depends largely on the read lengths considered. With increasing read length, the aligner is less likely to miss a potential alignment and therefore fewer alignments are rescued.

Handling AB SOLiD reads

AB SOLiD reads are represented in the colorspace rather than in the more conventional basespace. Most downstream applications do not support colorspace and thus alignments require conversion to basespace for maximum utility. MOSAIK is equipped to align colorspace reads against a colorspace reference and then convert the resulting alignments into basespace. The di-base quality conversion algorithm uses the minimum of the two qualities that overlap a nucleotide in basespace. This approach allows users to specify parameters, such as the maximum number of mismatches. Additionally, it enables users to merge aligned SOLiD datasets with datasets from other sequencing technologies.

Known-Sequence Insertion Detections

MOSAIK is aware of user-specified insertion sequences, e.g. mobile element insertions. When the insertion sequences are provided, the reference hashes are prioritized such that alignment to the given insertion sequences are attempted prior to alignment to the genome reference. An additional tag in the BAM file (the ZA tag) then indicates any alignments of a read hitting the given insertion sequences. Since MEIs are repetitive elements, a read from an MEI can be mapped to several locations within the genome (potentially hundreds of locations). The ZA tag then populated with valuable information about the reads mate, including location, mapping quality and number of mapping locations for the mate. This information ensures that multiple BAM search operations (which can be lengthy for large BAM files) can be avoided. The downstream MEI detector can detect MEI by using ZA tag easily.

Supporting Information

Figure S1 The distributions of alignments' softclips.

Figure S2 The complete information of Figure 1. The positive predictive value of aligners (the number of correctly mapped reads divided by the total number of mapped reads) as a function of mapping quality threshold. Datasets in (A) 100 bp and (B) 76 bp read lengths. PPV, TP, and FP stand for positive predictive value, true positive, and false positive, respectively.

Figure S3 The complete information of Figure 5. The receiver operating characteristic (ROC) curves of SNPs called by FREEBAYES and SAMTOOLS. The points on the curves are sorted by called qualities and the points closer to the upper-right corner have higher called qualities. The true positive (TP), false positive (FP), and false negative (FN) are calculated by intersecting SNPs called on each aligner's alignments and gold SNPs called on the simulated alignments.

Figure S4 The short INDELs that are inserted for investigating the aligners' abilities for them, and the read coverage for each length INDEL.

Figure S5 MOSAIK hash clustering. (A) The read uniquely aligns perfectly to the references, all hashes will succeed in finding the adjacent reference locations and the AVL tree will consolidate those hashes into one alignment candidate region. (B) However, if only one hash succeeds in finding the proper reference location because of sequencing errors, an alignment candidate region is still present in the AVL tree.

Figure S6 The distribution of candidate loci in the genome of reads. MOSAIK applies a Smith-Waterman algorithm to each candidate locus of a read to generate an alignment. Therefore, the number of candidate loci is equal to the number of executed the Smith-Waterman algorithm. The mhp of MOSAIK is the maximum number of investigated hash positions per 15-mer.

Method S1 The methods of (A) Retraining Mapping Quality Neural Network and (B) Detecting Specified Insertion Sequences.

Table S1 The runtime of each mapper for aligning six million 100 bp reads. The version without '*' are the exact version of each mapper for which we report performance comparisons. For up to date information, we also report speed for the current version (indicated by '*') of each software.

STAMPY is a single-threaded program and thus the runtime of using 4 cpus is not available.

Author Contributions

Conceived and designed the experiments: WL AW EG GM. Performed the experiments: WL. Analyzed the data: WL AW. Contributed reagents/materials/analysis tools: MS CS EG. Wrote the paper: WL AW. Started the project: MS.

References

1. Drmanac R, Sparks AB, Callow MJ, Halpern AL, Burns NL, et al. (2010) Human genome sequencing using unchained base reads on self-assembling DNA nanoarrays. Science 327: 78–81. doi:10.1126/science.1181498.

2. Eid J, Fehr A, Gray J, Luong K, Lyle J, et al. (2009) Real-time DNA sequencing from single polymerase molecules. Science 323: 133–138. doi:10.1126/science.1162986.

3. Rothberg JM, Hinz W, Rearick TM, Schultz J, Mileski W, et al. (2011) An integrated semiconductor device enabling non-optical genome sequencing. Nature 475: 348–352. doi:10.1038/nature10242.

4. Schneider GF, Dekker C (2012) DNA sequencing with nanopores. Nat Biotechnol 30: 326–328. doi:10.1038/nbt.2181.

5. Burrows M, Burrows M, Wheeler DJ (1994) A block-sorting lossless data compression algorithm.

6. Cox AJ, Bauer MJ, Jakobi T, Rosone G (2012) Large-scale compression of genomic sequence databases with the Burrows-Wheeler transform. Bioinformatics. doi:10.1093/bioinformatics/bts173.

7. Boytsov L (2011) Indexing methods for approximate dictionary searching. J Exp Algorithmics 16: 1.1. doi:10.1145/1963190.1963191.

8. Li H, Ruan J, Durbin R (2008) Mapping short DNA sequencing reads and calling variants using mapping quality scores. Genome Res 18: 1851–1858. doi:10.1101/gr.078212.108.

9. Alkan C, Kidd JM, Marques-Bonet T, Aksay G, Antonacci F, et al. (2009) Personalized copy number and segmental duplication maps using next-generation sequencing. Nat Genet 41: 1061–1067. doi:10.1038/ng.437.

10. Hach F, Hormozdiari F, Alkan C, Hormozdiari F, Birol I, et al. (2010) mrsFAST: a cache-oblivious algorithm for short-read mapping. Nat Methods 7: 576–577. doi:10.1038/nmeth0810-576.

11. Rumble SM, Lacroute P, Dalca A V, Fiume M, Sidow A, et al. (2009) SHRiMP: accurate mapping of short color-space reads. PLoS Comput Biol 5: e1000386. doi:10.1371/journal.pcbi.1000386.

12. David M, Dzamba M, Lister D, Ilie L, Brudno M (2011) SHRiMP2: sensitive yet practical SHort Read Mapping. Bioinformatics 27: 1011–1012. doi:10.1093/bioinformatics/btr046.

13. Lin H, Zhang Z, Zhang MQ, Ma B, Li M (2008) ZOOM! Zillions of oligos mapped. Bioinformatics 24: 2431–2437. doi:10.1093/bioinformatics/btn416.

14. Zhang Z, Lin H, Ma B (2010) ZOOM Lite: next-generation sequencing data mapping and visualization software. Nucleic Acids Res 38: W743–8. doi:10.1093/nar/gkq538.

15. Eaves HL, Gao Y (2009) MOM: maximum oligonucleotide mapping. Bioinformatics 25: 969–970. doi:10.1093/bioinformatics/btp092.

16. Campagna D, Albiero A, Bilardi A, Caniato E, Forcato C, et al. (2009) PASS: a program to align short sequences. Bioinformatics 25: 967–968. doi:10.1093/bioinformatics/btp087.

17. Kim YJ, Teletia N, Ruotti V, Maher CA, Chinnaiyan AM, et al. (2009) ProbeMatch: rapid alignment of oligonucleotides to genome allowing both gaps and mismatches. Bioinformatics 25: 1424–1425. doi:10.1093/bioinformatics/btp178.

18. Li R, Li Y, Kristiansen K, Wang J (2008) SOAP: short oligonucleotide alignment program. Bioinformatics 24: 713–714. doi:10.1093/bioinformatics/btn025.

19. Gontarz PM, Berger J, Wong CF (2013) SRmapper: a fast and sensitive genome-hashing alignment tool. Bioinformatics 29: 316–321. doi:10.1093/bioinformatics/bts712.

20. Lunter G, Goodson M (2011) Stampy: a statistical algorithm for sensitive and fast mapping of Illumina sequence reads. Genome Res 21: 936–939. doi:10.1101/gr.111120.110.

21. Li H, Durbin R (2009) Fast and accurate short read alignment with Burrows-Wheeler transform. Bioinformatics 25: 1754–1760. doi:10.1093/bioinformatics/btp324.

22. Langmead B (2010) Aligning short sequencing reads with Bowtie. Curr Protoc Bioinforma Ed board Andreas D Baxevanis al Chapter 11: Unit 11.7.

23. Langmead B, Salzberg SL (2012) Fast gapped-read alignment with Bowtie 2. Nat Methods 9: 357–360. doi:10.1038/nmeth.1923.

24. Hoffmann S, Otto C, Kurtz S, Sharma CM, Khaitovich P, et al. (2009) Fast mapping of short sequences with mismatches, insertions and deletions using index structures. PLoS Comput Biol 5: e1000502. doi:10.1371/journal.pcbi.1000502.

25. Li R, Yu C, Li Y, Lam T-W, Yiu S-M, et al. (2009) SOAP2: an improved ultrafast tool for short read alignment. Bioinformatics 25: 1966–1967. doi:10.1093/bioinformatics/btp336.

26. Ferragina P, Manzini G (2005) Indexing compressed text. J ACM 52: 552–581. doi:10.1145/1082036.1082039.

27. Ferragina P, Manzini G (2001) An experimental study of an opportunistic index: 269–278.

28. Mahmud MP, Wiedenhoeft J, Schliep A (2012) Indel-tolerant read mapping with trinucleotide frequencies using cache-oblivious kd-trees. Bioinformatics 28: i325–i332. doi:10.1093/bioinformatics/bts380.

29. Tipton KF (1994) Nomenclature Committee of the International Union of Biochemistry and Molecular Biology (NC-IUBMB). Enzyme nomenclature. Recommendations 1992. Supplement: corrections and additions. Eur J Biochem 223: 1–5.

30. The 1000 Genomes Project Consortium (2010) A map of human genome variation from population-scale sequencing. Nature 467: 1061–1073. doi:10.1038/nature09534.

31. The 1000 Genomes Project Consortium (2012) An integrated map of genetic variation from 1,092 human genomes. Nature 491: 56–65. doi:10.1038/nature11632.

32. Harris TD, Buzby PR, Babcock H, Beer E, Bowers J, et al. (2008) Single-molecule DNA sequencing of a viral genome. Science 320: 106–109. doi:10.1126/science.1150427.

33. Garrison E, Marth G (2012) Haplotype-based variant detection from short-read sequencing: 9.

34. Li H (2011) A statistical framework for SNP calling, mutation discovery, association mapping and population genetical parameter estimation from sequencing data. Bioinformatics 27: 2987–2993. doi:10.1093/bioinformatics/btr509.

35. Prak ET, Kazazian HH (2000) Mobile elements and the human genome. Nat Rev Genet 1: 134–144. doi:10.1038/35038572.

36. Griffiths D (2001) Endogenous retroviruses in the human genome sequence. Genome Biol 2: reviews1017.1–reviews1017.5. doi:10.1186/gb-2001-2-6-reviews1017.

37. Costantini M, Bernardi G (2009) Mapping insertions, deletions and SNPs on Venter's chromosomes. PLoS One 4: e5972. doi:10.1371/journal.pone.0005972.

38. Levy S, Sutton G, Ng PC, Feuk L, Halpern AL, et al. (2007) The diploid genome sequence of an individual human. PLoS Biol 5: e254. doi:10.1371/journal.pbio.0050254.

39. Osoegawa K, Mammoser AG, Wu C, Frengen E, Zeng C, et al. (2001) A bacterial artificial chromosome library for sequencing the complete human genome. Genome Res 11: 483–496. doi:10.1101/gr.169601.

40. Marth GT, Yu F, Indap AR, Garimella K, Gravel S, et al. (2011) The functional spectrum of low-frequency coding variation. Genome Biol 12: R84. doi:10.1186/gb-2011-12-9-r84.

41. Su X, Zhang L, Zhang J, Meric-Bernstam F, Weinstein JN (2012) PurityEst: estimating purity of human tumor samples using next-generation sequencing data. Bioinformatics 28: 2265–2266. doi:10.1093/bioinformatics/bts365.

42. Roberts KG, Morin RD, Zhang J, Hirst M, Zhao Y, et al. (2012) Genetic alterations activating kinase and cytokine receptor signaling in high-risk acute lymphoblastic leukemia. Cancer Cell 22: 153–166. doi:10.1016/j.ccr.2012.06.005.

43. Lin Y, Li Z, Ozsolak F, Kim SW, Arango-Argoty G, et al. (2012) An in-depth map of polyadenylation sites in cancer. Nucleic Acids Res 40: 8460–8471. doi:10.1093/nar/gks637.

44. Wang J, Mullighan CG, Easton J, Roberts S, Heatley SL, et al. (2011) CREST maps somatic structural variation in cancer genomes with base-pair resolution. Nat Methods 8: 652–654. doi:10.1038/nmeth.1628.

45. Chung CC, Ciampa J, Yeager M, Jacobs KB, Berndt SI, et al. (2011) Fine mapping of a region of chromosome 11q13 reveals multiple independent loci associated with risk of prostate cancer. Hum Mol Genet 20: 2869–2878. doi:10.1093/hmg/ddr189.

46. Goya R, Sun MGF, Morin RD, Leung G, Ha G, et al. (2010) SNVMix: predicting single nucleotide variants from next-generation sequencing of tumors. Bioinformatics 26: 730–736. doi:10.1093/bioinformatics/btq040.

47. Cridland JM, Thornton KR (2010) Validation of rearrangement break points identified by paired-end sequencing in natural populations of Drosophila melanogaster. Genome Biol Evol 2: 83–101. doi:10.1093/gbe/evq001.

48. Hillier LW, Marth GT, Quinlan AR, Dooling D, Fewell G, et al. (2008) Whole-genome sequencing and variant discovery in C. elegans. Nat Methods 5: 183–188. doi:10.1038/nmeth.1179.

49. Henn MR, Boutwell CL, Charlebois P, Lennon NJ, Power KA, et al. (2012) Whole genome deep sequencing of HIV-1 reveals the impact of early minor variants upon immune recognition during acute infection. PLoS Pathog 8: e1002529. doi:10.1371/journal.ppat.1002529.

50. Malboeuf CM, Yang X, Charlebois P, Qu J, Berlin AM, et al. (2012) Complete viral RNA genome sequencing of ultra-low copy samples by sequence-independent amplification. Nucleic Acids Res 41: e13. doi:10.1093/nar/gks794.

51. Campbell MS, Mullins JI, Hughes JP, Celum C, Wong KG, et al. (2011) Viral linkage in HIV-1 seroconverters and their partners in an HIV-1 prevention clinical trial. PLoS One 6: e16986. doi:10.1371/journal.pone.0016986.

52. Wilen CB, Wang J, Tilton JC, Miller JC, Kim KA, et al. (2011) Engineering HIV-resistant human CD4+ T cells with CXCR4-specific zinc-finger nucleases. PLoS Pathog 7: e1002020. doi:10.1371/journal.ppat.1002020.

53. Farrell A, Thirugnanam S, Lorestani A, Dvorin JD, Eidell KP, et al. (2012) A DOC2 protein identified by mutational profiling is essential for apicomplexan parasite exocytosis. Science 335: 218–221. doi:10.1126/science.1210829.

54. Dark MJ, Al-Khedery B, Barbet AF (2011) Multistrain genome analysis identifies candidate vaccine antigens of Anaplasma marginale. Vaccine 29: 4923–4932. doi:10.1016/j.vaccine.2011.04.131.

55. Dark MJ, Lundgren AM, Barbet AF (2012) Determining the repertoire of immunodominant proteins via whole-genome amplification of intracellular pathogens. PLoS One 7: e36456. doi:10.1371/journal.pone.0036456.

56. Iorizzo M, Senalik DA, Grzebelus D, Bowman M, Cavagnaro PF, et al. (2011) De novo assembly and characterization of the carrot transcriptome reveals novel genes, new markers, and genetic diversity. BMC Genomics 12: 389. doi:10.1186/1471-2164-12-389.

57. Neves L, Davis J, Barbazuk B, Kirst M (2011) Targeted sequencing in the loblolly pine (Pinus taeda) megagenome by exome capture. BMC Proc 5: O48. doi:10.1186/1753-6561-5-S7-O48.

58. Cannon CH, Kua C-S, Zhang D, Harting JR (2010) Assembly free comparative genomics of short-read sequence data discovers the needles in the haystack. Mol Ecol 19 Suppl 1: 147–161. doi:10.1111/j.1365-294X.2009.04484.x.

59. Aslam ML, Bastiaansen JW, Elferink MG, Megens H-J, Crooijmans RP, et al. (2012) Whole genome SNP discovery and analysis of genetic diversity in Turkey (Meleagris gallopavo). BMC Genomics 13: 391. doi:10.1186/1471-2164-13-391.

60. Fraser BA, Weadick CJ, Janowitz I, Rodd FH, Hughes KA (2011) Sequencing and characterization of the guppy (Poecilia reticulata) transcriptome. BMC Genomics 12: 202. doi:10.1186/1471-2164-12-202.

61. Stewart C, Kural D, Strömberg MP, Walker JA, Konkel MK, et al. (2011) A Comprehensive Map of Mobile Element Insertion Polymorphisms in Humans. PLoS Genet 7: 1.

62. Tae H, McMahon KW, Settlage RE, Bavarva JH, Garner HR (2013) ReviSTER: an automated pipeline to revise misaligned reads to simple tandem repeats. Bioinformatics 29: 1734–1741. doi:10.1093/bioinformatics/btt277.

63. David M, Mustafa H, Brudno M (2013) Detecting Alu insertions from high-throughput sequencing data. Nucleic Acids Res: gkt612–. doi:10.1093/nar/gkt612.

64. Xing J, Witherspoon DJ, Jorde LB (2013) Mobile element biology: new possibilities with high-throughput sequencing. Trends Genet 29: 280–289. doi:10.1016/j.tig.2012.12.002.

65. Zhao M, Lee W-P, Garrison EP, Marth GT (2013) SSW Library: An SIMD Smith-Waterman C/C++ Library for Use in Genomic Applications. PLoS One 8: e82138. doi:10.1371/journal.pone.0082138.

66. Farrar M (2007) Striped Smith-Waterman speeds database searches six times over other SIMD implementations. Bioinformatics 23: 156–161. doi:10.1093/bioinformatics/btl582.

67. Adel'son-Vel'skii GM, Landis EM (1962) An algorithm for the organization of information. Sov Math Dokl 3: 263–266.

68. Smith TF, Waterman MS (1981) Indentification of common molecular subsequences. J Mol Biol 147: 195–197.

69. Gotoh O (1982) An improved algorithm for matching biological sequences. J Mol Biol 162: 705–708.

70. Chao KM, Pearson WR, Miller W (1992) Aligning two sequences within a specified diagonal band. Comput Appl Biosci 8: 481–487.

23

The Role of Mutation Rate Variation and Genetic Diversity in the Architecture of Human Disease

Ying Chen Eyre-Walker, Adam Eyre-Walker*

School of Life Sciences, University of Sussex, Brighton, United Kingdom

Abstract

Background: We have investigated the role that the mutation rate and the structure of genetic variation at a locus play in determining whether a gene is involved in disease. We predict that the mutation rate and its genetic diversity should be higher in genes associated with disease, unless all genes that could cause disease have already been identified.

Results: Consistent with our predictions we find that genes associated with Mendelian and complex disease are substantially longer than non-disease genes. However, we find that both Mendelian and complex disease genes are found in regions of the genome with relatively low mutation rates, as inferred from intron divergence between humans and chimpanzees, and they are predicted to have similar rates of non-synonymous mutation as other genes. Finally, we find that disease genes are in regions of significantly elevated genetic diversity, even when variation in the rate of mutation is controlled for. The effect is small nevertheless.

Conclusions: Our results suggest that gene length contributes to whether a gene is associated with disease. However, the mutation rate and the genetic architecture of the locus appear to play only a minor role in determining whether a gene is associated with disease.

Editor: I. King Jordan, Georgia Institute of Technology, United States of America

Funding: These authors have no support or funding to report.

Competing Interests: The authors have declared that no competing interests exist.

* E-mail: a.c.eyre-walker@sussex.ac.uk

Introduction

Why do humans suffer from the diseases that we do? In part this is clearly due to our anatomy and physiology, and that of the organisms that infect us - we cannot have a disease of an organ that we do not possess. But why do we suffer from cystic fibrosis rather than some other disease of the lungs? One simple reason might be variation in the mutation rate. Those genes and genomic regions that have high mutation rates are more likely to generate disease mutations, and hence be associated with a disease. The rate of mutation of a locus will depend upon two factors: the rate of mutation per site and the number of sites at which a mutation can generate a disease phenotype. The per site mutation rate is known to vary across the human genome at a number of different scales such that some genes have mutation rates that are several fold higher than other genes (reviewed in Hodgkinson *et al.* [1]). Genes also vary considerably in their length, with some of the largest, such as the dystrophin gene, being association with disease.

A more subtle factor affecting the likelihood of a gene being associated with a disease is the genealogy. At each site in the genome there is an underlying genealogy whereby every chromosome in the population is related via a bifurcating tree to every other chromosome at that site. If there is no recombination between sites then sites share the same genealogy. The shape and depth of the genealogy depends on several factors. The first is chance; for example, the average total length of a genealogy for a neutral locus in a population of stationary size is expected to be

proportional to $4N$ generations in a diploid species, where N is the population size, but this is expected to have a variance of at least $(4N)^2$ generations [2]. Second, the genealogy depends on the effective population size of the locus (N_e). N_e is thought to vary across the human genome as a consequence of natural selection [3,4]. Selection can reduce the N_e of a genomic region through either a selective sweep caused by the passage of an advantageous mutation through the population [5], or via background selection caused by the removal of deleterious mutations [6]. Those regions of the genome with low rates of recombination or a high density of selected sites are expected to have low N_e, and this is expected to reduce the genetic diversity of neutral and weakly selected variants in these regions (reviewed in [7]). Analyses suggest that N_e varies across the human genome by a few-fold [4]. The effective population size is not expected to affect the frequency of deleterious mutations in which the product of N_e and the strength selection is greater than one. However, stochastic factors affecting the genealogy are expected to be important irrespective of the selection acting upon a mutation.

Previous analyses have shown that Mendelian disease genes are 30% longer than non-disease genes [8,9]. Comparative analyses have also shown that genes associated with Mendelian diseases have significantly, but only slightly higher rates of mutation per site, as inferred from levels of synonymous divergence between species [8,10]. The rather modest differences between disease and non-diseases genes in the inferred mutation rate might be due to time frame over which the mutation rate was inferred: Smith and

A)

B)

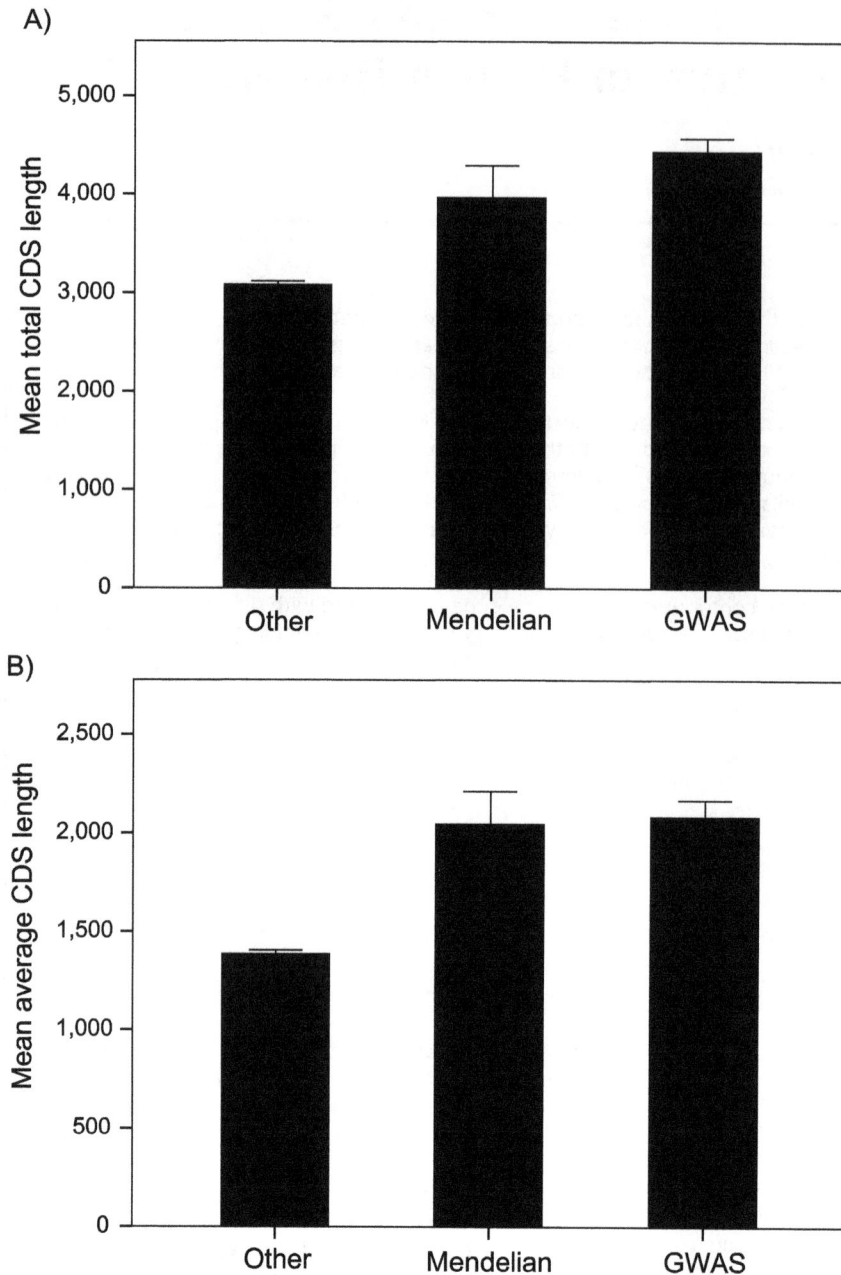

Figure 1. CDS length. (A) Mean total CDS length, and (B) Mean average CDS length. Total CDS length is the sum of all constitutive and alternately spliced exons; average CDS length is the average CDS length of each transcript. Error bars represent the 95% confidence intervals.

Eyre-Walker [8] considered the divergence between human and mouse, and Huang et al. [10] considered the divergence between mouse and rat. This will give a poor estimate of the current mutation rate at a locus in humans because the relative mutation rate of a locus appears to have evolved through time [1,11]. The mutation rate has also recently been predicted, based on a model fitted to the locations of *de novo* mutations in humans, to be slightly higher in disease associated genes [12], but the accuracy of this model is unproven, and they consider the total mutation rate of the exon, rather than the rate at non-synonymous sites. Here we

consider the divergence between humans and their most closely related extant relative, chimpanzee, as our measure of the mutation rate. We also consider whether the density of single nucleotide polymorphism (SNP) is greater in disease than non-disease genes.

Materials and Methods

To estimate mutation rate for each gene, we estimated their intron divergence between the human and chimpanzee genomes as follows. Alignments using the NCBI build 36 version of the

A)

B)

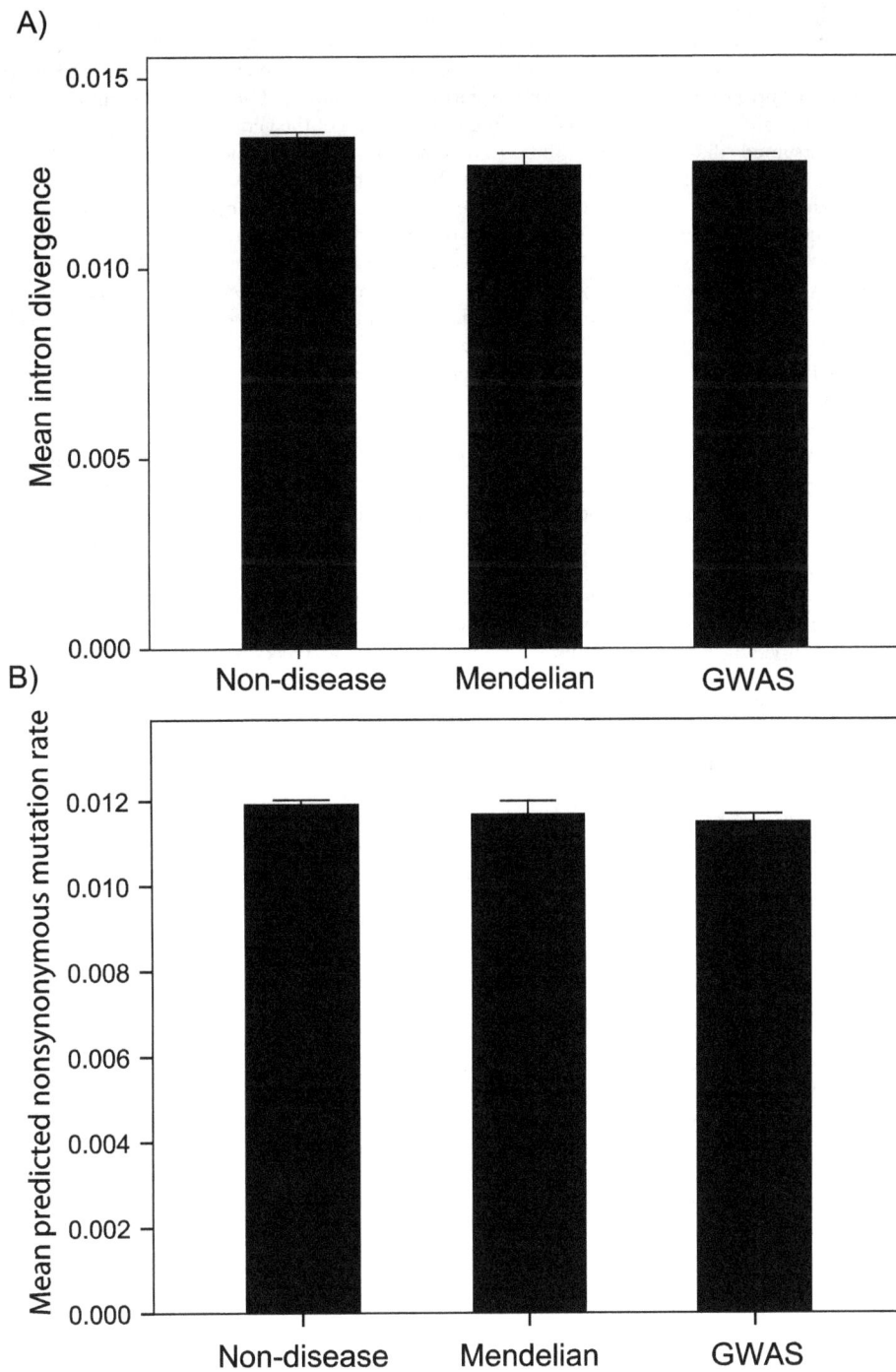

Figure 2. Mutation rates. The mutation rate per site, as inferred from intron divergence between human and chimpanzee. A) Intron divergence per site between human and chimpanzee; B) the predicted non-synonymous mutation rate per CDS site. Error bars represent the 95% confidence intervals.

human genome (hg18) and PanTro2 version of the chimp genome were downloaded from the UCSC website (http://genome.ucsc. edu/). Alignments were parsed into individual genic sequences and realigned with MAFFT version 6 (http://mafft.cbrc.jp/ alignment/software/). Exon sequences were masked according to exon annotation of the NCBI build 36 version of the human genome from the ensemble database (http://www.ensembl.org/). We did not correct for multiple hits; this is not necessary since the average intron divergence between human and chimpanzee sequences is 1.05% [13]. We calculated the rates of intron divergence for CpG and nonCpG sites separately since the former have much higher rates of mutation. We used these intron divergences to infer the rate of non-synonymous mutation in human exons, by calculating the number of CpG and non-CpG sites in each exon which when mutated would give a non-synonymous change; in this calculation we assumed that all mutations at CpGs are transitions, which is a good approximation [14], and that 60% of mutations at other sites were transitions. If a gene had multiple transcripts we made these calculations for each transcript and averaged the result.

DNA sequence diversity data were taken from the 1000 genome project [15].

Genes were designated as being associated with Mendelian disease based upon the compilation made by Blekhman *et al.* [16]. Genes associated with genome-wide association studies (GWAS) were obtained from the GWAS catalog (http://www.genome.gov/gwastudies/); a gene in which the strongest GWAS signal was found within the boundaries of a gene were designated as being a GWAS gene.

To investigate what factors might influence patterns of genic mutations, estimated by intron divergence, we considered a number of variables. Intron GC content, nucleosome occupancy, replication timing and male and female recombination rates were downloaded from the UCSC website (http://genome.ucsc.edu/). We used A365 values to study the influence of nucleosome occupancy on the distribution of genic mutations rate across the genome. Recombination rates per MB were from Kong *et al* [17]. Replication time data were from Chen *et al.* [18] and Hansen *et al.* [19]. Qualitatively similar results were obtained using each of four replication time datasets, so we only present the analysis using data from an embryonic stem cell line BG02 [19]. Germ-line expression data were from a study by McVicker and Green [20].

The dataset for this analysis is available as Table S1.

Results

We predict that unless all possible diseases with a genetic basis, and all the genes that can cause them, have already been discovered, then genes associated with diseases should have higher genic mutation rates than non-disease genes, where the genic mutation rate is determined by the product of gene length and the mutation rate per site. We also predict that disease genes should be in relatively diverse regions of the genome. To investigate these predictions we compiled data from 17577 nuclear genes with introns, of which 854 genes are known to cause a Mendelian

disease. We also analysed 1732 genes in which the strongest signal in a genomic region in a genome wide association study (GWAS) lay within the boundaries of the gene (i.e. all exons and introns between the start and stop codon). The presence of an association signal within the boundaries of the gene does not necessarily mean that the causative mutation is within the protein coding sequence or even within the boundaries of the gene, and many of these associations may be in regulatory sequences [21]. We subsequently excluded genes on the sex chromosomes since the Y-chromosome is known to have a higher mutation rate and the X-chromosome a lower mutation rate than the autosomes [22]. This yielded a dataset of 17062 autosomal genes including 820 associated with a Mendelian disease and 1726 with a GWAS signal. Details of the dataset are given in Table S1.

Gene Length

Consistent with the hypothesis that disease genes should have higher overall rates of mutation we find, as others have in the past for genes causing Mendelian disease [8,9], that genes associated with disease are significantly longer, in terms of their total coding sequence (CDS) length (i.e. the sum of all constitutive and alternatively spliced exons), than non-disease genes; Mendelian disease genes are ~28% and GWAS genes ~44% longer than non-disease genes (One-way ANOVA $p < 0.001$) (Figure 1a). A similar pattern is evident for average CDS length; both Mendelian and GWAS disease genes are 50% longer than non-disease genes (Figure 1b). The difference in average CDS length is greater than in previous studies [8,9], but this is likely to be due to the improvement in genome annotation; the average length of genes is slightly shorter than in previous analyses.

Strikingly, the difference in length is as great or greater for the GWAS than the Mendelian disease genes despite the fact that many of the GWAS signals are likely to be outside the protein coding sequence [21]. GWAS genes might have longer CDSs for three reasons. First, genes with longer CDSs have a greater chance of generating a disease mutation. Second, longer genes are more likely to have a non-causative marker SNP in the CDS that is associated with the disease. And finally since intron and total CDS lengths are correlated ($r = 0.36$, $p < 0.001$), genes with long CDSs have longer introns and hence an increase chance of having causative or non-causative SNPs in their introns. However, if we control for the correlation between intron and CDS length by regressing CDS length against intron length and taking the residuals, we find that GWAS genes have longer CDSs, than non-disease genes, even given their longer introns (t-test $p < 0.001$;

Table 1. Standardised regression coefficients from multiple regressions.

Factor	Intron Divergence	Predicted non-synonymous mutation rate	Intron SNP density	Average genealogy length
GC content	0.525***	0.325***	0.192***	−0.212***
Nucleosome occupancy	−0.396***	−0.167***	−0.412***	−0.035
Female recombination rate	−0.020*	0.018*	0.058***	0.042***
Male recombination rate	0.202***	0.143***	0.129***	−0.048***
Germ-line expression	−0.062***	−0.116***	−0.020*	0.032***
Replication time	−0.132***	−0.157***	−0.071***	0.038***
Distance to telomere	−0.158***	−0.097***	−0.117***	0.060***
Distance to centromere	−0.018*	−0.016	0.020*	0.014

Note that the replication time data is such that a negative slope indicates an increase in the variable through the cell cycle * $p < 0.05$, ** $p < 0.01$ and *** $p < 0.001$.

A)

B)

C)

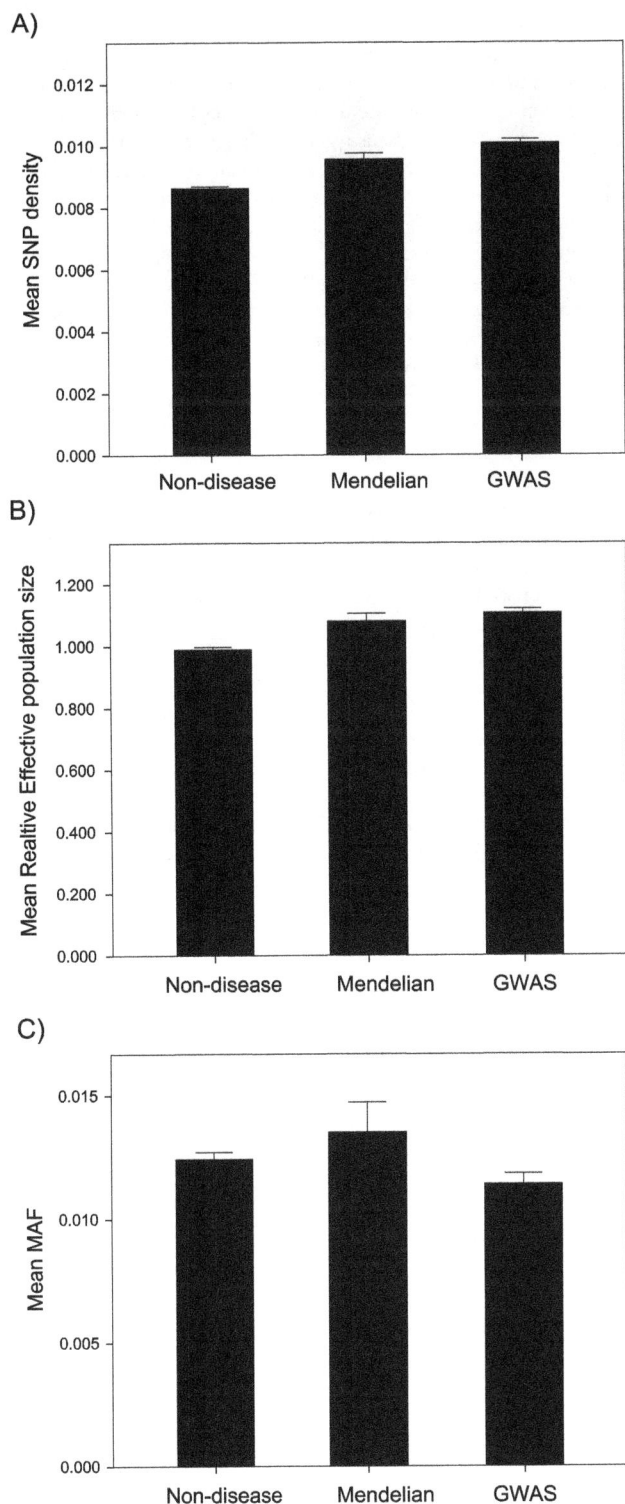

Figure 3. Diversity and genealogy estimates. The diversity in disease and non-disease genes measured as the A) average intron SNP density, B) the average intron SNP density divided by intron divergence, C) and the mean minor allele frequency (MAF). Error bars represent the 95% confidence intervals.

similar results are obtained if we regress log CDS length against

log intron length (p<0.001)). This suggests that GWAS genes are not simply longer because they have longer introns; it therefore seems that either GWAS genes are more likely to be associated with disease because some causative mutations are within their exons, or because there is a greater number of marker SNPs in exons.

Mutation Rates

However, contrary to our expectations, we find that disease genes are found in regions of the genome with significantly lower per site mutation rates, as measured by intron divergence between human and chimpanzee. The difference is highly significant (one-way ANOVA p<0.001), but the difference is small with disease genes having approximately 5% lower intron divergence than non-disease genes (Figure 2a). The pattern differs between CpG and non-CpG sites, with disease genes having lower divergence at CpG sites and either similar or higher divergence at non-CpG sites (results not shown). If we calculate the expected non-synonymous mutation rate in the CDS by multiplying the proportion of non-synonymous sites that are CpG and non-CpG in the CDS by the respective levels of intron divergence, we still find that both Mendelian and complex disease genes have slightly lower mutation rates per site than non-disease genes (p = 0.004) (Figure 2b). As expected, both Mendelian and complex disease genes have significantly higher overall predicted rates of non-synonymous mutation (p<0.001), driven by the fact that disease genes have longer CDSs.

The fact that disease genes have lower predicted rates of non-synonymous mutation per site is inconsistent with our hypothesis, but this might be due to the fact that they have features which predispose them to lower mutation rates - for example they might be transcribed at lower levels and hence have lower rates of mutation [23]. Divergence at intronic and intergenic sites is known to be significantly correlated to a number of other variables including GC-content [3,18,24,25], recombination rate [3,24,26,27,28], replication time [18,29,30], distance to the telomere and centromere [3,13,18,24], gene density [3,24], nucleosome occupancy [11] and expression level [23]. We confirm previous results and show that intron divergence is positive correlated to GC content and male recombination rate within a multiple regression; and that intron divergence is negatively correlated to replication time (later genes have higher divergence), distance to the telomere, distance to the centromere, female recombination rate, nucleosome occupancy and germ-line expression (Table 1). Similar patterns are evident for the predicted non-synonymous mutation rate (Table 1). If we take the residuals from a multiple regression of intron divergence against all the genomic variables above we find that intron divergence and the predicted rate of non-synonymous mutation do not differ significantly between disease and non-disease genes.

Genetic Diversity

Although, disease genes are found in regions of the genome with relatively low rates of intron mutation we find that disease genes have a significantly greater density of polymorphisms segregating in their introns than non-disease genes; the difference is 11% and 17% for the Mendelian and GWAS genes respectively (Figure 3a). If we divide the density of SNPs by the divergence of introns to estimate a quantity that is proportional to the average length of the genealogies at the locus, we find that Mendelian and GWAS genes have significantly longer average genealogy lengths that are 9% and 12% greater than non-disease genes (ANOVA p<0.001; t-test of Mendelian versus non-disease p<0.001; t-test of GWAS versus non-disease p<0.001). It is odd that the difference between disease and non-disease genes is less pronounced for average genealogy length than diversity given that disease genes have lower intron divergence than non-disease genes. This is probably due to non-linearities associated with ratios.

Although, we find that disease genes have higher diversities and average genealogy lengths than non-disease genes, we find no evidence that the predicted non-synonymous population mutation rate in the CDS (calculated as the proportion of non-synonymous sites that are CpG multiplied by the SNP density at CpG sites in introns plus the proportion of non-synonymous sites that are non-CpG multiplied by the SNP density at non-CpG in introns) differs between disease and non-disease genes. However, the calculation of the predicted non-synonymous population mutation rate is subject to considerable error because we have relatively few intron CpG sites and SNP density is very low in humans.

It is possible that disease genes have higher diversities and average genealogy lengths because disease genes have features that predispose them to higher values, not because by having higher values they are more likely to be associated with disease. We find that intron SNP density is positively correlated to GC content, female and male rates of recombination and distance to the centromere and negatively correlated to the time of replication (late genes have higher diversity), nucleosome occupancy, germ-line expression and distance to the telomere (Table 1). If control for these factors by taking the residuals from the multiple regression we find that SNP density is still significantly greater in both Mendelian and GWAS genes, than in non-disease genes (ANOVA $p<0.001$; individual t-tests $p<0.001$). Likewise we find the average genealogy length is positively correlated to all variables except GC content, nucleosome occupancy and male recombination rate (Table 1), and that after controlling for these associations, disease genes still have significantly greater average genealogy lengths than non disease genes (ANOVA $p = 0.019$; individual t-tests Mendelian versus non-disease $p = 0.21$, GWAS versus non-disease $p = 0.001$).

Although disease genes have a greater number of SNPs per bp than non-disease genes the distribution of the genetic variation varies in an inconsistent manner between categories of genes; the average minor allele frequency is ~10% greater in Mendelian, and ~10% lower in GWAS genes, than in non-disease genes (ANOVA $p<0.01$) (Figure 3b).

Discussion

We have found that genes associated with disease are longer and reside in regions of the genome with greater intron diversities and average genealogy lengths than non-disease genes. This is consistent with a role for mutation and genetic variation in determining whether a gene becomes associated with disease.

However, we find no evidence that the mutation rate per site is greater in disease than non-disease genes. Nevertheless, what is ultimately important is the mutation rate of the gene, and we find that the overall mutation rate of disease genes is greater than non-disease genes because disease genes are longer ($p<0.001$). The effect of gene length may be more conspicuous than for the other variables, because there is substantially more variation in CDS length per gene (coefficient of variation (CV) = 0.78) than in intron divergence (CV = 0.56), intron SNP density (CV = 0.42) and average genealogy length (CV = 0.47); in reality the differences in CV are even larger because intron divergence, and in particular SNP density and average genealogy length, are likely to be subject to large sampling error variances that CDS length is not.

We have interpreted the fact that disease genes are longer than non-disease genes as evidence that genes with higher mutation rates are more likely to generate disease mutations, however, it is possible that disease genes are longer simply because genes involved in particular processes that could cause disease to be longer. It is difficult to test this hypothesis without knowing all the genes that might cause disease. We have also interpreted the greater diversity in disease genes as being what causes them to be associated with disease. However, in the case of the complex disease genes this might simply reflect a bias towards a better ability to detect GWAS signals in regions of higher diversity.

Although, we have found that disease genes are longer than non-disease genes, and that they have greater diversity and average genealogy lengths, the differences are fairly small. It is therefore evident that either most disease associated genes have been discovered, which seems unlikely, or that the function of the gene is far more important in determining whether a gene causes disease than its effective mutation rate.

Supporting Information

Table S1 The data matrix used in the analysis. A description of column headings is provided as a separate worksheet within the Excel spreadsheet.

Acknowledgments

The authors are grateful to the comments of three anonymous referees.

Author Contributions

Conceived and designed the experiments: YEW AEW. Performed the experiments: YEW AEW. Analyzed the data: YEW AEW. Contributed reagents/materials/analysis tools: YEW AEW. Wrote the paper: YEW AEW.

References

1. Hodgkinson A, Eyre-Walker A (2011) Variation in the mutation rate across mammalian genomes. Nature Reviews Genetics 12: 756–766.
2. Charlesworth B, Charlesworth D (2010) Elements of evolutionary genetics. Greenwood Village: Ben Roberts.
3. Hellmann I, Prufer K, Ji H, Zody MC, Paabo S, et al. (2005) Why do human diversity levels vary at a megabase scale? Genome Res 15: 1222–1231.
4. Gossmann TI, Woolfit M, Eyre-Walker A (2011) Quantifying the variation in the effective population size within a genome. Genetics 189: 1389–1402.
5. Maynard Smith J, Haigh J (1974) The hitch-hiking effect of a favourable gene. Genet Res 23: 23–35.
6. Charlesworth B, Morgan MT, Charlesworth D (1993) The effect of deleterious mutations on neutral molecular variation. Genetics 134: 1289–1303.
7. Charlesworth B (2009) Fundamental concepts in genetics: effective population size and patterns of molecular evolution and variation. Nat Rev Genet 10: 195–205.
8. Smith NG, Eyre-Walker A (2003) Human disease genes: patterns and predictions. Gene 318: 169–175.
9. Kondrashov FA, Ogurtsov AY, Kondrashov AS (2004) Bioinformatical assay of human gene morbidity. Nucleic acids research 32: 1731–1737.
10. Huang H, Winter EE, Wang H, Weinstock KG, Xing H, et al. (2004) Evolutionary conservation and selection of human disease gene orthologs in the rat and mouse genomes. Genome biology 5: R47.
11. Hodgkinson A, Chen Y, Eyre-Walker A (2012) The large scale distribution of somatic mutations in cancer genomes. Human Mutation 33: 136–143.
12. Michaelson JJ, Shi Y, Gujral M, Zheng H, Malhotra D, et al. (2012) Whole-genome sequencing in autism identifies hot spots for de novo germline mutation. Cell 151: 1431–1442.
13. Chimpanzee-Sequencing-and-Analysis-Consortium (2005) Initial sequence of the chimpanzee genome and comparison with the human genome. Nature 437: 69–87.
14. Nachman MW, Crowell SL (2000) Estimate of the mutation rate per nucleotide in humans. Genetics 156: 297–304.
15. _Genomes_Project_Consortium (2012) An integrated map of genetic variation from 1,092 human genomes. Nature 491: 56–65.

16. Blekhman R, Man O, Herrmann L, Boyko AR, Indap A, et al. (2008) Natural selection on genes that underlie human disease susceptibility. Curr Biol 18: 883–889.
17. Kong A, Gudbjartsson DF, Sainz J, Jonsdottir GM, Gudjonsson SA, et al. (2002) A high-resolution recombination map of the human genome. Nat Genet 31: 241–247.
18. Chen CL, Rappailles A, Duquenne L, Huvet M, Guilbaud G, et al. (2010) Impact of replication timing on non-CpG and CpG substitution rates in mammalian genomes. Genome Res 20: 447–457.
19. Hansen RS, Thomas S, Sandstrom R, Canfield TK, Thurman RE, et al. (2010) Sequencing newly replicated DNA reveals widespread plasticity in human replication timing. Proceedings of the National Academy of Sciences of the United States of America 107: 139–144.
20. McVicker G, Green P (2010) Genomic signatures of germline gene expression. Genome research 20: 1503–1511.
21. Maurano MT, Humbert R, Rynes E, Thurman RE, Haugen E, et al. (2012) Systematic localization of common disease-associated variation in regulatory DNA. Science 337: 1190–1195.
22. Ellegren H (2007) Characteristics, causes and evolutionary consequences of male-biased mutation. Proc Biol Sci 274: 1–10.
23. Park C, Qian W, Zhang J (2012) Genomic evidence for elevated mutation rates in highly expressed genes. EMBO reports 13: 1123–1129.
24. Tyekucheva S, Makova KD, Karro JE, Hardison RC, Miller W, et al. (2008) Human-macaque comparisons illuminate variation in neutral substitution rates. Genome Biol 9: R76.
25. Wolfe KH, Sharp PM, Li W-H (1989) Mutation rates differ among regions of the mammalian genome. Nature 337: 283–285.
26. Duret L, Arndt PF (2008) The impact of recombination on nucleotide substitutions in the human genome. PLoS Genet 4: e1000071.
27. Hellmann I, Ebersberger I, Ptak SE, Paabo S, Przeworski M (2003) A neutral explanation for the correlation of diversity with recombination rates in humans. Am J Hum Genet 72: 1527–1535.
28. Lercher MJ, Hurst LD (2002) Human SNP variability and mutation rate are higher in regions of high recombination. Trends Genet 18: 337–340.
29. Pink CJ, Hurst LD (2010) Timing of replication is a determinant of neutral substitution rates but does not explain slow Y chromosome evolution in rodents. Mol Biol Evol 27: 1077–1086.
30. Stamatoyannopoulos JA, Adzhubei I, Thurman RE, Kryukov GV, Mirkin SM, et al. (2009) Human mutation rate associated with DNA replication timing. Nat Genet 41: 393–395.

24

A New Exhaustive Method and Strategy for Finding Motifs in ChIP-Enriched Regions

Caiyan Jia[1,2]*, Matthew B. Carson[3,4], Yang Wang[1], Youfang Lin[1], Hui Lu[2,5]*

1 School of Computer and Information Technology & Beijing Key Lab of Traffic Data Analysis, Beijing Jiaotong University, Beijing, China, 2 Department of Bioengineering/ Bioinformatics, University of Illinois at Chicago, Chicago, Illinois, United States of America, 3 Center for Healthcare Studies, Institute for Public Health and Medicine, Northwestern University Feinberg School of Medicine, Chicago, Illinois, United States of America, 4 Division of Health and Biomedical Informatics, Department of Preventive Medicine, Northwestern University Feinberg School of Medicine, Chicago, Illinois, United States of America, 5 Shanghai Institute of Medical Genetics, Shanghai Children's Hospital, Shanghai JiaoTong University, Shanghai, China

Abstract

ChIP-seq, which combines chromatin immunoprecipitation (ChIP) with next-generation parallel sequencing, allows for the genome-wide identification of protein-DNA interactions. This technology poses new challenges for the development of novel motif-finding algorithms and methods for determining exact protein-DNA binding sites from ChIP-enriched sequencing data. State-of-the-art heuristic, exhaustive search algorithms have limited application for the identification of short (l, d) motifs $(l \leq 10, d \leq 2)$ contained in ChIP-enriched regions. In this work we have developed a more powerful exhaustive method (FMotif) for finding long (l, d) motifs in DNA sequences. In conjunction with our method, we have adopted a simple ChIP-enriched sampling strategy for finding these motifs in large-scale ChIP-enriched regions. Empirical studies on synthetic samples and applications using several ChIP data sets including 16 TF (transcription factor) ChIP-seq data sets and five TF ChIP-exo data sets have demonstrated that our proposed method is capable of finding these motifs with high efficiency and accuracy. The source code for FMotif is available at http://211.71.76.45/FMotif/.

Editor: Ying Xu, University of Georgia, United States of America

Funding: This work was supported in part by National Nature Science Foundation of China (Grant No. 60905029, 61105055, 61105056, 81230086, and 31071167), the Beijing Natural Science Foundation (Grant No. 4112046), and the Fundamental Research Funds for the Central Universities. The funders had no role in study design, data collection and analysis, decision to publish, or preparation of the manuscript.

Competing Interests: The authors have declared that no competing interests exist.

* E-mail: cyjia@bjtu.edu.cn (CJ); huilu.bioinfo@gmail.com (HL)

Introduction

Protein-DNA interactions play key roles in several cellular processes and functions including DNA transcription, packaging, replication, and repair. Identification of regions such as transcription factor binding sites (TFBSs), which are targeted by proteins called transcription factors (TFs), is crucial for a better understanding of transcriptional regulation. Although traditional footprinting assays can accurately identify the precise binding sites of any factor, this low-throughput method is highly technical and can only be used to analyze a single small region (< 1 kilobase pairs (kb)) at a time. Chromatin immunoprecipitation followed by high-throughput deep sequencing (ChIP-seq) enables genome-wide detection of transcription factor binding sites as well as the localization of epigenetic regulatory markers on a genomic scale [1,2]. It typically returns millions of short (35–50 base pairs (bps)) sequence tags mapped onto a reference genome from a sample organism. Putative binding sites with high confidence can be extracted from peak-enriched regions in the genome by peak-calling programs [3]. However, the resolution of binding regions identified from ChIP-seq can be a few hundred base pairs and is one or two orders of magnitude larger than a typical TFBS. By using an exonuclease that trims DNA regions at a precise distance from binding sites, the novel ChIP-seq technique ChIP-exo is able to locate binding sites at high resolution [4]. However, according to the results in Rhee and Pugh [4], binding regions identified

from ChIP-exo experiments may be tens of bps away from the exact binding locations, although some of them at the location indicated by the experiments. Computational methods are still needed to identify the exact binding locations of a TF in ChIP-seq or ChIP-exo data sets.

Binding sites for a specific TF are often highly conserved and have strong evidence for sequence specificity [5]. An actual DNA region interacting with and bound by a single TF usually ranges in size from 8–10 to 16–20 bps. In the past two decades, numerous programs have been developed to identify over-represented DNA sequence motifs from the promoters of co-regulated or homologous genes [6]. These programs can be divided into two groups. The first includes profile-based methods such as CONSENSUS [7], MEME [8], Gibsampler [9], AlignACE [10], PROJECTION [11], and CRMD [12], each of which attempts to maximize a statistic- or entropy-related score from a profile matrix (also called a position weight matrix (PWM)). The second group is comprised of consensus-based methods, which include SPELLER [13], WEEDER [14,15], MITRA-count [16], Voting [17], PMSprune [18], WINNOWER [19], iTriplet [20], VINE [21], Stemming [22], and RecMotif [23]. These progams are designed to find potential (l,d) motifs within DNA sequences [19], where l is the length of a motif and d is the maximum number of mutations between a predicted binding site and the motif consensus. In most cases, profile-based methods are faster but suffer from lower accuracy due to their tendency to be trapped in a local optimum. Consensus-based methods are more accurate but slower due to the exponential growth of the search space with increasing values of l and d.

Consensus-based methods can be further divided into two categories: pattern-driven and sample-driven approaches [16]. A pattern-driven approach attempts to enumerate all possible 4^l l-mer motifs with lexical order, while a sample-driven approach tries to test all possible (l, d) motifs generated from real l-mers of input sequences. For the methods mentioned above, SPELLER, WEEDER, and MITRA-count are pattern-driven approaches and Voting, PMPprune, WINNOWER, iTriplet, VINE, Stemming, and RecMotif are sample-driven. By using pattern-driven approaches (with the exception of MITRA-count), one can automatically find planted (l, d) motifs without prior knowledge of the length l. On the contrary, sample-driven approaches require that l be specified for each run. In real applications, the exact length of motifs contained in a set of sequences is usually unknown. The pattern-driven algorithm WEEDER has been successful in real eukaryotic applications [24] but has not been improved upon to the best of our knowledge. In this study, we have developed a more powerful method to extract (l,d) motifs and their binding locations contained in DNA sequences without prior knowledge of motif length and have used this method to identify motifs and their binding locations in ChIP-enriched regions.

The pattern-driven approach MITRA-count builds a mismatch tree for all l-mers first, then traverses search space recursively from the root down in depth-first order. Therefore, the length l of a predicted motif must be specified in advance. SPELLER enumerates all possible motifs in a depth-first manner throughout the search space, then scans and counts all possible instances of the current motif with length i ($i \in \{1,2,\cdots,l\}$) from the suffix tree of input sequences. The algorithm can identify planted (l, d) motifs efficiently when $l \leq 13$ and $d \leq 3$ (see Table 1). In order to increase the speed of SPELLER, WEEDER includes an error ratio ε ($\varepsilon \simeq d/l$) for the algorithm that narrows the search space such that for all $i \in \{1,2,...,l\}$ the number of mismatches between the first i nucleotides of a candidate l-mer motif and the first i nucleotides of a valid instance of the motif is at most εi. The algorithm can accelerate SPELLER to some extent, especially when d/l is small (e.g., $d/l \leq 0.25$). Unfortunately, not all motif occurrences satisfy this restriction. WEEDER must lower the occurrence frequency $q \leq N$ to make sure exact motifs will not be missed. However, WEEDER's run time increases dramatically with the decrease of q. For instance, for (15, 4), q should be lowered to half the number of sequences at the OOPS constraint (one occurrence(s) of the motif instance(s) per sequence) to make sure that the true motif will be discovered [14]. However, WEEDER's run time may be even longer than SPELLER under the condition that the two algorithms use the same programming techniques. Thus, a more efficient method is needed to improve the efficiency of pattern-driven algorithms without knowledge of the length of predicted motifs under the ZOMOPS constraint (zero, one or multiple occurrence(s) of the motif instance(s) per sequence).

Additionally, the programs mentioned above are not computationally efficient enough to process a large number of ChIP-seq peaks. In recent years, several programs have been developed to cope with large-scale ChIP-seq data. Some are ChIP-tailored versions of previously-developed software (e.g., ChIP-MEME [25], DREME [26], and GimmeMotifs [27]). These typically restrict motif discovery to a few hundred peaks and usually ignore the remaining unselected sequences. Other programs are faster versions of previous software (e.g., STEME [28], ChIPMunk [29], and HMS [30]). STEME is a faster version of MEME and

Table 1. Comparisons between FMotif and other pattern-driven algorithms on (l, d) samples with $N = 20$, $L = 600$, and $\alpha = 0\%$ noise sequences.

(l,d); $\tau = d/l$	SPELLER	WEEDER(q)	MITRA	FMotif
(10, 2); $\tau = 0.2$	17.16s-1	7.47s (19)-1	1.83s-2	0.59s-1
(11, 2); $\tau = 0.18$	17.73s-1	32.53s (15)-1	1.82s-1	0.59s-1
(12, 3); $\tau = 0.25$	4.42m-1	9.35m (15)-1	21.22s-1	6.77s-1
(13, 3); $\tau = 0.23$	4.42m-1	2.80m (18)-1	21.25s-3	6.73s-1
(14, 4); $\tau = 0.28$	1.05h-1	2.41h (15)-1	3.94m-1	1.32m-1
(15, 4); $\tau = 0.27$	1.05h-1	1.08h (16)-1	3.93m-1	1.31m-1
(15, 5); $\tau = 0.33$	–	–	41.25m-2	15.51m−/
(16, 5); $\tau = 0.31$	–	–	41.19m-1	15.50m-1
(17, 6); $\tau = 0.35$	–	–	6.58h-1	3.17h−/
(18, 6); $\tau = 0.35$	–	–	6.84h-1	3.17h-1

'WEEDER(q)' indicates the execution time of WEEDER given the occurrence frequency threshold q. '–' indicates a run time of over 10 hours. s, m, and h are the units of a run time and denote seconds, minutes, and hours respectively. The number after each run time is the ranking number of a true planted motif among the top 25 predicted motifs. '/' after a run time indicates that the real motifs were not in the top 25.

involves indexing sequences with a suffix tree, which accelerates the expectation-maximization (EM) steps. ChIPMunk combines EM with a greedy approach similar to CONSENSUS and decreases the run time of the optimization procedure. HMS is an improved version of Gibbs Sampler and combines stochastic sampling with deterministic, greedy search steps. Another group of programs integrate other information such as TFBS positional priors [31] or transcription start sites [32] in order to optimize a PWM of ChIP-enriched regions. As mentioned above, these programs still have a local optimum problem. Similar to SPELLER and WEEDER, some of these programs are consensus-based methods (sometimes called word enumeration methods). These include RSAT [33], Cisfinder [34] and POSMO [35]. RSAT is a word enumeration method and has been developed to process whole ChIP-seq peak data sets, but is limited to short (l, d) motifs ($l \leq 10, d \leq 2$). Cisfinder is a word clustering method and combines short k-mer enumeration ($k = 7$, 8, or 9) with a clustering strategy. POSMO, also a word clustering method, uses TFBS positional bias information along with k-mer enumeration and clustering. However, both Cisfinder and POSMO use clustering methods to group short k-mers and therefore cannot find exact (l, d) motifs contained in sequences. Thus, finding exact (l, d) motifs with larger values of l and d in a large-scale sequence data set is still very difficult.

According to a previous study by Keich and Pevzner [36], real signals may be mixed with spurious motifs contained in background sequences under the OOPS constraint when the degenerative ratio $\tau = d/l > 0.25$. A larger τ makes it more difficult to discriminate between a real motif and spurious motifs. However, some sequences may not contain any occurrence of a motif. As previously mentioned, we have concentrated on a more generalized model (the ZOMOPS constraint). Under this constraint, we have found that, except for the degenerative ratio τ, the ratio of noise sequences $\alpha = (N - Q)/N$, where N is the number of sequences and Q is the number of sequences containing at least one variant of a motif, negatively affects (l, d) motif searches. A larger α leads to more spurious motifs in background sequences. It is suspected that 30% of factor-bound locations in

ChIP-seq data may be false positives [4]. Plus, there may be different versions of DNA-binding motifs for any given TF. A specified motif may only occur in 30% of binding regions. Although false positive rates in ChIP-seq data sets are low enough that statistical conclusions can be drawn in most cases, the noise (plus the diversity of DNA-binding modes) still interrupts the motif-finding process and alters motif-finding results. Thus, this may not be the best way to identify motifs in full-size ChIP-seq data sets. After running a peak-calling program on a raw ChIP-seq data set, peaks along with their ChIP enrichment values, p-values, or false discovery rates (FDRs) can be obtained. False positive peaks are those with low peak enrichment values, p-values, or FDRs. A better method may be to find motifs with a high confidence value (i.e., those that are plentiful enough to draw statistical conclusions) in peak-enriched regions and subsequently scan their binding locations with the degenerative value d in the remaining peak regions that have low peak enrichment values, p-values, or FDRs. This would not only exclude more noise and spurious motifs [37], but it would also take advantage of well-developed motif-finding tools with an acceptable level of scalability. A similar idea was used in MICSA and achieved good performance [38]. However, MICSA used the optimal method MEME (the accuracy of which is limited [12,19]) to get the PWM of a motif for only the first three hundred peak-enriched regions.

In this study we have found that for motifs with length l, both SPELLER and WEEDER have been designed to check each i-mer ($i \leq l$) in the pattern space with depth-first order and count the variants of the i-mer in the suffix tree of sequences from the root to layer i. The suffix tree is scanned one time for each i-mer pattern. Thus, as i increases, the algorithms scan the suffix tree an increasing number of times. In fact, the mismatch information in layer i of a suffix tree can be used to search for $(i+1)$-mers in the pattern space. For this reason, we constructed a new suffix tree structure with mismatch information (called a mismatched suffix tree) and developed a fast <u>motif</u> enumerative method (FMotif) under the ZOMOPS constraint. Using the newly constructed suffix trees, we incorporated the mismatch information in layer i of the mismatched suffix trees to verify $(i+1)$-mers in the pattern space. We then updated mismatch information in layer $i+1$ of the mismatched suffix trees. In this way we were able to implement a depth-first search within the pattern space and the mismatched suffix trees simultaneously. To process large-scale ChIP-seq data sets, we integrated the peak detection method MACS [39] with our motif-finding method and ChIP-enriched sampling strategy, which allowed us to locate the exact binding locations in ChIP-seq and ChIP-exo data sets. We chose MACS because it has been shown to perform well when compared to several other peak-calling programs [3].

Results

Experimental Results on Artificial Data Sets

We compared FMotif with the existing pattern-driving methods including SPELLER, WEEDER, and MITRA-count (MITRA for short) on synthetic samples to show the efficiency of our proposed method. All synthetic samples were generated following the method of Pevsner and Sze [19], where Q ($Q \leq N$) variants of an l-length motif were randomly planted into Q sequences selected randomly from a set of N sequences with length L. In this (l, d) model, each planted variant of the motif with length l had exactly d mismatches with the motif itself.

In the first group of experiments, we tested the performance of these algorithms on (l, d) sample sets without noise sequences (i.e., $Q = N$) at standard settings, where the number N and the length L

of sequences are set to 20 and 600, respectively [14,16,19]. These test results are shown in Table 1. 'WEEDER(q)' indicates the execution time of WEEDER given the occurrence frequency threshold q. '$-$' indicates a run time of over 10 hours. s, m, and h denote seconds, minutes, and hours respectively. The number after each run time is the ranking number of a true planted motif among the top 25 predicted motifs. '/' after a run time indicates that the real motifs were not in the top 25. In the second group of experiments, we first tested the influence of the ratio of noise sequences α ($\alpha = (N - Q)/N$) on (l, d) samples using FMotif with typical settings (i.e., $N = 20$ and $L = 600$). In order to provide a more comprehensive comparison of the calculation speed when noise was added, we compared FMotif to SPELLER and MITRA on $(10, 2)$, $(11, 2)$, $(12, 3)$, and $(13, 3)$ samples. We avoided comparisons over motifs more complicated than $(13, 3)$ because SPELLER lacked computational efficiency on these problems and WEEDER required tuning of parameter q. These test results are shown in Table 2, where α is set at 5%,10%,15%,$\cdots$, and 40%, '/' indicates that the real motifs were not in the top 25, and the first line for $(10, 2)$, $(11, 2)$, $(12, 3)$ and $(13, 3)$ is the FMotif result, the second line (denoted by 'M..') is the MITRA result, and the third line (denoted 'S..') is the SPELLER result. We then tested the influence of the noise ratio α on samples with $N = 1000$ and $L = 100$ to simulate ChIP-enriched regions because those regions are usually relatively short and the number of regions is usually large. We subsequently compared FMotif to SPELLER and MITRA on $(10, 2)$, $(11, 2)$, $(12, 3)$, and $(13, 3)$ samples as before. These test results are shown in Table 3, where α is set at 10%,20%,$\cdots$, and 80%. In the third group of experiments, we tested FMotif scalability using two groups of samples to see whether it was suitable for recognizing motifs in large-scale ChIP-enriched regions. The settings of the first group were $L = 100$, $N = 1000,2000,\cdots,8000$ and no noise sequences ($\alpha = 0\%$). These test results are shown in Table 4. The settings of the second group were $L = 100$, $N = 1000,2000,\cdots,8000$ and $\alpha = 30\%$ noise sequences in order to mimick ChIP-seq data. These test results are shown in Table 5. All experiments were performed on a computer with an Intel 2.99 GHz processor, 2.00GB of main memory, and the Windows XP operating system.

The results in Tables 1–5 lead to three observations. First, FMotif is a fast and exact algorithm and capable of finding (l, d) motifs in synthetic samples without being given the length l of a predicted motif. It performs faster than SPELLER, MITRA, and WEEDER without sacrificing accuracy. As mentioned above, WEEDER's efficiency suffers significantly (see (14, 4) in Table 1 for an example) when the occurrence frequency threshold q is too low, and MITRA requires that the length l be specified a priori. It should be noted that FMotif ranked all motifs with different lengths together by significance score. For the samples whose true motifs were not ranked in the top 25, the top motifs were usually $(l-1)$- or $(l-2)$- substrings of the true motifs with length l. In these cases the true motifs were still in the output list but were ranked below the top 25. Second, noise sequences have a strong effect on the results and the speed of the method. With an increase in α, the run time increases as well. Like the degenerative ratio $\tau = d/l$, the ratio of noise sequences α also weakens motifs, especially when background sequences are long (see Table 2). Spurious motifs in background sequences bury the authentic signals when either τ or α is large. For example, real motifs were difficult to filter by their significance score for the (15, 5) motif in Table 2, even when the noise sequence ratio was set to 5%. In this case, many spurious motifs of length 14–15 with a large significance score were ranked among the top 25. When the length of background sequences was shorter and the number of

Table 2. Results for noise-influenced models on (l, d) samples with $N = 20$, $L = 600$, and $\alpha = 5\%$, 10%, $\cdots$, 40% noise sequences.

(l,d)	5%	10%	15%	20%	25%	30%	35%	40%
(10, 2)	0.78s-1	0.95s-1	1.06s-1	1.19s-1	1.30s-1	1.44s-3	1.63s-5	/
M..	2.78s-1	3.05s-1	3.23s-1	3.44s-1	3.78s-1	4.83s-3	5.23s-15	/
S..	38.99s-1	1.09m-1	1.53m-1	1.96m-1	2.51m-1	3.34m-3	4.69m-21	/
(11, 2)	0.77s-1	0.94s-1	1.06s-1	1.17s-1	1.30s-1	1.44s-1	1.61s-1	1.83s-1
M..	2.77s-1	3.05s-1	3.23s-1	3.23s-1	3.44s-1	3.77s-1	4.38s-1	5.19s-1
S..	38.13s-1	1.08m	1.52m-1	1.95m-1	2.48m-1	3.31m-1	4.66m-1	6.53m-1
(12, 3)	9.16s-1	11.69s-1	13.92s-1	15.88s-1	17.61s-6	/	/	/
M..	33.13s-1	33.23s-1	39.28s-1	43.45s-1	46.64s-4	50.36s-20	/	/
S..	9.78m-1	17.69m-1	26.94m-1	36.16m-1	45.64m-2	58.11m-8	/	/
(13, 3)	9.17s-1	11.64s-1	13.95s-1	15.88s-1	17.63s-1	19.50s-5	21.83s-2	24.80s-1
M..	27.23s-1	34.38s-1	40.22s-1	43.44s-1	46.66s-1	50.58s-1	57.14s-1	1.12m-1
S..	9.80m-1	17.71m-1	27.07m-1	36.12m-1	45.80m-1	58.33m-1	1.30h-1	1.79h-1
(14, 4)	1.85m-1	2.33m-1	2.89m-9	3.47m-5	/	4.52m-20	/	/
(15, 4)	1.82m-1	2.32m-1	2.89m-1	3.48s-1	4.03m-1	4.52m-3	5.02m-3	5.66m-1
(15, 5)	/	/	/	/	/	/	/	/
(16, 5)	22.90m	29.52m-1	36.05m-1	/	/	/	/	/
(17, 6)	/	/	/	/	/	/	/	/
(18, 6)	5.18h	7.07h-1	/	/	/	/	/	/

The ratio of noise sequences α is set at $5\%, 10\%, 15\%, \cdots$, and 40%, '/' indicates that the real motifs were not in the top 25. The number after each run time is the ranking number of a true planted motif among the top 25 predicted motifs. The first line for (10, 2), (11, 2), (12, 3) and (13, 3) is the FMotif result, the second line (denoted by 'M..') is the MITRA result, and the third line (denoted 'S..') is the SPELLER result. s, m, and h denote seconds, minutes, and hours respectively.

sequences was larger, the signals were stronger and could be easily identified even if a large portion of noise sequences was added (see Table 3). This is consistent with the previous result that false-positives could be reduced by decreasing the sequence length or by adding more sequences to the data set [37]. Third, as shown in Tables 4 and 5, FMotif is capable of operating on a large scale even when there are 30% noise sequences in samples. This allowed us to use FMotif to process peak regions within ChIP-seq and ChIP-exo data sets.

Additionally, we compared FMotif with CisFinder, which uses k-mer enumeration with k-mer clustering to find motifs in large-scale ChIP-seq peak regions. Using both algorithms, we verified the accuracy of FMotif and CisFinder by searching for long (l, d) motifs in synthetic sample sets with 3000 sequences, each of which contained a planted variant of a parent motif. The experimental results are shown in Table 6, where 'Planted Motif' indicates a planted motif consensus in a set of sequences, 'FMotif (Top-1)' indicates the top ranked motif consensus found by FMotif in a sample set, 'CisFinder' indicates the closest matching motif consensus (described by IUPAC nucleotide codes) found by CisFinder in a sample set, '#' indicates the number of variants of a reported motif found by FMotif or CisFinder in a sample set, and 'Rank' after '# −' is the ranking number of the reported motif found by Cisfinder in Table 6.

We used the *site-level sensitivity* (sSn) and *positive predictive value* ($sPPV$) metrics described by Tompa [24] to statistically quantify the accuracy of the two methods, where $sSn = sTP/(sTP + sFN)$ and $sPPV = sTP/(sTP + sFP)$, sTP is the number of known sites overlapping predicted sites, sFN is the number of known sites not overlapping predicted sites, and sFP is the number of predicted sites not overlapping known sites. A predicted site overlaps a known site if they share at least a half of the length of known sites. In order to give a more comprehensive comparison of the

accuracy of the two methods on simulated ChIP-seq data sets, we added 30% noise sequences to samples with $N = 3000$ and $L = 100$ and performed the experiments again. These test results are shown in Table 7.

As evident from Tables 6 and 7, FMotif is an exact algorithm. It reported all true motif consensuses and their planted variants plus false positive variants in background sequences. CisFinder performed quickly but suffered from low accuracy (due to low *sensitivity*), especially when $\tau = d/l$ was large. FMotif and CisFinder both were robust after 30 noise sequences were added to the samples. It should be pointed out that, although there are various resources on CisFinder's website (http://lgsun.grc.nia.nih.gov/cis-finder/download.html), we used only the motif-finding program. There are other programs focused on motif clustering, motif improvement, motif comparison, and other tasks. If all programs were used together, a better motif and more of its binding sites may be identified. However, the CisFinder algorithm [34] was implemented in that motif-finding program and there was no direct way to use all these programs together based on our knowledge.

Experimental Results Using ChIP-seq Data Sets

We tested FMotif using 12 mouse ChIP-seq data sets for 12 DNA-binding TFs (CTCF, cMyc, Esrrb, Klf4, Nanog, nMyc, Oct4, Smad1, Sox2, STAT3, Tcfcp2I1, and Zfx) involved in mouse embryonic stem cell pluripotency and self-renewal [40]. These ChIP-seq data sets have been deposited in the GEO database with ID number GSE11431. We also tested FMotif using four widely used human ChIP-seq data sets for four DNA-binding TFs including CTCF (CCCTC-binding factor [41], named CTCF(h) in the paper), FoxA1 (hepatocyte nuclear factor 3α [42]), NRSF (neuron-restrictive silencer factor [2]), and STAT1 (signal transducer and activator of transcription protein [1]). The

Table 3. Results for noise-influenced models on (l, d) samples with $N = 1000$, $L = 100$, and $\alpha = 10\%$, 20%, $\cdots$, 80% noise sequences.

(l,d)	$\alpha = 10\%$	$\alpha = 20\%$	$\alpha = 30\%$	$\alpha = 40\%$	$\alpha = 50\%$	$\alpha = 60\%$	$\alpha = 70\%$	$\alpha = 80\%$
(10,2)	6.39s-1	7.63s-1	10.36s-1	12.95s-1	13.63s-1	14.64s-1	23.13s-1	24.84s-6
M..	12.44s-1	12.49s-1	19.38s-1	31.61s-1	31.97s-1	32.29s-1	1.47m-1	1.64m-1
S..	1.23m-1	2.29m-1	6.84m-1	10.68m-1	12.72m-1	15.03m-1	1.05h-1	1.23h-1
(11,2)	6.41s-1	7.70s-1	10.28s-1	12.59s-1	13.60s-1	14.66s-1	23.01s-1	24.86s-2
M..	12.41s-1	12.66s-1	18.58s-1	31.44s-1	31.86s-1	32.16s-1	1.41m-1	1.64m-1
S..	1.03m-1	2.09m-1	5.86m-1	10.64m-1	12.74m-1	15.08m-1	1.05h-1	1.23h-1
(12,3)	1.19m-1	1.44m-1	1.66m-1	2.40m-1	2.94m-1	3.25m-1	3.56m-1	/
M..	2.44m-1	2.84m-1	2.87m-1	4.55m-1	7.92m-1	7.98m-1	8.63m-1	25.91m-1
S..	15.45m-1	32.14m-1	45.25m-1	2.22h-1	3.80h-1	4.42h-1	6.00h-1	22.34h-1
(13,3)	1.18m-1	1.43m-1	1.66m-1	2.33m-1	2.91m-1	3.50m-1	3.56m-1	5.72m-10
M..	2.33m-1	2.84m-1	2.87m-1	4.31m-1	7.90m-1	7.96m-1	8.41m-1	25.90m-1
S..	15.39m-1	32.30m-1	46.20m-1	2.20h-1	3.82h-1	4.40h-1	5.92h-1	22.25h-1
(14,4)	10.03m-1	16.35m-1	19.17m-1	21.63m-1	30.65m-1	42.05m-1	45.05m-1	/
(15,4)	10.07m-1	16.36m-1	19.19m-1	21.78m-1	30.67m-1	42.02m-1	45.04m-1	/
(15,5)	1.69h-1	2.23h-1	3.84h-1	4.66h-1	5.18h-1	7.38h-1	10.60h-13	/
(16,5)	1.70h-1	2.24h-1	3.89h-1	4.67h-1	5.20h-1	7.38h-1	10.61h-3	/

The ratio of noise sequences α is set at $10\%, 20\%, \cdots$, and 80%. Row definitions are the same as those in Table 2.

raw sequence of the FoxA1 ChIP-seq data set was downloaded from http://liulab.dfci.harvard.edu/MACS/Sample.html. The bed format of the CTCF, NRSF and STAT1 ChIP-seq data sets was downloaded from http://dir.nhlbi.nih.gov/papers/lmi/epigenomes/sissrs/. These downloaded short reads were mapped onto the newest version of mouse genome assembly mm10 and human genome assembly hg19, respectively. The peak regions were extracted from these reads using the peak finding program MACS [39] with a false discovery rate (FDR) threshold of 0.2. The reads were ranked by their FDR if a negative control was available, or by p-value otherwise. To prepare the data sets for use with motif discovery algorithms, we mapped the summits of the ChIP-seq peaks and extracted the 100 bps of genomic sequence centered around each peak.

In order to facilitate a fast motif search, avoid the potential influence of false positive peaks, and reduce false positive motifs in background sequences [37], we ran FMotif on the first 3000 ChIP-enriched genomic sequences and then scanned for potential binding locations in the remaining genomic sequences with the degenerative value d. Since binding sites could exist on either DNA strand and CisFinder searched both, we counted the instances of a predicted motif and those of its reverse complement motif. We then compared FMotif with CisFinder and published motifs [2,40,42–44] in literature. The experimental results are shown in Figures 1 and 2, where 'Nb' indicates the number of peak-enriched regions predicted by the peak-calling program MACS with an FDR threshold of 0.2 or a p-value threshold of 10^{-5}, 'FMotif' and 'CisFinder' indicate the closest matching motif logos found by these programs (all motif logos were generated using the web-based tool Weblogo [45]), 'Literature' indicates the corresponding motif logos published in literature, '#' indicates the number of binding sites found by either FMotif or CisFinder, and 'Rank' after '# −' is the ranking number of a reported motif found by either FMotif or CisFinder.

We compared predicted and published motifs using a motif-level accuracy measure called the *performance coefficient*

$\|U \cap V\| / \|U \cup V\|$, where U is a predicted motif consensus and V is the motif consensus published in literature [19]. As shown in Figures 1 and 2, the motif logos found by FMotif were more accurate compared with published logos from literature than those found by CisFinder. Furthermore, FMotif identified more TFBS locations than CisFinder. As for the 12 mouse TFs DNA-binding logos in [40], Chen et al. used the motif discovery algorithm WEEDER and subsequently extended the motifs using an expectation-maximization method. This second step was necessary because the supplied version of the WEEDER algorithm limited the motif search to a maximum of 12 bps. As discussed in the previous sections, WEEDER operated with low efficiency for long motifs and was difficult to tune for the parameter q.

To estimate the robustness of our sampling strategy, we ran FMotif on the first 500, 1000, 1500, ..., and 5000 sequences and the full-size ChIP-enriched genomic sequences of TFs n-Myc, Oct4, and NRSF. For all subsets and the full-size data sets, each of the corresponding motifs in Figures 1 and 2 was ranked within the top 25 motifs predicted by FMotif. The ranking number of reported motifs increased with subset size and tended to be stable when the size was greater than 1000. All potential binding sites of reported motifs were obtained from subsets and discovered during the scanning step. Thus, it was not necessary to run a motif-finding algorithm on the whole ChIP-seq data sets, especially when data sets were very large. Additionally, we tested FMotif on N randomly selected sequences ($N = 500$, 1000, 1500, ..., and 3000). These experiments were repeated 10 times. We then compared these results to those of the first N sequences and those of the last N sequences for TFs n-Myc, Oct4, and NRSF. In general, motif consensuses predicted from the first N sequences were the most similar to published motifs and ranked highest in the final output. Those predicted from randomly selected N sequences tended to be ranked second, while those predicted from the last N sequences were usually ranked the lowest. Furthermore, for the same reported motif of a TF, the number of binding sites found in the first N sequences was significantly greater than that

Table 4. A demonstration of FMotif scalability on (l, d) samples for $N = 1000, 2000, \cdots, 8000$ sequences, $L = 100$, and no ($\alpha = 0\%$) noise sequences.

(l,d)	1000	2000	3000	4000	5000	6000	7000	8000
(10, 2)	3.34s-1	8.05s-1	11.90s-1	15.80s-1	19.58s-1	23.34s-1	26.83s-1	42.81s-1
(11, 2)	3.36s-1	8.08s-1	11.86s-1	15.61s-1	19.45s-1	23.16s-1	26.84s-1	39.25s-1
(12, 3)	33.41s-1	1.31m-1	1.84m-1	2.38m-1	2.78m-1	3.32m-1	4.03m-1	4.46m-1
(13, 3)	34.62s-1	1.30m-1	1.85m-1	2.40m-1	2.84m-1	3.32m-1	3.82m-1	4.33m-1
(14, 4)	4.51m-1	10.22m-1	15.14m-1	19.93m-1	24.85m-1	29.54m-1	34.25m-1	39.03m-1
(15, 4)	4.52m-1	10.30m-1	15.23m-1	20.05m-1	24.87m-1	29.16m-1	34.38m-1	39.26m-1
(15, 5)	35.06m-1	1.46h-1	2.15h-1	2.68h-1	3.21h-1	3.75h-1	4.27h-1	4.77h-1
(16, 5)	35.02m-1	1.47h-1	2.13h-1	2.70h-1	3.24h-1	3.84h-1	4.27h-1	4.75h-1

The number after each run time is the ranking number of a true planted motif among the top 25 predicted motifs. s, m, and h denote seconds, minutes, and hours respectively.

found in randomly selected N sequences. The number of predicted binding sites found in the last N sequences was the lowest. In some cases there was no corresponding motif in randomly selected N sequences or in the last N sequences when employing the same parameter settings. This situation occurred more often when using the last N sequences. Therefore, we decided that the first N sequences with the lowest p-value or FDR (i.e., the most ChIP-enriched sequences) were the best choice for drawing statistical conclusions about a corresponding motif. This is because, as discussed in the Introduction section, the first N sequences were the least affected by noise. We selected the first 3000 peak regions to be sure that the selected subsets were large enough to account for the specificity of TF-DNA binding and to exclude false positive motifs. The same results may be obtained by running the algorithm on the first 1000–2000 sequences and then scanning potential locations in the remaining sequences.

FMotif Sensitivity

To test the sensitivity of FMotif, we ran it on an NRSF-positive TFBS set (NRSF/qPCR), which was composed of 83 binding sites verified by qPCR [2]. We then ran FMotif on four yeast DNA-binding TFs (Reb1, Gal4, Phd1, and Rap1) and one human TF (CTCF) ChIP-exo data sets. Raw sequence of the five ChIP-exo data sets are available from the NCBI Sequence Read Archive with accession number SRA0044886 [4]. Since it is thought that >98% of ChIP-exo peak regions contain one recognizable

DNA-binding motif within tens of bps away from peak summits, these can be viewed as positive TFBS sets. We used the five ChIP-exo peaks reported in Data S1 from Rhee and Pugh [4]. Similarly, we mapped the summits of ChIP-exo peaks and extracted 50 bps of genomic sequence centered around each peak in yeast genome sacCer3 and human genome hg19, respectively. This allowed us to avoid peak regions overlapping with each other due to some of the summits of ChIP-exo peaks being very close together. Results from CisFinder and published motifs [2,4,43,46–48] in literature are shown for comparison (see Figure 3).

As shown in Figure 3, FMotif was capable of finding more matching motifs and true TF-binding locations when compared to CisFinder. For example, 76 true binding sites of NRSF/qPCR were predicted exactly by FMotif. On the same data set (NRSF/qPCR), MICSA [38] using MEME reported only 55 sites. This highlights the fact that FMotif is capable of identifying TF-binding locations with high sensitivity. It is well-established that specificity is an important consideration for this type of method. However, the ChIP-exo technique is a high-throughput approach, and the resolution of binding regions identified by ChIP-exo may still be tens of base pairs from where the true binding sites of between 8 and 25 base pairs are located. In addition, some of those binding regions are false positives, and it is difficult to say which ones are truly false positives without carefully designed wet-lab experiments. However, we show the specificity (i.e., $sPPV$) of FMotif for artificial samples in Tables 6 and 7. From this information we

Table 5. A demonstration of FMotif scalability on (l, d) samples for $N = 1000, 2000, \cdots, 8000$ sequences, $L = 100$, and $\alpha = 30\%$ noise sequences.

(l,d)	1000	2000	3000	4000	5000	6000	7000	8000
(10, 2)	10.33s-1	26.19s-1	45.16s-1	1.10m-1	1.63m-1	2.09m-1	2.70m-1	2.97m-1
(11, 2)	10.33s-1	25.89s-1	45.30s-1	1.09m-1	1.49m-1	1.90m-1	2.43m-1	2.88m-1
(12, 3)	1.66m-1	4.13m-1	7.26m-1	10.70m-1	14.07m-1	17.52m-1	21.40m-1	25.76m-1
(13, 3)	1.66m-1	4.15m-1	7.27m-1	10.72m-1	14.06m-1	17.56m-1	21.51m-1	25.94m-1
(14, 4)	19.14m-1	45.85m-1	1.28h-1	1.90h-1	2.59h-1	3.24h-1	3.88h-1	4.52h-1
(15, 4)	19.16m-1	45.81m-1	1.29h-1	1.91h-1	2.59h-1	3.25h-1	3.89h-1	4.54h-1
(15, 5)	3.86h-1	8.83h-1	14.70h-1	21.28h-1	28.40h-1	35.48h-1	43.08h-1	50.36h-1
(16, 5)	3.87h-1	8.61h-1	14.64h-1	20.81h-1	28.52h-1	35.15h-1	43.35h-1	51.22h-1

Row and column definitions are the same as those in Table 4.

Table 6. Comparisons between FMotif and CisFinder on large (l, d) samples with $L = 100$, $N = 3000$, and no ($\alpha = 0\%$) noise sequences.

(l,d)	Planted Motif	FMotif (*Top-1*)	CisFinder (#-Rank)
		(#)–(*sSn*)–(*sPPV*)	(#-Rank)–(*sSn*)–(*sPPV*)
(10, 2)	CACGAGAACC	CACGAGAACC	CACGANAACC
		(3108)–(1.0)–(0.97)	(66-1)–(0.02)–(0.98)
(11, 2)	TTGACAAGGAT	TTGACAAGGAT	TTVACAASGA
		(3026)–(1.0)–(0.99)	(186-1)–(0.06)–(0.96)
(12, 3)	TCCATTAGGTGG	TCCATTAGGTGG	CCWMCTAABKGAMC
		(3089)–(1.0)–(0.97)	(80-1)–(0.02)–(0.93)
(13, 3)	CGATAGGTCTATG	CGATAGGTCTATG	ATAGKYCTA
		(3026)–(1.0)–(0.99)	(148-1)–(0.05)–(0.96)
(14, 4)	AGCTATCTATTTAA	AGCTATCTATTTAA	TAAANWGATA
		(3161)–(1.0)–(0.95)	(75-1)–(0.02)–(0.85)
(15, 4)	GATCACACGGAAACC	GATCACACGGAAACC	CACACGGAAAC
		(3022)–(1.0)–(0.99)	(109-3)–(0.04)–(0.98)
(15, 5)	GGTGGGGCGGGCGAT	GGTGGGGCGGGCGAT	CMGGYYGGGKCG
		(3371)–(1.0)–(0.89)	(40-1)–(0.01)–(0.77)
(16, 5)	GAGGCTTGTAAACGTT	GAGGCTTGTAAACGTT	GGMGKGTAAAMGTTKC
		(3062)–(1.0)–(0.98)	(59-1)–(0.02)–(0.85)

'Planted Motif' indicates a planted motif consensus in a set of sequences, 'FMotif (*Top-1*)' indicates the top ranked motif consensus found by FMotif in a sample set, 'CisFinder' indicates the closest matching motif consensus (described by IUPAC nucleotide codes) found by CisFinder in a sample set, '#' indicates the number of variants of a reported motif found by FMotif or CisFinder in a sample set, and 'Rank' after '#−' is the ranking number of the reported motif found by Cisfinder. The *site-level sensitivity* (*sSn*) and *positive predictive value* (*sPPV*) metrics described by Tompa [24] were used to statistically quantify the accuracy of the two methods, where $sSn = sTP/(sTP + sFN)$ and $sPPV = sTP/(sTP + sFP)$, sTP is the number of known sites overlapping predicted sites, sFN is the number of known sites not overlapping predicted sites, and sFP is the number of predicted sites not overlapping known sites. A predicted site overlaps a known site if they share at least a half of the length of known sites.

Table 7. Comparisons between FMotif and CisFinder on large (l, d) samples with $L = 100$, $N = 3000$, and $\alpha = 30\%$ noise sequences.

(l,d)	Planted Motif	FMotif (*Top-1*)	CisFinder
		(#)–(*sSn*)–(*sPPV*)	(#-Rank)–(*sSn*)–(*sPPV*)
(10, 2)	TGACCCCACG	TGACCCCACG	YHGAYCHMACGSM
		(2192)–(1.0)–(0.96)	(65-2)–(0.03)–(0.89)
(11, 2)	GCGGTGTACCA	GCGGTGTACCA	GCGGTNTACC
		(2130)–(1.0)–(0.99)	(120-2)–(0.06)–(0.99)
(12, 3)	CACGGGCCTTAG	CACGGGCCTTAG	CAKSGGCCBBAG
		(2182)–(1.0)–(0.96)	(61-2)–(0.03)–(0.85)
(13, 3)	TTCAGTAAGCACG	TTCAGTAAGCACG	TTCRGTAARCAYG
		(2124)–(1.0)–(0.99)	(99-1)–(0.05)–(0.96)
(14, 4)	GCAAGTCACCGTGT	GCAAGTCACCGTGT	RVAAGTVVBNGTGT
		(2167)–(1.0)–(0.97)	(42-2)–(0.02)–(0.90)
(15, 4)	AAGGTGTTGGTATGG	AAGGTGTTGGTATGG	AARGTGTTGGTATGGG
		(2137)–(1.0)–(0.98)	(70-2)–(0.03)–(0.90)
(15, 5)	AATACTGTGCATGGA	AATACTGTGCATGGA	AATWCTGTSCA
		(2272)–(1.0)–(0.92)	(27-1)–(0.01)–(0.70)
(16, 5)	AGCTTGCCAGCGACGT	AGCTTGCCAGCGACGT	VGCTSKCCAGCWACGT
		(2145)–(1.0)–(0.98)	(51-1)–(0.02)–(0.90)

Column definitions are the same as those in Table 6.

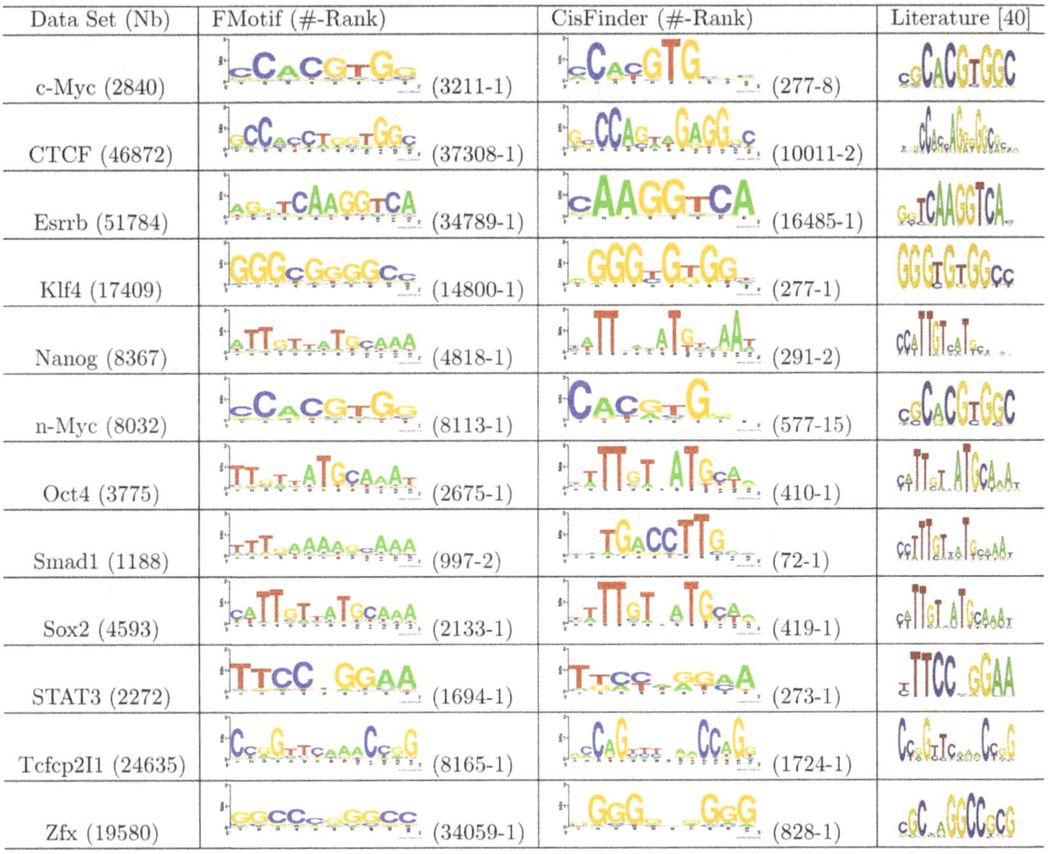

Data Set (Nb)	FMotif (#-Rank)	CisFinder (#-Rank)	Literature [40]
c-Myc (2840)	(3211-1)	(277-8)	
CTCF (46872)	(37308-1)	(10011-2)	
Esrrb (51784)	(34789-1)	(16485-1)	
Klf4 (17409)	(14800-1)	(277-1)	
Nanog (8367)	(4818-1)	(291-2)	
n-Myc (8032)	(8113-1)	(577-15)	
Oct4 (3775)	(2675-1)	(410-1)	
Smad1 (1188)	(997-2)	(72-1)	
Sox2 (4593)	(2133-1)	(419-1)	
STAT3 (2272)	(1694-1)	(273-1)	
Tcfcp2I1 (24635)	(8165-1)	(1724-1)	
Zfx (19580)	(34059-1)	(828-1)	

Figure 1. Motifs in 12 mouse ES Cell ChIP-seq data sets. FMotif was tested using mouse ChIP-seq data sets for 12 DNA-binding TFs (CTCF, cMyc, Esrrb, Klf4, Nanog, nMyc, Oct4, Smad1, Sox2, STAT3, Tcfcp2I1, and Zfx) involved in mouse embryonic stem cell pluripotency and self-renewal [40]. Results from CisFinder and published motifs in literature are shown for comparison. 'Nb' indicates the number of peak-enriched regions predicted by the peak-calling program MACS with an FDR threshold of 0.2 or a p-value threshold of 10^{-5}, 'FMotif' and 'CisFinder' indicate the closest matching motif logos found by these programs (all motif logos were generated using the web-based tool Weblogo [45]), 'Literature' indicates the corresponding motif logos published in literature, '#' indicates the number of binding sites found by either FMotif or CisFinder, and 'Rank' after '# −' is the ranking number of a reported motif found by either FMotif or CisFinder.

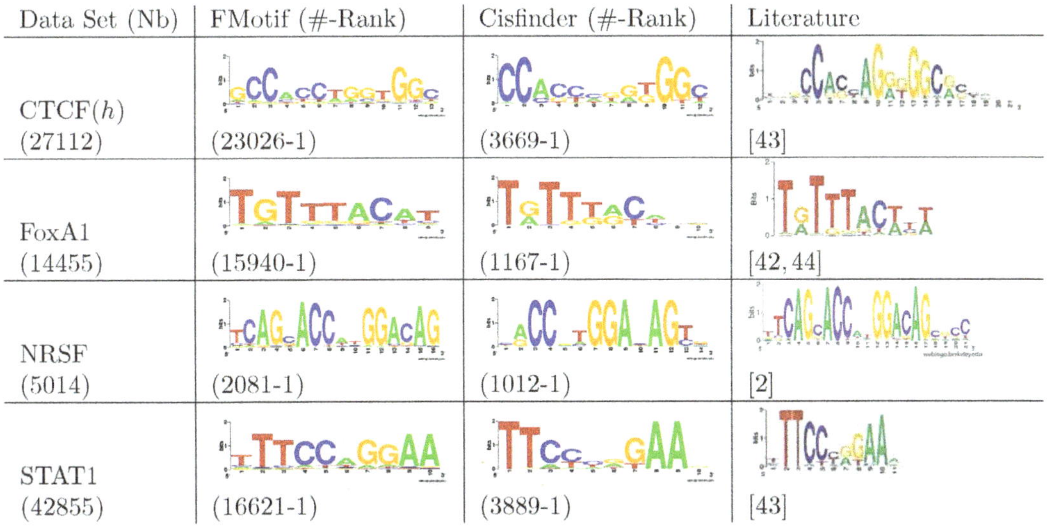

Data Set (Nb)	FMotif (#-Rank)	Cisfinder (#-Rank)	Literature
CTCF(h) (27112)	(23026-1)	(3669-1)	[43]
FoxA1 (14455)	(15940-1)	(1167-1)	[42, 44]
NRSF (5014)	(2081-1)	(1012-1)	[2]
STAT1 (42855)	(16621-1)	(3889-1)	[43]

Figure 2. Motifs in 4 human TF ChIP-seq data sets. FMotif was tested with four widely used human ChIP-seq data sets for four DNA-binding TFs including CTCF (CCCTC-binding factor [41], named CTCF(h)), FoxA1 (hepatocyte nuclear factor 3α [42]), NRSF (neuron-restrictive silencer factor [2]), and STAT1 (signal transducer and activator of transcription protein [1]). Results from CisFinder and published motifs in literature are shown for comparison. Column definitions are the same as those in Figure 1.

conclude that FMotif has both a higher sensitivity and a higher specificity than CisFinder.

Discussion

In this study, we have proposed a new and fast heuristic enumeration method, FMotif, for extracting motifs from sequences. We have used this method to identify motifs and their binding locations in widely-used large-scale ChIP-seq and ChIP-exo data sets by combining FMotif with a peak-enriched sampling strategy. Our empirical studies have shown that this algorithm is fast and exact when searching for motifs in (l, d) samples and has achieved good performance when identifying motifs in ChIP-enriched regions. In addition, the ChIP-enriched sampling strategy worked well on large-scale ChIP-seq and ChIP-exo data sets. It not only allowed us to exclude both noise occurring in lower ChIP-enriched peak regions and false positive motifs contained in background sequences, but also let us take advantage of well-developed motif-finding tools with low-level scalability. However, it should be pointed out that, in general, no method can outperform others under all conditions. FMotif performed faster than SPELLER, WEEDER, and MITRA but used more memory to store mismatched information in suffix trees, and FMotif was much more accurate but much slower than CisFinder. FMotif does, however, provide a good trade-off between time, space, and accuracy.

Motif discovery has been a popular area of study for more than two decades. Many successful motif-finding programs have been developed. The programs are ideal for finding motifs in tens or hundreds of promoters of co-regulated or homologous genes and for extracting motifs in genome-wide ChIP-enriched regions contained in large-scale ChIP-chip, ChIP-seq, and ChIP-exo data sets. Still, the problem is far from solved due to diversity in gene expression/regulation and the low specificity of binding sites. With the advance of high-throughput and high-resolution sequencing techniques like ChIP-exo, researchers have an increasing number of tools for studying gene regulation on a genomic scale. This will make the motif-finding problem easier to solve. Using advanced techniques such as ChIP-exo, it is possible to acquire new knowledge of regulatory binding sites. This will not only be beneficial for understanding the mechanisms of gene regulation, but also for creating a proper computational model that will replace (l, d) models and PWM matrix profiles for motif representation.

Materials and Methods

The (l, d) Motif Search and Suffix Tree

A transcription factor binds to specific DNA sequences and is involved in controlling the transcription of genetic information from DNA to mRNA. The actual DNA regions bound by a TF usually range in size from 8–10 to 16–20 bps and display a short motif, but differ by a few nucleotides from one another. The computational problem is to determine such a motif by analyzing a set of sequences that contain instances of the motif.

In current literature, there are two main approaches to motif representation. The first involves using a motif profile character-

Data Set (Nb)	FMotif (#-Rank)	CisFinder (#-Rank)	Literature
NRSF/qPCR (83)	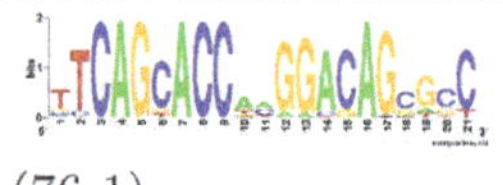 (76-1)	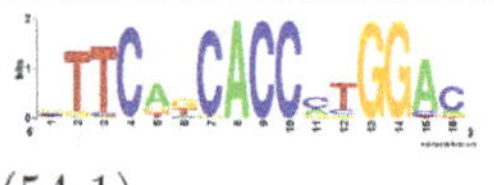 (54-1)	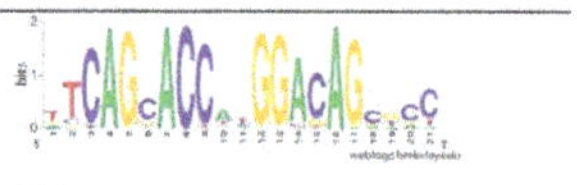 [2]
Reb1 (1776)	 (1579-1)	 (200-1)	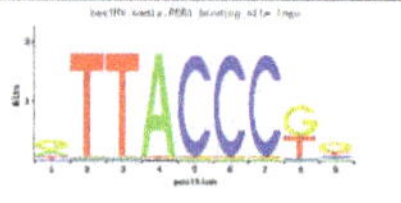 [4, 46, 47]
Gal4 (15)	 (14-1)	No-Motif-Found	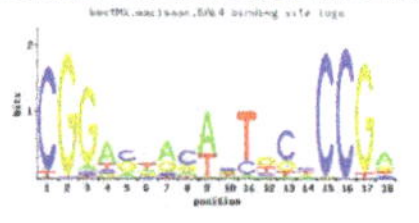 [4, 47, 48]
Phd1 (967)	 (1610-1)	 (40-1)	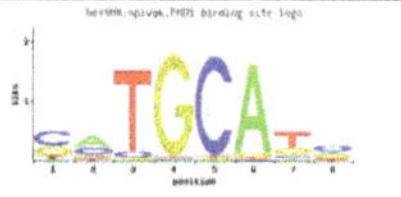 [4, 47]
Rap1 (576)	 (896-1)	 (113-1)	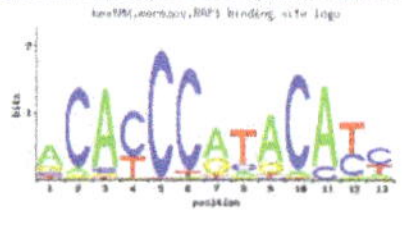 [4, 47]
CTCF(h-exo) (35161)	 (12919-2)	 (1633-1)	 [43]

Figure 3. FMotif sensitivity. FMotif sensitivity was measured using an NRSF-positive TFBS set (NRSF/qPCR), which was composed of 83 binding sites verified by qPCR [2], four yeast DNA-binding TFs (Reb1, Gal4, Phd1, and Rap1), and one human TF (CTCF) ChIP-exo data sets. Results from CisFinder and published motifs in literature are shown for comparison. Column definitions are the same as those in Figure 1.

ized by a PWM $[p_{j,k}]_{l\times 4}$. The PWM records the probability of an observed nucleotide k ($k\in\{A,C,G,T\}$) at position j ($j=\{1,2,\cdots,l\}$) for all aligned sites, where l is the length of the motif. Numerous programs have been developed to maximize the score of a PWM by measuring, for example, the information content of a PWM:

$$IC = -\sum_{i,j} p_{i,j}\cdot log\left(p_{i,j}/p_b\right)$$

where p_b is the background frequency for the nucleotide, which measures motif conservation [7]. Using the second approach, one can characterize a motif as an l-length consensus string and describe it using the most frequent nucleotide in each position of all aligned sites under the assumption that each sequence contains zero or one motif instance with up to d or exactly d mutations within the motif. Finding (l, d) motifs with exactly d mutations is more challenging than finding (l, d) motifs with up to d mutations, and algorithms designed for the former can usually be directly used to find the latter. In addition, profile-based optimization methods, e.g., CONSENSUS and MEME, have failed to find (l, d) motifs such as (15, 4), where a 15-bp motif is planted into 20 sequences, each 600 bps in length with exactly 4 mismatches [19]. Thus, in this study we focus on designing a fast and exact algorithm to find (l, d) motifs with exactly d mutations on (l, d) samples.

When searching for the exact motifs contained in an (l, d) sample, it is customary to perform an exhaustive search for all potential l-mers and verify their occurrence in the entire sample set. When using SPELLER and WEEDER to perform fast l-mer substring searching in a sequence set, a suffix tree structure is used to index sequences. A suffix tree presents the suffixes of a given string or a given set of strings in a way that allows for a very fast implementation of string operations. An example of a classic suffix tree for the string GAGAC is shown in Figure 4a. When the suffix tree of a string with length L is constructed, searching for a substring of length l ($l\le L$) in the string only requires time proportional to $O(l)$ instead of $O(L)$.

Nevertheless, it is still time consuming to perform an exhaustive motif search in a suffix tree of sequences because the search space (shown as a four-branch tree in Figure 4b) can be up to 4^l in size. With the increase in degenerative value d, the valid instances of a motif in the suffix tree will increase dramatically. Therefore, SPELLER can handle only short (l, d) motifs with $l\le 13$ and $d\le 3$. In order to increase the speed of SPELLER, WEEDER introduces an error ratio ε ($\varepsilon\simeq d/l$) to narrow the search space such that for all $i\in\{1,2,...,l\}$, the number of mismatches between the first i nucleotides of a candidate l-mer motif and the first i nucleotides of a valid instance of the motif is at most εi. The strategy can quickly discard l-mers in the search space that do not satisfy this restriction. However, not all motif occurrences satisfy this restriction, and therefore the real motif may be missed by the algorithm. WEEDER lowers the occurrence frequency $q\le Q\le N$ to make sure that the true motif will not be missed. Still, WEEDER is an almost exact algorithm. What's more, with the decrease of q, WEEDER's run time will increase dramatically [14]. Therefore, a fast and exact (l, d) motif search method is needed.

Mismatched Suffix Trees and FMotif

SPELLER and WEEDER use a depth-first search to scan the entire pattern space. If an i-mer along the pattern tree has enough instances in the suffix tree of sequences, the i-mer can grow up to 4

$(i+1)$-mers in the next layer of the pattern tree (see Figure 4b). Otherwise, the end node of an i-mer in the pattern tree will not be allowed to grow and the sibling nodes of the i-mer will be checked. If any of the sibling nodes can grow to the $(i+1)$-th layer in the pattern tree, the search process will go down to the $(i+1)$-th layer of the pattern tree in a depth-first manner. Otherwise, it will backtrack to the uncle nodes (the siblings of parent nodes) of the i-mer in the $(i-1)$-th layer of the pattern tree and so forth. The algorithms will end at the longest l-mer or l-mers in the pattern tree. The difference between SPELLER and WEEDER is that WEEDER reduces the number of possible instances of a motif by restricting its mutation locations such that the valid paths on the pattern tree are sharply reduced.

We discovered that for finding motifs with length l, both SPELLER and WEEDER must check each i-mer ($i\le l$) in the pattern space with depth-first order and count the variants of the i-mer in the suffix tree of sequences from the root to layer i. The suffix tree is scanned one time for each i-mer pattern. Thus, the algorithms scan the suffix tree an increasing number of times with the increase of i. Actually, the mismatch information in layer i of a suffix tree can be used to search $(i+1)$-mers in the pattern space. In this work we constructed a new suffix tree structure with mismatch information, called a mismatched suffix tree, for each sequence. Using these trees, we took advantage of the mismatch information in the i-th layer of the trees to verify $(i+1)$-mers in the pattern space and then updated the mismatch information in the $(i+1)$-th layer. In this way we were able to implement a depth-first search on the pattern space and mismatched suffix trees simultaneously, which avoided a large number of repeated scans on the suffix trees of sequences.

For instance, when searching occurrences of (4, 1) motifs in the sequence GAGAC, we started from the root of the pattern tree represented as $P_0=\ominus$ in Figure 5a and initialized the mismatched suffix tree for the sequence GAGAC. We then checked the occurrences of pattern $P_1=A$ with up to 1 mismatch in the mismatched suffix tree and found that all nodes in the first layer have 0 or 1 mismatch(es) with P_1. Next, we updated the mismatch value along the valid nodes and linked all of these nodes by points (see Figure 5b). We subsequently performed a depth-first search again and arrived at the pattern $P_2=AA$. We updated mismatch information for all child nodes of the nodes in the link set in the first layer by using the mismatch information of those nodes in the link set and found all nodes in the second layer had 1 mismatch with the pattern $P_2=AA$. We updated the mismatch value along the valid nodes in the second layer and linked all of these nodes by points to form a new link set (see Figure 5c). Then, we moved to the pattern $P_3=AAA$ in a depth-first manner and updated mismatch information of all child nodes of the nodes in the newly generated link set by using the mismatch information of those nodes in the new link set. We found that only the child node A, representing the 3-mer AGA from the root to node A in the third layer of the suffix tree, had 1 mismatch with the pattern $P_3=AAA$ (see Figure 5d). Other nodes with 2 mismatches did not need to be updated and checked for the longer pattern $AAAA$. We found that the child node of the node A in the third layer did not satisfy the 1-mismatch restriction with the pattern $AAAA$, so we looked at the pattern $P_4=AAAC$ and found a (4, 1) occurrence of P_4 (see Figure 5e). We then went to the patterns $AAAG$ and $AAAT$ and found no occurrence of these patterns in the sequence GAGAC. We backtracked to the pattern $P_3'=AAC$ and updated the mismatch information in the third layer by using the mismatch information of their parent nodes in link set of the second layer. There we found that only node C in the third layer satisfies the restriction (see Figure 5f), but that node C has no child. We then

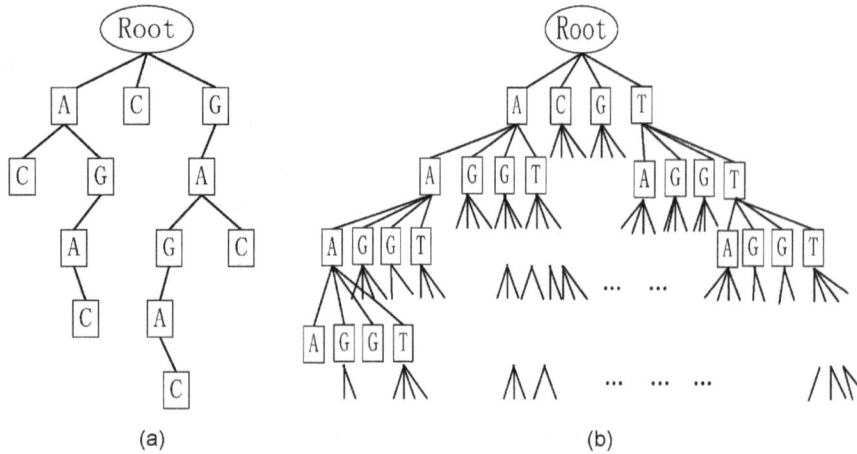

Figure 4. An example of a suffix tree and a tree representation of pattern space. (a) The suffix tree of the sequence GAGAC. (b) A tree representation of pattern space in the search for an (l, d) motif.

backtracked to pattern AAG and continued the process as before until we found all occurrences for each $(4, 1)$ motif. The details of the pattern search and mismatched suffix tree construction are shown in the subroutine $PatternOnTree(P_k,d,T,List(P_{k-1},T))$, where T is the mismatched suffix tree for a sequence, P_k is the node currently being processed (representing a k-mer pattern) in the k-th layer of the pattern tree, $List(P_{k-1},T)$ is the link set representing all valid occurrences of the $(k-1)$-mer pattern represented by the node P_{k-1} in the $(k-1)$-th layer of the pattern tree, and $MMC(n_{jk})$ is mismatch value of the pattern represented by P_k compared with the substring represented by the node n_{jk} in the tree T.

$PatternOnTree(P_k,d,T,List(P_{k-1},T))$
   Initialize $List(P_k,T)=\varnothing$;
   for $n_{i,k-1}=$ head node to tail node of $List(P_{k-1},T)$ **do**
      for each $n_{jk}\in NS$ **do** (NS is the child node set of the node $n_{i,k-1}$)
         if $n_{jk}=P_k$, **then**
            $MMC(n_{jk})=MMC(n_{i,k-1})$
         else
            $MMC(n_{jk})=MMC(n_{i,k-1})+1$;
         if $MMC(n_{jk})\leq d$ **then**
            add n_{jk} to $List(P_k,T)$;
  return $List(P_k,T)$;

   $MotifFinding(P_k,TS,List(P_k,TS),q,d,l_{max})$
   Initialize $str=ACGT$;
   if $k>l_{max}$ or $List(P_k,TS)=\varnothing$ **then**
   return;
   for j$=1\cdots 4$ **do**
      $P_{k+1}=P_k+str[j]$;
      $FailureCount=0$;
      for each tree T_i in the tree set TS **do**
         $List(P_{k+1},T_i)=PatternOnTree(P_{k+1},d,T_i,List(P_k,T_i))$;
         if $List(P_{k+1},T_i)=\varnothing$ **then**
            $FailureCount++$;
   if $FailureCount>|TS|-q$ **then**
   break;
   if $FailureCount\leq|TS|-q$ **then**
      $Output(P_{k+1})$;
   $MotifFinding(P_{k+1},TS,List(P_{k+1},TS),q,d,l_{max})$

For each set of sequences, we counted the number of occurrences of a potential pattern in all sequences instead of just one sequence shown in subroutine $PatternOnTree(P_k, d,T,List(P_{k-1},T))$. If the number of occurrences was larger than the threshold of occurrence frequency q, it was reported as a potential pattern. The subroutine for counting occurrences of a $(k+1)$-mer pattern, represented by the node P_{k+1} in the $(k+1)$-th layer of the pattern tree, is shown in $MotifFinding(P_k, TS,List(P_k,TS), q,d,l_{max})$, where l_{max} is the maximum length allowed for a motif, TS is the set of mismatched suffix trees for all sequences, $List(P_k,TS)=\bigcup_{i=0}^{N}List(P_k,T_i)$, and N is the number of total sequences.

The entire process of finding motifs with at least q occurrences in a set of sequences is shown below. Additionally, since there may be many motifs that satisfy the quorum restriction q, we sorted all potential motifs according to their statistical significance using the method in [14,15]. We reported the top n significant motifs and their occurrences as output, where (i, j) indicates an instance of a motif starting at the j-th position of the i-th sequence s_i.

The FMotif Algorithm

1) Initialize a mismatched suffix tree T_i like the one shown in Figure 5a, $i=1,2,\cdots,N$;
2) Initialize $List(\ominus,T_i)=$&tempnode, where &tempnode is the pointer of a temporary node, $i=1,2,\cdots,N$;
3) Input q,d,l_{max},n;
4) $MotifFinding(\ominus,TS,List(\ominus,TS),q,d,l_{max})$;
5) Rank the found motifs according to their significance scores;
6) Output the top n motifs, their instances, and the positions (i,j) of these instances.

According to our empirical study, FMotif is capable of increasing the speed of the algorithms SPELLER and WEEDER without loss of accuracy. In addition, we used the WEEDER strategy to further decrease the search space by allowing mismatches occurring at most εi times with an increase in i. This strategy decreased FMotif's run time but caused problems during the tuning of the parameter q and resulted in a loss of accuracy.

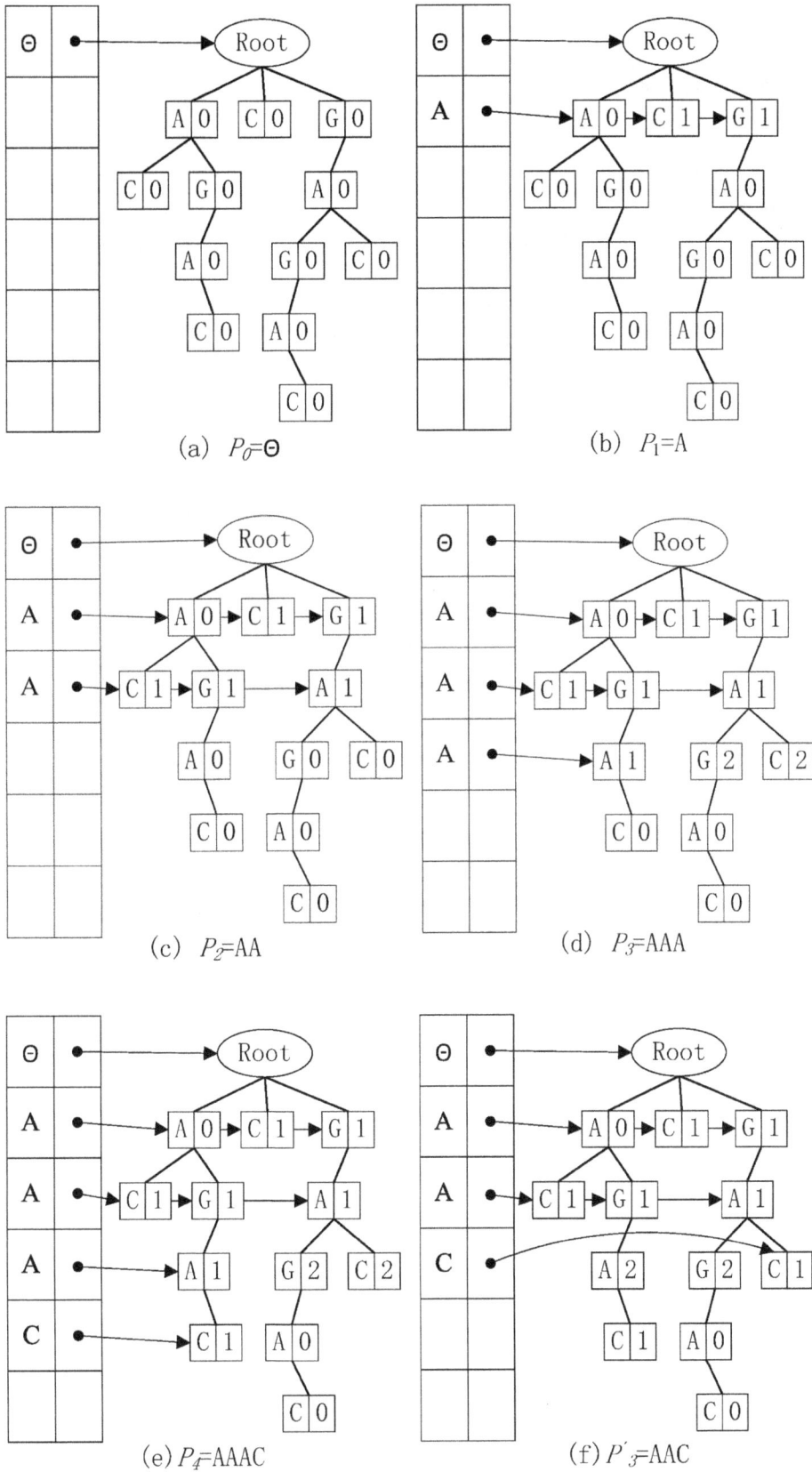

Figure 5. An example of a (4,1) motif search using FMotif. Figures (a)–(f) illustrate the search process of (4, 1) motifs on the mismatched suffix tree of the sequence GAGAC.

Author Contributions

Conceived and designed the experiments: CJ YW. Performed the experiments: CJ YW. Analyzed the data: CJ MBC. Contributed reagents/materials/analysis tools: YL HL. Wrote the paper: CJ MBC.

References

1. Robertson G, Hirst M, Bainbridge M, Bilenky M, Zhao Y, et al. (2007) Genome-wide profiles of STAT1 DNA association using chromatin immunoprecipitation and massively parallel sequencing. Nat Methods 4: 651–657.

2. Johnson DS, Mortazavi A, Myers RM, Wold B (2007) Genome-wide mapping of in vivo protein- DNA interactions. Science 316: 1497–1502.

3. Wilbanks EG, Facciotti MT (2010) Evaluation of algorithm performance in ChIP-seq peak detection. PLoS One 5: e11471.

4. Rhee, S H, Pugh BF (2011) Comprehensive genome-wide protein-DNA interactions detected at single-nucleotide resolution. Cell 147: 480–483.

5. Zhao Y, Stormo GD (2011) Quantitative analysis demonstrates most transcription factors require only simple models of specificity. Nat Biotechnol 29: 1408–1419.

6. Zambelli F, Pesole G, Pavesi G (2013) Motif discovery and transcription factor binding sites before and after the next-generation sequencing era. Brief Bioinform 14: 225–237.

7. Hertz GZ, Stormo GD (1999) Identifying DNA and protein patterns with statistically significant alignments of multiple sequences. Bioinformatics 15: 563–577.

8. Bailey TL, Elkan C (1994) Fitting a mixture model by expectation maximization to discover motifs in biopolymers. In: Proceedings of the Second International Conference on Intelligent Systems for Molecular Biology, 1994, Menlo Park, CA. 28–36.

9. Lawrence CE, Altschul SF, Boguski MS, Liu JS, Neuwald AF, et al. (1993) Detecting subtle sequence signals: a Gibbs sampling strategy for multiple alignment. Science 262: 208–214.

10. Hughes JD, Estep PW, Tavazoie S, Church GM (2000) Computational identification of cisregulatory elements associated with groups of functionally related genes in saccharomyces cerevisiae. Journal of Molecular Biology 296: 1205–1214.

11. Buhler J, Tompa M (2002) Finding motifs using random projections. Journal of Computational Biology 9: 225–242.

12. Gang L, Chan TM, Leung KS, Lee KH (2010) A cluster refinement algorithm for motif discovery. IEEE/ACM Trans on Computational Biology and Bioinformatic 7: 654–668.

13. Sagot MF (1998) Spelling approximate repeated or common motifs using a suffix tree. Proceedings of LATIN'98: Theoretical Informatics, LNCS 1380: 111–127.

14. Pavesi G, Mauri G, Pesole G (2001) An algorithm for finding signals of unknown length in DNA sequences. Bioinformatics 17: 207–214.

15. Pavesi G, Mereghetti P, Mauri G, Pesole G (2004) Weeder Web: discovery of transcription factor binding sites in a set of sequences from co-regulated genes. Nucleic Acids Research: Web Server Issue 32: W199–W203.

16. Eskin E, Pevzner P (2002) Finding composite regulatory patterns in DNA sequences. Bioinformatics 18: 354–363.

17. Chin YL, Leung CM (2005) Voting algorithms for discovering long motifs. In: Proceedings of the Third Asia-Pacific Bioinformatics Conference, 2005, Singapore. 261–271.

18. Davila J, Balla S, Rajasekaran S (2007) Fast and practical algorithms for planted (l, d) motif search. IEEE/ACM Trans On Computational Biology and Bioinformatics 4: 544–552.

19. Pevzner P, Sze S (2000) Combinatorial approaches to finding subtle signals in DNA sequences. In: Proceedings of the Eighth International Conference on Intelligent Systems for Molecular Biology, 2000, California, USA. 269–278.

20. Ho ES, Jakubowski CD, Gunderson SI (2009) iTriplet, a rule-based nucleic acid sequence motif finder. Algorithms Molecular Biology 4: doi:10.1186/1748-7188-4-14.

21. Huang CW, Lee WS, Hsieh SY (2010) An improved heuristic algorithm for finding motif signals in DNA sequences. IEEE/ACM Trans On Computational Biology and Bioinformatics 8: 959–975.

22. Kuksa PP, Pavlovic V (2010) Efficient motif finding algorithms for large-alphabet inputs. BMC Bioinformatics 1: S1.

23. Sun HQ, Low MYH, Hsu WJ, Rajapakse JC (2010) RecMotif: a novel fast algorithm for weak motif discovery. BMC Bioinformatics 11: S8.

24. Tompa M (2005) Assessing computational tools for the discovery of transcription factor binding sites. Nature Biotechnology 23: 137–144.

25. Machanick P, Bailey TL (2011) MEME-ChIP: motif analysis of large DNA datasets. Bioinformatics 27: 1696–1697.

26. Bailey TL (2011) DREME: motif discovery in transcription factor ChIP-seq data. Bioinformatics 27: 1653–1659.

27. Heeringen SJV, Veenstra GJ (2011) GimmeMotifs: a de novo motif prediction pipeline for ChIPsequencing experiments. Bioinformatics 27: 270–271.

28. Reid JE, Wernisch L (2011) STEME: efficient EM to find motifs in large data sets. Nucleic Acids Res 38: e126.

29. Kulakovskiy IV, Boeva VA, Favorov AV, Makeev VJ (2010) Deep and wide digging for binding motifs in ChIP-seq data. Bioinformatics 26: 2622–2623.

30. Hu M, Yu J, Taylor JM, Chinnaiyan AM, Qin ZS (2010) On the detection and refinement of transcription factor binding sites using ChIP-seq data. Nucleic Acids Res 38: 2154–2167.

31. Guo Y, Mahony S, Gifford DK (2012) High resolution genome wide binding event finding and motif discovery reveals transcription factor spatial binding constraints. PLoS Comput Biol 8: e1002638.

32. He Y, Zhang Y, Zheng G, Wei C (2012) CTF: a CRF-based transcription factor binding sites finding system. BMC Genomics 13: S18.

33. Thomas-CholllierM, Herrmann C, Defrance M, Sand O, Thieffry D, et al. (2012) RSAT peak-motifs: motif analysis in full-size ChIP-seq datasets. Nucleic Acids Res 40: e31.

34. Sharov AA, Ko MSH (2009) Exhaustive search for over-represented DNA sequence motif with CisFinder. DNA Research 16: 261–273.

35. Ma X, Kulkarni A, Zhang Z, Xuan Z, Serfling R, et al. (2012) A highly efficient and effective motif discovery method for ChIP-seq/ChIP-chip data using positional information. Nucleic Acids Res 40: e50.

36. Keich U, Pevzner P (2002) Subtle motif: defining the limits of finding algorithms. Bioinformatics 18: 1382–1390.

37. Zia A, Moses AM (2012) Towards a theoretical understanding of false positives in DNA motif finding. BMC Bioinformatics 13: doi: 10.1186/1471-2105-13-151.

38. Boeva V, Surdez D, Guillon N, Tirode F, Fejes A, et al. (2010) De novo motif identification improves the accuracy of predicting transcription factor binding sites in ChIP-Seq data analysis. Nucleic Acids Res 38: e126.

39. Zhang Y, Liu T, Meyer CA, Eeckhoute J, Johnson D, et al. (2008) Model-based analysis of ChIPSeq (MACS). Genome Biology 9: R137.

40. Chen X, Xu H, Yuan P, Fang F, Huss M, et al. (2008) Integration of external signaling pathways with the core transcriptional network in embryonic stem cells. Cell 133: 1106–1117.

41. Barski A, Cuddapah S, Cui K, Roh T, Schones D, et al. (2007) High-resolution profiling of histone methylations in the human genome. Cell 129: 823–837.

42. Lupien M, Eeckhoute J, Meyer CA, Wang Q, Zhang Y, et al. (2008) FoxA1 translates epigenetic signatures into enhancer-driven lineage-specific transcription. Cell 132: 958–970.

43. Jothi R, Cuddapah S, Barski A, Cui K, Zhao K (2008) Genome-wide identification of in vivo protein-DNA binding sites from ChIP-Seq data. Nucleic Acids Res 36: 5221–5231.

44. Jothi R, Cuddapah S, Barski A, Cui K, Zhao K (2012) Breast cancer risk-associated SNPs modulate the affinity of chromatin for FOXA1 and alter gene expression. Nat Genet 44: 1191–1198.

45. Crooks GE, Hon G, Chandonia JM, Brenner SE (2004) WebLogo: a sequence logo generator. Genome Research 14: 1188–1190.

46. Badis G, Chan ET, van Bakel H, Pena-Castillo L, Tillo D, et al. (2008) A library of yeast transcription factor motifs reveals a widespread function for Rsc3 in targeting nucleosome exclusion at promoters. Mol Cell 32: 878–887.

47. Spivak AT, Stormo GD (2012) ScerTF: a comprehensive database of benchmarked position weight matrices for Saccharomyces species. Nucleic Acids Res 40: D162–168.

48. Pachkov M, Erb I, Molina N, van Nimwegen E (2007) SwissRegulon: a database of genome-wide annotations of regulatory sites. Nucleic Acids Res 35: D127–131.

Permissions

The contributors of this book come from diverse backgrounds, making this book a truly international effort. This book will bring forth new frontiers with its revolutionizing research information and detailed analysis of the nascent developments around the world.

We would like to thank all the contributing authors for lending their expertise to make the book truly unique. They have played a crucial role in the development of this book. Without their invaluable contributions this book wouldn't have been possible. They have made vital efforts to compile up to date information on the varied aspects of this subject to make this book a valuable addition to the collection of many professionals and students.

This book was conceptualized with the vision of imparting up-to-date information and advanced data in this field. To ensure the same, a matchless editorial board was set up. Every individual on the board went through rigorous rounds of assessment to prove their worth. After which they invested a large part of their time researching and compiling the most relevant data for our readers.

The editorial board has been involved in producing this book since its inception. They have spent rigorous hours researching and exploring the diverse topics which have resulted in the successful publishing of this book. They have passed on their knowledge of decades through this book. To expedite this challenging task, the publisher supported the team at every step. A small team of assistant editors was also appointed to further simplify the editing procedure and attain best results for the readers.

Apart from the editorial board, the designing team has also invested a significant amount of their time in understanding the subject and creating the most relevant covers. They scrutinized every image to scout for the most suitable representation of the subject and create an appropriate cover for the book.

The publishing team has been an ardent support to the editorial, designing and production team. Their endless efforts to recruit the best for this project, has resulted in the accomplishment of this book. They are a veteran in the field of academics and their pool of knowledge is as vast as their experience in printing. Their expertise and guidance has proved useful at every step. Their uncompromising quality standards have made this book an exceptional effort. Their encouragement from time to time has been an inspiration for everyone.

The publisher and the editorial board hope that this book will prove to be a valuable piece of knowledge for researchers, students, practitioners and scholars across the globe.

List of Contributors

Rakesh K. Bhat, Wallis Rudnick, Joseph M. Antony, Ferdinand Maingat, Kristofor K. Ellestad and Christopher Power
Department of Medicine (Neurology), University of Alberta, Edmonton, Alberta, Canada

Blaise M. Wheatley
Department of Surgery (Neurosurgery), University of Alberta, Edmonton, Alberta, Canada

Ralf R. Tönjes
Division of Medical Biotechnology, Paul-Ehrlich-Institut, Langen, Germany

Tiffany Renee Oliver
Department of Human Genetics, Emory University School of Medicine, Atlanta, Georgia, United States of America
Department of Biology, Spelman College, Atlanta, Georgia, United States of America

Candace D. Middlebrooks, Stuart W. Tinker, Emily Graves Allen, Lora J. H. Bean and Stephanie L. Sherman
Department of Human Genetics, Emory University School of Medicine, Atlanta, Georgia, United States of America

Ferdouse Begum
Department of Biostatistics, Graduate School of Public Health, University of Pittsburgh, Pittsburgh, Pennsylvania, United States of America

Eleanor Feingold
Department of Biostatistics, Graduate School of Public Health, University of Pittsburgh, Pittsburgh, Pennsylvania, United States of America
Department of Human Genetics, Graduate School of Public Health University of Pittsburgh, Pittsburgh, Pennsylvania, United States of America

Reshmi Chowdhury
Department of Human Genetics, Graduate School of Public Health University of Pittsburgh, Pittsburgh, Pennsylvania, United States of America

Vivian Cheung
Howard Hughes Medical Institute, University of Michigan, Ann Arbor, Michigan, United States of America
Department of Human Genetics, University of Michigan, Ann Arbor, Michigan, United States of America

Shaoli Das and Suman Ghosal
Computational Biology Group, Indian Association for the Cultivation of Science, Kolkata, West Bengal, India

Rituparno Sen
Gyanxet, Kolkata, West Bengal, India

Jayprokas Chakrabarti
Computational Biology Group, Indian Association for the Cultivation of Science, Kolkata, West Bengal, India

Luis Santana-Quintero and Vahan Simonyan
Center for Biologics Evaluation and Research, US Food and Drug Administration, Rockville, Maryland, United States of America

Hayley Dingerdissen
Center for Biologics Evaluation and Research, US Food and Drug Administration, Rockville, Maryland, United States of America
Department of Biochemistry and Molecular Biology, George Washington University Medical Center, Washington, DC, United States of America

Jean Thierry-Mieg
National Center for Biotechnology Information, U.S. National Library of Medicine, National Institutes of Health, Bethesda, Maryland, United States of America

Raja Mazumder
Department of Biochemistry and Molecular Biology, George Washington University Medical Center, Washington, DC, United States of America

Hsueh-Ting Chu
Department of Biomedical informatics, Asia University, Taichung, Taiwan
Department of Computer Science and Information Engineering, Asia University, Taichung, Taiwan

William W.L. Hsiao
British Columbia Public Health Microbiology and Reference Laboratory, Vancouver, British Columbia, Canada
Department of Pathology and Laboratory Medicine, University of British Columbia, Vancouver, British Columbia, Canada

Chaur-Chin Chen
Department of Computer Science, National Tsing Hua University, Hsinchu, Taiwan

Theresa T.H. Tsao and Cheng-Yan Kao
Department of Computer Science and Information Engineering, National Taiwan University, Taipei, Taiwan

D. Frank Hsu
Department of Computer and Information Science, Fordham University, New York, New York, United States of America

Sheng-An Lee
Department of Information Management, Kainan University, Taoyuan, Taiwan

Eric S. Ho
Department of Biology, Lafayette College, Easton, Pennsylvania, United States of America

Joan Kuchie
New Jersey City University, Jersey City, New Jersey, United States of America

Siobain Duffy
Department of Ecology, Evolution and Natural Resources, Rutgers University, New Brunswick, New Jersey, United States of America

Jiang Bian and Mathias Brochhausen
Division of Biomedical Informatics, University of Arkansas for Medical Sciences, Little Rock, AR 72205, United States of America

Mengjun Xie
Department of Computer Science, University of Arkansas at Little Rock, Little Rock, AR 72204, United States of America

Teresa J. Hudson
Department of Psychiatry and Behavioral Sciences, University of Arkansas for Medical Sciences, Little Rock, AR 72205, United States of America
Department of Veterans Affairs HSR&D Center for Mental Healthcare and Outcomes Research, Central Arkansas Veterans Healthcare System, Little Rock, AR 722205, United States of America

Hari Eswaran
Division of Biomedical Informatics, University of Arkansas for Medical Sciences, Little Rock, AR 72205, United States of America
Department of Obstetrics & Gynecology Research, University of Arkansas for Medical Sciences, Little Rock, AR 72205, United States of America

Josh Hanna
Clinical and Translational Science Informatics and Technology, University of Florida, Gainesville, FL 32610, United States of America

William R. Hogan
Department of Health Outcomes & Policy, University of Florida, Gainesville, FL 32610, United States of America
Clinical and Translational Science Institute, University of Florida, Gainesville, FL 32610, United States of America

Michael E. Sparks, Daniel Kuhar and Dawn E. Gundersen-Rindal
USDA-ARS Invasive Insect Biocontrol and Behavior Laboratory, Beltsville, Maryland, United States of America

Kent S. Shelby
USDA-ARS Biological Control of Insects Research Laboratory, Columbia, Missouri, United States of America

Weizhao Yang
Chengdu Institute of Biology, Chinese Academy of Sciences, Chengdu, China
University of Chinese Academy of Sciences, Beijing, China

Yin Qi
Chengdu Institute of Biology, Chinese Academy of Sciences, Chengdu, China

Jinzhong Fu
Chengdu Institute of Biology, Chinese Academy of Sciences, Chengdu, China

Department of Integrative Biology, University of Guelph, Guelph, Ontario, Canada

Xiangkai Zhu Ge, Zihao Pan, Lin Hu, Haojin Wang, Jianjun Dai and Hongjie Fan
College of Veterinary Medicine, Nanjing Agricultural University, Nanjing, China

Jingwei Jiang and Frederick C. Leung
Bioinformatics Center, Nanjing Agricultural University, Nanjing, China,
School of Biological Sciences, University of Hong Kong, Hong Kong SAR, China

Shaohui Wang
Shanghai Veterinary Research Institute, Chinese Academy of Agricultural Sciences, Shanghai, China

Matthew McKnight Croken
Department of Microbiology and Immunology, Albert Einstein College of Medicine, Bronx, New York, United States of America

Yanfen Ma
Department of Pathology, Albert Einstein College of Medicine, Bronx, New York, United States of America

Lye Meng Markillie and Galya Orr
Environmental Molecular Sciences Laboratory, Pacific Northwest National Laboratory, Richland, Washington, United States of America

Ronald C. Taylor
Computational Biology and Bioinformatics Group, Biological Sciences Division, Pacific Northwest National Laboratory, Richland Washington, United States of America

Louis M. Weiss
Department of Pathology, Albert Einstein College of Medicine, Bronx, New York, United States of America Department of Medicine, Albert Einstein College of Medicine, Bronx, New York, United States of America

Kami Kim
Department of Microbiology and Immunology, Albert Einstein College of Medicine, Bronx, New York, United States of America
Department of Pathology, Albert Einstein College of Medicine, Bronx, New York, United States of America Department of Medicine, Albert Einstein College of Medicine, Bronx, New York, United States of America

Haluk Dogan, Handan Can and Hasan H. Otu
Department of Genetics and Bioengineering, Istanbul Bilgi University, Istanbul, Turkey

Wonseok Shin, Jungnam Lee and yudong Han
Department of Nanobiomedical Science and WCU Research Center, Dankook University, Cheonan, Republic of Korea

Seung-Yeol Son
Department of Microbiology, College of Advanced Science, Dankook University, Cheonan, Republic of Korea

Kung Ahn and Heui-Soo Kim
Department of Biological Sciences, College of Natural Sciences, Pusan National University, Busan, Republic of Korea

Quan Li
Endocrine Genetics Lab, The McGill University Health Center (Montreal Children's Hospital), Montréal, Québec, Canada

Hui-Qi Qu
Division of Epidemiology, Human Genetics and Environmental Sciences, The University of Texas School of Public Health, Houston, Texas, Unite States of America

Yaoliang Chen, Ji Hong, Wanyun Cui, Wei Wang and Yanghua Xiao
School of Computer Science, Fudan University, Shanghai, China

Jacques Zaneveld, Richard Gibbs and Rui Chen
Human Genome Sequencing Center, Department of Molecular and Human Genetics, Baylor College of Medicine, Houston, Texas, United States of America

Vikas Bansal
Department of Cardiovascular Genetics, Experimental and Clinical Research Center, Charité-Universitätsmedizin Berlin and Max Delbrück Center (MDC) for Molecular Medicine, Berlin, Germany
Department of Mathematics and Computer Science, Free University of Berlin, Berlin, Germany

Cornelia Dorn and Silke R. Sperling
Department of Cardiovascular Genetics, Experimental and Clinical Research Center, Charité-Universitätsmedizin Berlin and Max Delbrück Center (MDC) for Molecular Medicine, Berlin, Germany
Department of Biology, Chemistry, and Pharmacy, Free University of Berlin, Berlin, Germany

Marcel Grunert
Department of Cardiovascular Genetics, Experimental and Clinical Research Center, Charité-Universitätsmedizin Berlin and Max Delbrück Center (MDC) for Molecular Medicine, Berlin, Germany

Sabine Klaassen
For the National Register for Congenital Heart Defects, Berlin, Germany
Experimental and Clinical Research Center, Charité-Universitätsmedizin Berlin and Max Delbrück Center (MDC) for Molecular Medicine, Berlin, Germany
Department of Pediatric Cardiology, Charité-Universitätsmedizin Berlin, Berlin, Germany

Roland Hetzer
Department of Cardiac Surgery, German Heart Institute Berlin, Berlin, Germany

Felix Berger
Department of Pediatric Cardiology, Charité-Universitätsmedizin Berlin, Berlin, Germany
Department of Pediatric Cardiology, German Heart Institute Berlin, Berlin, Germany

Sumit Middha, Saurabh Baheti, Steven N. Hart and Jean-Pierre A. Kocher
Division of Biomedical Statistics and Informatics, Department of Health Sciences Research, Mayo Clinic, Rochester, Minnesota, United States of America

Yongchao Liu and Bertil Schmidt
Institut für Informatik, Johannes Gutenberg Universität Mainz, Mainz, Germany

Bernt Popp
Institute of Human Genetics, University of Erlangen-Nuremberg, Erlangen, Germany

Hao Wu
Department of Biostatistics and Bioinformatics, Emory University, Atlanta, Georgia, United States of America

Hongkai Ji
Department of Biostatistics, Johns Hopkins University, Baltimore, Maryland, United States of America

Lun-Ching Chang and Chien-Wei Lin
Department of Biostatistics, University of Pittsburgh, Pittsburgh, Pennsylvania, United States of America

Stephane Jamain
Inserm U955, Psychiatrie Génétique, Créteil, France
Université Paris Est, Créteil, France
Fondation FondaMental, Créteil, France

Dan Rujescu
Department of Psychiatry, University of Halle, Halle, Germany

George C. Tseng
Department of Biostatistics, University of Pittsburgh, Pittsburgh, Pennsylvania, United States of America
Department of Human Genetics, University of Pittsburgh, Pittsburgh, Pennsylvania, United States of America

Etienne Sibille
Department of Psychiatry, Center For Neuroscience, University of Pittsburgh, Pittsburgh, Pennsylvania, United States of America

Guénola Drillon and Gilles Fischer
Sorbonne Universités, UPMC Univ Paris 06, UMR 7238, Biologie Computationnelle et Quantitative, Paris, France
CNRS, UMR7238, Biologie Computationnelle et Quantitative, Paris, France
Institut Universitaire de France, Paris, France

Alessandra Carbone
Sorbonne Universités, UPMC Univ Paris 06, UMR 7238, Biologie Computationnelle et Quantitative, Paris, France

CNRS, UMR7238, Biologie Computationnelle et Quantitative, Paris, France
Institut Universitaire de France, Paris, France

Wan-Ping Lee, Michael P. Stromberg, Alistair Ward, Erik P. Garrison and Gabor T. Marth
Department of Biology, Boston College, Chestnut Hill, Massachusetts, United States of America

Chip Stewart
Department of Biology, Boston College, Chestnut Hill, Massachusetts, United States of America
Broad Institute of Harvard and Massachusetts Institute of Technology, Cambridge, Massachusetts, United States of America

Ying Chen Eyre-Walker and Adam Eyre-Walker
School of Life Sciences, University of Sussex, Brighton, United Kingdom

Caiyan Jia
School of Computer and Information Technology & Beijing Key Lab of Traffic Data Analysis, Beijing Jiaotong University, Beijing, China
Department of Bioengineering/Bioinformatics, University of Illinois at Chicago, Chicago, Illinois, United States of America

Matthew B. Carson
Center for Healthcare Studies, Institute for Public Health and Medicine, Northwestern University Feinberg School of Medicine, Chicago, Illinois, United States of America
Division of Health and Biomedical Informatics, Department of Preventive Medicine, Northwestern University Feinberg School of Medicine, Chicago, Illinois, United States of America

Yang Wang and Youfang Lin
School of Computer and Information Technology & Beijing Key Lab of Traffic Data Analysis, Beijing Jiaotong University, Beijing, China

Hui Lu
Department of Bioengineering/Bioinformatics, University of Illinois at Chicago, Chicago, Illinois, United States of America
Shanghai Institute of Medical Genetics, Shanghai Children's Hospital, Shanghai JiaoTong University, Shanghai, China

Index

www.ingramcontent.com/pod-product-compliance
Lightning Source LLC
Chambersburg PA
CBHW080523200326
41458CB00012B/4314